BENDING COMBINED w/ TENSION

$$\text{STRESS} = \frac{M_{\text{moment}}\, c_{\text{centroid distance}}}{I_{\text{inertia}}} - \frac{P_{\text{load}}}{A_{\text{area}}} =$$

STATICS AND STRENGTH OF MATERIALS

STATICS AND STRENGTH OF MATERIALS

CHARLES O. HARRIS

JOHN WILEY & SONS
New York Chichester Brisbane Toronto Singapore

Library of Congress Cataloging in Publication Data:

Harris, Charles Overton, 1909–
Statics and strength of materials.

Includes index.
1. Statics. 2. Strength of materials.
I. Title.

TA351.H34 620.1′053 81-16237
ISBN 0-471-08293-7 AACR2

Printed in the United States of America

10 9 8 7 6 5 4 3 2

PREFACE

This book is a basic treatment of statics and strength of materials for students in engineering technology; thus the mathematics used is below the level of calculus.

I have presented the material in the same way that I would present it in the classroom. The first paragraph of each chapter explains the purpose of the chapter, telling the student what he or she is going to learn from studying the chapter. Then, each topic is introduced with a definition and illustration of terms as well as an explanation of method. Next, a simple Illustrative Example is used to show how the method is applied. This is followed by an explanation of some expansion or variation of the topic, another Illustrative Example, and more explanation and examples until the exposition is complete. Finally, there is a set of Practice Problems for the student to solve—this is where the subject is really learned. At the end of each chapter, there is a Summary, a set of Review Questions, and a set of Review Problems, all of which will be helpful to the student.

I have not spent much time in proving theorems and deriving formulas. I believe that the plausibility of a theorem is more important than the proof, so I have emphasized the physical aspects of the methods and their uses. Throughout the book, I have provided a review of material presented in earlier chapters when it is necessary for the study of a particular chapter.

It is evident that the metric system is used quite a bit in American industry, but it also seems evident that it will be a long time, if ever, before the traditional system of units is abandoned. Therefore, I have provided much explanation and many problems in each system of units.

Now for a few words about subject matter.

I don't believe in lengthy abstract discussions when introducing a subject to students; thus, in Chapter 1, I have started right in with what I consider to be the elements of statics—for example, topics such as force, components of a force, moment of a force, and the principle of moments.

I have organized Chapter 2 in terms of methods of solving equilibrium problems, rather than in terms of types of force systems, because I believe that the emphasis on methods is the more efficient way for the student to learn to solve problems. I have included a section on the equilibrium of shafts as preliminary material for the study of torsion in Chapter 10. I have put three special topics in equilibrium—trusses, friction, and parabolic cables—in a separate part of Chapter 2, with the thought that some might wish to omit these topics in order to get to strength of materials more quickly.

Stress and strain are introduced in Chapter 3, and I have extended the treatment to include calculating the size of a member.

Chapter 4 covers the centroid of an area, and Chapter 5 discusses moment of inertia of an area. The explanation is thorough, and there are many Illustrative Examples.

Chapter 6 has a few tables of properties of standard beam sections, just enough for the student to learn that this sort of thing is available, along with an explanation of how to use such tables. I have also put special emphasis on the selection of a beam section for a given loading.

I have included stress, stress concentration, and special topics in torsion; these subjects are too important to leave out.

In Chapter 7, which discusses bending combined with tension or compression, I have included many problems in which it is necessary to locate the centroid of an area and calculate the moment of inertia, because this is a common problem in real engineering work.

I think that the only practical way to calculate beam deflections is to have a collection of a few formulas for special cases, and then to use the method of superposition. This is the way I have organized Chapter 8, which is on beam deflection.

There is a brief treatment of statically indeterminate beams in Chapter 9. Here, I have emphasized the conditions that make a beam statically indeterminate.

Chapter 10 is a thorough treatment of torsion. I felt that it was necessary to include quite a bit on noncircular sections, such as rectangles and thin-walled sections, both open and closed.

In Chapter 11, I decided to use the maximum shear stress as the criterion of failure in presenting bending combined with torsion. It is the simplest criterion, and it represents the behavior of most commonly used materials. Here I have also shown the use of torque diagrams.

It does not seem possible that all can agree on what single type of column formula is the best, so I have explained the use of several different types in Chapter 12. I hope that the student will then be able to use whatever type he or she needs in subsequent engineering work.

Chapter 13, which discusses riveted, bolted, and welded joints, shows a personal view of mine. I believe that the huge riveted joint with many repeating sections is obsolete; therefore, I have concentrated on the simple type of joint that is used to fasten one member to another.

Most failures of members in machines and structures occur under repetition of loads and at points of stress concentration. This is why I have included Chapter 14—Repeated Stress and Stress Concentration.

Now, here is the book.

Charles O. Harris

CONTENTS

LIST OF TABLES

STATICS AND STRENGTH OF MATERIALS

1 INTRODUCTION TO STATICS

PURPOSE OF THIS CHAPTER. The purpose of this chapter is to introduce you to the subject of *statics*. Here, we will study some of the fundamental ideas and principles of statics that are used in all problems of *strength of materials*. New words such as *force* and *moment* will be introduced, but don't be afraid of them, because each term will be explained when we are ready for it. The basic principles and methods of statics are easy to understand if taken one at a time and studied thoroughly so that each principle is understood before going on to the next. You must master the fundamental principles of this chapter so you will be ready for the chapters which follow.

You must learn what force is (Section 1-3) and how to make calculations with forces. You must also learn what the moment of a force is (Section 1-7), why it is important, and how to calculate it. These basic ideas form the introduction to the subject of statics. You must learn statics to be able to study strength of materials.

Now it is time for you to learn what strength of materials is, and then you will want to know how strength of materials is used in different branches of engineering. Knowing these things will make it easier and more interesting for you to study this book.

1-1. WHAT STRENGTH OF MATERIALS IS

Strength of materials is the study of the relations between (1) the forces that are applied to a member of a structure, (2) the size and shape of the member, and (3) the internal forces and deformations of the member. It is a basic subject. Its principles and methods apply to any case in which the function of an object is to withstand forces. So you can apply what you will learn here whether you are going to design machine parts, members of buildings and bridge structures, components of airplanes and boats, transmission line towers, or whatever.

1-2. HOW THE DESIGNER USES STRENGTH OF MATERIALS

The engineering designer uses his or her knowledge of strength of materials to design each part of a machine or structure in such a manner

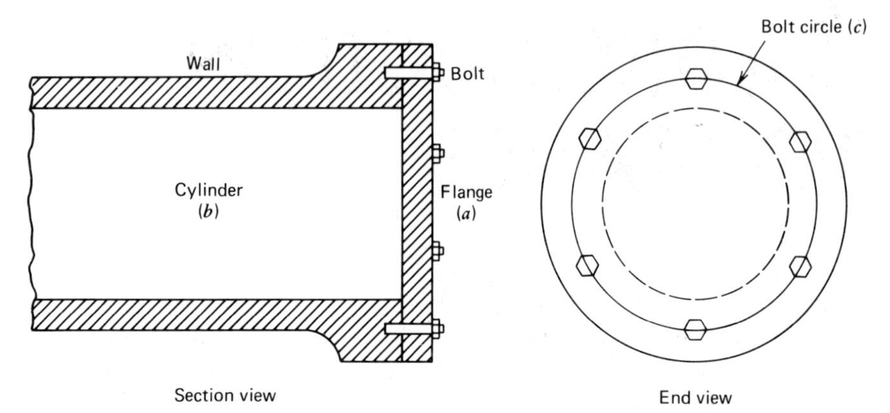

Fig. 1-1. Part of cylinder of air compressor.

that the part will have the necessary strength and stiffness, in this way making sure that it will neither break nor deform in ordinary use or under normal conditions of service. Your job in studying this book is to learn how to do this.

Here is a simple example of the application of strength of materials. Figure 1-1 shows a cross section and end view of part of the cylinder of an air compressor; it holds air under pressure (say, 120 pounds per square inch). The flange (*a*) is fastened to the cylinder (*b*) by the bolts in the bolt circle (*c*). The machine designer faces certain problems in designing a unit even as simple as this. In this case the problems are: (1) How thick shall I make the wall of the cylinder so that it won't break at this pressure? (2) How many bolts do I need in the bolt circle, and what size bolts shall I use? (3) How deep should the bolt holes be tapped in the cylinder so that the bolts won't pull out? (4) How tightly should the bolts be drawn so the cylinder will be airtight under pressure?

Figure 1-2 shows another example. This is the part of a building structure. A steel beam is welded to two steel columns. The person who designs this system must decide: (1) what size and shape of beam to use so that it will be strong enough to support the loads and so that it won't bend too much; (2) what size and arrangement of welds to use to fasten the beam to the column so that the welds will be strong enough to

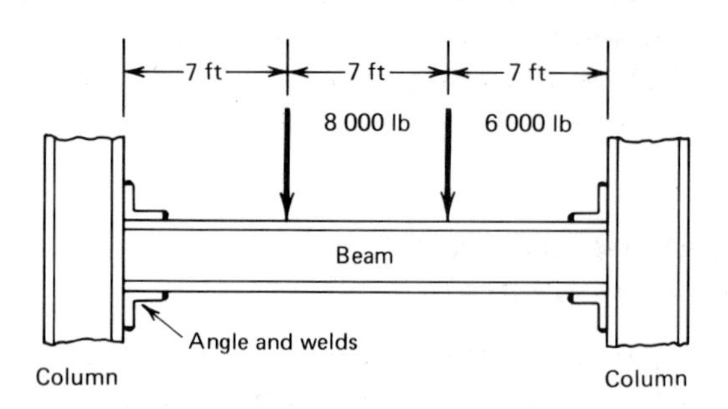

Fig. 1-2. Beam and columns.

support the beam; and (3) what size and shape of columns to use so that they will be strong enough to support the beam.

These are only two of a great many examples to which knowledge of strength of materials can be applied. As we go through this book you will encounter many others, but first we must go back to the beginning and learn the simple elements of statics.

1-3. FORCE

Force is our starting point. Practically all problems in strength of materials deal with forces; therefore, it is necessary to get a clear idea of what a force is. A *force* is a push or pull exerted by one body on another; we represent a force by an arrow, as in Fig. 1-3, which shows forces P, Q, and F acting on a bell crank (a *bell crank* is a bent lever). The forces P and F are pulls exerted by parts of the machine that are fastened to the bell crank; the force P is exerted by a link that pulls on the bell crank to move it, and the force F is exerted by a member that is moved by the bell crank. The force Q is a push exerted by a bearing that supports the bell crank and holds it so that it turns about the center O. In this case the bearing is just a block of steel with a circular hole. A pin in the bell crank goes through the hole.

One of the most commonly encountered forces is weight; the *weight* of an object is the force of gravity that is exerted on the object by the earth. Weight is a downward force that tends to pull the body to the center of the earth. We can show it as W, as in Fig. 1-4.

Figure 1-5 shows forces acting on a piston of the type used in a gas engine. All forces here are pushes. The force P is exerted on the piston by the connecting rod; the force Q is exerted by the combustion gases on the piston, and the force N is exerted on the piston by the wall of the cylinder.

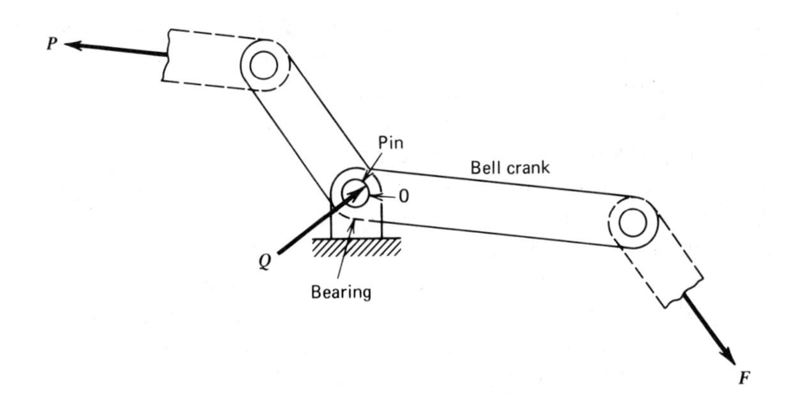

Fig. 1-3. Forces on a bell crank.

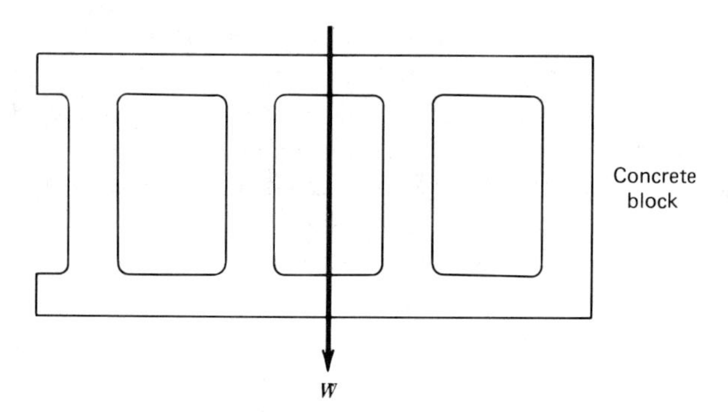

Fig. 1-4. Weight of an object.

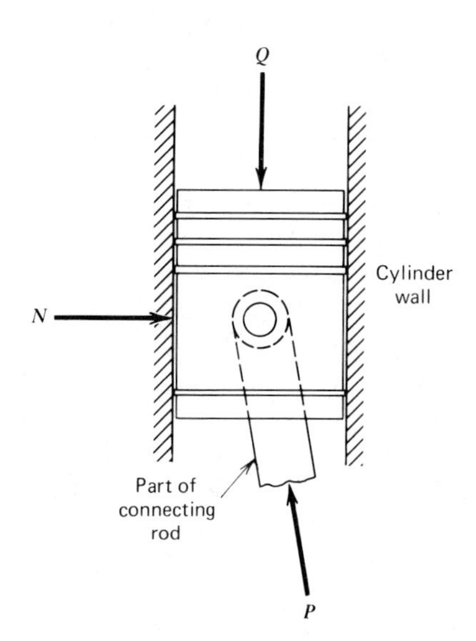

Fig. 1-5. Forces on piston.

1-4. EFFECTS OF A FORCE

Force has two effects: (1) the *external effect,* which is the change in motion of the body acted upon: starting it or stopping it, speeding it up or slowing it down; and (2) the *internal effect,* which is the development of internal forces and deformation within the body acted on. (*Deformation* is a bending, twisting, or stretching action that alters the form or shape of a body.) Sometimes the external effect of a force is the most dominant, as when an automobile is stopped by applying the brakes. In other cases the internal effect is more noticeable; for example, the forming of a flat piece of steel into an automobile fender or in compressing a helical spring, as shown in Fig. 1-6. It is the internal effect with which we are concerned in pursuing the study of strength of materials. The external effect is treated in the subject of dynamics.

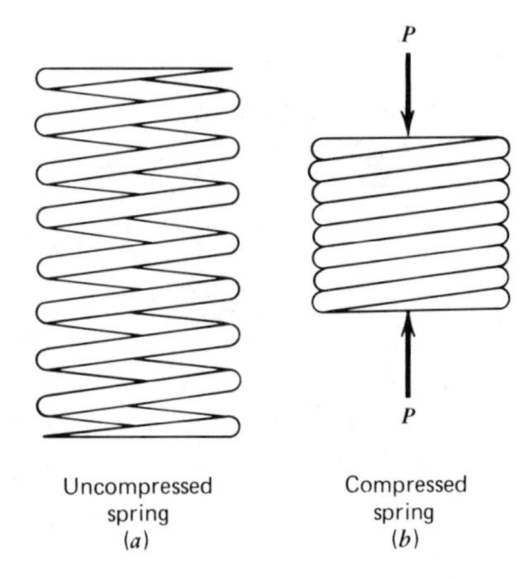

Fig. 1-6. Helical spring.

1-5. CHARACTERISTICS OF A FORCE

The characteristics of a force are the properties by which it is described. They are its *magnitude* (size), its *direction,* and its *location.* These are the properties that must be known about a force in order to calculate the effect it will produce when it is exerted on a body. (*Body* is the term used to designate an object.)

The magnitude of a force is simply the amount or size of the force. There are three units of force that have considerable use in American engineering practice.

The first of these units is the pound. Thus a force might be expressed as 100 pounds or 8 000 pounds; pounds is often abbreviated as lb.

The second unit in which force is expressed is the kip. A kip is equivalent to 1 000 pounds. The kip is used mostly in structural work. Its main advantage is that it saves writing zeros in designating large forces. Thus a force that has a magnitude of 116 000 lb could be written as 116 kips.

The third unit for force is the newton, which is part of the metric system, also known as *Système International d' Unités* (SI). One pound is equivalent to 4.448 newtons. Newton is abbreviated as N.

The United States is in the process of converting to the metric system, but the process has not yet been completed and, in fact, may never be completed. So a person who is going to apply a knowledge of strength of materials may need to work with each of these force units at one time or another. For this reason we are going to give you some problems in which the forces are expressed in pounds, some in which the forces are expressed in kips, and some in which the forces are expressed in newtons. Then you will be ready to use whatever units you may need to use in practice.

The *direction* of a force is the angle the force makes with some reference line. The best way to describe this is to show the direction in

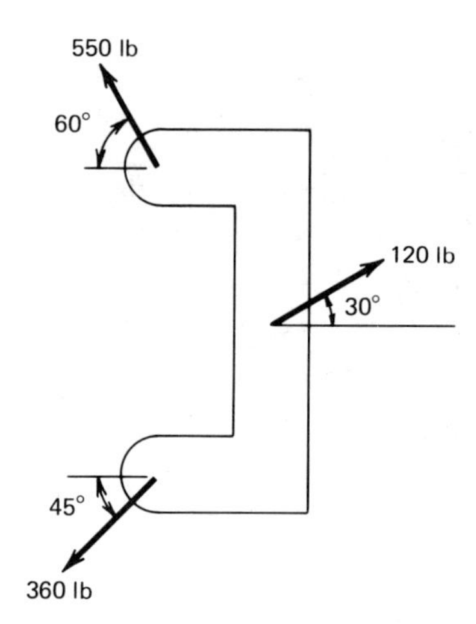

Fig. 1-7. Directions of forces on a machine part.

a sketch giving the acute angle the force makes with a horizontal line. For example, the 120-lb force acting on the machine part in Fig. 1-7 is upward to the right at an angle of 30° with the horizontal, and the 360-lb force is downward to the left at an angle of 45° with the horizontal.

The *location* of a force is the place (a line) along which the force acts. The effect of a force depends a great deal on its location. For example, it is easier to bend a pipe, such as the one shown in Fig. 1-8, by applying the force P at A than by applying it along the line B because of the greater leverage of the force at A.

Force is a vector. This means that it has direction as well as magnitude. The direction of a force exerted by one body on another is specified by giving the angle the force makes with a horizontal line, and the magnitude or amount of the force is usually given in pounds, kips, or newtons. Anything that has direction as well as magnitude is called a *vector*. Anything that has magnitude only and that can be completely described without stating a direction is called a *scalar*. Temperature is a good example of a scalar.

1-6. COMPONENTS OF A FORCE

If we could have our way, every force applied to a part of a machine or structure would be parallel to the length of the part or perpendicular

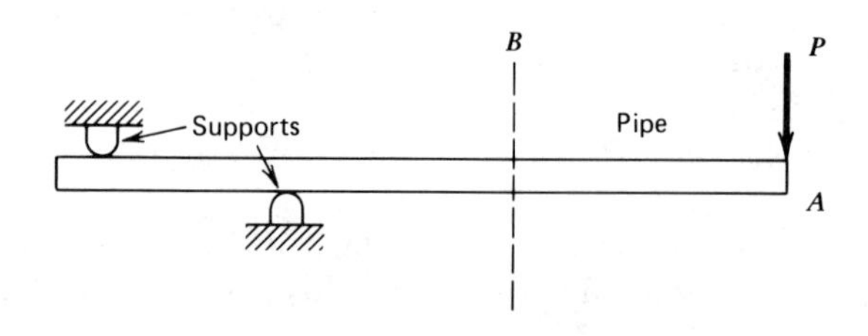

Fig. 1-8. Location of force on a pipe.

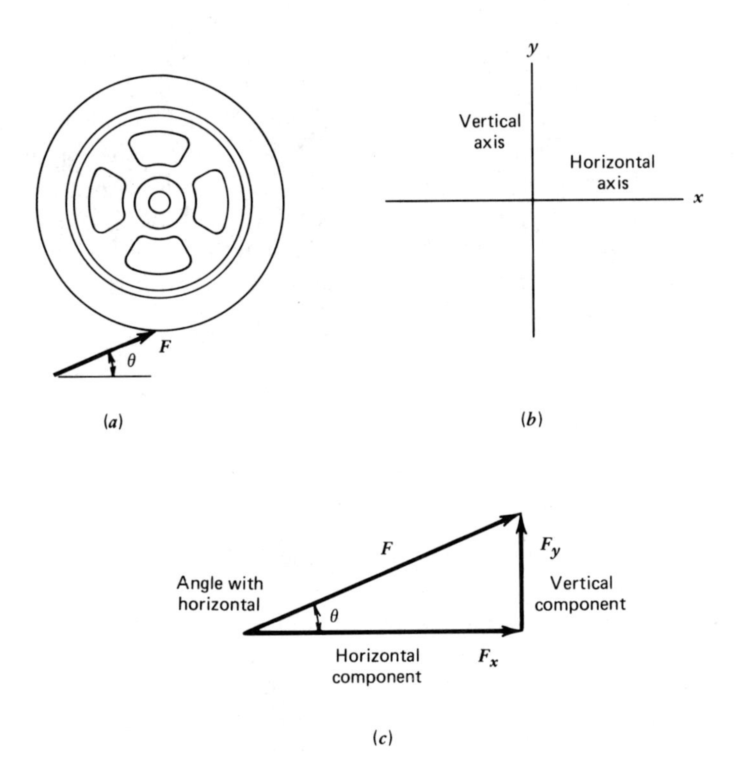

Fig. 1-9. Components of a force. (*a*) Force on wheel. (*b*) *xy* Coordinate system. (*c*) Force triangle.

to it, because it is easier to make calculations with forces so exerted. It doesn't always work out this way, unfortunately, so what we must do in many problems is to resolve each force into two components, one parallel to a line we select and one perpendicular. This means that we replace the force with two other forces that are easier to work with and will produce the same effect.

A *component* of a force is one of two or more forces that can replace the given force. A force can be resolved into components in any direction whatever, but here we are interested only in components that are perpendicular to each other. This is the way we go about it. Let us say that we are given a force of magnitude F acting on a wheel, as in Fig. 1-9*a*. The force is upward to the right at an angle θ (Greek letter *theta*) with the horizontal. We want to resolve this force into two components, one horizontal and one vertical. It is convenient to work in terms of an xy coordinate system like that shown in Fig. 1-9*b*. (You probably remember this from your study of algebra; x is horizontal and y is vertical). We can designate the horizontal component by F_x (F sub x) and call it the x component, and designate the vertical component by F_y (F sub y) and call it the y component. Then in Fig. 1-9*c* we draw a right triangle with the force F as the hypotenuse, making one leg horizontal and the other vertical. If this is drawn to scale, the horizontal leg represents the x

component of the force F_x, and the vertical leg represents the y component of the force F_y. Notice that the arrowheads on F_x and F_y are so directed that if you follow them from the beginning of F you arrive at the end of F. The relation between the force F and its components is the same as the relation between the sides of a triangle. Now you should remember from trigonometry that the sine of the angle θ is equal to the vertical leg of the triangle divided by the hypotenuse. In this case, where the vertical leg represents the y component of the force F_y and the hypotenuse represents the force F, we have

$$\frac{F_y}{F} = \sin\theta$$

from which

$$F_y = F\sin\theta$$

This equation says that the vertical component of the force is the product of the force and the sine of the angle the force makes with the horizontal. Again you should remember from trigonometry that the cosine of the angle θ is equal to the horizontal leg of the triangle divided by the hypotenuse. Here the horizontal leg of the triangle represents the horizontal component of the force F_x, so we have for the cosine

$$\frac{F_x}{F} = \cos\theta$$

or

$$F_x = F\cos\theta$$

This equation says that the horizontal component of the force is the product of the force and the cosine of the angle the force makes with the horizontal. Now we are ready for some examples.

From this point on we are going to show you many illustrative examples with numerical calculations, and we are going to ask you to work many practice problems and review problems, also with numerical calculations. We are going to work on the basis that you will be using an electronic calculator from which you can find the sine or cosine of any angle, and with which you can carry out all of the arithmetic, such as addition, subtraction, multiplication, division, square, square root, and so on. But, as we go through these calculations, we are going to keep in mind that the calculator may display as many as eight significant figures in the result of a calculation and that not all of these figures are meaningful as engineering results. There is likely to be so much uncertainty in the initial data of an engineering problem that it is not useful to express more than three or four significant figures in the result of a calculation. So we are going to round off our answers to three or four significant figures. Now, punch your calculator along with us as we go through many illustrative examples in this book.

Illustrative Example 1. Calculate the horizontal and vertical components of the 200-lb force of Fig. 1-10*a*.

Solution:
The force has a magnitude of 200 lb and the angle θ is 60°. Fig. 1-10*b* shows the triangle of forces.

1. The horizontal component is

$$F_x = F \cos \theta = 200 \cos 60°$$

You can use your calculator to determine that the cosine of 60° is 0.5. Then, using this value,

$$F_x = 200 \times 0.5 = 100 \text{ lb}$$

2. The vertical component of the force is

$$F_y = F \sin \theta = 200 \sin 60°$$

Again, you can use your calculator to find that the sine of 60° is 0.866, using three significant figures. Using this value,

$$F_y = 200 \times 0.866 = 173.2 \text{ lb}$$

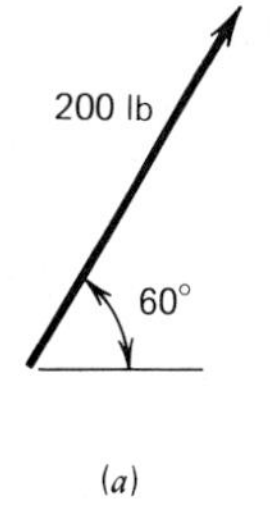

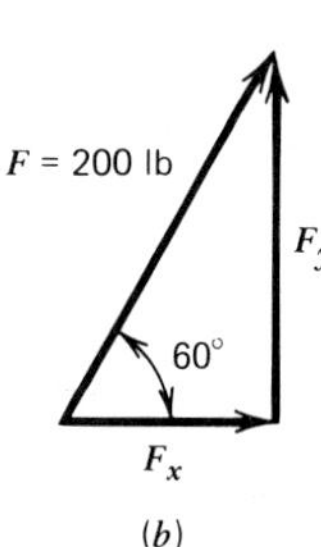

Fig. 1-10. Force for Example 1. (*a*) Force. (*b*) Force triangle.

Illustrative Example 2. Calculate the horizontal and vertical components of the 165-N force in Fig. 1-11*a*.

Solution:
The magnitude of the force is 165 N. The direction is given by the angle of 35°.

1. The horizontal component is

$$F_x = F \cos \theta = 165 \cos 35°$$

You can punch all of this through on your calculator to get

$$F_x = 135 \text{ N}$$

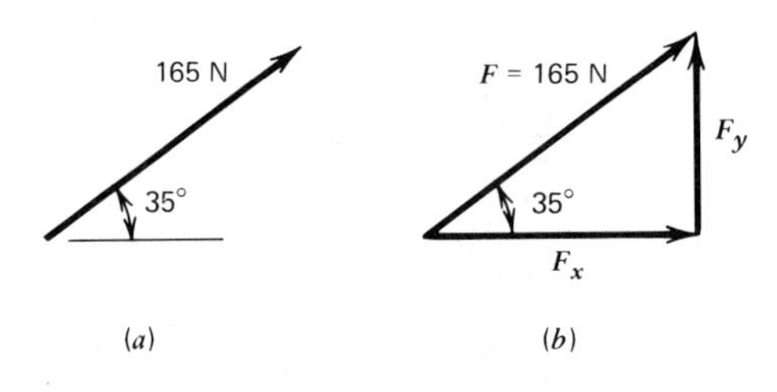

Fig. 1-11. Force for Example 2. (*a*) Force. (*b*) Force triangle.

(You might read the calculator as 135.16009 for this answer instead of 135, but we round this off to 135.)

2. The vertical component is

$$F_y = F \sin \theta = 165 \sin 35°$$

and punching the calculator and rounding off,

$$F_y = 94.6 \text{ N}$$

OTHER DIRECTIONS

There is a little more to it than this. Forces aren't always directed upward to the right; they may instead be upward to the left, downward to the left, or downward to the right, as are the forces acting on the *bent lever* in Fig. 1-12. If you think of the horizontal component of a force as the horizontal projection of the force, you can tell by looking at it whether the horizontal component is directed toward the right or toward the left. Also, if you think of the vertical component of a force as the vertical projection of the force, you can tell at once whether the vertical component is directed upward or downward. Then we use the scheme shown in Fig. 1-13. We say that a horizontal component is positive (plus) if it is directed toward the right and negative (minus) if it is directed toward the left. A vertical component is positive if it is directed upward and negative if it is directed downward. (This is about the same scheme as the one you used in plotting curves in algebra when you plotted an x coordinate to the right if it was positive and to the left if negative, a y coordinate upward if it was positive and downward if negative). Now let's try a few examples.

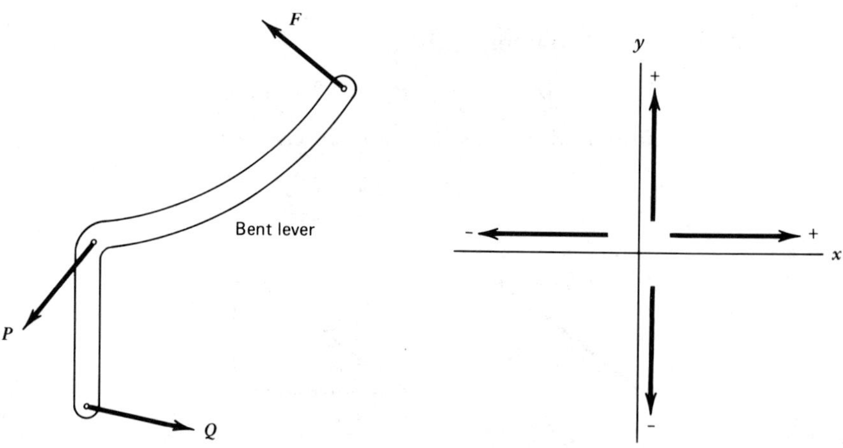

Fig. 1-12. Bent lever.

Fig. 1-13. Signs of components.

Illustrative Example 3. Find the horizontal and vertical components of the force shown in Fig. 1-14.

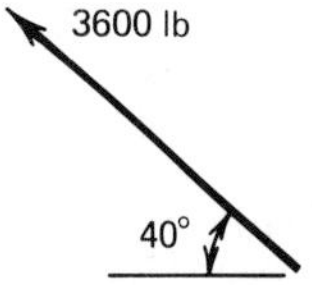

Fig. 1-14. Force for Example 3.

Solution:

1. The force is directed upward to the left, so the x component is directed to the left and is, therefore, negative. Thus

$$F_x = -F \cos \theta = -3\,600 \cos 40°$$
$$= -2\,758 \text{ lb}$$

2. The y component is directed upward and is, therefore, positive

$$F_y = F \sin \theta = 3\,600 \sin 40° = 2\,314 \text{ lb}$$

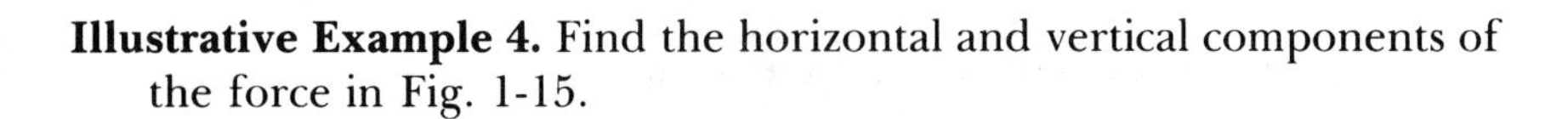

Illustrative Example 4. Find the horizontal and vertical components of the force in Fig. 1-15.

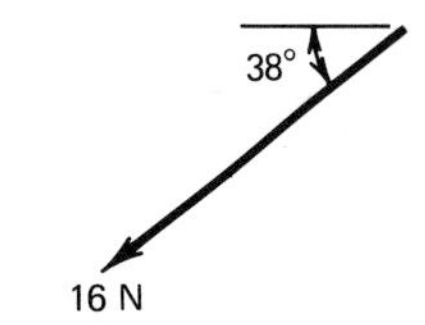

Fig. 1-15. Force for Example 4.

Solution:

The horizontal component is directed to the left and the vertical component is directed downward, so both are negative. Then,

1. $F_x = -F \cos \theta = -16 \cos 38° = -12.6$ N
   and
2. $F_y = -F \sin \theta = -16 \sin 38° = -9.85$ N

Illustrative Example 5. Find the x and y components of the force in Fig. 1-16.

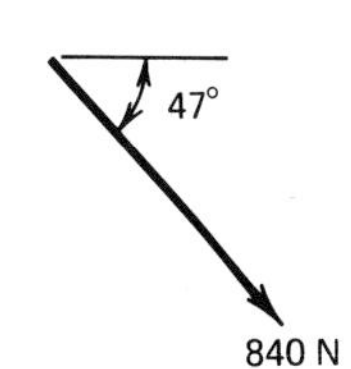

Fig. 1-16. Force for Example 5.

Solution:

1. The x component is seen to be directed to the right, so it is positive. Then,

$$F_x = 840 \cos 47° = 573 \text{ N}$$

2. The y component is directed downward, so it is negative. Then,

$$F_y = -840 \sin 47° = -614 \text{ N}$$

Practice Problems (Section 1-6).[1] Find the horizontal and vertical component of each of the forces described in the following problems. Indicate directions by giving a positive sign to a horizontal component that is directed to the right or a vertical component that is

[1]You can check your answers to Practice Problems by referring to the section at the back entitled Answers to Problems.

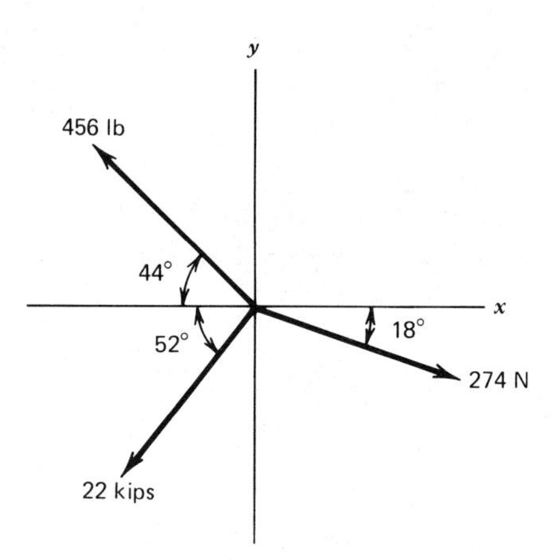

Fig. 1-17. Forces for Problems 8, 9, and 10.

directed upward and by giving a negative sign to a horizontal component that is directed to the left or a vertical component that is directed downward.

1. A force of 86 lb that is directed upward to the right at an angle of 50° with the horizontal.
2. A force of 27.5 N that is directed upward to the right at an angle of 20° with the horizontal.
3. A force of 7.3 lb that is directed downward to the left at an angle of 32° with the horizontal.
4. A force of 165 kips that is directed upward to the left at an angle of 60° with the horizontal.
5. A force of 668 N directed upward to the left at 37.5° with the horizontal.
6. A force of 1 170 N directed downward to the left at 29° with the horizontal.
7. A force of 9.3 lb directed downward to the left at 65.4° with the horizontal.
8. The 274-N force in Fig. 1-17.
9. The 456-lb force in Fig. 1-17.
10. The 22-kip force in Fig. 1-17.

CALCULATING A FORCE WHEN ITS COMPONENTS ARE KNOWN

Sometimes it is necessary to turn the problem around. You may be able to calculate the components of a force from other data, and then you may want to find its magnitude F and its direction θ. This can be done by drawing a triangle like the one in Fig. 1-18. Here the relation between the force F and its components F_x and F_y is the same as between the sides of a triangle. You should remember the Pythagorean theorem from plane geometry. This theorem states that the square of the hypotenuse of a right triangle is equal to the sum of the squares of the other two sides. It can be applied to give

$$F^2 = F_x^2 + F_y^2$$

Fig. 1-18. Force triangle.

Taking the square root of each side of the equation,

$$F = \sqrt{F_x^2 + F_y^2}$$

If you know the values of F_x and F_y you can calculate F from this equation. Then you can calculate

$$\frac{F_y}{F_x} = \tan\theta$$

This gives the value of the tangent of θ, and you can find the angle θ with your calculator.

The best way to show the answers is to make a sketch such as that shown in Fig. 1-18.

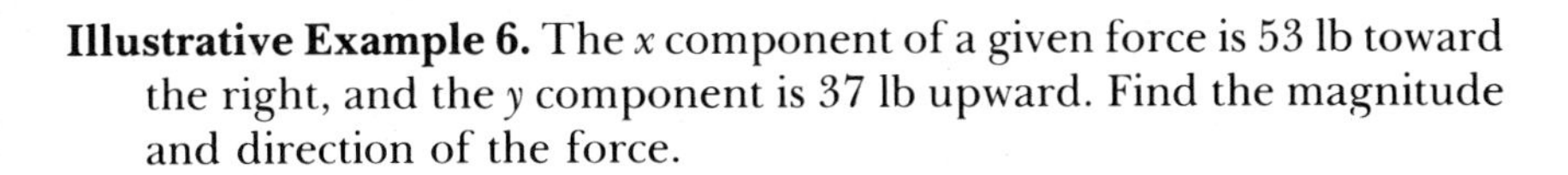

Illustrative Example 6. The x component of a given force is 53 lb toward the right, and the y component is 37 lb upward. Find the magnitude and direction of the force.

Solution:
Figure 1-19 shows the triangle from which F and θ are calculated. The force F must be directed upward to the right, because its x component is toward the right and its y component is upward.

1. The Pythagorean theorem gives

$$F = \sqrt{F_x^2 + F_y^2} = \sqrt{(53)^2 + (37)^2}$$

$$= \sqrt{2\,809 + 1\,369} = \sqrt{4\,178} = 64.6 \text{ lb}$$

2. Then for the angle θ

$$\tan\theta = \frac{F_y}{F_x} = \frac{37}{53} = 0.698$$

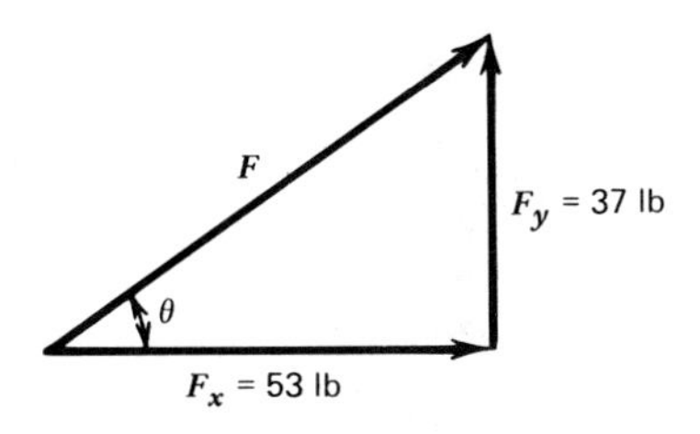

Fig. 1-19. Force triangle for calculating F and θ.

Your calculator will give you $\theta = 34.92°$

Illustrative Example 7. Find the magnitude and direction of a force that has a horizontal component of 440 N toward the left and a vertical component of 580 N downward.

Solution:

Figure 1-20 shows the triangle of forces. The horizontal component of the force is directed toward the left and the vertical component is directed downward, so the force F must be directed downward to the left.

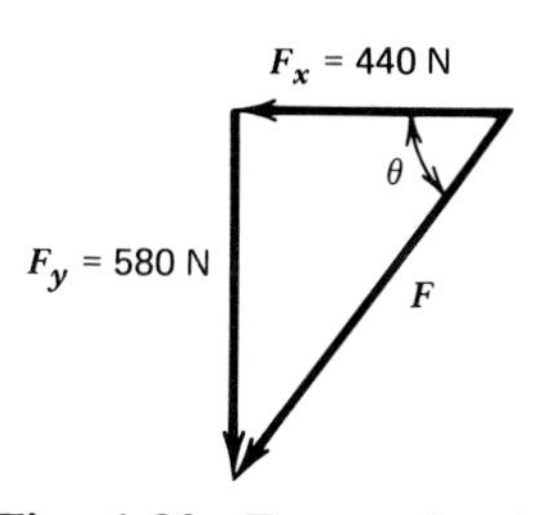

Fig. 1-20. Force triangle for magnitude and direction of force.

1. The magnitude of the force is

$$F = \sqrt{F_x^2 + F_y^2} = \sqrt{(440)^2 + (580)^2}$$

$$= \sqrt{193\,600 + 336\,400} = \sqrt{530\,000} = 728 \text{ N}$$

2. The angle θ can be found from

$$\tan\theta = \frac{F_y}{F_x} = \frac{580}{440} = 1.318$$

Then,

$$\theta = 52.82°$$

Illustrative Example 8. Find the magnitude and direction of a force that has an x component of 170 lb to the left and a y component of 215 lb upward.

Solution:

The triangle of forces is shown in Fig. 1-21. The force F must be directed upward to the left, because its x component is directed toward the left and its y component is directed upward.

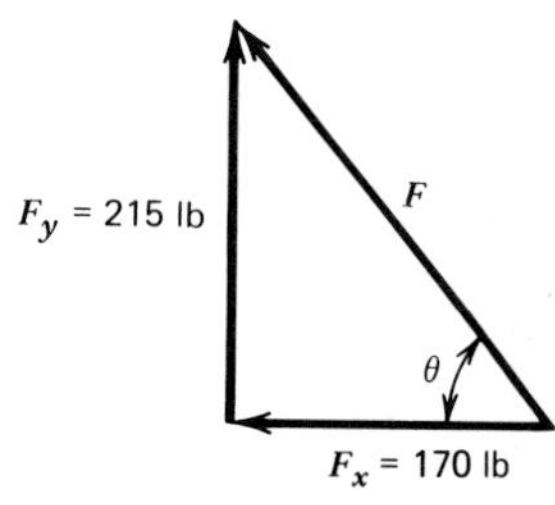

Fig. 1-21. Force directed upward to left.

1. F is given by

$$F = \sqrt{F_x^2 + F_y^2} = \sqrt{(170)^2 + (215)^2}$$

$$= \sqrt{28\,900 + 46\,225} = \sqrt{75\,125} = 274.1 \text{ lb}$$

2. The tangent of the angle θ is

$$\tan\theta = \frac{F_y}{F_x} = \frac{215}{170} = 1.265$$

Then,

$$\theta = 51.7°$$

Practice Problems (Section 1-6 *continued*). In each of the following problems the two components of a force are given. Find the magnitude and direction of the force. Each force is a push or a pull exerted by one body on another.

1. The horizontal component is 40 lb to the right; the vertical component is 26 lb upward.
2. The horizontal component is 5 700 lb to the left; the vertical component is 1 200 lb upward.
3. The horizontal component is 280 lb to the left; the vertical component is 760 lb downward.
4. The horizontal component is 108 lb to the right; the vertical component is 62 lb downward.
5. $F_x = 16\ 000$ N; $F_y = 9\ 000$ N.
6. $F_x = -195$ N; $F_y = 240$ N.
7. $F_x = 11$ N; $F_y = -4.9$ N.
8. $F_x = -310$ N; $F_y = -430$ N.

1-7. MOMENT OF A FORCE

The moment (M) of a force with respect to a point is the product of the force and the perpendicular distance from the point to the line along which the force acts. Thus in Fig. 1-22 the moment of the force F with respect to the point O is

$$M_o = Fd$$

The term M_o *means the moment of the force with respect to point* O. *Point* O *is called the* ***moment center****. The perpendicular distance* d *is called the* ***moment arm.***

The moment of a force is an important quantity, because the moment is a measure of the tendency of the force to turn or bend the body on which it acts. In Fig. 1-23*a*, where the wheel is mounted on an axle, the

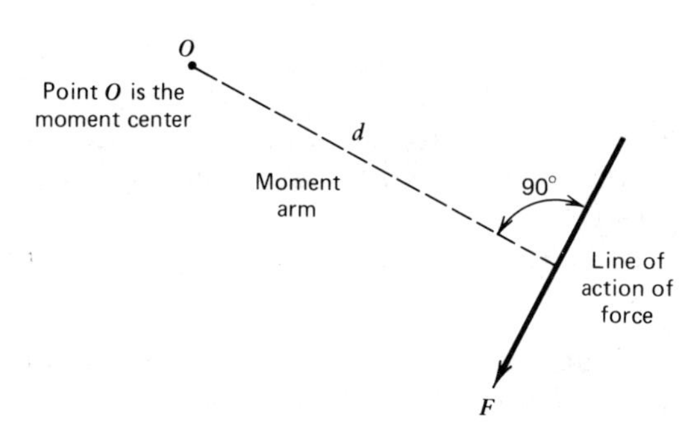

Fig. 1-22. Moment of a force.

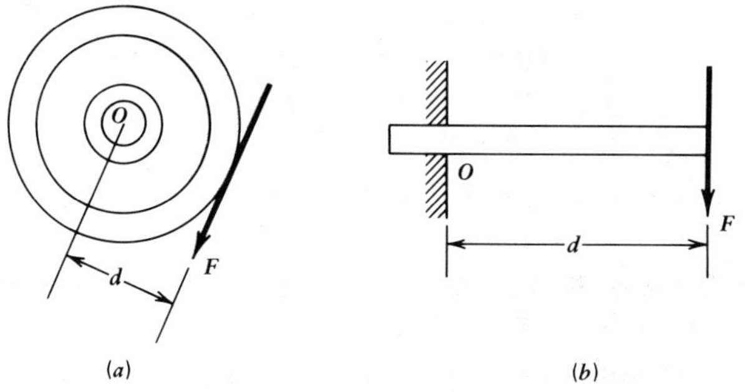

Fig. 1-23. Forces and moment arms. (*a*) Force on wheel. (*b*) Force on Cantilever beam.

force F tends to turn the wheel, and the moment of the force with respect to the center of the wheel, Fd, is a measure of this tendency. The greater the moment, the greater the tendency to turn. In Fig. 1-23*b*, where the force F is applied to the end of a cantilever beam, the force tends to bend the beam, and the moment of the force with respect to point O (Fd) is a measure of the tendency. We shall have many occasions to find the moment of a force in later chapters, and the full significance of moment will be brought out then.

Looking again at Fig. 1-23*a* you can see that the force F tends to turn the wheel in a clockwise direction, which is the direction in which the hands of a clock move. If the force acted in the opposite direction, it would tend to turn the wheel in a counterclockwise direction, which is opposite to the way the hands of a clock move. You can tell at once by inspection whether the moment of a force with respect to a certain point is clockwise or counterclockwise. It is necessary for us to adopt some simple way to distinguish between clockwise and counterclockwise moments, and so we will follow the rule that a clockwise moment is positive (plus) and a counterclockwise moment is negative (minus). If you ever have trouble deciding whether a moment is clockwise or counterclockwise, just imagine that the force is applied to the rim of a wheel that is mounted on an axle through the hub, and visualize which way the wheel would be turned by the force. The direction of turning is the direction of the moment.

UNITS

If we are using the English gravitational system units for force and length, the force will be expressed in pounds or kips, and the moment arm will be expressed in feet or inches. Then, when we calculate the moment of a force, the units of the moment will be a unit of force times a unit of length. So

1. If the force is in pounds and the moment arm in feet, the moment will be in pound feet, abbreviated as lb ft.

2. If the force is in kips and the moment arm in feet, the moment will be in kip feet, abbreviated as kip ft.
3. If the force is in pounds and the moment arm in inches, the moment will be in pound inches, abbreviated as lb in.
4. If the force is in kips and the moment arm in inches, the moment will be in kip inches, abbreviated as kip in.

If we are using the metric system, the force will be expressed in newtons, and the moment arm will be expressed in meters, centimeters, or millimeters. Thus

1. If the force is in newtons and the moment arm in meters, the moment will be in newton meters, abbreviated as N·m.
2. If the force is in newtons and the moment arm in centimeters, the moment will be in newton centimeters, abbreviated as N·cm.
3. If the force is in newtons and the moment arm is in millimeters, the moment will be in newton millimeters, abbreviated as N·mm.

Remember, 1 in. is equivalent to 0.0254 m or 2.54 cm; 1 m is equivalent to 39.47 in.

Now let's try an example or two.

Illustrative Example 1. Figure 1-24 shows a bar that is subjected to two forces. Each space is 1 ft in width. What is the moment of the 400-lb force with respect to point *O*?

Solution:
The moment arm of the force is 3 ft. The moment of the force with respect to point *O* is clockwise and is, therefore, positive. Then,

$$M_o = Fd = 400 \times 3 = 1\ 200 \text{ lb ft}$$

Illustrative Example 2. Calculate the moment of the 250-lb force in Fig. 1-24 with respect to point *O*.

Solution:
The moment arm of the force is 8 ft. The moment of the force is

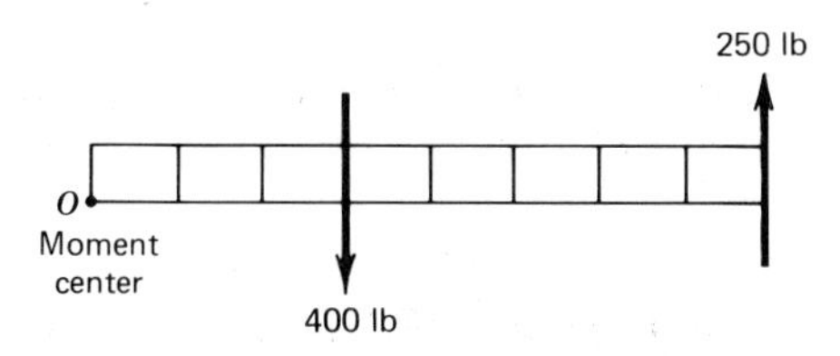

Fig. 1-24. Forces acting on a bar.

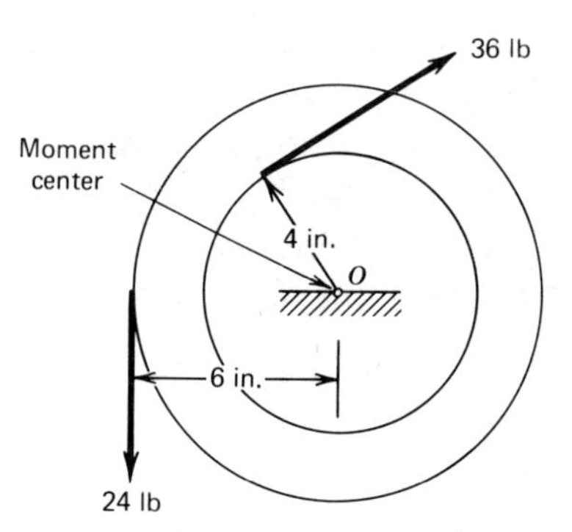

Fig. 1-25. Compound pulley.

counterclockwise with respect to point O and is, therefore, negative. So

$$M_o = -Fd = -250 \times 8 = -2\,000 \text{ lb ft}$$

MOMENT ARM IN INCHES

Sometimes it is more convenient for us to take the moment arm in inches. Then we are multiplying a force in pounds by a distance in inches, and the result is in pound inches (lb in.). Let's try it.

Illustrative Example 3. The compound pulley in Fig. 1-25 is acted on by the two forces shown. Calculate the moment of the 24-lb force with respect to the center point O.

Solution:

The moment arm of the force is 6 in. The force tends to turn the wheel in the counterclockwise direction, and therefore, the moment of the force is negative. Thus

$$M_o = -Fd = -24 \times 6 = -144 \text{ lb in.}$$

METRIC UNITS

Now let's try an example or two with metric units.

Illustrative Example 4. What is the moment of the 460-N force in Fig. 1-26 with respect to point O?

Solution:

The moment arm of the force is 0.2 m. It is seen that the force tends

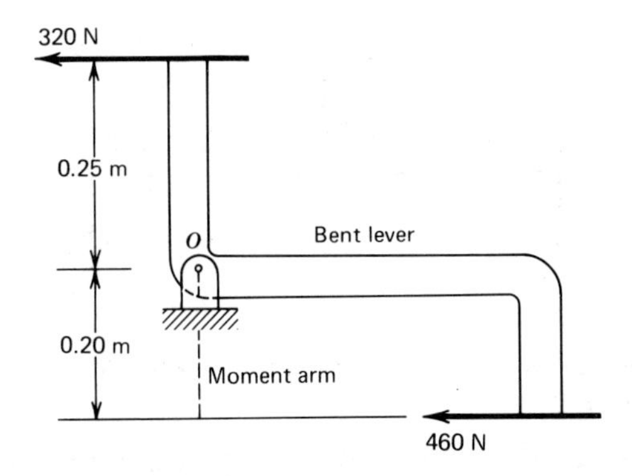

Fig. 1-26. Bent lever.

to turn the lever clockwise, and therefore, its moment is positive. Then,

$$M_o = Fd = 460 \times 0.2 = 92 \text{ N·m}$$

Illustrative Example 5. Figure 1-26 shows a bent lever. What is the moment of the 320-N force with respect to point O?

Solution:
The moment arm is 0.25 m. The moment of the force is counterclockwise with respect to point O and is, therefore, negative. Then,

$$M_o = -Fd = -320 \times 0.25 = -80 \text{ N·m}$$

Now you are ready to try it for yourself.

Practice Problems (Section 1-7). Working the following problems will help you to remember how to find the moment of force with respect to a point.

1. What is the moment of the 310-lb force in Fig. 1-27 with respect to point O?

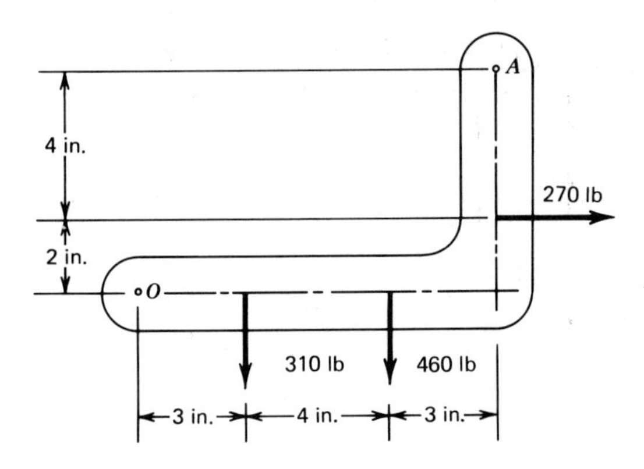

Fig. 1-27. Forces acting on a machine link.

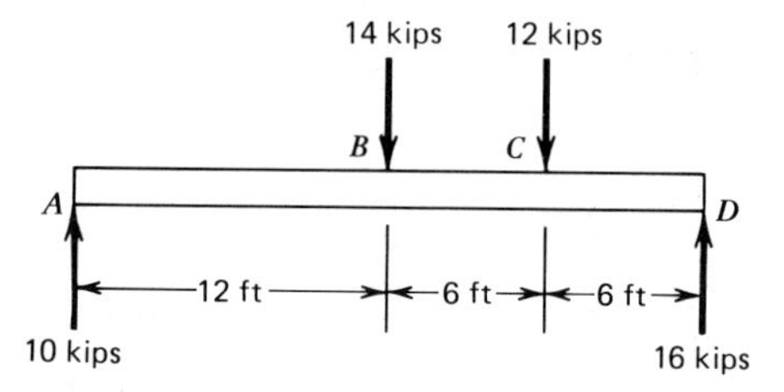

Fig. 1-28. Forces on a beam.

2. Find the moment of the 460-lb force in Fig. 1-27 with respect to point *O*.
3. In Fig. 1-27, what is the moment of the 270-lb force with respect to point *O*?
4. Calculate the moment of the 460-lb force in Fig. 1-27 with respect to point *A*.
5. Find the moment of the 310-lb force in Fig. 1-27 with respect to point *A*.
6. What is the moment of the 270-lb force with respect to point *A* in Fig. 1-27?
7. Figure 1-28 shows a beam in a building structure. Calculate the moment of the 14-kip force with respect to point *A*.
8. In Fig. 1-28, calculate the moment of the 12-kip force with respect to point *D*.
9. In Fig. 1-29, what is the moment of the 1 600-N force with respect to point *B*?
10. Find the moment of the 2 200-N force with respect to point *B* in Fig. 1-29.
11. Find the moment of the 1 100-N force at point *C* with respect to point *B* in Fig. 1-29.

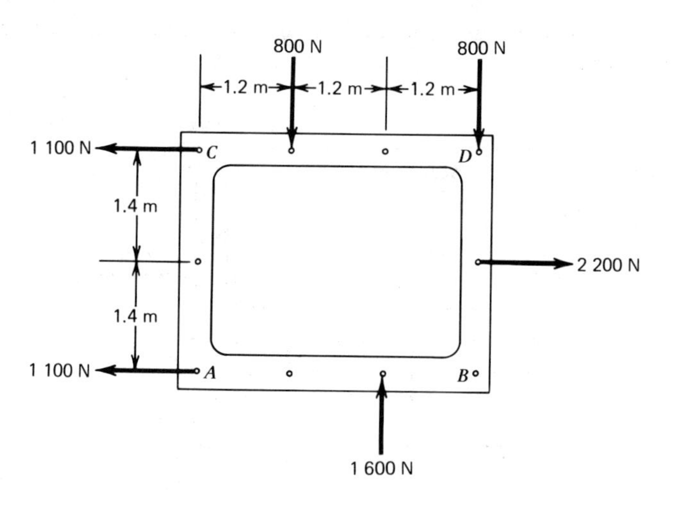

Fig. 1-29. Forces on a frame.

12. In Fig. 1-29, what is the moment of the 800-N force at point D with respect to point A?

1-8. PRINCIPLE OF MOMENTS

We could always calculate the moment of a force with respect to a point by multiplying the force by the perpendicular distance from the point to the line of action of the force, but there are many problems in which this procedure is not practical, because it is too difficult to calculate the perpendicular distance. It is often better to use the *principle of moments.* This important principle states that the moment of a force with respect to any point is equal to the sum of the moments of the two components of the force with respect to the point, provided the components are placed so they intersect on the line of action of the force. Let's see how it works. In Fig. 1-30*a* we would like to calculate the moment of the force F with respect to the point O. This is

$$M_o = Fd$$

But suppose that the distance d is hard to find. Then we can resolve the

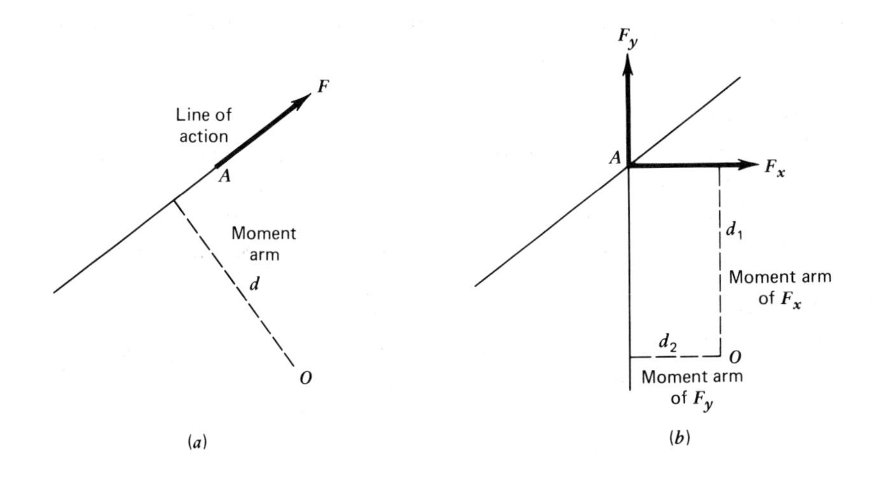

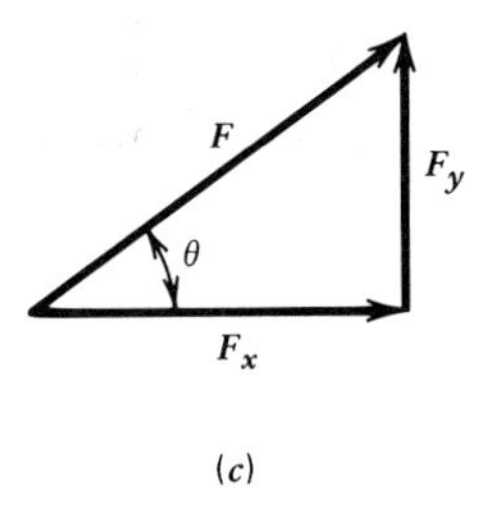

Fig. 1-30. Principle of moments. (*a*) Force. (*b*) Components of force. (*c*) Force triangle.

force into x and y components (F_x and F_y) at any point such as A, Fig. 1-30b, and the moment of F with respect to point O is equal to the moment of F_x (its x component) with respect to point O plus the moment of F_y (its y component) with respect to point O. Fig. 1-30b shows the components F_x and F_y, which have replaced the force F at A. The moment of F_x is ($F_x d_1$), and the moment of F_y is ($F_y d_2$), so

$$M_o = F_x d_1 + F_y d_2$$

It is often so much easier to find the distances d_1 and d_2 than to find the distance d that using the principle of moments is the best method.

In applying the principle of moments, you must remember to place the components of the force so they intersect on the line of action of the force. This may seem to be in conflict with the triangle of forces shown in Fig. 1-30c, but it isn't at all. The triangle of forces shows only the magnitude and direction of the force and its components and does not represent their locations. The only requirement with respect to location is that the components must intersect on the line of action of the force. You are free to choose the point of intersection of the components. Now let's try an example.

Illustrative Example 1. Calculate the moment of the 300-lb force in Fig. 1-31a with respect to the point O.

Solution:
We could, of course, find the perpendicular distance from point O to the line of action of the force, but we can solve the problem more easily by using the principle of moments. Let's resolve the force into components at point A where it crosses the x axis. The components are:

1. $F_x = F \cos \theta = 300 \cos 60° = 300 \times 0.5 = 150$ lb.
2. $F_y = F \sin \theta = 300 \sin 60° = 300 \times 0.866 = 259.8$ lb. Figure 1-31b shows the components at point A.

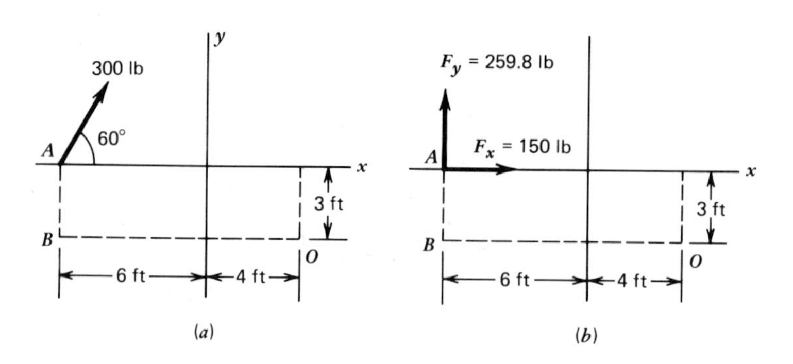

Fig. 1-31. Application of principle of moments. (a) Force. (b) Components.

3. The moment arm of F_x is 3 ft and the moment arm of F_y is 10 ft. Each has a clockwise moment with respect to point O, so the moment of each is positive. Then,

$$M_o = 3F_x + 10F_y = 3 \times 150 + 10 \times 259.8$$

$$= 450 + 2598 = 3\,048 \text{ lb ft}$$

Illustrative Example 2. Find the moment of the force in Fig. 1-31 with respect to point B.

Solution:
This is an easy one. The force is already resolved into components. The moment arm of F_x is 3 ft, and the moment is clockwise and, therefore, positive. The moment arm of F_y is zero (force through B), so its moment with respect to point B is zero. Then the moment of F with respect to point B is equal to the moment of F_x with respect to point B. This is

$$M_B = 3F_x = 3 \times 150 = 450 \text{ lb ft}$$

MOMENT SIGNS
Sometimes the moment of one component is positive, and the moment of the other component is negative. Then we just take their algebraic sum. Don't try to get the algebraic sign (positive or negative) of the moment from the algebraic sign of a component. The moment of a component is always positive if the moment is clockwise, and it is always negative if the moment is counterclockwise.

Illustrative Example 3. What is the moment of the force in Fig. 1-32*a* with respect to point O?

Solution:
The first step is to resolve the force into x and y components. The force is directed downward to the right, so the x component is toward the right and the y component is downward. The components of the force are shown in Fig. 1-32*b*. Their values are:

1. $F_x = F \cos \theta = 224 \cos 15° = 224 \times 0.966 = 216.4$ N.
2. $F_y = -F \sin \theta = -224 \sin 15° = -224 \times 0.259 = 57.98$ N.
3. The moment arm of F_x is 0.2 m; the moment is counterclockwise with respect to point O, and therefore it is negative. The moment arm of F_y is 0.3 m; this moment is clockwise with respect to point O, and therefore it is positive. Then,

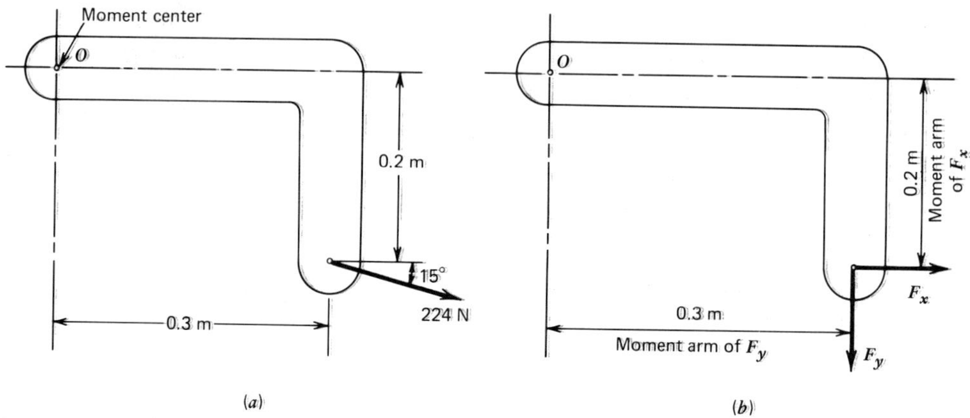

Fig. 1-32. Bent lever. (*a*) Force. (*b*) Components.

$$\begin{aligned} M_o &= -0.2F_x + 0.3F_y = -0.2 \times 216.4 + 0.3 \times 57.98 \\ &= -43.28 + 17.39 = -25.89 \text{ N·m} \end{aligned}$$

Illustrative Example 4. Find the moment of the 150-lb force with respect to point O, in Fig. 1-33*a*, by resolving it into components at point A where it crosses the x axis.

Solution:
First we need to know something about the angle θ so we can find the components. We can draw the triangle ABC in Fig. 1-33*b*. This has AB for its hypotenuse. The base AC is 4 ft, and the altitude BC is 3 ft. If we apply the Pythagorean theorem, the hypotenuse AB is

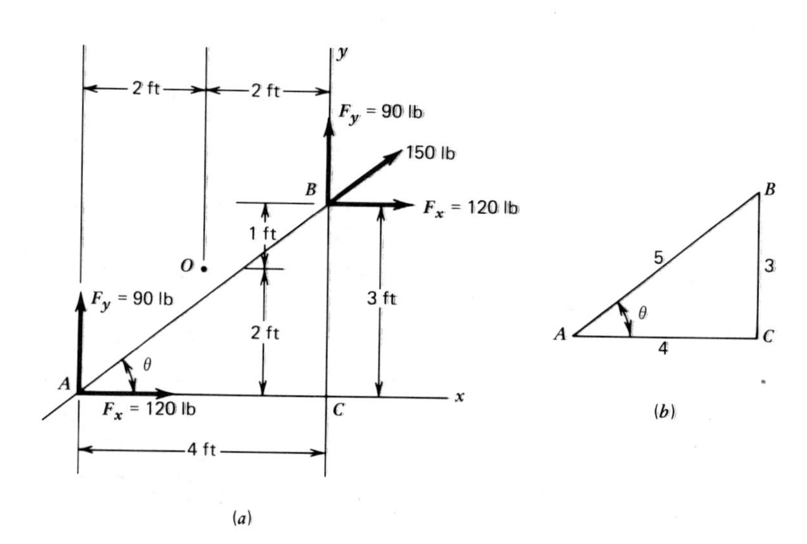

Fig. 1-33. Resolution of force into components. (*a*) Force and components. (*b*) Triangle.

$$AB = \sqrt{(4)^2 + (3)^2} = \sqrt{16 + 9} = \sqrt{25} = 5 \text{ ft}$$

The sine and cosine of θ can be obtained from the triangle without bothering to find the angle in degrees. The sine is equal to the altitude divided by the hypotenuse and is

$$\sin \theta = \frac{3}{5}$$

The cosine is equal to the base divided by the hypotenuse and is

$$\cos \theta = \frac{4}{5}$$

Then, the components of the force are

1. $F_x = F \cos \theta = 150 \times \frac{4}{5} = 120$ lb.
2. $F_y = F \sin \theta = 150 \times \frac{3}{5} = 90$ lb. These components are shown at point A.
3. The moment arm of F_x is 2 ft and its moment with respect to point O is counterclockwise, and therefore, it is negative. The moment arm of F_y is 2 ft, and its moment with respect to point O is clockwise; therefore, it is positive. The moment of the 150-lb force with respect to point O is the sum of the moments of F_x and F_y. It is

$$\begin{aligned} M_o &= -2F_x + 2F_y = -2 \times 120 + 2 \times 90 \\ &= -240 + 180 = -60 \text{ lb ft} \end{aligned}$$

Illustrative Example 5. Find the moment of the 150-lb force with respect to point O in Fig. 1-33*a* by resolving the force into components at point B.

Solution:
We already have the components from Example 4. They are shown at point B. The moment arm of F_x is 1 ft, and the moment is clockwise; therefore, it is positive. The moment arm of F_y is 2 ft, and the moment is counterclockwise; therefore, it is negative. Then,

$$\begin{aligned} M_o &= 1 \times F_x - 2F_y = 1 \times 120 - 2 \times 90 \\ &= 120 - 180 = -60 \text{ lb ft} \end{aligned}$$

Practice Problems (Section 1-8). The purpose of the following problems is to give you practice in applying the principle of moments. You can't learn the principle without practice.

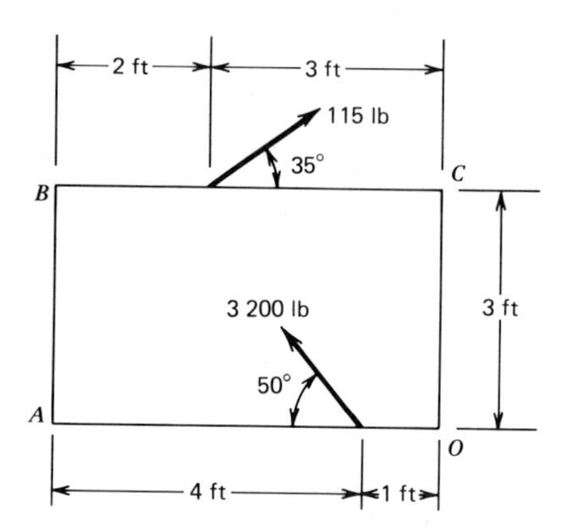

Fig. 1-34. Forces on a plate.

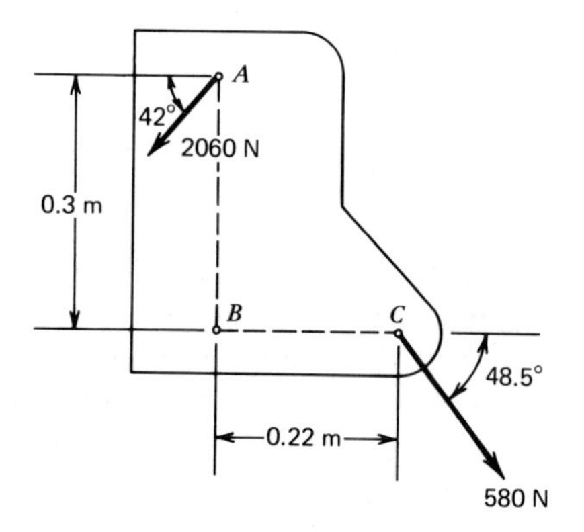

Fig. 1-35. Forces on a machine part.

1. Find the moment of the 115-lb force in Fig. 1-34 with respect to point O.
2. Calculate the moment of the 115-lb force in Fig. 1-34 with respect to point A.
3. In Fig. 1-34, what is the moment of the 3 200-lb force with respect to point B?
4. Find the moment of the 3 200-lb force in Fig. 1-34 with respect to point C.
5. Find the moment of the 2 060-N force in Fig. 1-35 with respect to point C.
6. Find the moment of the 580-N force in Fig. 1-35 with respect to point A.
7. Find the moment of the 16 000-N force in Fig. 1-36 with respect to point B.
8. In Fig. 1-36, what is the moment of the 17 000-N force with respect to point A?

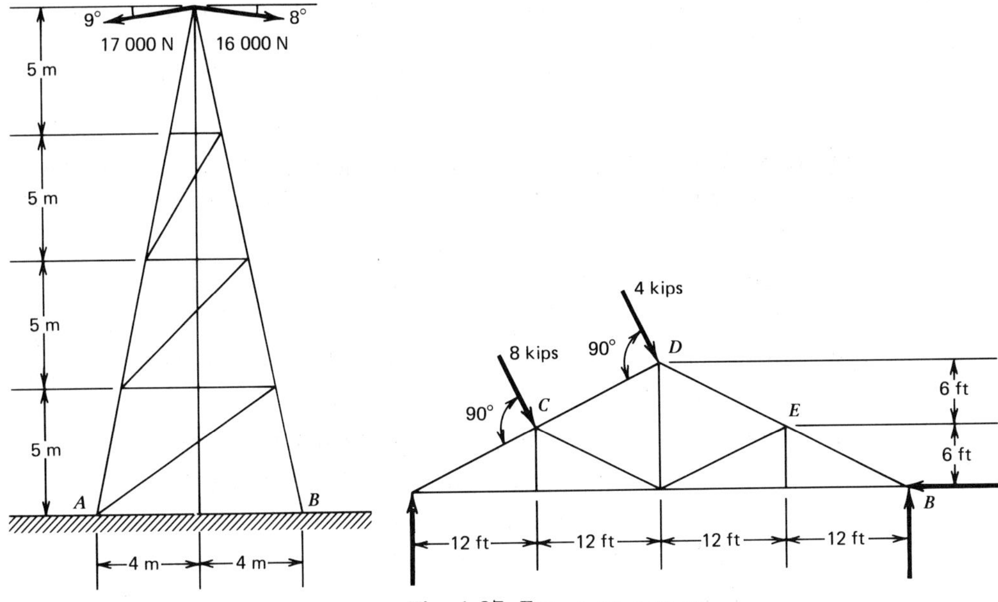

Fig. 1-36. Forces on a tower.

Fig. 1-37. Forces on a truss.

9. Calculate the moment of the force at D in Fig. 1-37 with respect to point A.
10. Find the moment of the force at C in Fig. 1-37 with respect to point B.

SUMMARY Here is a brief summary of what you should have learned in this chapter.

1. The components of a force F are given by
   (a) $F_x = F \cos \theta$
   (b) $F_y = F \sin \theta$
   where θ is the angle the force makes with the horizontal. You are to tell the direction of each component by inspection. Then F_x is positive if it is directed toward the right and negative if directed toward the left. F_y is positive if it is directed upward and negative if directed downward.
2. The moment of a force with respect to a point is equal to the product of the force and its perpendicular distance from the point. A clockwise moment is positive, and a counterclockwise moment is negative.
3. The principle of moments states that the moment of a force with

respect to any point is equal to the sum of the moments of the two components of the force with respect to that point.

REVIEW QUESTIONS These review questions are designed to test your mastery of this chapter. They are to be answered without looking at the preceding pages.

1. What is *strength of materials*?
2. What is a *force*?
3. What do you multiply a force by to get its x component?
4. How do you state the direction of a force?
5. When is an x component positive and when is it negative?
6. When is a y component positive and when is it negative?
7. In what units may the magnitude of a force be given?
8. What is the moment of a force with respect to a point?
9. What is a *moment arm*?
10. When is a moment positive and when is it negative?
11. What is the *principle of moments*?
12. In what units can the moment of a force be given?

REVIEW PROBLEMS[2] By now you probably understand the material of Chapter 1. Now do these review problems to fix in mind what you have learned.

Find the horizontal and vertical components of each of the forces described below.

1. A force of 18 lb directed upward to the right at an angle of 50° with the horizontal.
2. A force of 66 N directed downward to the left at an angle of 36.9° with the horizontal.
3. A force of 12.6 kips directed upward to the left at an angle of 66° with the horizontal.
4. A force of 244 lb directed downward to the right at an angle of 20.3° with the horizontal.
5. A force of 1760 N directed downward to the left at an angle of 45° with the horizontal.

[2]To check your answers to Review Problems, refer to the section at the back entitled Answers to Problems.

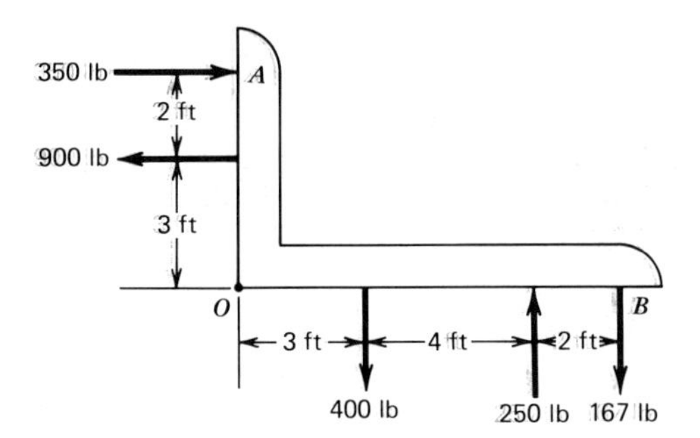

Fig. 1-38. Forces on a lever.

In each of the following problems the components of a force are given. Find the magnitude and direction of the force.

6. $F_x = 8\ 200$ lb; $F_y = 3\ 700$ lb.
7. $F_x = -736$ N; $F_y = 348$ N.
8. $F_x = 3.82$ kips; $F_y = -2.04$ kips.
9. $F_x = -1\ 070$ lb; $F_y = -893$ lb.
10. $F_x = -2\ 300$ N; $F_y = 5\ 900$ N.
11. Find the moment of the 900-lb force in Fig. 1-38 with respect to point *O*.
12. Find the moment of the 350-lb force in Fig. 1-38 with respect to point *B*.
13. Find the moment of the 34 000-N force in Fig. 1-39 about point *B*.
14. Find the moment of the 7 400-N force in Fig. 1-39 about point *A*.

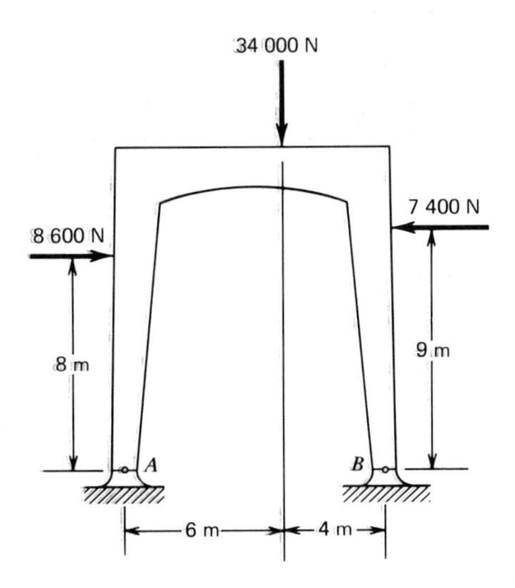

Fig. 1-39. Forces on a frame.

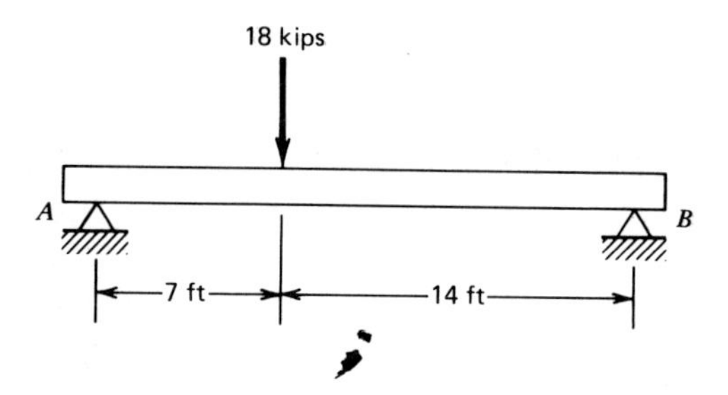

Fig. 1-40. Forces on a beam.

15. Find the moment of the force in Fig. 1-40 about point *A*.
16. Find the moment of the 260-lb force in Fig. 1-41 with respect to point *B*.
17. Find the moment of the 180-lb force in Fig. 1-41 with respect to point *A*.
18. Find the moment of the 42-lb force in Fig. 1-41 with respect to point *C*.
19. Find the moment of the 2 300-N force in Fig. 1-42 with respect to point *A*.
20. Find the moment of the 3 900-N force in Fig. 1-42 with respect to point *B*.

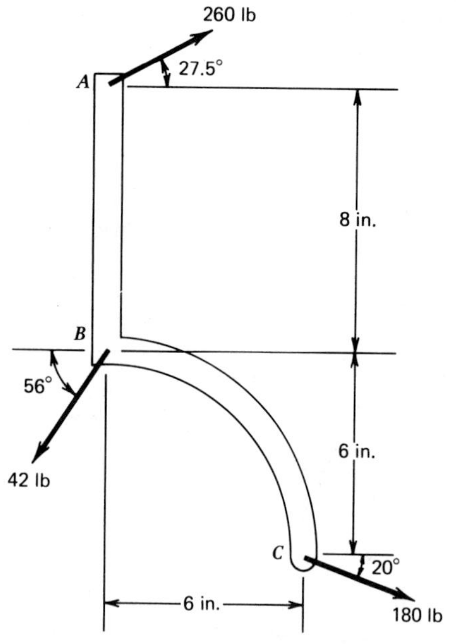

Fig. 1-41. Forces on a link.

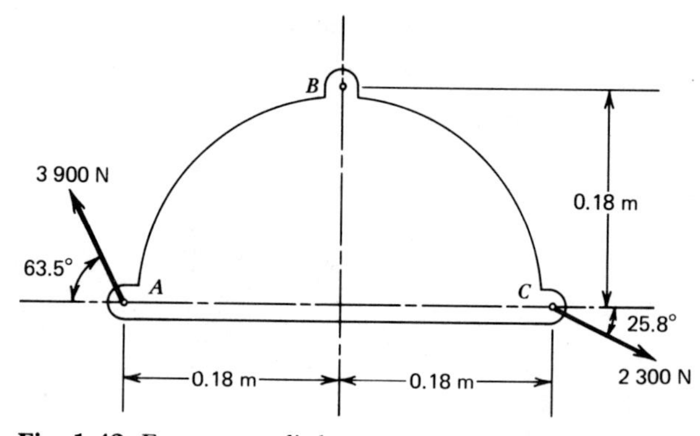

Fig. 1-42. Forces on a link.

2 EQUILIBRIUM OF BODIES

Part A BASIC METHODS AND PROBLEMS

PURPOSE OF PART A. The purpose of this part of the chapter is to show you how to find the forces acting on machine parts and other bodies. The design of each part of a machine or structure is a problem that resolves itself into a study of the forces acting on the part and the effects of these forces. To make such a study we must know all the forces exerted on each part, because the part must be strong and stiff enough to withstand all the forces exerted on it. The first step in the design of a member of a machine or structure, then, is to determine the properties of all the forces acting on the part, and that is what you are going to learn to do in this chapter. When you can do this, you will be prepared to study material presented in later chapters, where you are shown how forces affect the member and how the part should be designed to give it adequate strength for the use intended.

2-1. WHAT EQUILIBRIUM IS

Equilibrium is a state of balance between forces in which the tendency of one force to move a body in one direction is balanced or canceled by the tendencies of other forces to move the body in other directions. One of the simplest examples of a body in equilibrium is the link shown in Fig. 2-1. Here the 100-lb force at A tends to pull the body to the left, and this tendency is balanced by the tendency of the 100-lb force at B to pull the body to the right. We are interested in equilibrium, because most members of machines and structures are in a state of equilibrium. The

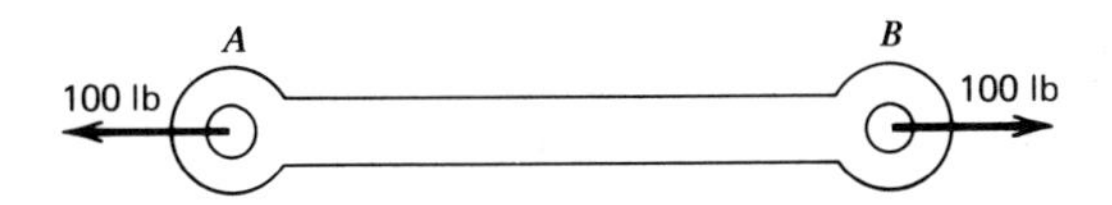

Fig. 2-1. Link in equilibrium.

forces acting on the part are balanced, so they do not move it. Some of the forces acting on the body are known forces, and these are called *loads*. There are unknown forces, exerted by bodies to which the member is fastened, and these forces are called *reactions*. The reactions balance the loads and hold the member in equilibrium. We must learn to find reactions so we will know all the forces acting on the member. Each reaction is a force, and we must find all of its characteristics (remember from Chapter 1 that the characteristics of a force are its *magnitude*, its *direction*, and its *location*). Usually the location of each reaction is known, and we have to find its magnitude (size) and direction. We will study the different methods of finding reactions so that you will be able to apply them as you need to. We will have to state the conditions necessary for the equilibrium of a body in terms of the forces applied to that body and we will do this, one at a time, as we come to them.

2-2. THE FREE-BODY DIAGRAM

The first thing to do in solving a problem in equilibrium is to draw the free-body diagram. This is a picture of the body showing all the forces exerted on it. The loads, or known forces, are shown first, then the reactions are put in. This places all the forces in the problem squarely in front of you so that you won't overlook any of them. The free-body diagram is one of the most important parts of the problem. Never try to solve an equilibrium problem without drawing the free-body diagram.

The free-body diagram doesn't have to show all of the details of the body. It can be a simple line drawing, offering just enough detail to show the characteristics of the forces. Remember that the characteristics of a force are its magnitude, direction, and location. Now let's look at some free-body diagrams.

Figure 2-2*a* shows a beam supported by walls at A and B. There are the two known forces (loads) of 3 000 lb and 2 400 lb (we call known force *loads*). Also, the walls must exert forces on the beam to hold it up. The forces exerted by the walls are unknown, and these forces, remember, are called reactions. The next thing is the free-body diagram and it is shown in Fig. 2-2*b*. The beam is represented by the heavy horizontal line. The loads of 3 000 lb and 2 400 lb are drawn first, and then the reactions of the walls are drawn at A and B. These are simply vertical forces and are designated by the R_1 and R_2 (R stands for reaction). All the forces acting on the beam are shown in this picture, and the next step will be to calculate the reactions. (See Section 2-3 to learn the method of doing this.)

Let's look at reactions in a little more detail and see how each type of support acts, so that we will know how to show the reactions. For one thing, a smooth bearing, as in Fig. 2-3*a*, or a knife-edge, as in Fig. 2-3*b*,

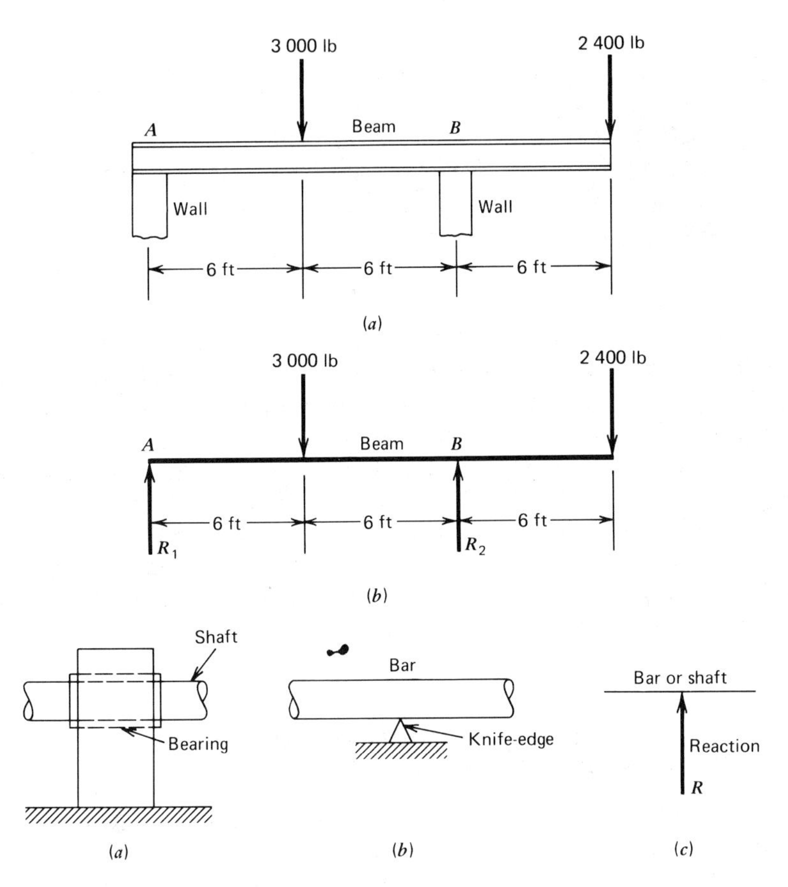

Fig. 2-2. A beam problem. (*a*) Beam. (*b*) Free-body diagram.

Fig. 2-3. Reactions on shaft or bar. (*a*) Shaft in smooth bearing. (*b*) Bar on knife-edge. (*c*) Reaction on bar or shaft.

can only exert a reaction perpendicular to the bar which it supports. In the free-body diagram the reaction can be drawn like the force R in Fig. 2-3*c*. A bearing can exert a force either upward or downward, because it can keep the bar from moving either upward or downward, but a knife-edge can only exert a force pushing against the bar, because it can't keep the bar from moving away. A knife-edge below the bar can only exert an upward reaction, and a knife-edge above the bar can only exert a downward reaction.

A pin connection, such as that shown in Fig. 2-4*a*, where the lower end of a member is fastened by a pin, can exert a reaction in any direction. Usually, then, the direction of the reaction of a pin is unknown. The reaction can be shown as a force R at an angle θ with the horizontal, as in Fig. 2-4*b*, or the two components R_x and R_y can be shown, as in Fig. 2-4*c*. If we know R and θ, we can find R_x and R_y. (Remember from Chapter 1, $R_x = R \cos \theta$ and $R_y = R \sin \theta$.) Also, if we know R_x and R_y, we can find R and θ. (From Chapter 1 $R = \sqrt{R_x^2 + R_y^2}$ and $\tan \theta = \frac{R_y}{R_x}$.)

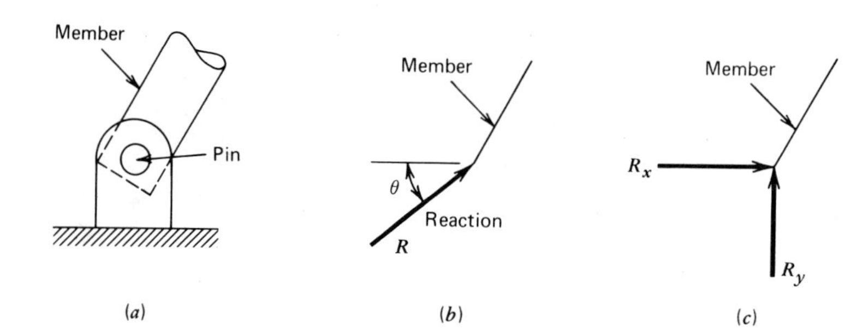

Fig. 2-4. Pin connection. (*a*) Bar supported by pin. (*b*) Reaction on bar. (*c*) Components of reaction.

Now look at the bell crank in Fig. 2-5*a*. There is a load (known force) of 120 N at *B*. Then there is a connecting link at *A*, and the bell crank is supported by a pin at *C*. The free-body diagram of the bell crank is shown in Fig. 2-5*b*, where the line *ACB* represents the bell crank. First, we draw the load of 120 N and then the reactions. They include the reaction R_1 at *A*, and the reaction of the pin at *C*. We don't know the direction of the reaction at *C*, so we show the two components R_x and R_y. This completes the free-body diagram, and the only thing left to do is to calculate the reactions. (This will be done in Section 2-3.)

2-3. SOLUTION OF EQUILIBRIUM PROBLEMS BY USE OF MOMENT EQUATIONS

One of the simplest ways of solving equilibrium problems is by use of moment equations. We know how to find the moment of a force with respect to a point. (You remember from Chapter 1 that the moment of a force with respect to a point is the product of the force and the perpendicular distance from the point to the line of action of the force.) When there are several forces acting on a body, we can choose a moment

Fig. 2-5. Bell crank. (*a*) Loading on bell crank. (*b*) Free-body diagram.

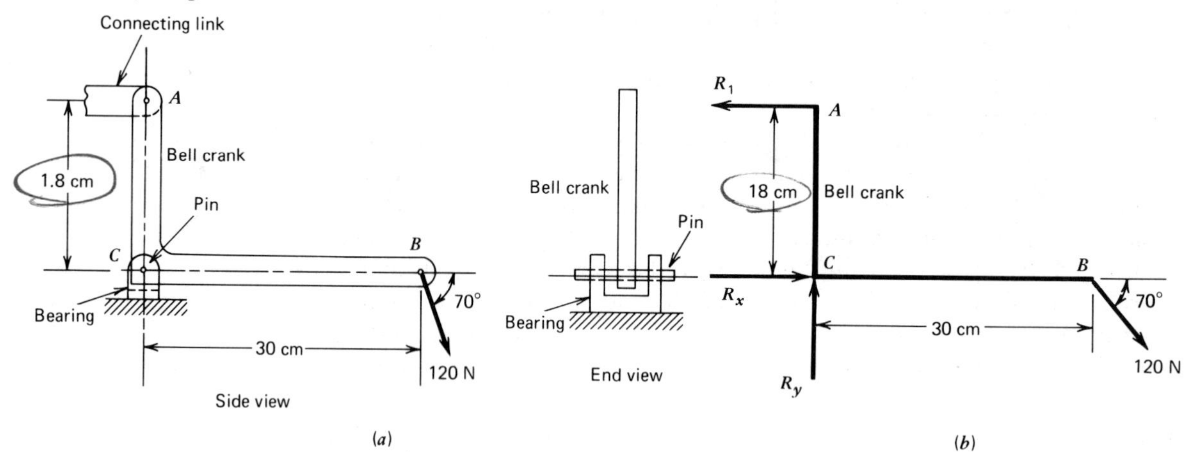

center, say, point O, and calculate the moment of each force with respect to this moment center. Next we can add the moments of the forces to obtain a moment sum. This moment sum is designated by ΣM_o and is to be read as "the sum of the moments with respect to point O." (The symbol Σ is the Greek letter sigma, which here stands for *sum*.) The idea of the sum of the moments is important because: If a body is in equilibrium, the sum of the moments of the forces acting on it is zero, with respect to any point whatever. Each force of the system tends to turn the body, and the moment of the force is a measure of the tendency to turn, but the tendency of one force is balanced or offset by the tendencies of the other forces, so that no turning occurs. We have to have equations to solve problems, and when we write this statement as an equation, it is

$$\Sigma M_o = 0$$

This is called a *moment equation* and is to be read, "the sum of the moments with respect to point O is zero." Point O can be any point, and, of course, the equation still holds true if we call the point A or B. Thus we could write

$$\Sigma M_A = 0 \qquad \text{or} \qquad \Sigma M_B = 0$$

We are going to use moment equations to solve equilibrium problems. We will pick a moment center and write the moment of each force with respect to that center point. If we select the moment center with care, the equation will contain only one unknown force, and we can solve the equation for that force. Remember to designate clockwise moments as positive and counterclockwise moments as negative.

The first step in solving an equilibrium problem is to draw the free-body diagram. This is true no matter what the method used. The remaining steps, in using moment equations, are to write the equations and solve them. Now let's try it.

Illustrative Example 1. Find the reactions on the beam in Fig. 2-6*a*. (This is the same beam as that shown in Fig. 2-2. The reactions are exerted by the walls at A and B.)

Solution:

1. The free-body diagram is drawn. This is shown in Fig. 2-6*B*. The loads of 3 000 lb and 2 400 lb are drawn first, and then the reactions at A and B are drawn. These are designated as R_1 and R_2.
2. We write a moment equation with point A as a center. The forces are taken in order from left to right, and the moment of R_1 with respect to point A is zero because its moment arm is zero. Clock-

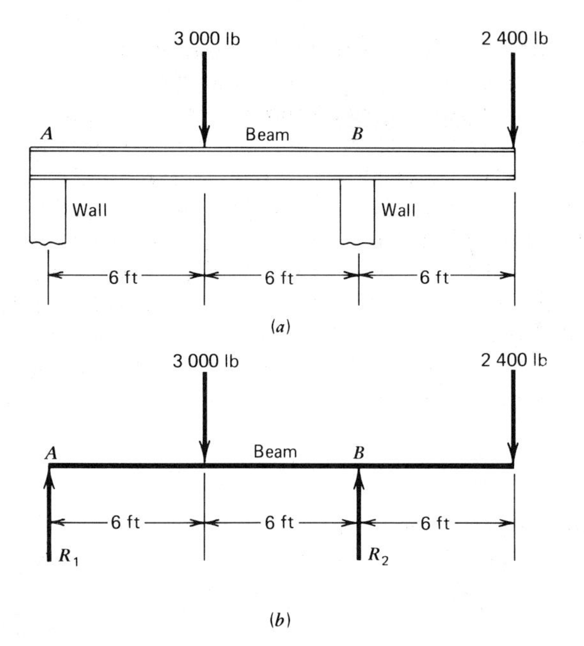

Fig. 2-6. A beam problem. (*a*) Beam. (*b*) Free-body diagram.

wise moments are positive and counterclockwise moments are negative. Then,

$$\Sigma M_A = 0$$

$$3\ 000 \times 6 - 12R_2 + 2\ 400 \times 18 = 0$$

(The term 3 000 × 6 is the moment of the 3 000-lb force, because the moment arm of this force is 6 ft. $12R_2$ is the moment of R_2, and it is negative because the moment of the force is counterclockwise with respect to point A. The term 2 400 × 18 is the moment of the 2 400-lb force; the moment arm is 18 ft.) Continuing, the term containing R_2 can be transposed, and

$$12R_2 = 3\ 000 \times 6 + 2\ 400 \times 18$$

$$12R_2 = 18\ 000 + 43\ 200 = 61\ 200$$

Then we divide by 12 to get R_2

$$R_2 = 5\ 100 \text{ lb}$$

3. Next we write a moment equation with point B as a center. The forces are taken in order from left to right, and the moment of R_2 is zero with respect to point B because its moment arm is zero. Then,

$$\Sigma M_B = 0$$

$$12R_1 - 3\ 000 \times 6 + 2\ 400 \times 6 = 0$$

The second and third terms in the equation can be transposed to give

$$12R_1 = 3\,000 \times 6 - 2\,400 \times 6 = 18\,000 - 14\,400 = 3\,600$$

$$R_1 = 300 \text{ lb}$$

SELECTION OF MOMENT CENTERS

The moment centers were selected with care in the foregoing example. It is always a good idea to choose a moment center on the line of action of an unknown force. Then this unknown force has no moment and doesn't appear in the equation; when there are two unknown forces, only one of them appears in the equation, and you can solve for it easily. The best procedure is to choose a moment center on the line of action of the first unknown force to find the second unknown force, and then to choose a moment center on the line of action of the second unknown force to find the first unknown force. Let's try it again, and use the metric system.

Illustrative Example 2. Figure 2-7*a* shows a beam in a weighing machine. The beam is subjected to loads of 3 000 N and supported at the knife-edges *A* and *B*. Find the reactions at *A* and *B*.

Solution:

1. Draw the free-body diagram. This is shown in Fig. 2-7*b*. The beam is represented by a line and the reactions at *A* and *B* are designated by R_1 and R_2.
2. Write a moment equation with point *A* as a center to find R_2.

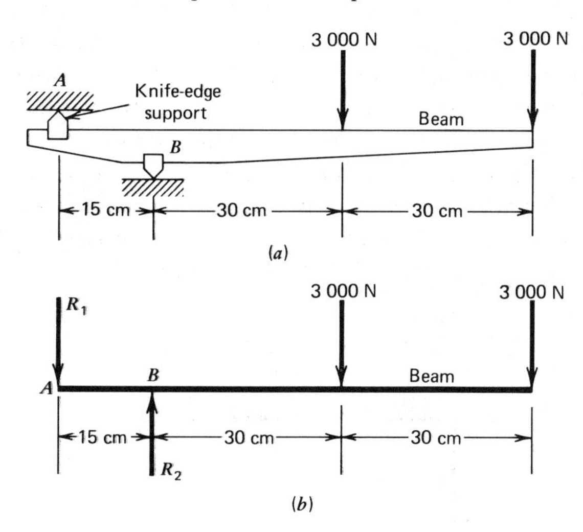

Fig. 2-7. Beam in weighing machine. (*a*) Loads on bar. (*b*) Free-body diagram.

Thus

$$\Sigma M_A = 0$$

$$-15R_2 + 3\,000 \times 45 + 3\,000 \times 75 = 0$$

Transpose the term containing R_2 to get

$$15R_2 = 135\,000 + 225\,000 = 360\,000$$

$$R_2 = 24\,000 \text{ N}$$

3. Write a moment equation with point B as a center to find R_1. Thus

$$\Sigma M_B = 0$$

$$-15R_1 + 3\,000 \times 30 + 3\,000 \times 60 = 0$$

Transpose the term containing R_1 to get

$$15R_1 = 90\,000 + 180\,000 = 270\,000$$

$$R_1 = 18\,000 \text{ N}$$

WEIGHT OF A MEMBER

Up to this point we haven't considered the weight of the member as a load. Most of the time you can ignore the weight because it is so small compared to the other forces in the problem. Frequently, a machine part weighing less than 10 lb carries forces of several thousand pounds; the weight of the part doesn't make enough difference to be worth calculating in such a case. However, there are a few cases in which the weight is a large part of the total load, and then it can't be ignored. The weight of a bar or beam or shaft is a force distributed along the bar or beam or shaft. It is the force of gravity acting on the material, and so it is distributed in the same manner as the material. If the bar has the same cross section throughout, the weight is evenly distributed along the length and is spoken of as a *uniformly distributed load*.

Figure 2-8 shows the free-body diagram of a beam that weighs 30 lb per foot of length. The uniformly distributed load is represented by the crosshatched rectangle, and the intensity of the load is written above it (30 lb/ft). The reactions at A and B, where the bar is supported,

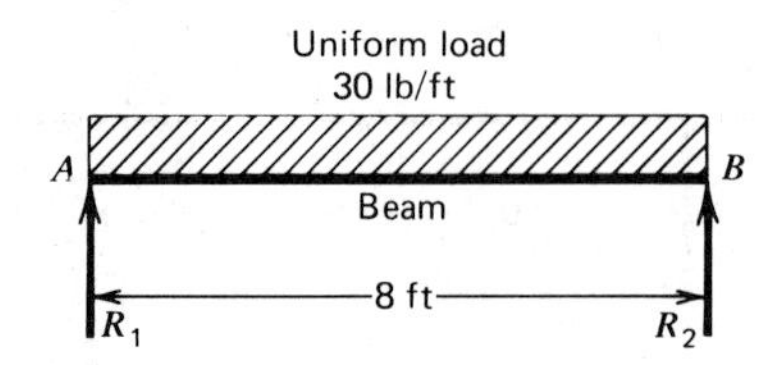

Fig. 2-8. Beam with uniform load.

complete the free-body diagram. The total amount of the distributed force is simply the number of pounds per foot multiplied by the number of feet. Thus the total load in this case is

$$W = (30 \text{ lb/ft}) \times (8 \text{ ft}) = 240 \text{ lb}$$

When a load is uniformly distributed, the moment is the same as it would be if all of the load were concentrated at the center of the length over which the load is distributed. The best way to calculate such a moment is to multiply the load in pounds per foot by the number of feet to get the total load, and then to multiply this by the number of feet from the center of the load to the moment center. The moment of the distributed load with respect to point A in Fig. 2-8 is

$$M_a = (30 \text{ lb/ft}) \times (8 \text{ ft}) \times (4 \text{ ft})$$

$$= 30 \times 8 \times 4 = 960 \text{ lb ft}$$

Let's try a problem of a beam with a distributed load.

Illustrative Example 3. Figure 2-9*a* shows a beam that weighs 40 lb/ft and is supported on walls at A and B. A bin rests on the beam and contains material weighing 100 lb/ft of length of the beam. Calculate the reactions at A and B.

Solution:

1. Figure 2-9*b* shows the free-body diagram of the beam. The heavy line AB represents the beam, the lower crosshatched rectangle

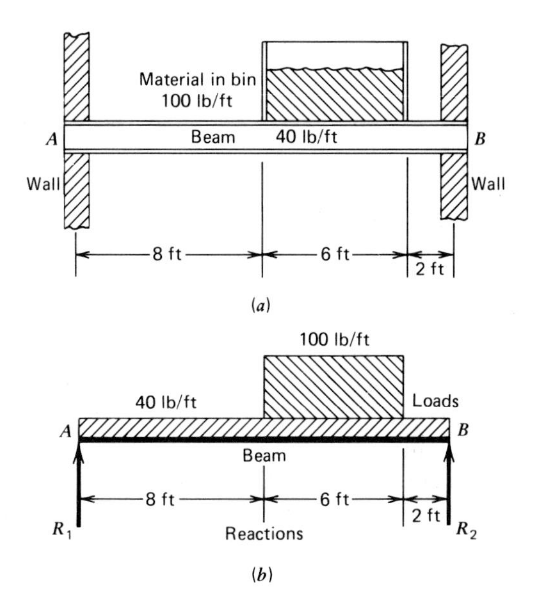

Fig. 2-9. Beam supporting a bin. (*a*) Picture of beam. (*b*) Free-body diagram.

represents the weight of the beam, and the upper crosshatched rectangle represents the weight of the material. The reactions are designated as R_1 (at the center of the left-hand wall) and R_2 (at the center of the right-hand wall).

2. We can find the reaction R_2 by writing a moment equation with center at A (because R_1 passes through A). Thus

$$\Sigma M_A = 0$$

$$40 \times 16 \times 8 + 100 \times 6 \times 11 - 16R_2 = 0$$

(Here $40 \times 16 \times 8$ is the moment of the beam weight of 40 lb/ft with respect to point A; the total of the load is 40×16 and the moment arm is 8 ft. The term $100 \times 6 \times 11$ is the moment of the weight of the material; this is 100×6 for the total amount of this load; and the moment arm is 11 ft, which is the distance from A to the center of this load.) We transpose the term $16R_2$ to get

$$16R_2 = 40 \times 16 \times 8 + 100 \times 6 \times 11$$

$$= 5\,120 + 6\,600 = 11\,720$$

$$R_2 = 732.5 \text{ lb}$$

3. We can find the reaction R_1 by writing a moment equation with point B as a center. Thus

$$\Sigma M_B = 0$$

$$16R_1 - 40 \times 16 \times 8 - 100 \times 6 \times 5 = 0$$

We can transpose the second and third terms in this equation to get

$$16R_1 = 40 \times 16 \times 8 + 100 \times 6 \times 5 = 5\,120 + 3\,000$$

$$= 8\,120$$

$$R_1 = 507.5 \text{ lb}$$

MORE GENERALLY

So far we have been considering problems of beams, in which all of the forces are vertical. Now we are ready to look at a little more general type of problem, one in which the forces are not all parallel. We can still solve each problem by moment equations, but there will usually be three unknown forces. We can choose a moment center at the intersection of two of the unknown forces; the moment equation can then be solved for the third unknown force. The two forces that intersect at the moment center have no moment, and so they do not appear in the moment equation. Let's try it.

Illustrative Example 4. Figure 2-10*a* shows the same bell crank as that shown in Fig. 2-5*a*. The bell crank is subjected to the 120-N load at *B*, and there are reactions at *A* and *C*. Find the reactions.

Solution:

1. The free-body diagram is shown in Fig. 2-10*b*. (This free-body diagram is also shown in Fig. 2-5*b*.) The line *ACB* represents the bell crank. The 120-N load is drawn first, then the reaction R_1, and finally the reaction at *C*. The reaction at *C* is exerted by a pin, and its direction is unknown, so it is represented by its two components, R_x and R_y.
   When moment equations are to be used it is best to replace the load by its two components, F_x and F_y. These are (remember from Section 1-6, Chapter 1):
   (a) $F_x = 120 \cos 70° = 120 \times 0.342 = 41.0 \text{ N}$
   (b) $F_y = 120 \sin 70° = 120 \times 0.9396 = 112.8 \text{ N}$

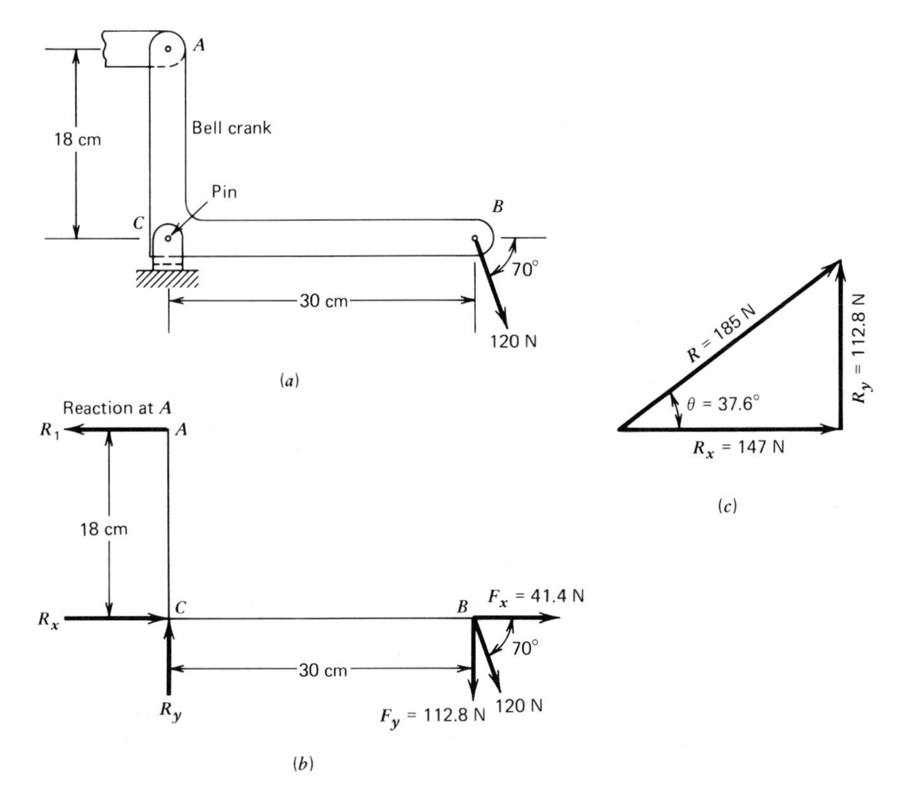

Fig. 2-10. Bell crank problem. (*a*) Loading on bell crank. (*b*) Free-body diagram. (*c*) Force triangle.

Then we will use the principle of moments, which states that the moment of a force is equal to the sum of the moments of its components.

2. The force R_1 is found by writing a moment equation with point C as a center. (The forces R_x and R_y go through C and have no moment, so R_1 is the only unknown force in the equation. Notice also that the x component of the 120-N force passes through C.) Then,

$$\Sigma M_c = 0$$

$$-18R_1 + 112.8 \times 30 = 0$$

(The term $18R_1$ is the moment of the force R_1, and the term 112.8×30 is the moment of the y component of the 120-N force.) Transposing

$$18R_1 = 3\ 384 \qquad R_1 = 188 \text{ N}$$

3. The component R_x is found by writing a moment equation with point A as a center. (The forces R_1 and R_y intersect at A. Neither component of the 120-N load passes through A.) Thus

$$\Sigma M_A = 0$$

$$-18R_x + 112.8 \times 30 - 41.0 \times 18 = 0$$

$$18R_x = 3\ 384 - 738 = 2\ 646$$

$$R_x = 147 \text{ N}$$

4. The component R_y is found by writing a moment equation with point B as a center. (The forces R_1 and R_x are parallel, so they don't intersect, but we know both of them by now and can find R_y by using any moment center which isn't on the line of action of R_y.)

$$\Sigma M_B = 0$$

$$30R_y - 18R_1 = 0$$

$$30R_y = 18R_1$$

We substitute the value already found for R_1, which is 188 N

$$30R_y = 18 \times 188 = 3\ 384$$

$$R_y = 112.8 \text{ N}$$

5. The magnitude and direction of the reaction at C are found by combining R_x and R_y by means of the triangle in Fig. 2-10c (remember from Section 1-6, Chapter 1). Thus

$$R = \sqrt{R_x^2 + R_y^2} = \sqrt{(147)^2 + (112.8)^2}$$
$$= \sqrt{21\,609 + 12\,724} = \sqrt{34\,333}$$
$$R = 185 \text{ N}$$

Also,

$$\tan\theta = \frac{R_y}{R_x} = \frac{112.8}{147} = 0.7673 \qquad \theta = 37.5°$$

Practice Problems (Section 2-3). It is now time for you to try some problems on your own. Here is a set of practice problems. Remember that a bearing or a knife-edge exerts a reaction perpendicular to the member, but a pin can exert a force in any direction.

1. The shaft in Fig. 2-11 is supported in bearings at A and B. Find the reactions.
2. The shaft and hanger in Fig. 2-12 is supported in bearings at A and B. Find the reactions.
3. The beam in Fig. 2-13 is subjected to a uniformly distributed

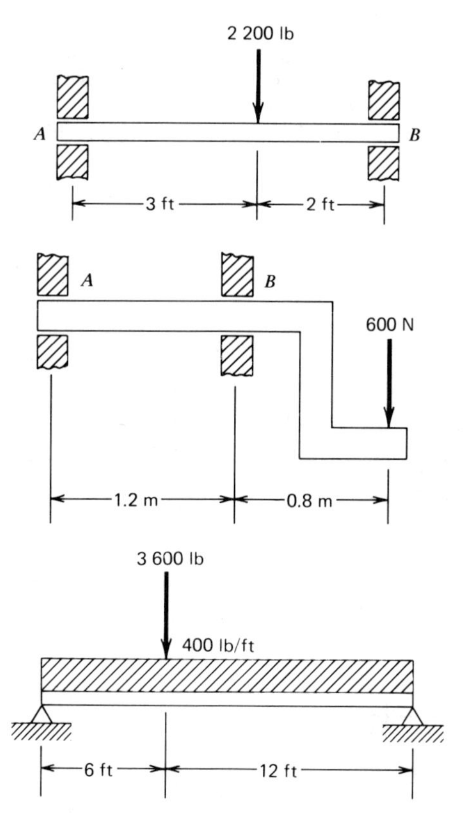

Fig. 2-11. Shaft in bearings.

Fig. 2-12. Shaft and hanger.

Fig. 2-13. Beam problem.

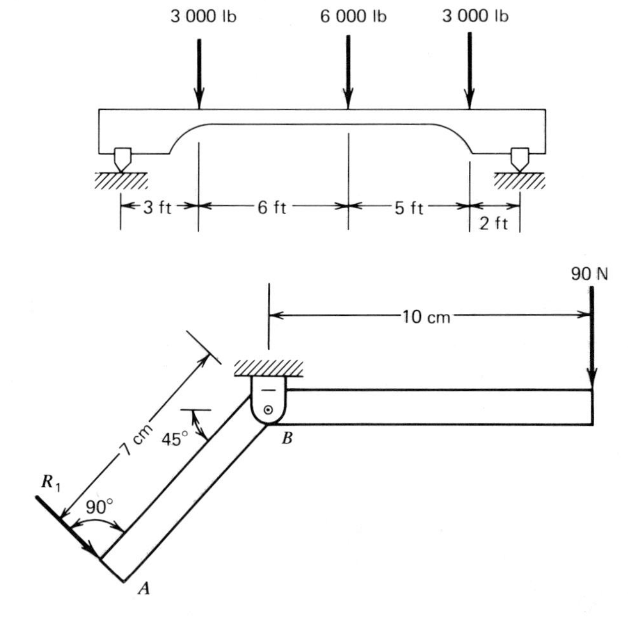

Fig. 2-14. Beam supported on knife-edges.

Fig. 2-15. Bell crank.

load and a concentrated load as shown. Find the reactions at the supports.

4. Find the reactions on the beam in Fig. 2-14.
5. Figure 2-15 shows a bell crank that is subjected to a load of 90 N as shown. A reaction R_1 is applied at A, and the bell crank is supported by a pin at B. Find the values of R_1 and the reaction at B.
6. The bracket in Fig. 2-16 is supported by the bearings on the

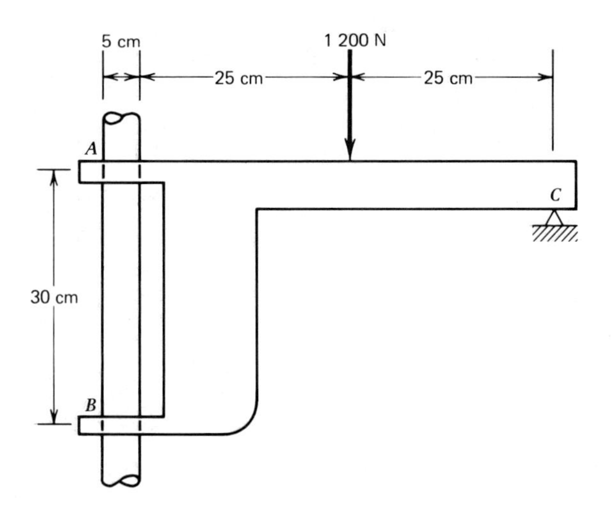

Fig. 2-16. Bracket.

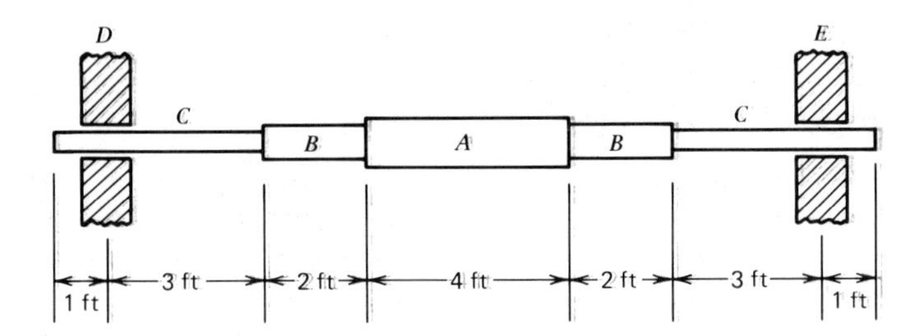

Fig. 2-17. Turbine shaft.

vertical rod at A and B and by the knife-edge at C. Find the reactions.

7. Figure 2-17 shows a shaft of the type used in large turbines. The central part A weighs 3 000 lb/ft, the parts B weigh 2 000 lb/ft, and the parts C weigh 1 000 lb/ft. The shaft is supported by bearings at D and E. Find the reactions.
8. The pulley in Fig. 2-18 is subjected to the belt pulls of 120 lb and 80 lb and is supported by a shaft through the center. Find the force R_1 and the reaction of the shaft.
9. Find the reactions at A and B in Fig. 2-19.
10. The frame in Fig. 2-20 is supported by knife-edges at A and B. Find the reactions.

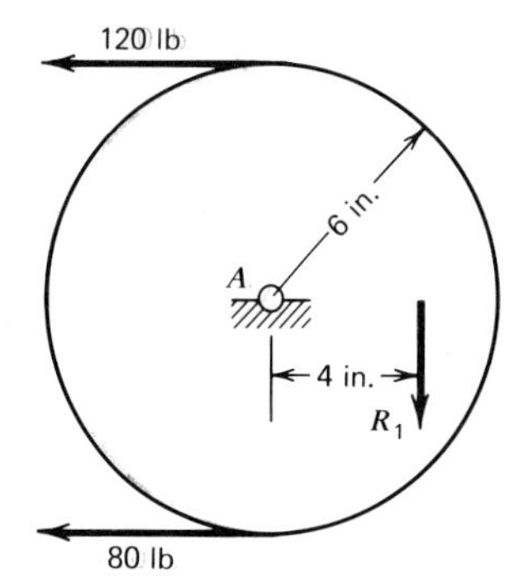

Fig. 2-18. Pulley.

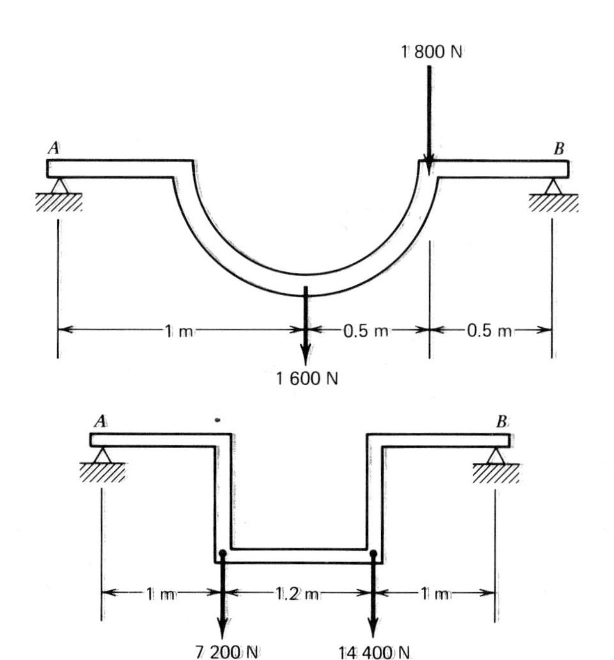

Fig. 2-19. Machine part.

Fig. 2-20. Frame.

2-4. SOLUTION OF EQUILIBRIUM PROBLEMS BY USE OF EQUATIONS FOR SUMS OF COMPONENTS

You are going to learn another method of solving equilibrium problems in this article. This method is often much simpler and easier to use than the method of writing moment equations, and every designer needs to know it. You remember how you learned to find the components of a force in Chapter 1. Given a force F, at an angle θ with the horizontal, the components are

$$F_x = F \cos \theta$$

$$F_y = F \sin \theta$$

The x components are positive if directed toward the right and negative if directed toward the left; the y components are positive if directed upward and negative if directed downward. Now we are going to use the fact that if the forces acting on a body are in equilibrium, the sum of the components of the forces in any direction is equal to zero. The forces tending to push the body in one direction are balanced by forces tending to push the body in the opposite direction. This leads to two equations, one for x components and one for y components. They are

$$\Sigma F_x = 0$$

$$\Sigma F_y = 0$$

The Σ (Greek letter sigma) still stands for summation and the equation, $\Sigma F_x = 0$, is to be read as, "The sum of the x components of the forces acting on the body is equal to zero." The other equation, $\Sigma F_y = 0$, is to be read as, "The sum of the y components of the forces acting on the body is equal to zero." When these equations are written for the forces acting on a body, they contain both the known and unknown forces and can be solved for the unknown forces.

Illustrative Example 1. Figure 2-21*a* shows a pulley that is subjected to two equal, vertical belt pulls of 200 lb and is supported by a shaft through the center. Find the reaction of the shaft.

Solution:

1. Figure 2-21*b* shows the free-body diagram of the pulley. The loads of 200 lb are drawn first, and then the reaction R is drawn.
2. The reaction R can be found by writing the equation for the sum of the y components. Thus

$$\Sigma F_y = 0$$

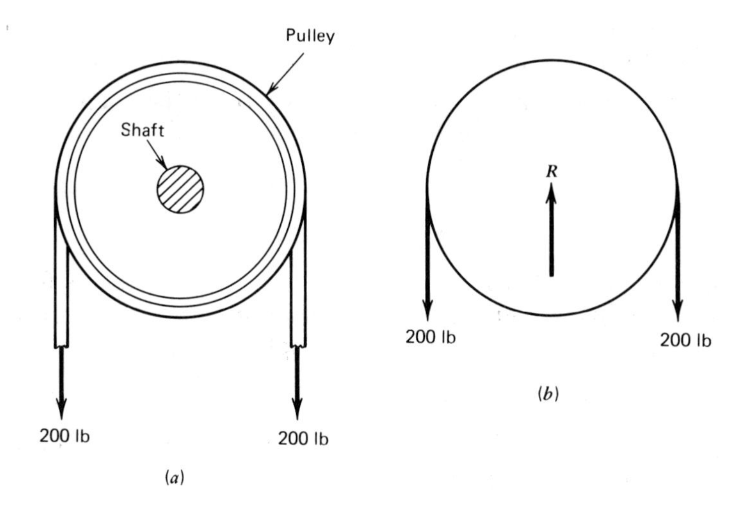

Fig. 2-21. Pulley. (*a*) Pulley and belt. (*b*) Free-body diagram.

$$R - 200 - 200 = 0$$

The force R is positive in this equation because it is directed upward. The 200-lb forces are directed downward so they are negative. Continuing,

$$R = 200 + 200 = 400 \text{ lb}$$

Illustrative Example 2. Figure 2-22*a* shows a pulley that is subjected to two equal belt pulls of 600 N and is supported by a shaft through the center. Find the reaction of the shaft.

Solution:

1. The free-body diagram is shown in Fig. 2-22*b*. The loads of 600 N are drawn first and then the two components of the shaft

Fig. 2-22. Another pulley. (*a*) Pulley and belt. (*b*) Free-body diagram. (*c*) Force triangle.

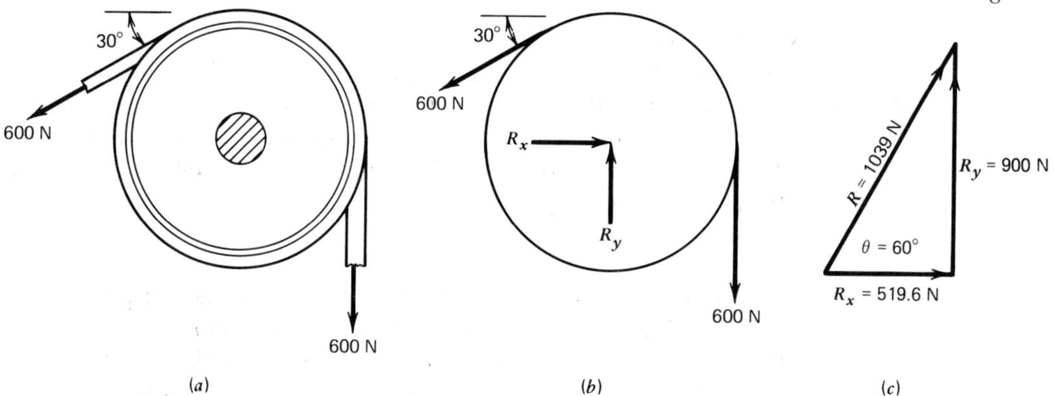

reaction. The direction of the shaft reaction is unknown, so it is best to represent the force by the components R_x and R_y.

2. The component R_x can be found by writing the equation for the sum of the x components. Thus

$$\Sigma F_x = 0$$

$$R_x - 600 \cos 30° = 0$$

(The term 600 cos 30° is the x component of the 600-N force and is negative because it is directed toward the left; R_x is positive because it is directed toward the right.) Continuing

$$R_x = 600 \cos 30° = 600 \times 0.866 = 519.6 \text{ N}$$

3. The component R_y can be found by writing the equation for the sum of the y components. So,

$$\Sigma F_y = 0$$

$$R_y - 600 \sin 30° - 600 = 0$$

$$R_y = 600 \sin 30° + 600 = 600 \times 0.5 + 600$$

$$R_y = 300 + 600 = 900 \text{ N}$$

4. The magnitude and direction of the reaction R can be found from the triangle of forces in Fig. 2-22c. Thus

$$R = \sqrt{R_x^2 + R_y^2} = \sqrt{(519.6)^2 + (900)^2}$$

$$= \sqrt{270\,000 + 810\,000}$$

$$R = \sqrt{1\,080\,000} = 1\,039 \text{ N}$$

$$\tan \theta = \frac{R_y}{R_x} = \frac{900}{519.6} = 1.732 \qquad \theta = 60°$$

COMBINATIONS OF EQUATIONS

Probably the most effective use of equations for the sums of components is in combination with moment equations. Sometimes they are used to check answers found by writing moment equations, and sometimes they are used to find one of the reactions after the others have been found by moment equations.

Illustrative Example 3. Figure 2-23a shows a beam that is supported on knife-edges at A and B. The beam carries two concentrated loads and a uniformly distributed load over part of its length. Calculate the reactions by moment equations and check the results.

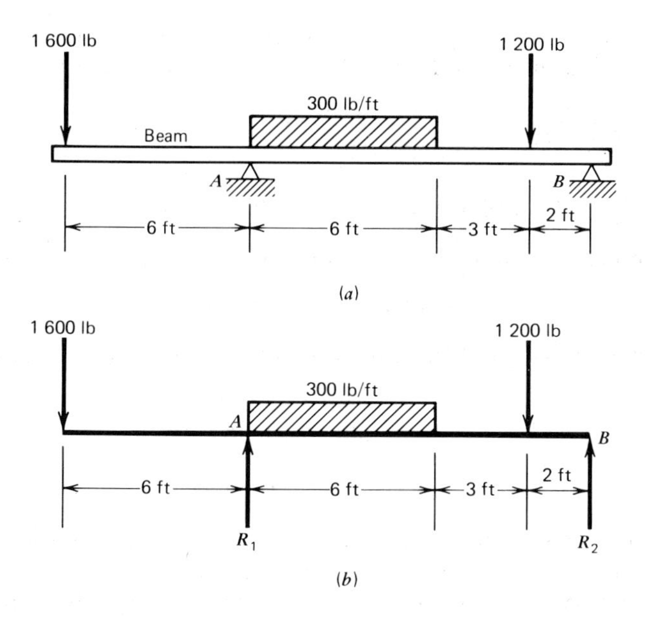

Fig. 2-23. Beam. (*a*) Beam. (*b*) Free-body diagram.

Solution:

1. The free-body diagram of the beam is shown in Fig. 2-23*b*. The heavy line represents the beam. The loads are drawn first and then the reactions, which are designated as R_1 and R_2.
2. We can find the reaction R_2 by writing a moment equation with center at A. (We choose A because R_1 goes through A and has no moment.) Then, we take the forces in order from left to right,

$$\Sigma M_A = 0$$

$$-1\,600 \times 6 + 300 \times 6 \times 3 + 1\,200 \times 9 - 11R_2 = 0$$

$$11R_2 = -9\,600 + 5\,400 + 10\,800 = 6\,600$$

$$R_2 = 600 \text{ lb}$$

3. We can find the reaction R_1 by writing a moment equation with center at B. Taking the forces in order from left to right,

$$\Sigma M_B = 0$$

$$-1\,600 \times 17 + 11R_1 - 300 \times 6 \times 8 - 1\,200 \times 2 = 0$$

$$11R_1 = 27\,200 + 14\,400 + 2\,400 = 44\,000$$

$$R_1 = 4\,000 \text{ lb}$$

4. We can check these results for R_1 and R_2 by writing an equation for the sum of the y components of the forces on the beam. This

sum of the y components must be zero, because the beam is in equilibrium. The values of the reactions must satisfy the equation. So

$$\Sigma F_y = 0$$

$$-1\,600 + R_1 - 300 \times 6 - 1\,200 + R_2 = 0$$

We can substitute $R_1 = 4\,000$ lb and $R_2 = 600$ lb to get

$$-1\,600 + 4\,000 - 1\,800 - 1\,200 + 600 = 0$$

Then, we can add the positive terms and the negative terms. This gives

$$4\,600 - 4\,600 = 0$$

$$0 = 0$$

This check shows us that the values we calculated for the reactions are correct. We found the reactions separately by writing moment equations; if we made an error in calculating R_2, it wouldn't affect the value of R_1. Then we checked the values of the reactions by writing the equation $\Sigma F_y = 0$, and this equation was satisfied. We can say that either the reactions are correct or that the value of R_1 is just as much too small as R_2 is too large. (If R_1 was 200 lb too small and R_2 was 200 lb too large, the sum of R_1 and R_2 would still check.) This is not likely, so we can be reasonably sure that our answers are correct.

Illustrative Example 4. Figure 2-24*a* shows a machine part which is subjected to loads of 100 N and 80 N and a reaction R_1. In addition it is supported by a pin at A. Find the reactions.

Solution:

1. The free-body diagram is shown in Fig. 2-24*b*. A line is enough to represent the body. The loads are drawn first, then the reaction R_1, and finally, the two components of the reaction at A.
2. The reaction R_1 is obtained from a moment equation with point A as a center. (The components R_x and R_y intersect at A.) Thus,

$$\Sigma M_A = 0$$

$$-100 \times 18 - 80 \times 9 + 12R_1 = 0$$

$$12R_1 = 1\,800 + 720 = 2\,520$$

$$R_1 = 210 \text{ N}$$

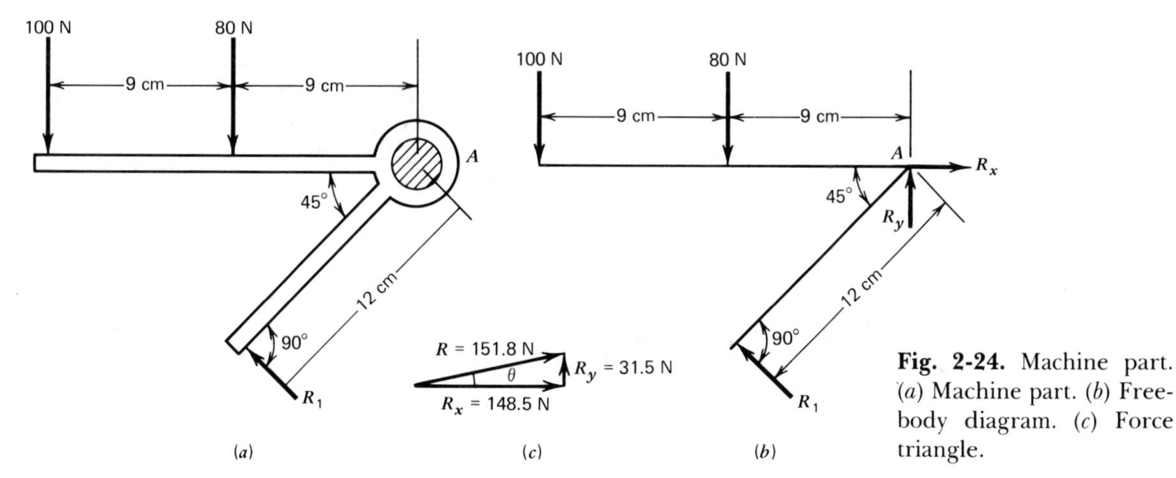

Fig. 2-24. Machine part. (*a*) Machine part. (*b*) Free-body diagram. (*c*) Force triangle.

3. The easiest way to find R_x is to write the equation for the sum of the x components. This is

$$\Sigma F_x = 0$$

$$R_x - R_1 \cos 45° = 0$$

Transposing and substituting give

$$R_1 = 210 \text{ N}$$

$$R_x = 210 \cos 45° = 210 \times 0.707 = 148.5 \text{ N}$$

4. The component R_y is found by writing the equation for the sum of the y components. So

$$\Sigma F_y = 0$$

$$-100 - 80 + R_y + R_1 \sin 45° = 0$$

Transposing and substituting give

$$R_1 = 210 \text{ N}$$

$$R_y = 100 + 80 - 210 \sin 45° = 180 - 210 \times 0.707$$

$$= 180 - 148.5 = 31.5 \text{ N}$$

5. The magnitude and direction of the reaction at A are found from the force triangle in Fig. 2-24*c*.

$$R = \sqrt{R_x^2 + R_y^2} = \sqrt{(148.5)^2 + (31.5)^2}$$

$$= \sqrt{22\,052 + 992} = \sqrt{23\,044} = 151.8 \text{ N}$$

$$\tan \theta = \frac{R_y}{R_x} = \frac{31.5}{148.5} = 0.212 \qquad \theta = 11.98°$$

STRATEGY

There isn't any set way in which an equilibrium problem must be solved. Usually it is best to start with a moment equation. If there are three unknown forces, choose the moment center at the intersection of two of them and write a moment equation to find the third unknown force. Then you might find the other two unknown forces by writing

$$\Sigma F_x = 0 \qquad \text{and} \qquad \Sigma F_y = 0$$

Illustrative Example 5. Figure 2-25*a* shows a brake lever. It is subjected to the loads of 340 lb and 40 lb and the reaction R_1. Also, it is supported by a pin at A. Find the reactions.

Solution:

1. The free-body diagram is shown in Fig. 2-25*b*. The loads are drawn first, then the reaction R_1, and, last, the components of the reaction at A. A line is enough to represent the brake arm.
2. The reaction R_1 is found by writing a moment equation with point A as a center. (Point A is chosen because R_x and R_y intersect at A.)

$$\Sigma M_A = 0$$

$$340 \times 1 - 40 \times 0.5 - 5R_1 = 0$$

$$5R_1 = 340 \times 1 - 40 \times 0.5 = 340 - 20 = 320$$

$$R_1 = 64 \text{ lb}$$

3. Then R_x is found by writing the sum of the x components. Thus

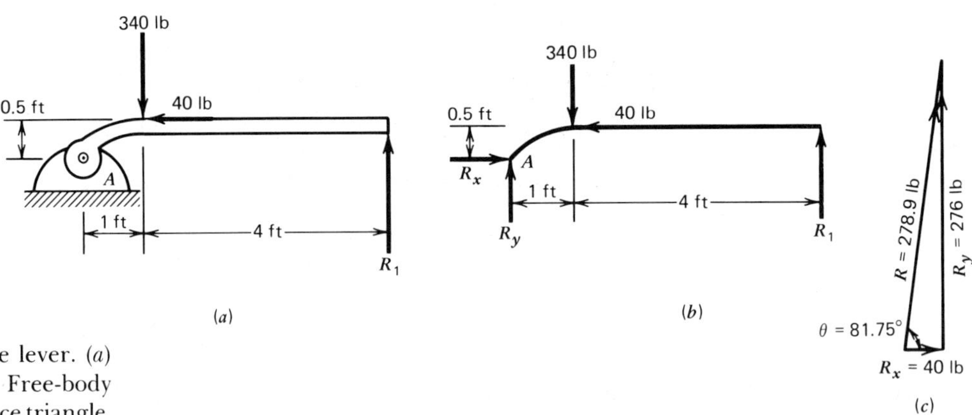

Fig. 2-25. Brake lever. (*a*) Brake lever. (*b*) Free-body diagram. (*c*) Force triangle.

$$\Sigma F_x = 0$$

$$R_x - 40 = 0 \qquad R_x = 40 \text{ lb}$$

4. Next R_y is found by writing the sum of the y components. This is

$$\Sigma F_y = 0$$

$$R_y - 340 + R_1 = 0$$

Transposing and substituting give

$$R_1 = 64 \text{ lb}$$

$$R_y = 340 - 64 = 276 \text{ lb}$$

5. The magnitude and direction of the reaction at A are found from the force triangle in Fig. 2-25c. Thus

$$R = \sqrt{R_x^2 + R_y^2} = \sqrt{(40)^2 + (276)^2} = \sqrt{1\,600 + 76\,176}$$

$$= \sqrt{77\,776} = 278.9 \text{ lb}$$

$$\tan\theta = \frac{R_y}{R_x} = \frac{276}{40} = 6.9 \qquad \theta = 81.75°$$

Practice Problems (Section 2-4). You are on your own again with this set of practice problems. Be sure to begin by drawing the free-body diagram. Then find the reactions by writing equations of equilibrium, either moment equations or equations for the sum of components, whichever you prefer.

1. A rope passes over the pulley in Fig. 2-26 and exerts the forces shown; one is horizontal and one is vertical. The pulley is supported by a shaft through the center. Find the reaction of the shaft.

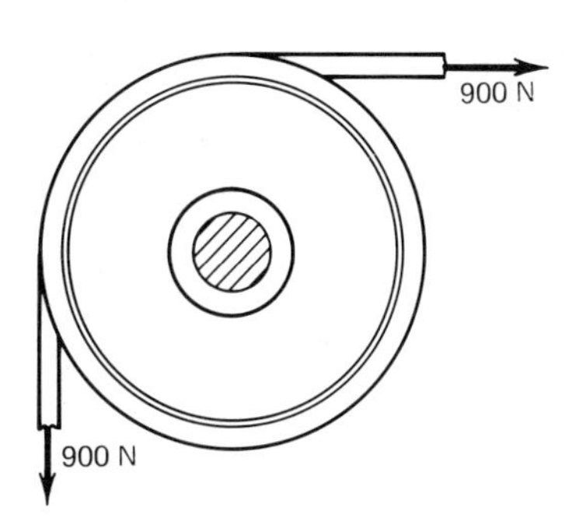

Fig. 2-26. Pulley and belt.

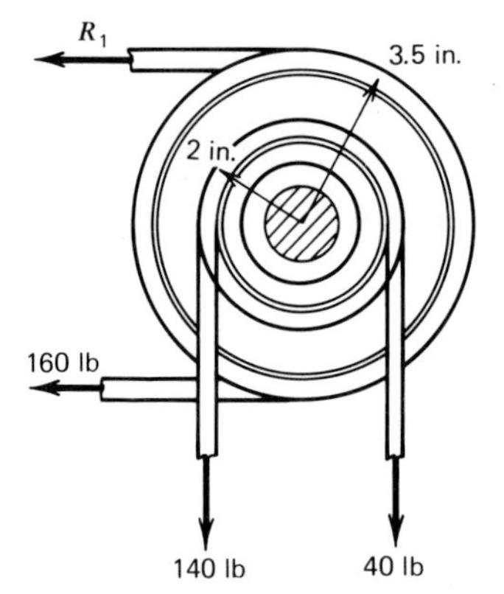

Fig. 2-27. Compound pulley.

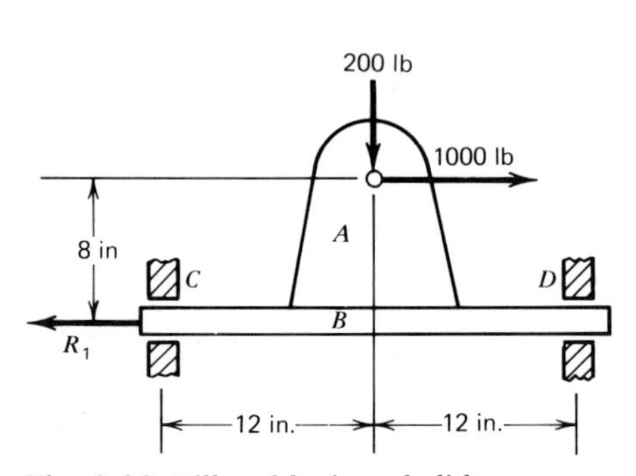

Fig. 2-28. Pillow block and slide.

2. The compound pulley in Fig. 2-27 is supported by a shaft through the center. Find the reaction R_1 and the reaction of the shaft.
3. In Fig. 2-28, the pillow block A is fastened to the slide B, which can slide in the smooth guide bearings C and D. Two loads are applied to the pillow block A. Find the reaction R_1 and the reactions at C and D.
4. The bar in Fig. 2-29 is part of a slider-crank mechanism. It is supported by a pin at A. Find the reactions.
5. Find the reactions R_1 and R_2 on the bracket shown in Fig. 2-30.
6. The frame in Fig. 2-31 is supported by a pin at A and the knife-edge at B. Find the reactions.

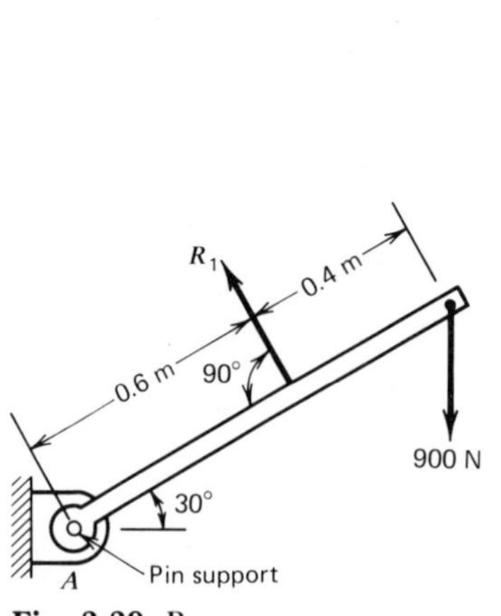

Fig. 2-29. Bar.

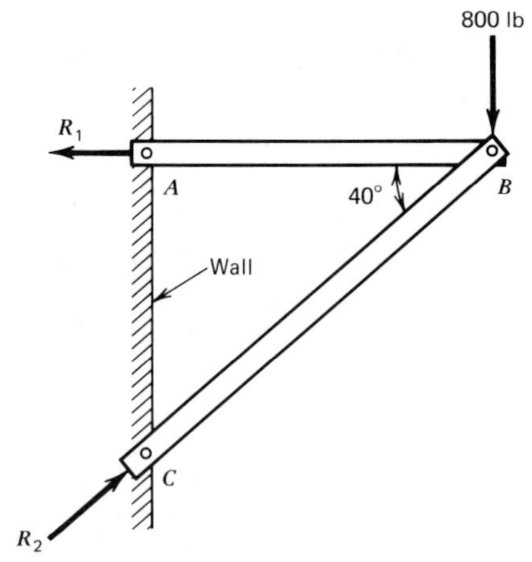

Fig. 2-30. Bracket.

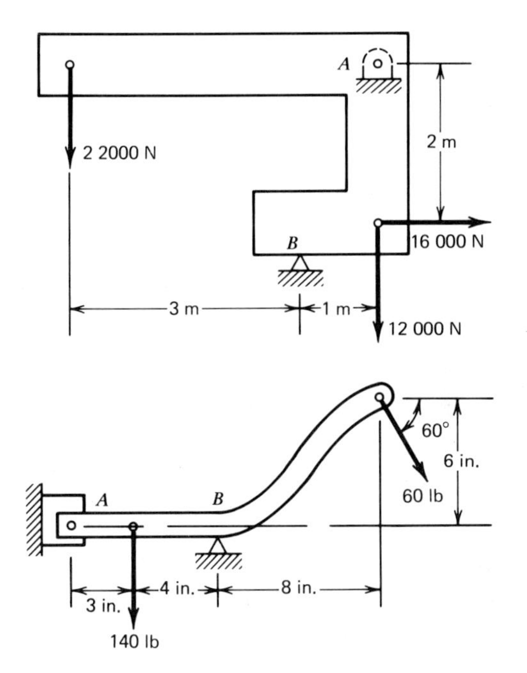

Fig. 2-31. Frame.

Fig. 2-32. Bent lever.

7. The bent lever in Fig. 2-32 is supported by a knife-edge at B and a pin at A. Find the reactions.
8. The compound pulley in Fig. 2-33 is subjected to the four belt pulls shown and is supported by a shaft through the center. Find the force R_1 and the reaction of the shaft.

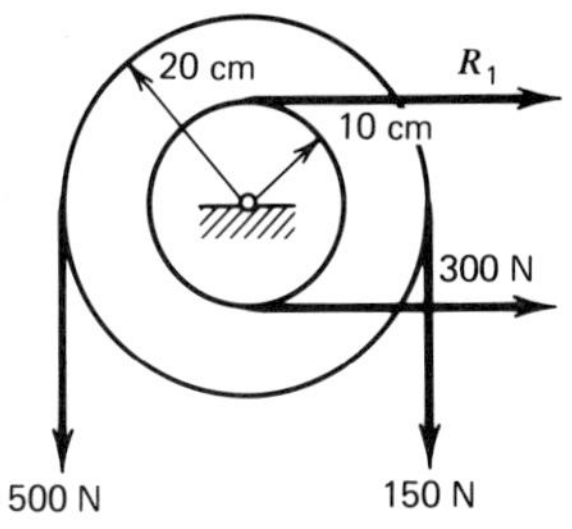

Fig. 2-33. Compound pulley.

2-5. SOLUTION OF EQUILIBRIUM PROBLEMS BY USE OF FORCE POLYGONS

One useful method of solving equilibrium problems is by means of the force polygon. In this, the forces acting on the body are laid off in a polygon, of which each side represents a force. For example, the forces F, P, and Q, acting on the bar in Fig. 2-34*a*, are drawn to form the polygon in Fig. 2-34*b*. Starting at point O, the force F is laid off to *scale* (this means that its length represents its magnitude) and in its true direction; from the end of F the force P is drawn, also to scale and in its true direction; finally, the force Q is drawn from the end of P, and it comes back to the starting point O. This construction is called a *force polygon*, and when the last force comes back to the starting point of the first one, it is said to *close*. If a body is in equilibrium, the force polygon of the forces acting on it must close. In using this method, the known force or forces are laid off first, and then the unknown forces are drawn in such a way that the force polygon closes, and the arrows on the forces

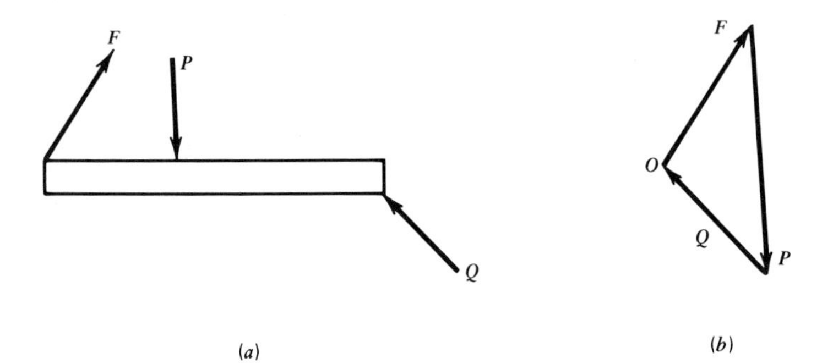

Fig. 2-34. Bar and force polygon. (*a*) Free-body diagram. (*b*) Force polygon.

can be followed continuously around the polygon to return to the starting point. Then the unknown forces are found by measuring. One of the easiest problems is that in which there is only one unknown force, but its magnitude and direction are both unknown. They can be found by closing the force polygon. Be sure to draw the free-body diagram as the start of the problem.

Force polygon work is graphical (*graphical* means by construction or drawings) work, and you must be equipped to do it properly. You need drawing instruments to do this.

Illustrative Example 1. A rope passes over the pulley in Fig. 2-35*a* and exerts the two forces of 360 lb each. The pulley is supported by a shaft through the center. Find the reaction of the shaft.

Solution:

1. Figure 2-35*a* can serve as the free-body diagram. The magnitude and direction of the reaction R are unknown and are to be found by closing the force polygon.
2. The two known forces are drawn in the force polygon in Fig. 2-35*b*. (A scale of 200 lb to the inch is suitable for this problem.)

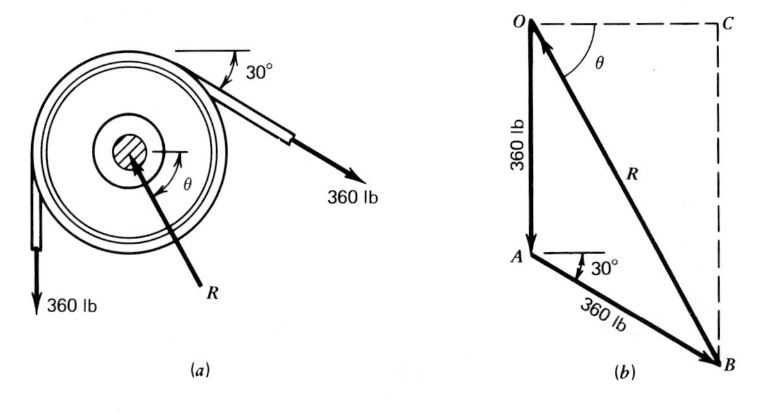

Fig. 2-35. Pulley problem. (*a*) Forces on pulley. (*b*) Force polygon.

This means that each inch of length of a force represents 200 lb; the proper length of a 360-lb force is (360 lb) ÷ (200 lb/in.) = 1.8 in. Starting at point O, the vertical force of 360 lb is drawn to scale, and from its end, point A, the oblique force of 360 lb is drawn to scale, ending at point B.

3. The reaction R is drawn from B to O to complete the polygon. This closes the polygon, and the arrowhead on R is directed toward point O, so that the arrows can be followed continously around the polygon.
4. The magnitude and direction of R are measured from the force polygon. Using the 20 scale on a decimal scale (this is the professional way to do it), the magnitude of R is measured to be $R = 624$ lb. Otherwise, the length of the force in inches is 3.12; the scale is 200 lb/in., which means that each inch of length of the force represents 200 lb, so the magnitude of R is

$$R = (3.12 \text{ in.}) \times (200 \text{ lb/in.}) = 624 \text{ lb}$$

A good way to determine the angle θ is to measure the lengths BC and OC in Fig. 2-35*b*. Then the tangent of the angle θ is

$$\tan \theta = \frac{BC}{OC}$$

In this problem BC is the vertical component of R and is measured as 540 lb; OC is the horizontal component and is measured as 316 lb. Then,

$$\tan \theta = \frac{540}{312} = 1.732$$

from which the angle θ is found to be 60°. The angle θ could also be measured with a protractor.

THREE FORCES IN EQUILIBRIUM

The force polygon is most useful when there are only three forces acting on the body, because then the directions of the unknown forces can be found by means of a special theorem which states: Three forces in equilibrium must be parallel or concurrent. You already know what parallel means, and *concurrent* simply means that the forces have a common point of intersection; that is, they all intersect at one point. Usually, the directions of two of the forces are known, and it remains to establish the point of intersection of these two forces. Then we know that the third force must also go through this point. This gives us one point on the line of action of the force and the point at which the force is applied is another. Two points on its line of action are enough to determine the direction of the force.

The procedure in drawing the force polygon is, first, to lay off the known force or forces to scale and in the proper direction. Second, draw one of the unknown forces from one end of the known force. Third, draw the other unknown force from the other end of the known force until it intersects the first unknown force; this completes the force polygon, and it is closed. Fourth, measure the values of the unknown forces. Remember that the known force is laid off to scale, so that its length represents its magnitude; naturally, the same scale is used to measure the magnitudes of the unknown forces. Let's see how it works.

Illustrative Example 2. The bell crank in Fig. 2-36*a* has the load of 160 N and the reaction R_1. It is supported by a pin at *A*. Find the reactions.

Solution:

1. The free-body diagram of the bell crank is shown in Fig. 2-36*b*.

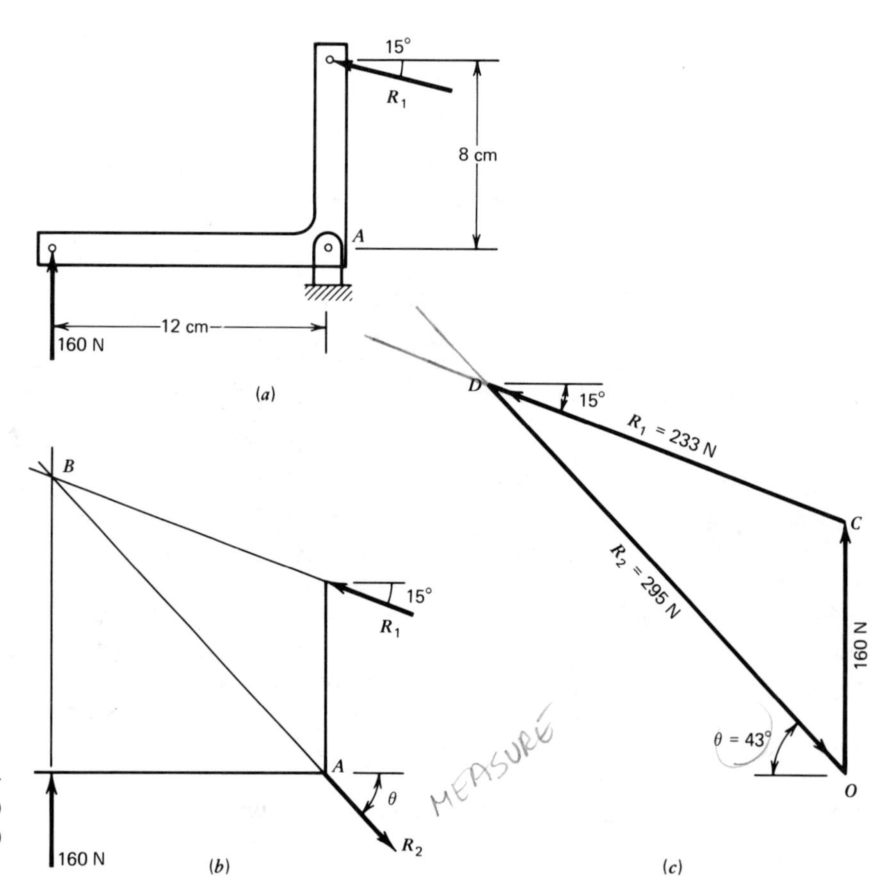

Fig. 2-36. Bell crank problem. (*a*) Bell crank. (*b*) Free-body diagram. (*c*) Force polygon.

It is drawn to scale (1 cm = 4 cm). There are only three forces acting, and, to be in equilibrium, they must be parallel or concurrent; it is clear that they are not parallel, because the reaction R_1 is not parallel to the 160 N force; hence they must be concurrent. The lines of action of the 160 N force and the reaction R_1 are extended to their point of intersection, which is at B. Then B is the common point of intersection for the forces and is one point on the line of action of the reaction at A. This force R_2 must pass through A, where it is applied, and with two points on its line of action known, its direction is determined. It is drawn through A and B as shown.

2. The known force of 160 N is laid off to scale (100 N/cm in this case) to start the force polygon in Fig. 2-36*c*. It starts at O and ends at C.
3. The reaction R_1 is drawn through point C in its proper direction.
4. The reaction R_2 is drawn through point O in its true direction (this can be done by sliding a triangle over from Fig. 2-36*b*) and extended until it intersects the force R_1 at point D. This closes the force polygon, and the arrowheads on R_1 and R_2 are directed so that the arrows on the three forces can be followed continuously around the polygon.
5. The magnitudes of R_1 and R_2 are measured from the force polygon in Fig. 2-36*c* (using the scale of 100 N/cm). They are

$$R_1 = 223 \text{ N}$$

$$R_2 = 295 \text{ N}$$

The direction of R is given by the angle θ. It is measured in either Fig. 2-36*b* or 2-36*c*.

$$\theta = 43°$$

Illustrative Example 3. The bar in Fig. 2-37*a* is part of a slider-crank mechanism. It carries the load of 150 lb and the reaction R_1 and is supported by a pin at A. Find the reactions.

Solution:

1. The free-body diagram of the bar is shown in Fig. 2-37*b*. Here a line is enough to represent the bar. There are only three forces acting on the bar, so they must be concurrent or parallel. The reaction R_1 is certainly not parallel to the 150-lb force, so they are concurrent. The 150-lb force is extended until it intersects the reaction R_1 at point B. Point B is then the common point of intersection of the forces, so the reaction at A must pass through

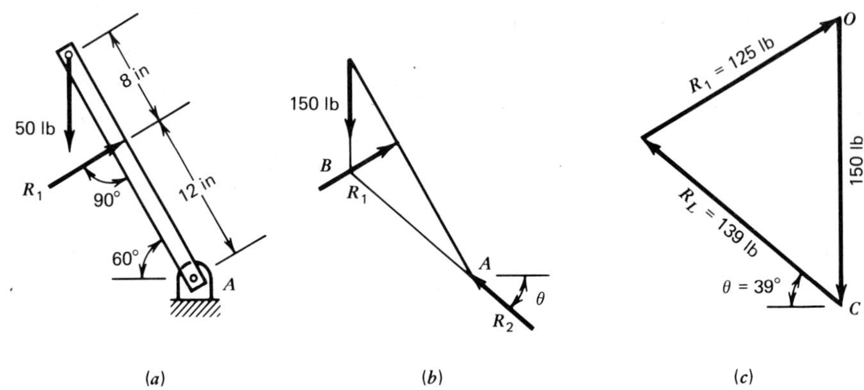

Fig. 2-37. Slider crank bar. (*a*) Bar. (*b*) Free-body diagram. (*c*) Force polygon.

B. This reaction R_2 must also pass through point *A*, because it is applied at *A*. Then *A* and *B* are two points on the line of action of R_2 and determine its direction. This line is drawn as shown in Fig. 2-37*b* and is in its correct direction, because the figure is drawn to scale (16 in./in.).

2. The force polygon is started in Fig. 2-37*c* by drawing the known force of 150 lb to scale (100 lb = 1 in.) and in its true direction. The point where it starts is lettered *O*, and the point where it ends is lettered *C*.
3. The reaction R_1 is drawn through point *O* in the force polygon.
4. The reaction R_2 is drawn through point *C*, in the direction found in Fig. 2-37*b*, and extended until it intersects R_1. This closes the force polygon. Then the arrowheads are placed on R_2 and R_1 and directed so that all of the arrowheads can be followed continuously around the polygon.
5. The magnitudes of R_1 and R_2 are measured from the force polygon in Fig. 2-37*c*. They are

$$R_1 = 125 \text{ lb}$$

$$R_2 = 139 \text{ lb}$$

The angle θ, for the direction of R_2, is measured either from Fig. 2-37*b* or 2-37*c*.

$$\theta = 39°$$

MORE THAN THREE FORCES

When there are more than three forces, or when the forces are parallel, ordinarily it doesn't pay to use the force polygon. The problem can usually be solved more simply by moment equations.

Practice Problems (Section 2-5). You can gain experience in drawing force polygons in the following problems. Be sure to draw the free-body diagram at the beginning of each. Also, be careful about scales. Use the same scale to measure results that you use to lay off the known forces.

1. Figure 2-38 shows an idler pulley. A belt passes over it and exerts the two forces shown. The pulley is supported by a shaft through its center. Find the reaction of the shaft.
2. The bar in Fig. 2-39 is a connecting rod in a pump mechanism. It is supported by a smooth bearing at *B* (which can only exert a force perpendicular to the rod) and by a pin on the pump rod at *A*. Find the reactions at *A* and *B*.
3. The machine part in Fig. 2-40 has the reaction R_1 and is supported by a pin at *A*. Find the reactions.
4. Figure 2-41 shows a cam follower. The reaction R_1 is exerted by the cam, and the body is supported by a pin at *A*. Find the reactions.

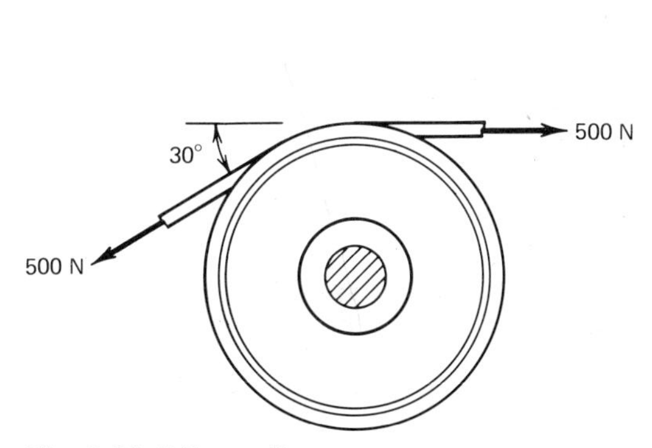

Fig. 2-38. Idler pulley.

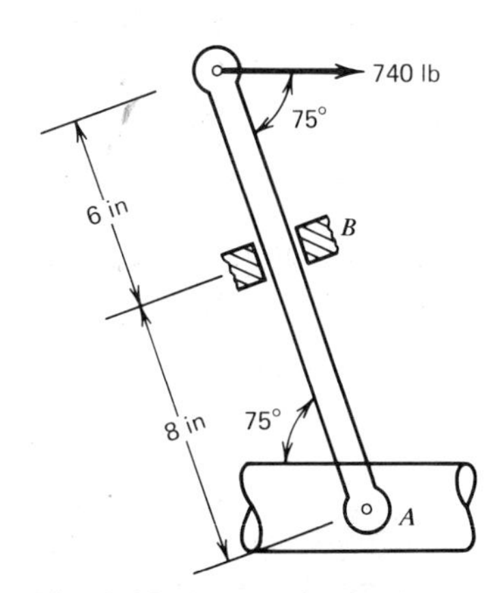

Fig. 2-39. Connecting rod.

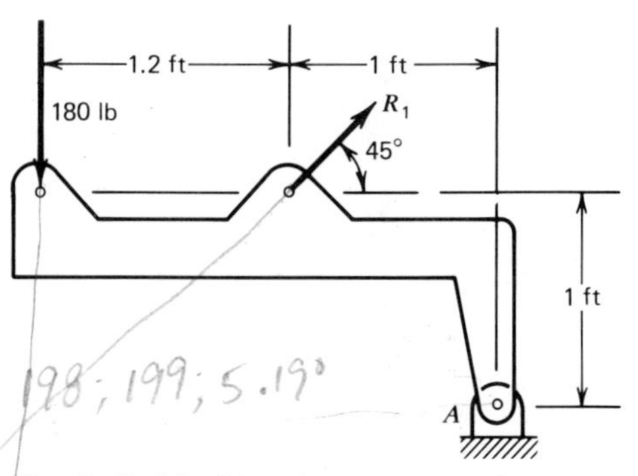

Fig. 2-40. Machine part.

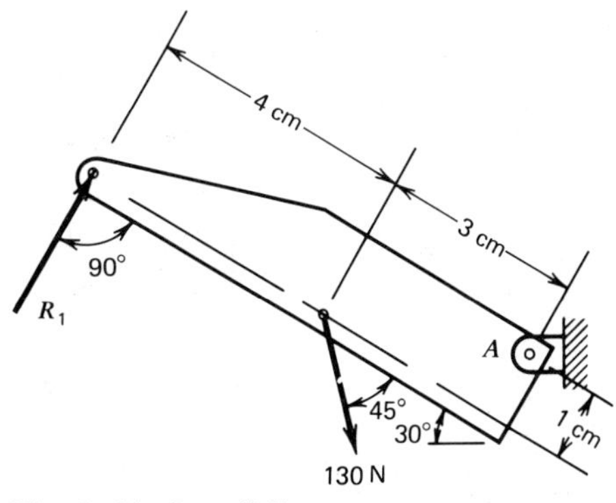

Fig. 2-41. Cam follower.

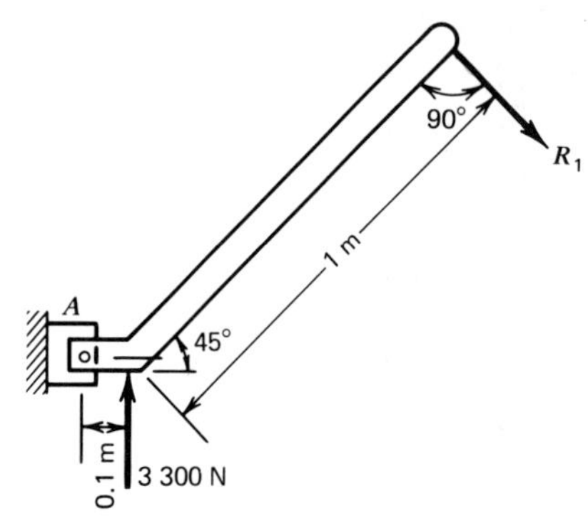

Fig. 2-42. Lever on a hand press.

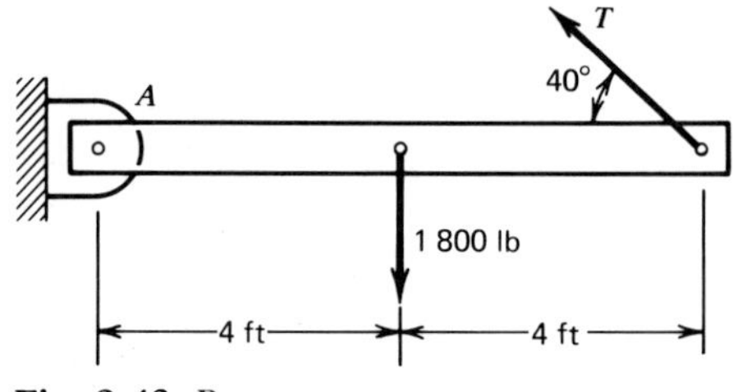

Fig. 2-43. Beam.

5. The bent bar in Fig. 2-42 is a lever on a hand press. The force of 3 300 N is exerted by the workpiece in the press, and the lever is actuated by the reaction R_1. In addition, it is supported by a pin at A. Find the reactions.
6. The beam in Fig. 2-43 is used to support a chain hoist which exerts the load of 1 800 lb. The reaction T is exerted by a tie rod, and the left end of the beam is supported at A. Find the reactions.

2-6. EQUILIBRIUM OF SHAFTS

The problem of the shaft is an important one, because shafts are commonly used in machines for transmitting power and motion. We must study the equilibrium of shafts now and learn to find the reactions so that we can calculate twisting moments and bending moments later and find stresses in shafts. The forces on a shaft are not usually in a single plane and cannot be shown satisfactorily in a single orthographic projection. Instead, an isometric drawing, such as Fig. 2-44 is the best way

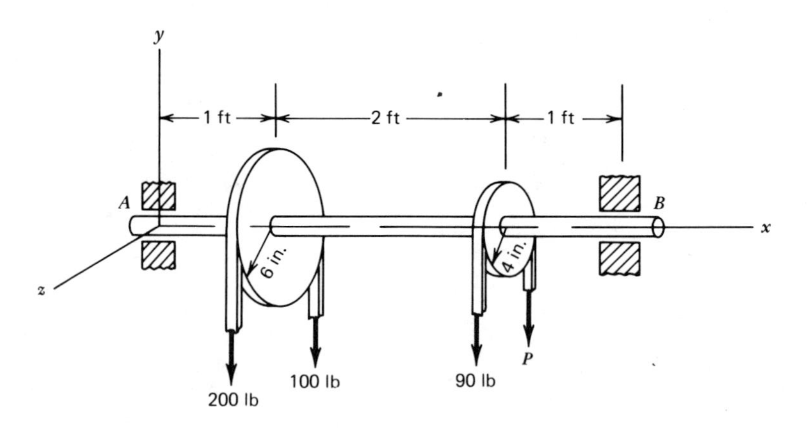

Fig. 2-44. Shaft and pulleys.

to show the forces. This shows a shaft that is supported in bearings at A and B and that carries two pulleys. The larger pulley has belt pulls of 200 and 100 lb, and the smaller pulley has belt pulls of P and 90 lb. The problem here is to find the force P and the reactions of the bearings. We are going to do this by introducing a three-dimensional coordinate system and expanding on notions of equilibrium equations. So, in Fig. 2-44, we have three coordinate axes. The origin of these axes is at point A.

1. The x axis is horizontal and coincides with the centerline of the shaft.
2. The y axis is vertical.
3. The z axis is horizontal and perpendicular to the plane of the paper.

We have already said that the sum of the moments about any point must be zero if a body is in equilibrium. But, the moment of a force with respect to a point, as we wrote it, is really the moment of the force about an axis that passes through the point and is perpendicular to the plane of the figure. Now, we are going to write moment equations about specific axes. The moment of the force with respect to an axis is just the product of the force and the perpendicular distance from the axis to the force. So we can say that

$$\Sigma M_x = 0 \qquad \Sigma M_y = 0 \qquad \Sigma M_z = 0$$

As we look from the positive end of the axis toward the origin, we will say that the moment of a force is positive if we see it as clockwise and negative if we see it as counterclockwise. The positive end of the axis is the end where the letter x, y, or z appears.

Next, the sum of the forces must be zero in any direction we choose, so

$$\Sigma F_x = 0 \qquad \Sigma F_y = 0 \qquad \Sigma F_z = 0$$

Now, we are ready to try an example.

Illustrative Example 1. Find the force P and the bearing reactions at A and B for the shaft in Fig. 2-44.

Solution:

1. The free-body diagram of the shaft is shown in Fig. 2-45. This is quite similar to Fig. 2-44. Notice the bearing reactions R_1 and R_2.
2. We write an equation for the sum of the moments about the x axis. The forces R_1 and R_2 go through the x axis so they have no moment about it. Thus

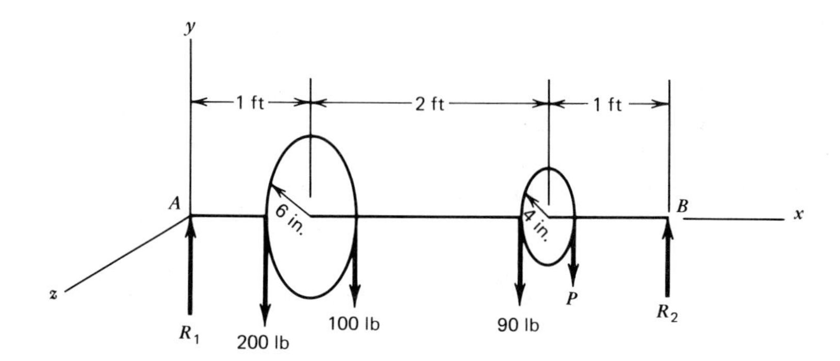

Fig. 2-45. Free-body diagram of shaft.

$$\Sigma M_x = 0 \qquad -200 \times 6 + 100 \times 6 - 90 \times 4 + 4P = 0$$

$$4P = 1\,200 - 600 + 360 = 960 \qquad P = 240 \text{ lb}$$

3. Next, we write an equation for the sum of the moments about the z axis. The force R_1 goes through the z axis. We have found the force P to be 240 lb. So

$$\Sigma M_z = 0 \qquad 200 \times 1 + 100 \times 1 + 90 \times 3 + 240 \times 3 - 4R_2 = 0$$

$$4R_2 = 200 + 100 + 270 + 720 = 1\,290$$

$$R_2 = 322.5 \text{ lb}$$

4. Finally, we write an equation for the sum of the forces in the direction of the y axis. It is

$$\Sigma F_y = 0 \qquad R_1 - 200 - 100 - 90 - 240 + 322.5 = 0$$

$$R_1 = 200 + 100 + 90 + 240 - 322.5 = 307.5 \text{ lb}$$

Illustrative Example 2. Figure 2-46*a* shows a shaft that is supported in bearings at A and B and carries the gears C and D. The gear D is subjected to a vertical load of 1 500 N, and the gear C has a reaction P. Find the reaction P and the bearing reactions at A and B.

Solution:

1. Figure 2-46*b* shows the free-body diagram of the shaft. There are four forces: the load of 1 500 N, the reaction P, and the bearing reactions which are designated as R_1 and R_2. Notice the x, y, and z axes.
2. First we write a moment equation about the x axis like this:

$$\Sigma M_x = 0 \qquad -4P + 1\,500 \times 8 = 0 \qquad P = 3\,000 \text{ N}$$

3. Next, we write a moment equation about the z axis. Thus

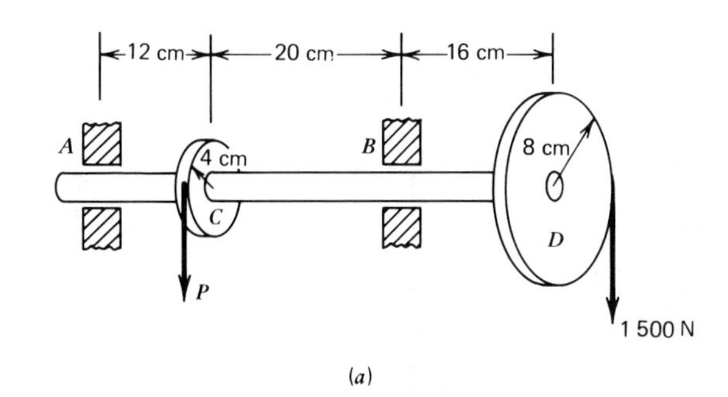

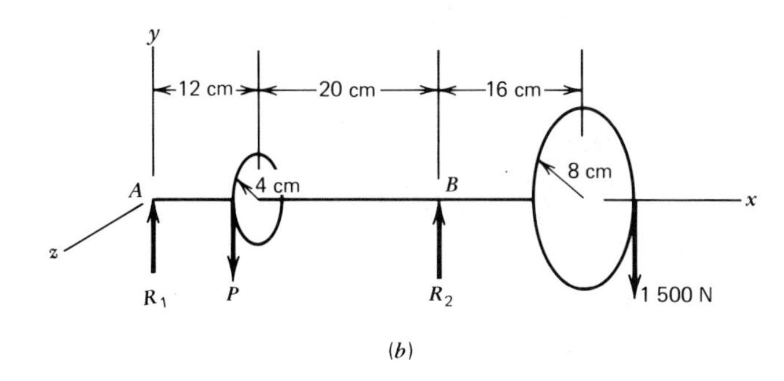

Fig. 2-46. Shaft and gears. (*a*) Shaft and gears. (*b*) Free-body diagram.

$$\Sigma M_z = 0 \qquad 3\,000 \times 12 - 32R_2 + 1\,500 \times 48 = 0$$

$$32R_2 = 36\,000 + 72\,000 = 108\,000 \qquad R_2 = 3\,375 \text{ N}$$

4. Then, we write an equation for the sum of the vertical forces. So,

$$\Sigma F_y = 0 \qquad R_1 - 3\,000 + 3\,375 - 1\,500 = 0$$

$$R_1 = 1\,125 \text{ N}$$

Illustrative Example 3. The shaft in Fig. 2-47*a* is supported in bearings at *A* and *B*. It has the pulley *C* and the gear *D*, with forces as shown. Find the reaction *P* and the reactions of the bearings at *A* and *B*.

Solution:

1. The free-body diagram of the shaft is shown in Fig. 2-47*b*. The forces acting on the pulley are vertical, so there must be vertical components of the reactions at the bearings; they are designated as R_{1y} and R_{2y}. The force *P* is horizontal, so there must be horizontal components of the reactions at the bearings; they are designated as R_{1z} and R_{2z}.

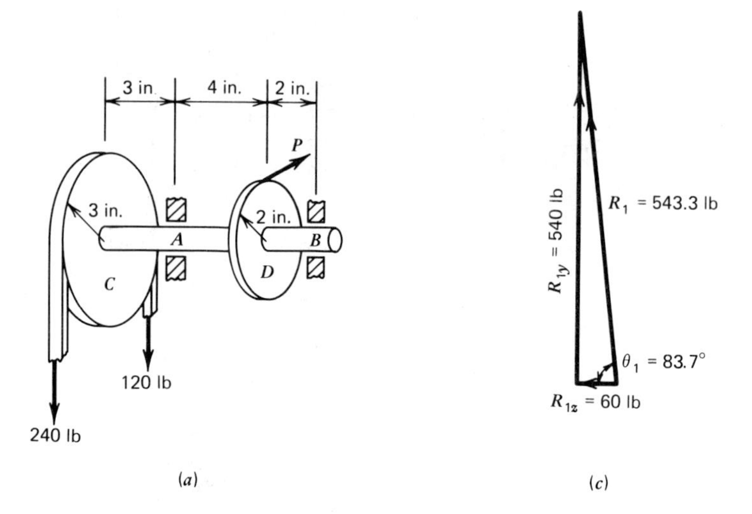

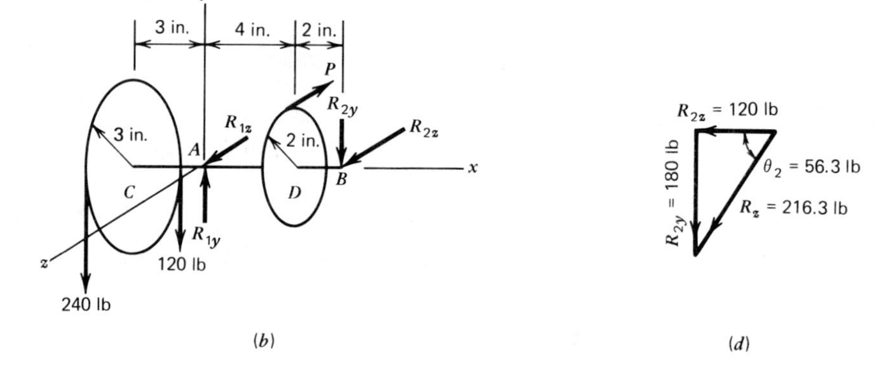

Fig. 2-47. Shaft with pulley and gear. (*a*) Shaft. (*b*) Free-body diagram. (*c*) Force triangle for R_1. (*d*) Force triangle for R_2.

2. First, we write a moment equation about the x axis. This is

$$\Sigma M_x = 0 \qquad -240 \times 3 + 120 \times 3 + 2P = 0$$

$$2P = 720 - 360 = 360 \qquad P = 180 \text{ lb}$$

3. Next, we write a moment equation about the z axis. Thus

$$\Sigma M_z = 0 \qquad -240 \times 3 - 120 \times 3 + 6R_{2y} = 0$$

$$6R_{2y} = 720 + 360 = 1\,080 \qquad R_{2y} = 180 \text{ lb}$$

4. Then we write an equation for the sum of the forces parallel to the y axis. So,

$$\Sigma F_y = 0 \qquad -240 - 120 + R_{1y} - 180 = 0 \qquad R_{1y} = 540 \text{ lb}$$

5. Now, we write a moment equation about the y axis. Remember, we know that $P = 180$ lb

$$\Sigma M_y = 0 \qquad -4 \times 180 + 6R_{2z} = 0$$

$$6R_{2z} = 720 \qquad R_{2z} = 120 \text{ lb}$$

6. This time we write an equation for the sum of the forces parallel to the z axis. Thus

$$\Sigma F_z = 0 \qquad R_{1z} - 180 + 120 = 0 \qquad R_{1z} = 60 \text{ lb}$$

7. The next step is to combine the components of R_1 by using the triangle in Fig. 2-47c. Then,

$$R_1 = \sqrt{(60)^2 + (540)^2} = \sqrt{3\,600 + 291\,600}$$

$$= \sqrt{295\,200} = 543.3 \text{ lb}$$

$$\tan \theta_1 = \frac{540}{60} = 9 \qquad \theta_1 = 83.7°$$

8. Finally, we combine the components of R_2 by using the triangle in Fig. 2-47d. This way,

$$R_2 = \sqrt{(120)^2 + (180)^2} = \sqrt{14\,400 + 32\,400}$$

$$= \sqrt{46\,800} = 216.3 \text{ lb}$$

$$\tan \theta_2 = \frac{180}{120} = 1.5 \qquad \theta_2 = 56.3°$$

Practice Problems (Section 2-6). Try these shaft problems for practice. Don't hurry, but take time to draw free-body diagrams carefully.

1. The shaft in Fig. 2-48 is supported in bearings at A and B. The two pulleys are subjected to vertical belt pulls as shown. Find the reaction P and the bearing reactions.

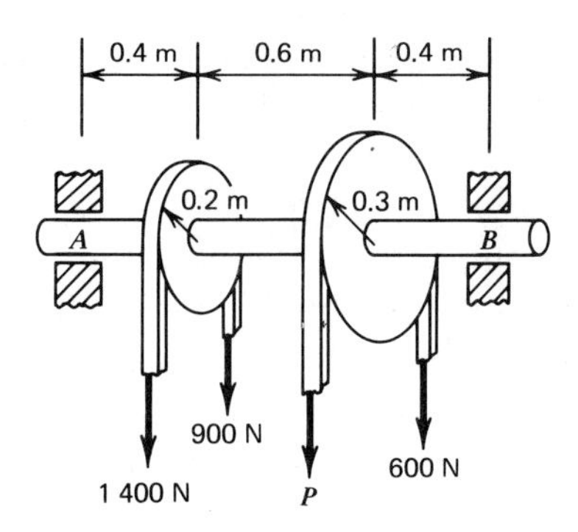

Fig. 2-48. Shaft with pulleys.

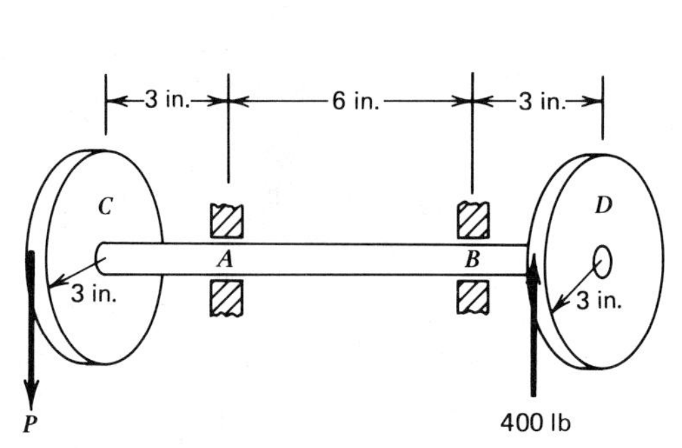

Fig. 2-49. Shaft with gears.

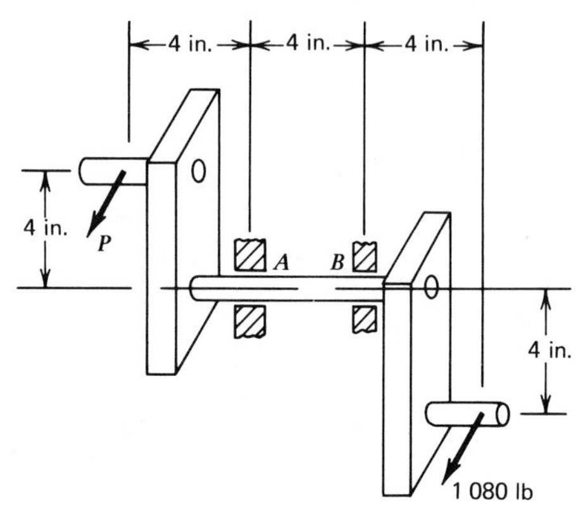

Fig. 2-50. Part of a crankshaft.

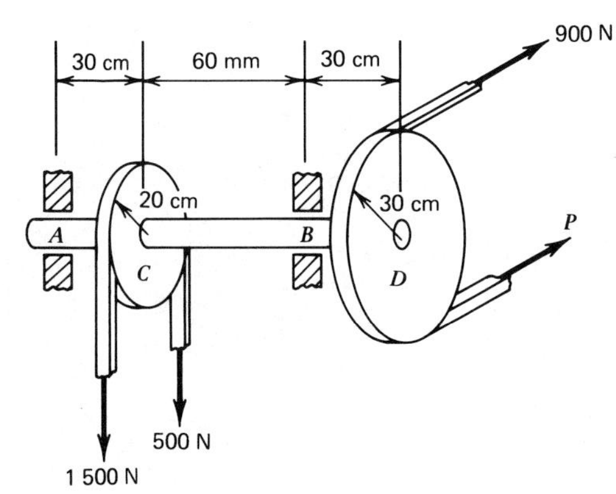

Fig. 2-51. Shaft with pulleys.

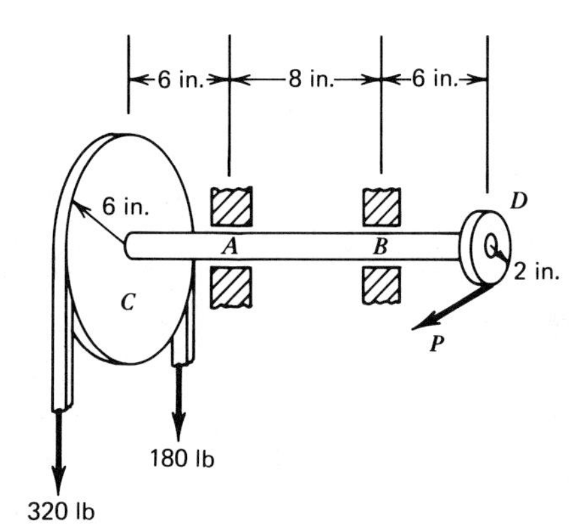

Fig. 2-52. Shaft with pulley and gear.

2. The shaft in Fig. 2-49 carries the gears C and D that are subjected to the vertical forces shown. Find the force P and the bearing reactions at A and B.
3. Figure 2-50 shows part of a crankshaft. The force P and the 1 080-lb load are horizontal. Find P and the bearing reactions at A and B.
4. The shaft in Fig. 2-51 is supported in bearings at A and B. The belt pulls on pulley C are vertical and those on pulley D are horizontal. Find the reaction P and the bearing reactions.
5. The shaft in Fig. 2-52 carries the pulley C, which is subjected to vertical belt pulls. The gear D is subjected to the horizontal force P. Find P and the bearing reactions at A and B.

SUMMARY Here is a summary of what you should have learned about equilibrium in Part A of this chapter.

1. Equilibrium is a state of balance between the forces acting on a body. The tendency of one force to cause motion is balanced by the tendencies of the other forces on the body.
2. The free-body diagram is a picture of the body showing all of the forces that act on the body. The first step in any equilibrium problem is to draw the free-body diagram.
   The procedure in drawing the free-body diagram is:
   (a) Draw the body (a line is often sufficient to represent it, and you needn't show a lot of detail).
   (b) Draw the loads (known forces).
   (c) Draw the reactions (unknown forces).
3. If a body is in equilibrium, the sum of the moments of the forces acting on the body is zero with respect to any point (for example, $\Sigma M_o = 0$). The most efficient way to use a moment equation is to choose a moment center on the line of action of one or two unknown forces. Then the other unknown force is found by solving the moment equation.
4. If a body is in equilibrium, the sum of the components of the forces acting on the body is zero in any direction (for example, $\Sigma F_x = 0$). This type of equation is useful in finding unknown forces and in checking results obtained with moment equations.
5. Three forces in equilibrium must be parallel or concurrent. This theorem is useful in finding the direction of an unknown force.
6. If a force system is in equilibrium, the force polygon must close. This offers a graphical method of solving equilibrium problems. In many cases the graphical method is the most efficient.
7. Shaft problems can be solved efficiently by using a three-dimensional coordinate system, thus x, y, z. Then, the sum of the moments of the forces about any axis is zero; that is, $\Sigma M_x = 0$, $\Sigma M_y = 0$, and $\Sigma M_z = 0$. Also, the sum of the forces in any direction is zero, so $\Sigma F_x = 0$, $\Sigma F_y = 0$, and $\Sigma F_z = 0$.

REVIEW QUESTIONS These review questions are on methods and principle and are to be answered without looking at the preceding pages.

1. What is equilibrium?
2. What is a free-body diagram?
3. What is the difference between a load and a reaction?
4. What is a moment equation?

5. What is the advantage of choosing a moment center on the line of action of an unknown force?
6. What is a force polygon?
7. What is the rule for directing the arrowheads on the forces in a force polygon?
8. Which forces are drawn first in a force polygon?
9. What is an equation for the sum of the components?
10. What is the direction of a reaction exerted on a shaft by a smooth bearing?

REVIEW PROBLEMS These problems offer you a chance to test yourself on what you have learned. If there are any problems you cannot solve, you need to study this chapter more.

1. The beam in Fig. 2-53 is supported at the ends. Find the reactions.
2. Find the reactions at *A* and *B* for the bar in Fig. 2-54.
3. Figure 2-55 shows a large shaft that is supported in bearings at *A* and *B*. Part *C* weighs 12 000 N/m and Part *D* weighs 5 000 N/m. Find the reactions.

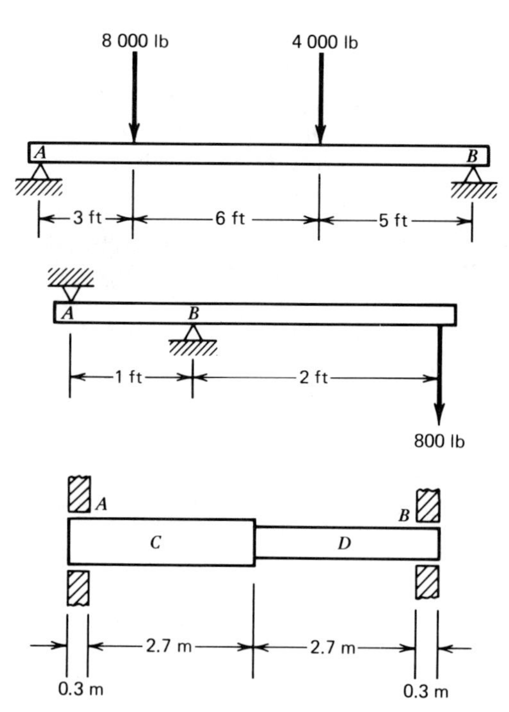

Fig. 2-53. Beam on knife-edges.

Fig. 2-54. Overhanging beam.

Fig. 2-55. Shaft supported in bearings.

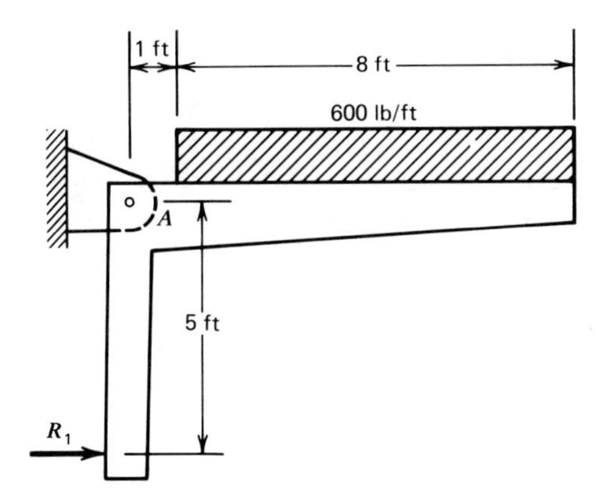

Fig. 2-56. Part of a loading device.

4. Figure 2-56 shows part of a loading device. It is supported by a pin at A. Find the reaction R_1 and the reaction at A.
5. The bent bar in Fig. 2-57 is supported by a pin at A and is subjected to the two vertical forces shown. Find the reaction R_1 and the reaction at A.
6. Figure 2-58 shows the piston rod of a pump. The force R_1 is exerted on the rod by the piston. Find the force R_1 and the bearing reactions.

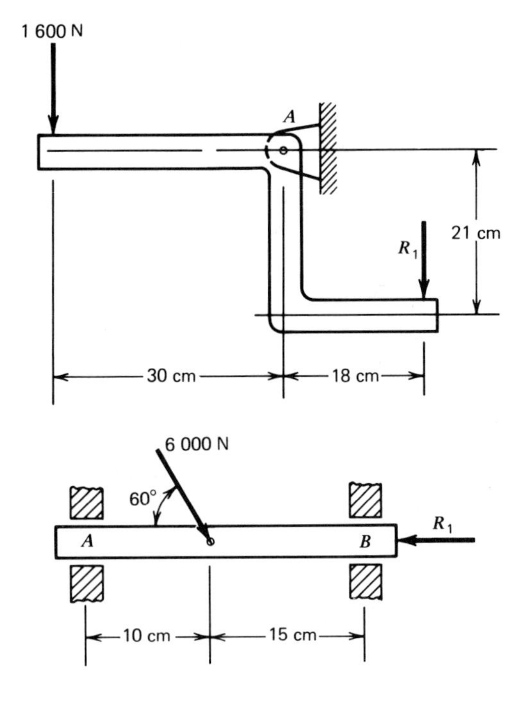

Fig. 2-57. Bent lever.

Fig. 2-58. Piston rod of pump.

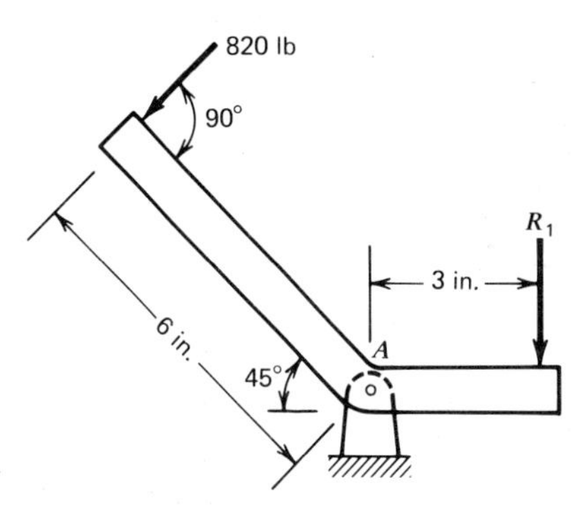

Fig. 2-59. Lever supported by a pin.

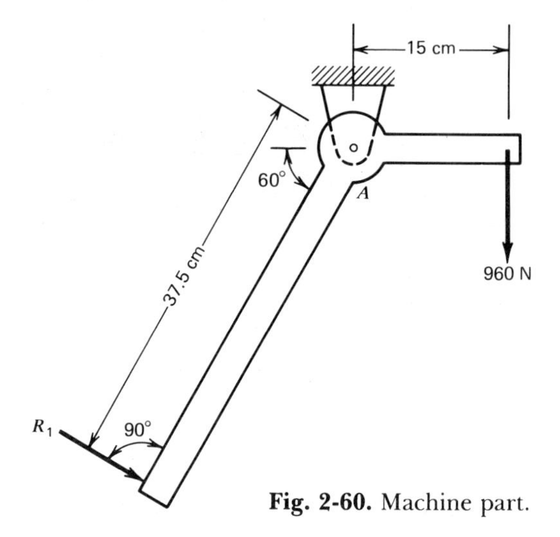

Fig. 2-60. Machine part.

7. The lever in Fig. 2-59 is supported by a pin at A and is subjected to the other forces shown. Find the reaction R_1 and the reaction at A.
8. Find the reaction R_1 and the reaction at A for the machine part in Fig. 2-60.
9. The shaft in Fig. 2-61 is supported in bearings at A and B. A vertical force of 270 lb is exerted on the gear C, and a horizontal force P is exerted on the gear D. Find the force P and the bearing reactions.
10. The shaft in Fig. 2-62 is supported in bearings at A and B, and the two pulleys are subjected to vertical belt pulls as shown. Find the force P and the reactions of the bearings.

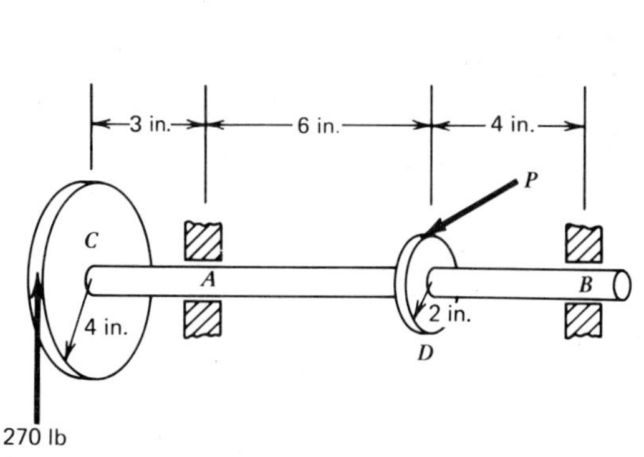

Fig. 2-61. Shaft with gears.

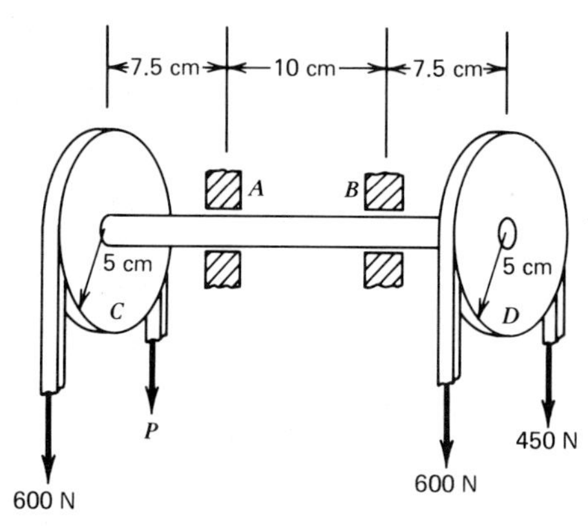

Fig. 2-62. Shaft with pulleys.

Part B THREE SPECIAL TOPICS IN EQUILIBRIUM

PURPOSE OF PART B. The purpose of this part of the chapter is to study three special topics in equilibrium. These special topics are (1) trusses, (2) flexible cables, and (3) friction between dry surfaces. Each of these special topics is a distinct class of equilibrium problem. We will take them one at a time, with enough explanation, illustrative examples, and practice problems for you to learn how to work with them.

TRUSSES

2-7. WHAT A TRUSS IS

A truss is a special type of framed structure. It consists of bars fastened together at their ends in such a manner that the spaces between bars are triangular. The triangular arrangement of the spaces between the bars is especially important, because it makes a truss stable. Now let's look at two kinds of trusses.

Figure 2-63*a* shows a type of truss that has been used frequently in

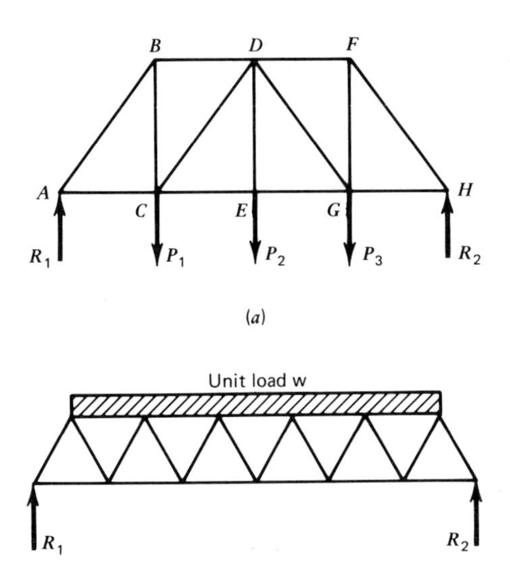

Fig. 2-63. Trusses. (*a*) Bridge truss. (*b*) Special roof truss.

highway bridges. Each line represents a bar, that is, a member of the truss. The members *AB*, *BD*, *DF*, and *FH* are the upper chord of the truss. The line *AH* is the lower chord. The members *BC*, *CD*, *DE*, *DG*, and *FG* are web members. Each point at which members are fastened together is a joint. The members might be fastened together by rivets, by bolts, or by welds. The analysis is the same for all methods of fastenening.

The weights of the truss, roadway, and vehicular traffic are assumed to be applied at the lower chord joints. These are the forces P_1, P_2, and P_3. The truss is supported at the ends where the reactions R_1 and R_2 are applied.

Figure 2-63*b* shows a special shape of truss that is often used to support the roofs of such buildings as supermarkets and automobile dealer garages. The members of this sort of truss are usually welded together.

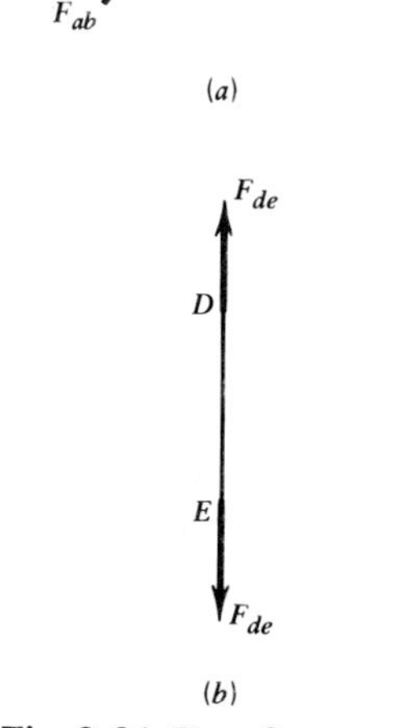

Fig. 2-64. Two-force members. (*a*) Free-body diagram of member *AB*. (*b*) Free-body diagram of member *DE*.

2-8. TWO-FORCE MEMBERS

Because truss members are fastened together at the ends, the ends are the only places at which forces are applied to the members. The joint at each end of a member exerts a force on the member. Figure 2-64*a* shows the free-body diagram of the member *AB* of the bridge truss in Fig. 2-63*a*. Here the forces F_{ab} are applied by the joints *A* and *B*. Each of these forces must act along the line *AB* for the sum of the moments about point *A* or point *B* to be zero. Furthermore, the two forces must be equal for the sum of the forces in the direction *AB* to be zero. We say that F_{ab} is the force in the member. We also say that F_{ab} is a compressive force, because it acts to shorten the member, and that the member is in compression. We call each truss member a two-force member, because there are only two forces applied to it.

Figure 2-64*b* shows the free-body diagram of member *DE* of the truss in Fig. 2-63*a*. Here, we say that F_{de} is a tensile force, because it acts to stretch the member, and that the member is in tension.

2-9. TRUSS ANALYSIS BY THE METHOD OF JOINTS

The object of truss analysis is to find the magnitude of the force in each member of the truss and to determine whether the force is tension or compression. This is the information a structural engineer needs to design the truss.

We are going to show you how to calculate the forces in truss members by means of the method of joints. This is a method in which we draw a free-body diagram of each joint of the truss and use equilibrium equations to find the forces exerted on the joint by the members that are fastened together there. Let's look at the free-body diagram of the joint A in the truss of Fig. 2-63a. Figure 2-65 shows this free-body diagram. The first force that we would draw here is the reaction R_1, which would be calculated by writing a moment equation about point H in Fig. 2-63a. Then, we could draw the force F_{ab}, which must be in the same direction as the member AB, because AB is a two-force member. We can see that F_{ab} must have a downward vertical component, to balance R_1, so F_{ab} must be downward to the left. Finally, we would draw the force F_{ac}, which must be directed to the right to balance the horizontal component of F_{ab}.

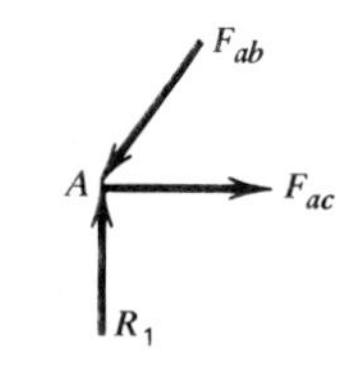

Fig. 2-65. Free-body diagram of joint.

We can reason our way to the conclusion that the force in AB is compression from the fact that the force F_{ab} pushes against the joint A. This force is the equal and opposite reaction to the compressive force F_{ab} which is exerted by the joint A on the member AB in Fig. 2-64a.

Also, we can decide that the force in AC is tensile, because the force F_{ac} pulls away from the joint A. This force is the equal and opposite reaction to the tensile force that is exerted by the joint A on the member AC in Fig. 2-64b.

The argument of these last two paragraphs leads to a pair of rules you must remember.

1. If the force exerted by a member on a joint is pushing against the joint, the member is in compression.
2. If the force exerted by a member on a joint is pulling away from the joint, the member is in tension.

Now, we are going to find the forces in the members of a truss by using the free-body diagrams of one joint after another until we know all the forces.

Illustrative Example 1. Calculate the force in each member of the truss in Fig. 2-66a. Notice that the loads and reactions are in kips. The reactions are given to save your time.

Solution:

1. Figure 2-66b shows the free-body diagram of the joint at A. The force F_{ab} must be directed downward to the left so that its vertical component can balance the reaction R_1. The force F_{ac} must be directed to the right so that it can balance the horizontal component of the force F_{ab}.
2. We need the sine and cosine of the angle θ. Let's use the triangle ABC in Fig. 2-66c. The length AB is

$$AB = \sqrt{(12)^2 + (16)^2} = 20 \text{ ft}$$

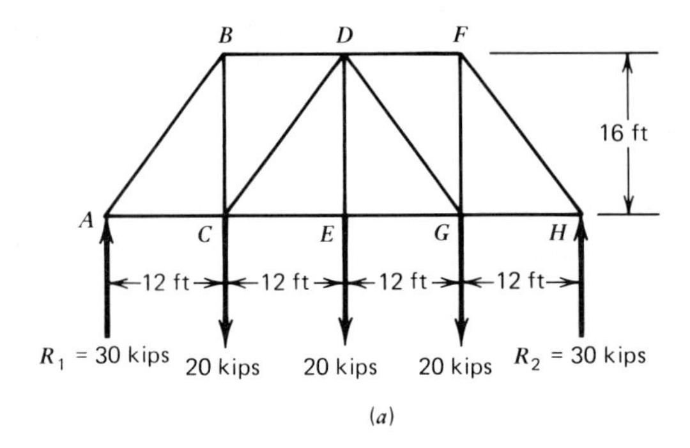

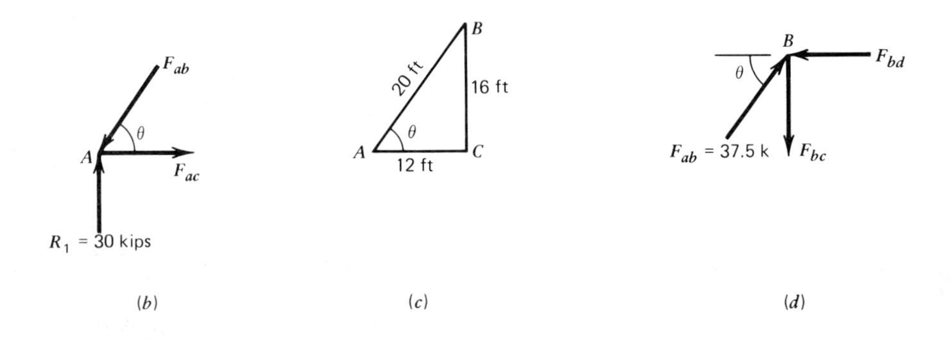

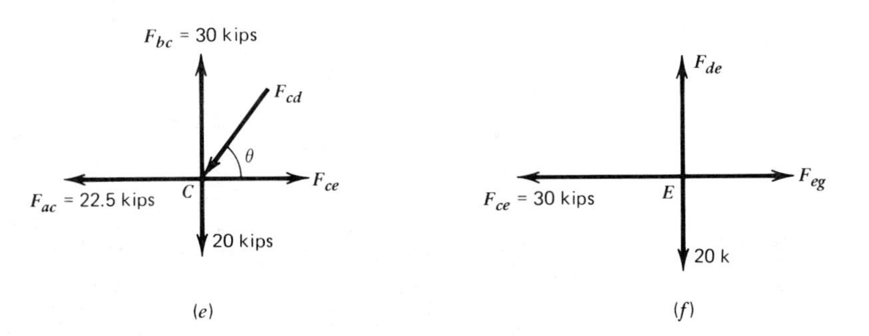

Fig. 2-66. Illustration for method of joints. (*a*) Free-body diagram of truss. (*b*) Free-body diagram of joint *A*. (*c*) Triangle for sin θ and cos θ. (*d*) Free-body diagram of joint *B*. (*e*) Free-body diagram of joint *C*. (*f*) Free-body diagram of joint *E*.

Then,

$$\sin\theta = \frac{16}{20} = 0.8 \qquad \cos\theta = \frac{12}{20} = 0.6$$

3. We can write $\Sigma F_y = 0$ to get F_{ab}. Thus

$$\Sigma F_y = 0 \qquad 30 - F_{ab}\sin\theta = 0$$

$$30 - 0.8\,F_{ab} = 0 \qquad F_{ab} = 37.5 \text{ kips compression}$$

We know that the force F_{ab} is compression because it pushes against the joint.

4. Now that we know F_{ab}, we can find F_{ac} by writing $\Sigma F_x = 0$. So,

$$\Sigma F_x = 0 \qquad F_{ac} - F_{ab} \cos \theta = 0$$

$$F_{ac} = 37.5 \times 0.6 = 22.5 \text{ kips tension}$$

We know that the force F_{ac} is tension because it pulls away from the joint.

5. Let's move on to the joint B, where we know that the force F_{ab} is 37.5 kips compression, so we show it pushing against the joint in Fig. 2-66*d*.
6. We can see that the force F_{bd} must be directed to the left so that it can balance the horizontal component of F_{ab}. Let's find F_{bd}. Thus

$$\Sigma F_x = 0 \qquad 37.5 \cos \theta - F_{bd} = 0$$

$$F_{bd} = 37.5 \cos \theta = 37.5 \times 0.6 = 22.5 \text{ kips compression}$$

We know that the force F_{bd} is compression because it pushes against the joint.

7. We can see that F_{bc} must be downward so that it can balance the vertical component of F_{ab}. We can write

$$\Sigma F_y = 0 \qquad 37.5 \sin \theta - F_{bc} = 0$$

$$F_{bc} = 37.5 \times 0.8 = 30 \text{ kips tension}$$

We know that F_{bc} is in tension because it pulls away from the joint.

8. We go next to joint C, Fig. 2-66*e*, where the forces F_{ac} and F_{bc} are known by this time. These are both tensile forces, and we show them as pulling away from the joint.
9. We show F_{cd} as downward to the left because its vertical component must balance the difference between F_{bc} and the 20-kip load. Then,

$$\Sigma F_y = 0 \qquad 30 - 20 - F_{cd} \sin \theta = 0$$

$$0.8\,F_{cd} = 10 \qquad F_{cd} = 12.5 \text{ kips compression}$$

10. It is easy to see that F_{ce} must be directed to the right so that it can balance F_{ac} and the horizontal component of F_{cd}.

$$\Sigma F_x = 0 \qquad -22.5 - 12.5 \cos \theta + F_{ce} = 0$$

$$F_{ce} = 22.5 + 12.5 \times 0.6 = 22.5 + 7.5 = 30 \text{ kips tension}$$

11. Now, in Fig. 2-66*f*, we draw the free-body diagram of the joint at E, where we know the force F_{ce} is 30 kips tension. We can see that the force F_{de} must be upward to balance the 20-kip load. So, let's write

$$\Sigma F_y = 0 \qquad F_{de} - 20 = 0 \qquad F_{de} = 20 \text{ kips tension}$$

12. We can finish this example by reasoning that the truss and loads are symmetrical, so the forces in the members must also be symmetrical. Thus

$$F_{fh} = F_{ab} = 37.5 \text{ kips compression}$$

$$F_{gh} = F_{ac} = 22.5 \text{ kips tension}$$

$$F_{fg} = F_{bc} = 30 \text{ kips tension}$$

$$F_{df} = F_{bd} = 22.5 \text{ kips compression}$$

$$F_{dg} = F_{cd} = 12.5 \text{ kips compression}$$

$$F_{eg} = F_{ce} = 30 \text{ kips tension}$$

LOOK BACK FOR A MOMENT

You may wonder how we know the order in which to take the joints as we move through the truss. The answer is that we must always go to a new joint where there are no more than two unknown forces, because we only have two equations of equilibrium for a joint, namely, $\Sigma F_x = 0$ and $\Sigma F_y = 0$.

Notice that we found each upper chord member (*AB*, *BD*, *DF*, and *FH*) to be in compression, and each lower chord member (*AC*, *CE*, *EG*, and *GH*) to be in tension. This is usually true if the truss is supported at the ends and the loads are downward. A truss is a special form of beam, and you will see compression in the top of a beam and tension in the bottom when you study beams in Chapter 6.

Also, we found that the web members (*BC*, *CD*, *DE*, *DG*, and *FG*) seem to alternate between tension and compression. This also usually occurs.

Illustrative Example 2. Calculate the forces in the members *AB*, *AC*, *BC*, *BD*, *CD*, and *CE* in the special roof truss in Fig. 2-67*a*. Every web member is at an angle of 60° with the horizontal. The reactions are given.

Solution:

1. We start with the free-body diagram of the joint at A. It is shown

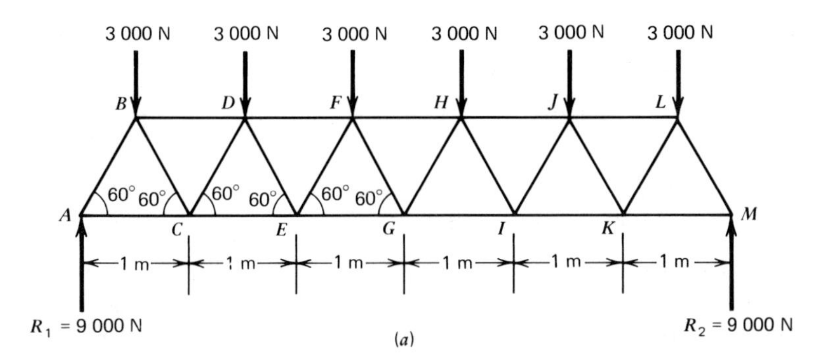

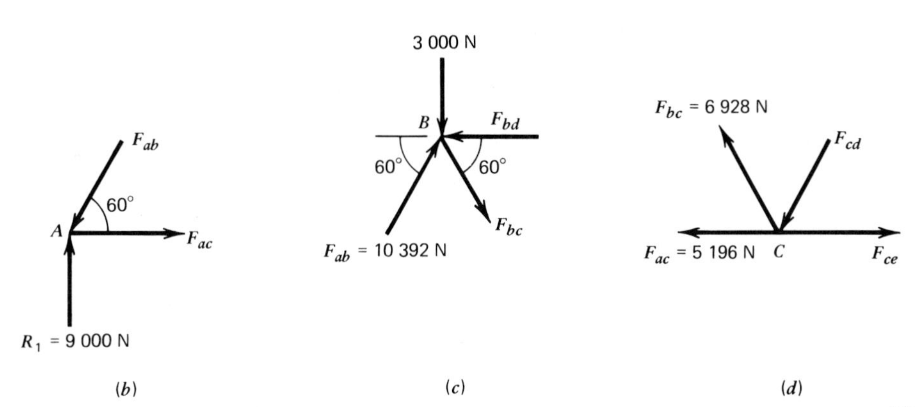

Fig. 2-67. Truss for Illustrative Example 2. (*a*) Free-body diagram of truss. (*b*) Free-body diagram of joint *A*. (*c*) Free-body diagram of joint *B*. (*d*) Free-body diagram of joint *C*.

in Fig. 2-67*b*. Here the force F_{ab} must be downward to the left so that its vertical component can balance the reaction R_1. Then, the force F_{ac} must be directed to the right so that it can balance the horizontal component of F_{ab}. Let's find F_{ab}:

$$\Sigma F_y = 0 \qquad 9\,000 - F_{ab} \sin 60° = 0$$

$$0.866\, F_{ab} = 9\,000 \qquad F_{ab} = 10\,392 \text{ N compression}$$

2. Next let's find F_{ac}:

$$\Sigma F_x = 0 \qquad F_{ac} - 10\,392 \cos 60° = 0$$

$$F_{ac} = 10\,392 \times 0.5 = 5\,196 \text{ N tension}$$

3. We move now to the free-body diagram of the joint at *B*, which is shown in Fig. 2-67*c*. We can find F_{bc} by summing the vertical components of the forces. Thus

$$\Sigma F_y = 0 \qquad 10\,392 \sin 60° - 3\,000 - F_{bc} \sin 60° = 0$$

$$0.866\, F_{bc} = 10\,392 \times 0.866 - 3\,000 = 9\,000 - 3\,000 = 6\,000$$

$$F_{bc} = 6\,928 \text{ N tension}$$

4. Now we can find F_{bd} by taking the sum of the horizontal components of the forces. So,

$$\Sigma F_x = 0 \qquad 10\,392 \cos 60° + 6\,928 \cos 60° - F_{bd} = 0$$

$$F_{bd} = 10\,392 \times 0.5 + 6\,928 \times 0.5 = 5\,196 + 3\,464$$

$$= 8\,660 \text{ N compression}$$

5. Next we draw the free-body diagram of the joint C, as shown in Fig. 2-67*d*. You can surely see for yourself how we decide the direction of the arrow on the force F_{cd} and then the direction of the arrow on the force F_{ce}. Let's find F_{cd}. This way,

$$\Sigma F_y = 0 \qquad 6\,928 \sin 60° - F_{cd} \sin 60° = 0$$

$$F_{cd} = 6\,928 \text{ N compression}$$

6. Then we calculate F_{ce}. So

$$\Sigma F_x = 0 \qquad -5\,196 - 6\,928 \cos 60° - 6\,928 \cos 60° + F_{ce} = 0$$

$$F_{ce} = 5\,196 + 6\,928 \times 0.5 + 6\,928 \times 0.5$$

$$F_{ce} = 5\,196 + 3\,464 + 3\,464 = 12\,124 \text{ N tension}$$

A FINAL WORD

You must determine whether the force in a truss member is tension or compression, in addition to finding the magnitude of the force. Remember that the force is compression if it pushes against the joint. Then, when you go on to the next joint, the force must be shown as pushing against that joint. Also, the force is tension if it pulls away from the joint and must be shown as pulling away from the joint in the next free-body diagram.

Practice Problems (Section 2-9). Here are some problems to help you really learn how to use the method of joints in truss analysis. You can calculate the reactions for yourself.

1. Calculate the force in each member of the truss in Fig. 2-68.

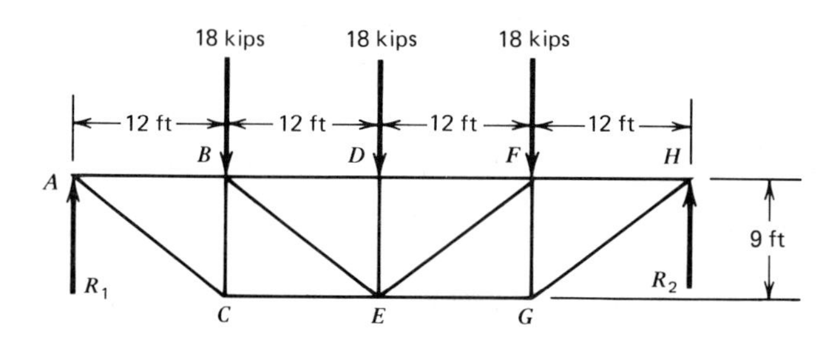

Fig. 2-68. Truss for Problem 1.

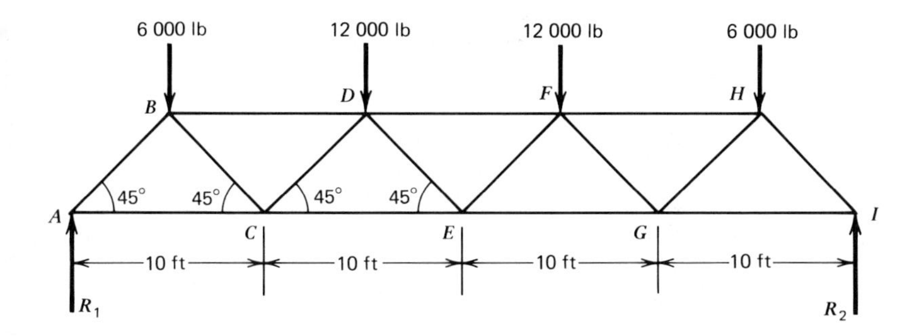

Fig. 2-69. Truss for Problem 2.

2. Calculate the force in each member of the truss in Fig. 2-69.
3. Figure 2-70 shows a cantilever truss of the sort that might be used to support the roof over a loading platform. Calculate the force in each member of the truss. (Hint: you can do this without calculating any reactions.)
4. Calculate the force in each member of the truss in Fig. 2-71.

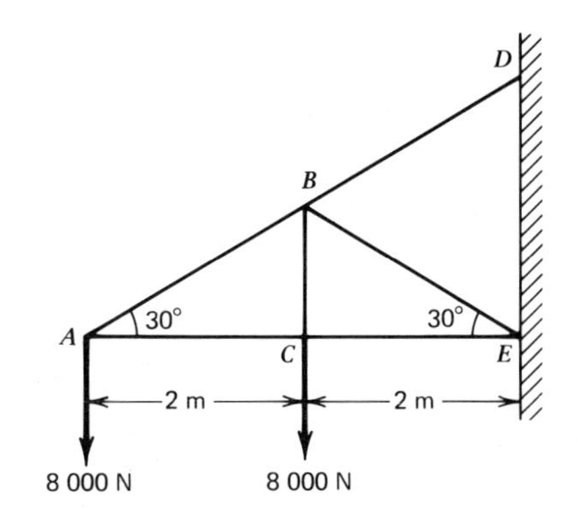

Fig. 2-70. Truss for Problem 3.

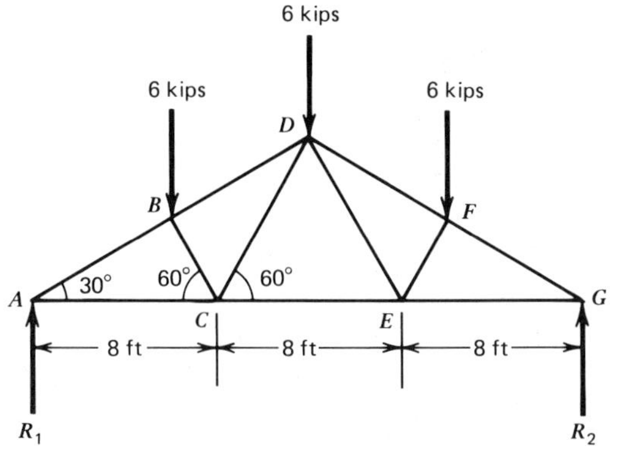

Fig. 2-71. Truss for Problem 4.

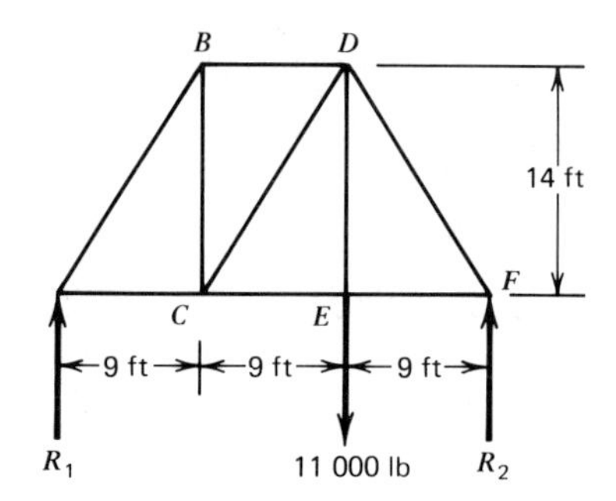

Fig. 2-72. Truss for Problem 5.

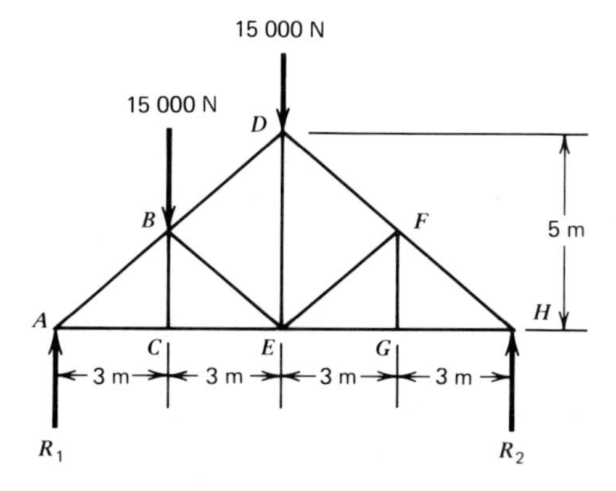

Fig. 2-73. Truss for Problem 6.

5. Figure 2-72 shows an unsymmetrical truss with unsymmetrical loads. Find the force in each member.
6. Calculate the force in each member of the truss in Fig. 2-73. Notice that the loading is not symmetrical.

FLEXIBLE CABLES

2-10. WHAT A FLEXIBLE CABLE IS

A flexible cable is a wire cable, a rope, or even a chain that is very slender; that is, the length is very large compared with the diameter. Such a slender cable has so little resistance to bending that it is assumed to have none. This means that the only force a flexible cable can withstand is a tensile force such as you see in Fig. 2-74*a*, where the force T is applied to a short piece of cable. The cable will always be in the direction of the force T.

Some engineering examples of flexible cables are electric and telephone wires, cables for suspension bridges, and cables for aerial tramways. Cables of this sort would probably be made of metal wire.

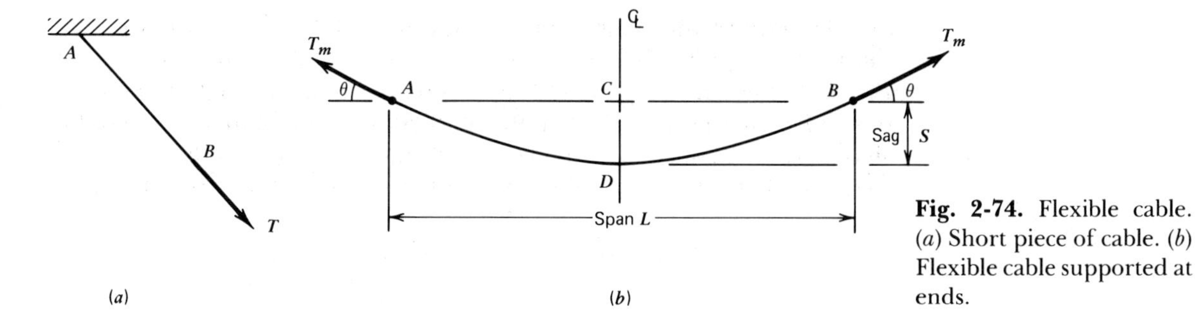

Fig. 2-74. Flexible cable. (*a*) Short piece of cable. (*b*) Flexible cable supported at ends.

Figure 2-74*b* shows a symmetrical cable of span L. The cable is supported at points A and B. The vertical distance CD is called the sag s. The maximum tension in the cable is T_m, and it occurs at the ends of the cable. The direction of T_m is given by the angle θ. The load on the cable is only its weight.

2-11. PARABOLIC CABLES

It can be shown that a cable will be in the shape of a parabola if the load on the cable is uniformly distributed over the span of the cable. (We are not going to prove this statement here.) This sort of weight distribution is not likely to occur, but it is easy to analyze for forces, and the results are very good approximations if the sag s is no more than about 10% of the span L.

Now, let's say that the weight of the cable in Fig. 2-74*b* is uniformly distributed over the span L and make an analysis. We will do this by drawing the free-body diagram of half the cable and applying equilibrium equations. Figure 2-75 shows this free-body diagram. The unit load w is the weight of the cable. The lowest point of the cable is at the center D, and the slope of the cable at this point is horizontal. Then, the tension T_o at the center must also be horizontal, because the tension is always in

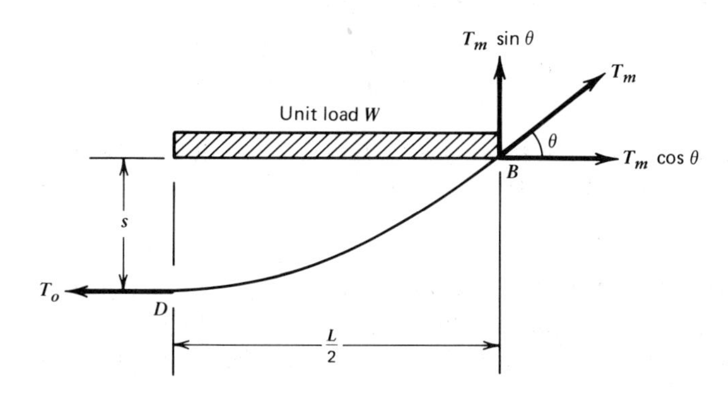

Fig. 2-75. Analysis of parabolic cable.

the direction of the cable. The maximum tension is T_m at point B. We will resolve T_m into its horizontal and vertical components $T_m \cos \theta$ and $T_m \sin \theta$. Let's find T_o by writing a moment equation about point B, remembering that we calculate the moment of the unit load w by taking all of it as concentrated at the center of this half of the cable. The total distributed load is $wL/2$, and its moment arm is half of $L/2$, or $L/4$. Then,

$$\Sigma M_b = 0 \qquad T_o \times s - \frac{wL}{2} \times \frac{L}{4} = 0$$

$$T_o = \frac{wL^2}{8s}$$

Next, let's find $T_m \cos \theta$ by using $\Sigma F_x = 0$.

$$\Sigma F_x = 0 \qquad T_m \cos \theta - \frac{wL^2}{8s} = 0$$

$$T_m \cos \theta = \frac{wL^2}{8s}$$

Then, we can get $T_m \sin \theta$ by using $\Sigma F_y = 0$.

$$\Sigma F_y = 0 \qquad T_m \sin \theta - \frac{wL}{2} = 0$$

$$T_m \sin \theta = \frac{wL}{2}$$

Now, we can use the Pythagorean theorem to calculate T_m. So,

$$T_m = \sqrt{(T_m \cos \theta)^2 + (T_m \sin \theta)^2}$$

$$= \sqrt{\left(\frac{wL^2}{8s}\right)^2 + \left(\frac{wL}{2}\right)^2} = \frac{wL}{2}\sqrt{\frac{L^2}{16s^2} + 1}$$

The angle θ is found by dividing $T_m \sin \theta$ by $T_m \cos \theta$.

$$\tan \theta = \frac{T_m \sin \theta}{T_m \cos \theta} = \frac{wL/2}{wL/8s^2} = \frac{4s}{L}$$

We can always find θ if we know $\tan \theta$. Now, it is time to try these formulas in some examples.

Illustrative Example 1. A flexible cable has a span of 200 ft and a sag of 8 ft. The weight of the cable is 2 lb/ft. What is the maximum tension in the cable and the direction of the cable at the supports?

Solution:

1. The span L is 200 ft.
2. The sag s is 8 ft.
3. The unit load w is 2 lb/ft.
4. Then,

$$T_m = \frac{wL}{2}\sqrt{\frac{L^2}{16s^2} + 1} = \frac{2 \times 200}{2}\sqrt{\frac{(200)^2}{16(8)^2} + 1}$$

$$= 100\sqrt{39.06 + 1} = 633 \text{ lb}$$

5. Finally,

$$\tan\theta = \frac{4s}{L} = \frac{4 \times 8}{200} = 0.16 \qquad \theta = 9.09°$$

Illustrative Example 2. A flexible cable has a span of 300 m and a sag of 20 m. The weight of the cable is 48 N/m. Calculate the maximum tension in the cable and the direction of the cable at the supports.

Solution:

1. The span L is 300 m.
2. The sag s is 20 m.
3. The unit load w is 48 N/m.
4. We calculate T_m:

$$T_m = \frac{wL}{2}\sqrt{\frac{L^2}{16s^2} + 1} = \frac{48 \times 300}{2}\sqrt{\frac{(300)^2}{16(20)^2} + 1}$$

$$= 7\,200\sqrt{14.06 + 1} = 27\,900 \text{ N}$$

5. And,

$$\tan\theta = \frac{4s}{L} = \frac{4 \times 20}{300} = 0.267 \qquad \theta = 14.95°$$

Practice Problems (Section 2-11). Now you should work a few problems of parabolic cables to get some experience in using these formulas. Just calculate the maximum tension and the direction of the cable at the supports for each of these cables.

1. A wire that weighs 0.1 lb/ft has a span of 24 ft and a sag of 3 in.
2. A cable that weighs 0.73 lb/ft has a span of 160 ft and a sag of 5 ft.

3. A cable that weighs 2 lb/ft has a span of 1 760 ft and a sag of 42 ft.
4. A cable that weighs 17 N/m has a span of 190 m and a sag of 12 m.
5. A cable that weighs 3 N/m has a span of 90 m and a sag of 4 m.
6. A cable that weighs 0.4 N/m has a span of 38 m and a sag of 2 m.

FRICTION BETWEEN DRY SURFACES

2-12. WHAT FRICTION IS

Friction is a force that develops in resistance when one body slides, or tends to slide, over another. The force is tangent to the surface of contact of the two bodies and is directed so that it opposes the tendency to slide.

Let's see how a frictional force might develop. Figure 2-76*a* shows a block A resting on a larger body B. The force W is the weight of body A. This force is balanced by the normal force N which is exerted by B on A, so the body A is in equilibrium. In fact, we have the free-body diagram of body A in Fig. 2-76*a*. There is no friction, because there is no tendency for body A to slide over body B.

Now, suppose we apply a small horizontal force P to body A, as in Fig. 2-76*b*. The force P tends to make the body A slide over body B, and the frictional force F develops in resistance to the tendency to slide. Notice that the body A tends to slide to the right and that F is directed to the left to oppose this tendency to slide, so that body A remains in equilibrium.

Friction between dry surfaces can be quite useful; in fact, it is only in useful circumstances that it is ordinarily allowed to occur in modern industry. Typical applications of friction are the transmission of power by belts on pulleys, belt conveyors, brakes, clamping devices, and friction between automobile tires and pavements.

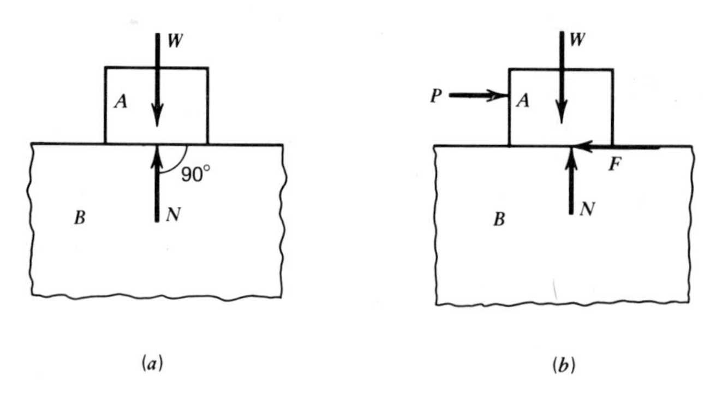

Fig. 2-76. Illustration of friction between bodies. (*a*) Bodies in contact. (*b*) Bodies with friction between them.

2-13. COEFFICIENT OF FRICTION

If the force P in Fig. 2-76*b* is increased gradually, the frictional force F will increase with it for a while, still holding body A in equilibrium. If we continue to increase the force P, the force F will increase with it until F reaches the maximum value that it can develop. If we increase P further, F cannot increase more to balance P and the body A will slide over the body B.

The condition when F is at its maximum value and body A is on the point of sliding is called *impending motion*. The magnitude of F when motion impends is equal to a certain constant times the normal force N. This constant is called the *coefficient of friction* and is represented here by f. So, $F = fN$ when motion impends. The magnitude of f depends on the nature of the surfaces in contact—both the structure of each material in contact and the smoothness or roughness of each surface. Table 2.1 gives you some values of the coefficient of friction.

Impending motion is a critical condition in which the maximum frictional resistance has been developed and the body is on the point of sliding. If the propulsive force is increased, even by a very small amount, the frictional force cannot increase with it, and the body will slide. For practical purposes, then, the force that develops impending motion is the force that causes the body to slide.

We must distinguish between (1) *static friction*, which is friction up to and including impending motion and in which sliding does not occur, and (2) *kinetic friction*, which is friction in which one body is actually sliding over another. The cóefficient of kinetic friction is smaller than the coefficient of static friction. Once the initial resistance to sliding is overcome, the frictional resistance is somewhat smaller.

It is important for you to realize that the frictional force F is at its maximum value fN only when motion impends, that is, when one body is on the point of sliding over another. If sliding can be prevented by a frictional force that is smaller than fN, then the frictional force will be smaller than fN.

TABLE 2.1 VALUES OF THE COEFFICIENT OF FRICTION

Rubber-covered conveyor belt on clean, dry iron pulley	0.25[a]
Rubber-covered conveyor belt on rubber-covered pulley	0.35[a]
Automobile tire on dry concrete pavement at 10 mph	0.90
Automobile tire on wet concrete pavement at 10 mph	0.60
Automobile tire on icy pavement at 10 mph	0.20
Wood on wood	0.25 to 0.50
Wood on metal	0.20 to 0.60
Metal on metal	0.15 to 0.30

[a]These appear to be numbers used in design calculations. It might be expected that the actual coefficient of friction is somewhat higher.

2-14. SIMPLE PROBLEMS

The way to solve a simple problem in friction is to draw the free-body diagram of the body and to use equilibrium equations to find the unknown forces. We would first put in the known forces, such as the weight of the body. Then, there is always a normal force N between two bodies in contact, so we would put in this normal force. If there is a separate propulsive force, such as the force P in Fig. 2-76*b*, we would show it. Finally, we would put in the frictional force F, tangent to the surface of contact, and in such a direction as to oppose sliding. Now, let's go through some examples.

Illustrative Example 1. Figure 2-77*a* shows a box that rests on a horizontal surface and weighs 300 lb. The coefficient of static friction between the box and the horizontal surface is 0.2. How large a horizontal force is required to make the box slide over the surface?

Solution:

1. Figure 2-77*b* shows the free-body diagram of the box when motion impends. First, we put in the weight of 300 lb.
2. Next, we put in the normal reaction N of the supporting surface.
3. Then, we put in the propulsive force P. This force is horizontal.
4. The last force is the frictional force F. It is in the surface of contact and is directed to the left so that it opposes sliding. The box is on the point of sliding so that $F = fN = 0.2N$.
5. We use the equilibrium equation $\Sigma F_y = 0$ to calculate N. Thus

$$\Sigma F_y = 0 \qquad N - 300 = 0 \qquad N = 300 \text{ lb}$$

6. We use the equilibrium equation $\Sigma F_x = 0$ to find the force P. Like this,

$$\Sigma F_x = 0 \qquad P - 0.2 \times 300 = 0 \qquad P = 60 \text{ lb}$$

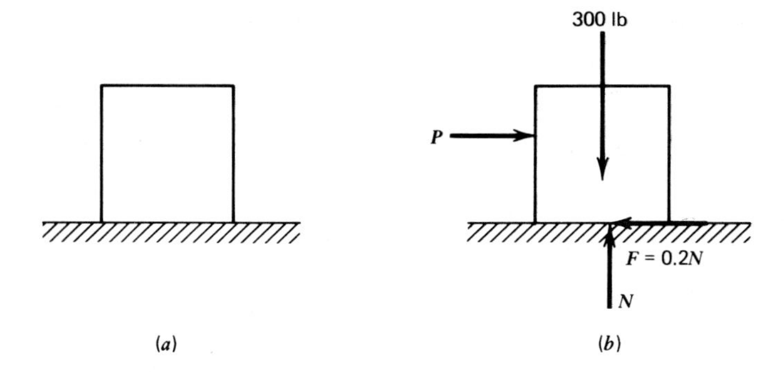

Fig. 2-77. Development of frictional force. (*a*) Block. (*b*) Free-body diagram of block when motion impends.

COMPLICATIONS

The problems are not always this simple. The equations of equilibrium can be more complicated. Let's see an example.

Illustrative Example 2. Figure 2-78*a* shows a crate that weighs 600 N and rests on an inclined plane. The coefficient of static friction between the crate and the plane is 0.3. How large a horizontal force must be applied to the crate to start it sliding up the plane?

Solution:

1. Figure 2-78*b* shows the free-body diagram of the crate when motion impends. We draw the weight of 600 N first.
2. Then, we draw the normal force N. It is at an angle of 75° with the horizontal.
3. Next, we draw the propulsive force P.
4. The last force is the frictional force. It is directed downward along the plane so that it opposes the tendency to slide upward. Since motion impends, the frictional force is $fN = 0.3N$.
5. We use the equilibrium equation $\Sigma F_y = 0$ to find the force N. This way,

$$\Sigma F_y = 0 \qquad N \sin 75° - 0.3N \sin 15° - 600 = 0$$

$$0.966N - 0.3N \times 0.259 = 600$$

$$0.966N - 0.078N = 600$$

$$0.888N = 600 \qquad N = 675.7 \text{ N}$$

6. Finally, we use the equilibrium equation $\Sigma F_x = 0$ to find P. Thus

$$\Sigma F_x = 0 \qquad P - 675.7 \cos 75° - 0.3 \times 675.7 \cos 15° = 0$$

$$P = 675.7 \times 0.259 + 0.3 \times 675.7 \times 0.966$$

$$= 175.0 + 195.8 = 370.8 \text{ N}$$

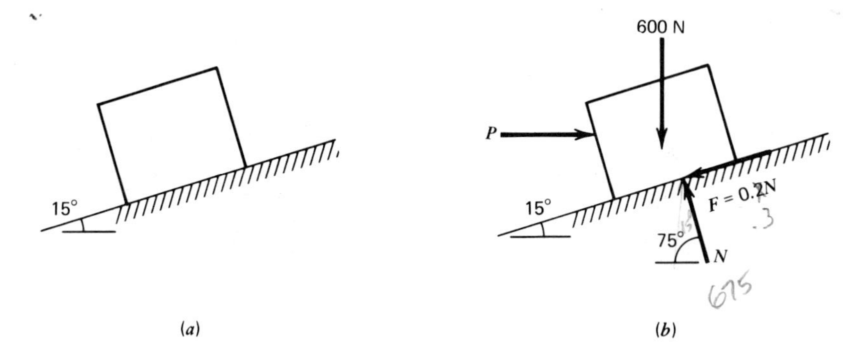

Fig. 2-78. Illustration for box problem. (*a*) Box on inclined plane. (*b*) Free-body diagram of box when motion impends.

MOMENT EQUATIONS

Sometimes we need to use a moment equation to solve a friction problem. Let's see how it goes.

Illustrative Example 3. Figure 2-79*a* shows a cylinder that weighs 400 lb and is in a corner between a vertical wall and a floor. The coefficient of static friction between the cylinder and the floor is 0.25. The wall is smooth, which means that there is no friction between the cylinder and the wall. A horizontal force, directed to the left, is applied at A. How large must this force be to cause the cylinder to rotate?

Solution:

1. Figure 2-79*b* shows the free-body diagram of the cylinder when motion impends. First, we put in the weight of 400 lb.
2. Then, we draw the normal force N the floor exerts on the cylinder.
3. The third force we show is the propulsive force P.
4. Next, we draw the frictional force $0.25N$. We can see that the force P will cause a clockwise rotation of the cylinder. The frictional force must oppose this rotation, so it must have a counterclockwise moment about the center of the cylinder and must be directed to the right.
5. The force P will push the cylinder against the wall so the wall must exert a normal force on the cylinder. We will designate this normal force as N_1. The wall is smooth, so there is no frictional force exerted by the wall on the cylinder.
6. We're ready to start calculating. Let's use $\Sigma F_y = 0$ to find N. Thus

$$\Sigma F_y = 0 \qquad N - 400 = 0 \qquad N = 400 \text{ lb}$$

7. The easiest way to calculate P is to use a moment equation about

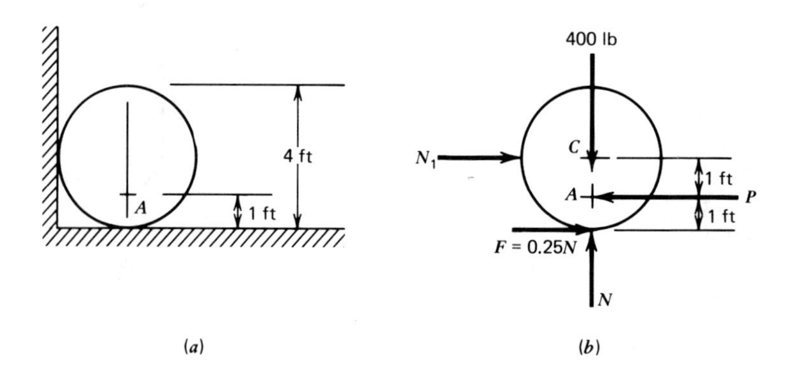

Fig. 2-79. Illustration for cylinder problem. (*a*) Cylinder. (*b*) Free-body diagram of cylinder when motion impends.

a point on the line of action of N_1. Let's choose point C, at the center of the cylinder, because the force N goes through C. So,

$$\Sigma M_c = 0 \qquad 1 \times P - 2 \times 0.25N = 0$$

Then, we'll use $N = 400$ lb

$$P = 0.5 \times 400 = 200 \text{ lb}$$

ANGLE OF FRICTION

We should tell you about one other topic at this time. This is the *angle of friction*. There are some problems, especially if you are using graphic methods, in which it is convenient to replace the normal and frictional forces at a surface by the single force of which they are the components. When motion impends, the frictional force is proportional to the normal force, and because it is, the direction of the resultant force can be determined from the coefficient of friction. In Fig. 2-80*a*, the force R is the resultant of N and F, and from the triangle in Fig. 2-80*b*, $\tan \phi = fN/N = f$. The angle ϕ between R and the normal to the surface is called the angle of friction. The direction in which R is inclined from the normal can be determined from the direction of the frictional force. When motion does not impend, the angle between R and the normal to the surface is less than ϕ.

Practice Problems (Section 2-14). The next thing for you to do is to solve these problems. The first step in each is to draw the free-body diagram.

1. A box rests on a horizontal surface. The box weighs 70 lb. The coefficient of friction between the box and surface is 0.2. How large a horizontal force must be applied to the box to make it slide across the surface?
2. A crate weighs 360 N and rests on an inclined plane that is at

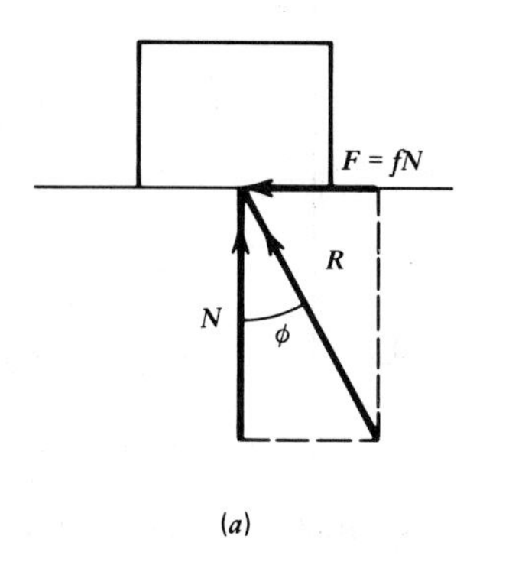

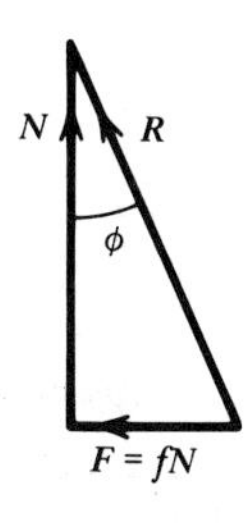

Fig. 2-80. Illustration for angle of friction. (*a*) Force exerted by surface on block. (*b*) Force triangle.

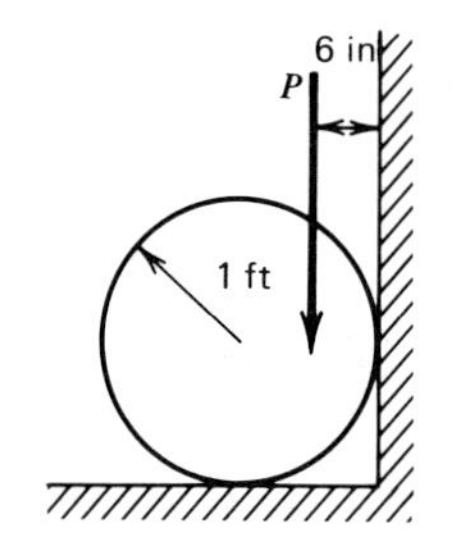

Fig. 2-81. Cylinder for Problem 5.

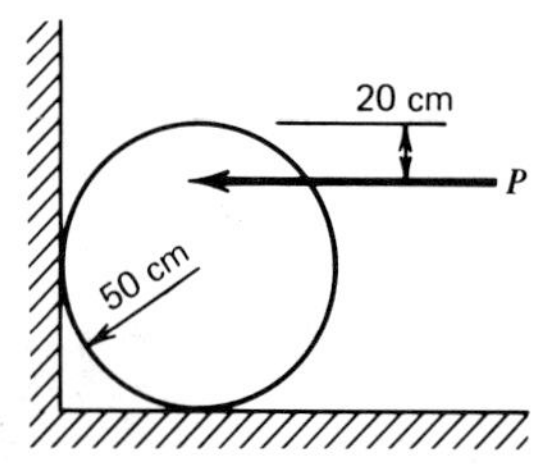

Fig. 2-82. Cylinder for Problem 6.

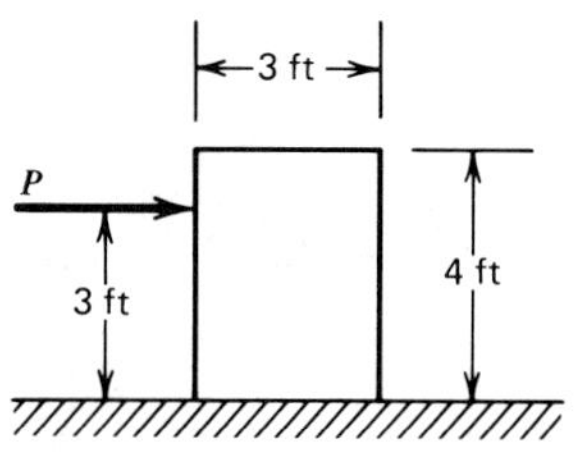

Fig. 2-83. Block for Problem 7.

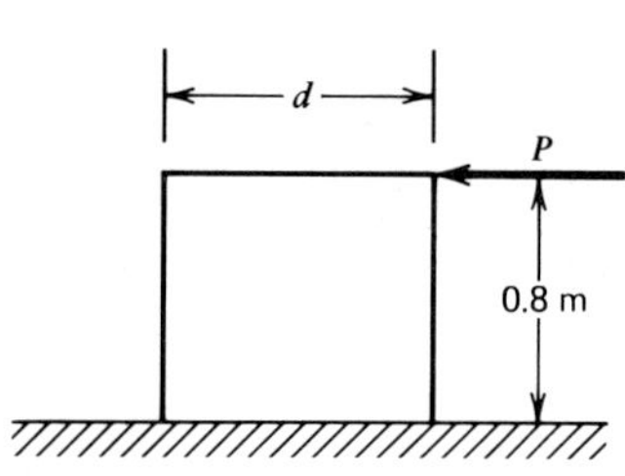

Fig. 2-84. Block for Problem 8.

an angle of 10° with the horizontal. The coefficient of friction between the crate and plane is 0.22. How large must a horizontal force be to cause the crate to slide up the plane?

3. A square box is placed on a 20° inclined plane. The box weighs 240 lb. The coefficient of friction between the box and plane is 0.25. What magnitude of horizontal force must be applied to the box to prevent it from sliding down the plane?
4. A rectangular block is placed on a 30° inclined plane. The block weighs 520 N. The coefficient of friction between the block and plane is 0.3. What magnitude of horizontal force must be applied to the block to keep it from sliding down the plane?
5. The cylinder in Fig. 2-81 weighs 210 lb. The coefficient of friction between the cylinder and floor is 0.1, and the wall is smooth. What magnitude of the force P will cause the cylinder to rotate?
6. The cylinder in Fig. 2-82 weighs 360 N. The coefficient of friction between the floor and cylinder is 0.2 and the wall is smooth. What magnitude of the force P will cause the cylinder to rotate?
7. The coefficient of friction between the block and plane in Fig. 2-83 is 0.25. The block weighs 200 lb. The force P is just large enough to cause impending motion. How far from the left-hand side of the block is the normal reaction of the plane?
8. The coefficient of friction between the block and plane in Fig. 2-84 is 0.3. The block weighs 600 N. The normal reaction of the plane on the block is located at the left-hand corner of the block when motion impends. What is the distance d?
9. The uniform ladder in Fig. 2-85 weighs 40 lb. The wall is smooth. What coefficient of friction between the ladder and floor is needed to keep the ladder from sliding?
10. The uniform ladder in Fig. 2-86 weighs 350 N. The wall is

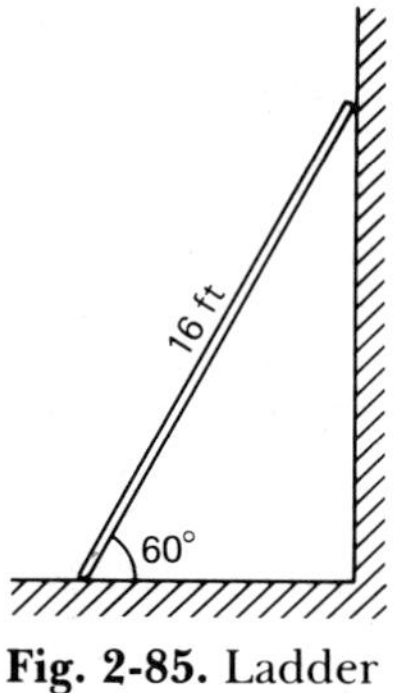

Fig. 2-85. Ladder for Problem 9.

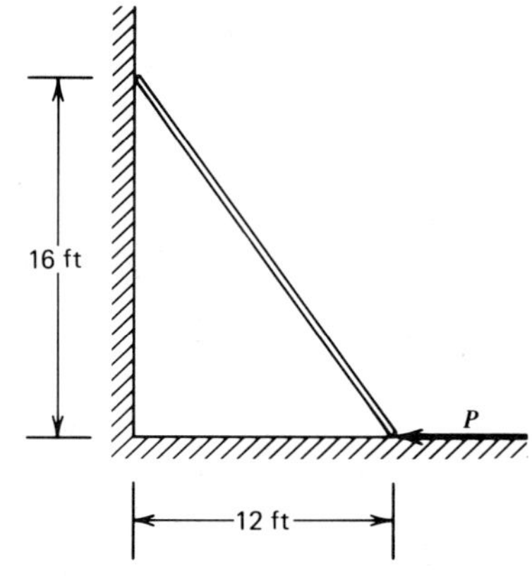

Fig. 2-86. Ladder for Problem 10.

smooth. The coefficient of friction between the wall and floor is 0.25. How large must the force P be to keep the ladder from sliding?

2-15. BELT FRICTION

One of the most useful applications of dry friction is in the transmission of power by belts on pulleys. It would be important to keep the belt from sliding on the pulley (to prevent wear of the belt) so we would want to be able to calculate the conditions when slipping would occur. Figure 2-87 shows a pulley with a belt. Here we have a belt tension T_1 that is larger than the belt tension T_2, so there is a tendency for the belt to slide counterclockwise around the pulley. The coefficient of friction between the belt and pulley is f. An important quantity in this problem is the angle of contact between the belt and pulley. This angle is shown as α (Greek letter alpha) in Fig. 2-87, and it must be expressed in radians in our calculations.

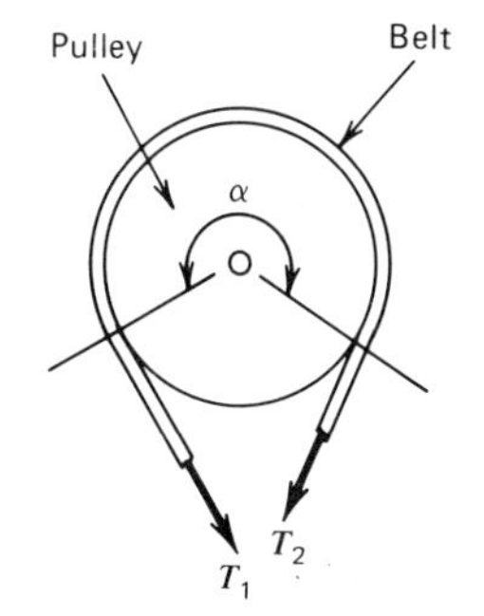

Fig. 2-87. Belt and pulley.

The kind of mathematics needed to derive a formula for belt slipping is higher than the level we are using in this book, so we are going to give you the formula. Here it is. The belt will slip when,

$$\frac{T_1}{T_2} = e^{f\alpha}$$

where e is the Naperian constant 2.718. . . . This is one equation, so we can solve it for one unknown quantity. For example, if we know the coefficient of friction f, the angle of contact α, and one belt tension, we can calculate the other belt tension. This sort of calculation can be done handily with a small electronic calculator. Be sure that you take the angle α in radians.

One special point must be mentioned here. The method given for converting degrees to radians on my calculator is limited to angles of 90° or less. Since most belts on pulleys have an angle of contact of more than 90°, it is probably best to make this conversion by dividing the angle in degrees by $180/\pi$, which is the number of degrees in a radian. Or, multiply by $\pi/180$. Now, let's try an example.

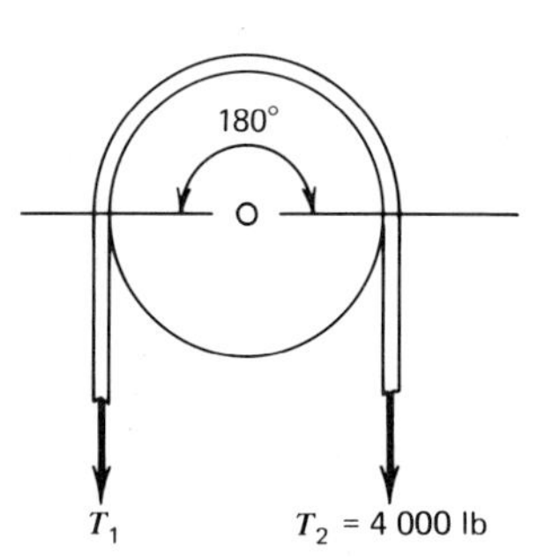

Fig. 2-88. Sheave and cable for Illustrative Example 1.

Illustrative Example 1. Figure 2-88 shows the traction sheave of a passenger elevator. The coefficient of friction between the cable and sheave is 0.5. What is the magnitude of T_1 when the cable is about to slide counterclockwise on the sheave?

Solution:

1. The coefficient of friction f is 0.5.

2. The angle of contact between the cable and sheave is 180°. Let's convert this to radians, like this,

$$\alpha = 180 \times \frac{\pi}{180} = 3.14$$

3. Now, we'll calculate $f\alpha$.

$$f\alpha = 0.5 \times 3.14 = 1.57$$

4. Then,

$$\frac{T_1}{T_2} = e^{f\alpha} = e^{1.57} = 4.807 \qquad \text{using the calculator}$$

5. We know $T_2 = 4\,000$ lb, so

$$\frac{T_1}{4\,000} = 4.807 \qquad T_1 = 19\,200 \text{ lb}$$

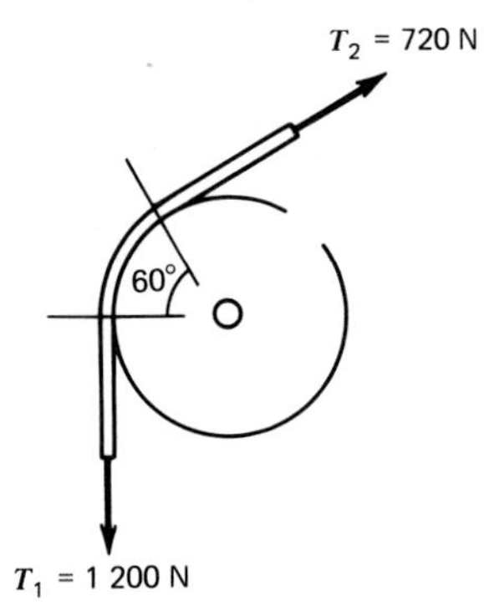

Fig. 2-89. Belt and pulley for Illustrative Example 2.

Illustrative Example 2. Figure 2-89 shows a belt and pulley. Both belt tensions are known. What coefficient of friction is needed to keep the belt from sliding on the pulley?

Solution:

1. The angle α is 60°. In radians this is,

$$\alpha = 60 \times \frac{\pi}{180} = 1.047$$

2. Let's calculate $e^{f\alpha}$. Thus

$$e^{f\alpha} = \frac{T_1}{T_2} = \frac{1\,200}{720} = 1.667$$

3. Now, we'll calculate $f\alpha$. This is just the natural logarithm of 1.667, so we use the *ln x* button and get,

$$f\alpha = 0.511$$

4. Finally, using $\alpha = 1.047$,

$$f\alpha = 0.511$$

$$1.047 f = 0.511 \qquad f = 0.488$$

Practice Problems (Section 2-15). The best way to learn how to use this formula is to use it in working problems. Here is your chance.

1. Figure 2-90 shows a driven pulley A and a snub pulley B. The

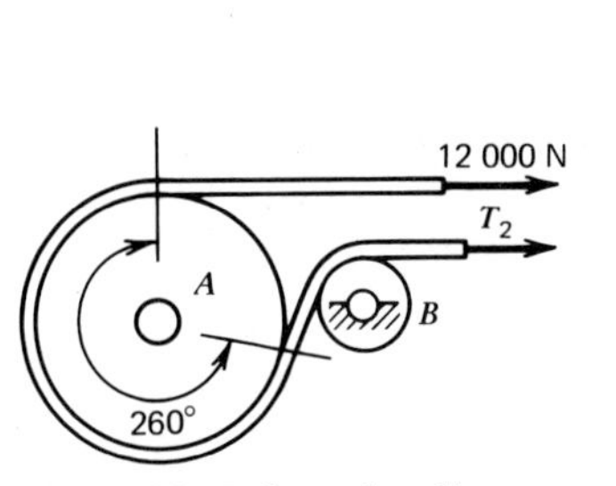

Fig. 2-90. Belt and pulleys for Problem 1.

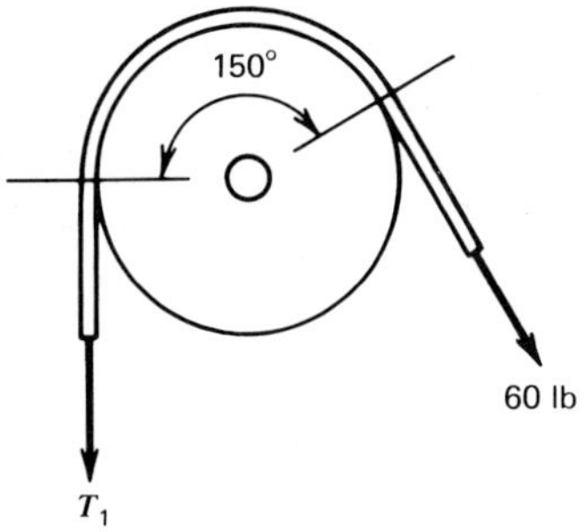

Fig. 2-91. Belt and pulley for Problem 2.

snub pulley can turn freely on its shaft; its only function is to increase the angle of contact of the belt on A. The coefficient of friction between the belt and pulley A is 0.4. Calculate the tension T_2 when the belt is about to slide clockwise on the pulley.

2. The coefficient of friction between the belt and pulley in Fig. 2-91 is 0.52. What is the magnitude of T_1 when the belt is about to slide counterclockwise on the pulley?
3. How large a coefficient of friction is needed to develop the condition shown in Fig. 2-92?
4. There was a time when people made practical use of the phenomenon of belt friction by snubbing a rope around a post to hold a fractious animal. What coefficient of friction is needed to cause the rope tension to increase to five times the initial value in one complete wrap around a post?

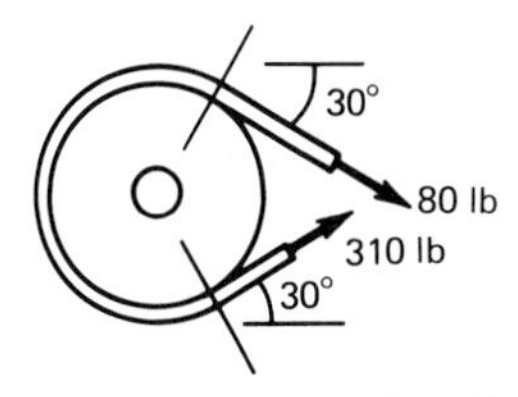

Fig. 2-92. Belt and pulley for Problem 3.

SUMMARY

Here is a summary of Part B of this chapter.

Trusses

1. A truss is a structure composed of slender bars fastened together at the ends and arranged in such a manner that the spaces between the members are triangular.
2. Each truss member is a two-force member; that is, there are only two forces exerted on a member, and these forces are applied at the ends of the member.
3. Tension is a condition in which the forces applied to the ends of a member tend to stretch it. A member in tension pulls away from the joints to which it is fastened.
4. Compression is a condition in which the forces applied to the ends of a member tend to shorten it. A member in compression pushes against the joints to which it is fastened.

5. The force in each member of a simple truss can be found by drawing free-body diagrams of the joints of the truss. The forces in the members appear in these free-body diagrams and are calculated by writing and solving equilibrium equations. It is necessary to work on the joints in such an order that there are no more than two unknown forces at a joint at the time that it is being analyzed.
6. The nature of the force in each truss member, that is, whether it is tension or compression, must be determined as well as the magnitude of the force.

Flexible Cables

1. A flexible cable is a thin member that has no bending stiffness. It can only withstand a tensile force in the direction of the cable.
2. A parabolic cable is a cable that has its load uniformly distributed along the span. It is a good approximation to consider any cable to be parabolic if the sag s is no more than 10% of the span L.
3. The maximum tension in a parabolic cable occurs at the support of the cable. Its magnitude is:

$$T_m = \frac{wL}{2}\sqrt{\frac{L^2}{16s^2} + 1}$$

   where w is the unit load.
4. The direction of the tension at the support is given by,

$$\tan\theta = \frac{4s}{L}$$

Friction between Dry Surfaces

1. Friction is a force that develops in resistance whenever one body tends to slide over another.
2. A frictional force is always tangent to the surface of contact between the bodies.
3. A frictional force is always directed so that it opposes the tendency to slide.
4. There is always a normal force N between two bodies in contact.
5. The maximum frictional force that can be developed between two bodies is fN where f is the coefficient of friction.
6. The magnitude of the coefficient of friction f depends on the nature of the surfaces in contact.
7. The condition when the maximum frictional force has been developed is called impending motion, because one body is just ready to slide over the other.

8. Simple problems in friction are solved by drawing free-body diagrams and applying equations of equilibrium.
9. A belt on a pulley is on the point of sliding relative to the pulley when the belt tensions fit this formula:

$$\frac{T_1}{T_2} = e^{f\alpha}$$

   (a) T_1 is the larger tension.
   (b) T_2 is the smaller tension.
   (c) e is the Naperian constant 2.718. . . .
   (d) f is the coefficient of friction.
   (e) α is the angle of contact between the belt and pulley, and is expressed in radians.

REVIEW QUESTIONS Now, see if you can answer these review questions.

1. How can you tell, either from a drawing or from seeing the actual structure, whether a framed structure is a truss?
2. What is a two-force member?
3. What is tension?
4. What is compression?
5. How do you determine that a truss member is in tension?
6. How do you determine that a truss member is in compression?
7. What is the method of joints for truss analysis?
8. How do you choose the order of analysis of joints in finding the forces in the members of a truss?
9. What is a flexible cable?
10. Under what conditions is it reasonable to consider a flexible cable to be a parabolic cable?
11. Where does the maximum tension occur in a parabolic cable?
12. What factors does the maximum tension in a parabolic cable depend on?
13. What is friction?
14. How can you determine the direction of a frictional force?
15. What is the meaning of the term *impending motion*?
16. What is the coefficient of friction?
17. How are friction problems analyzed?

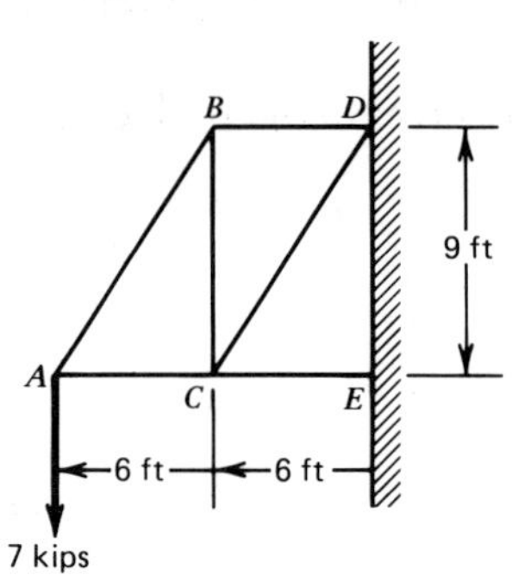

Fig. 2-93. Truss for Problem 1.

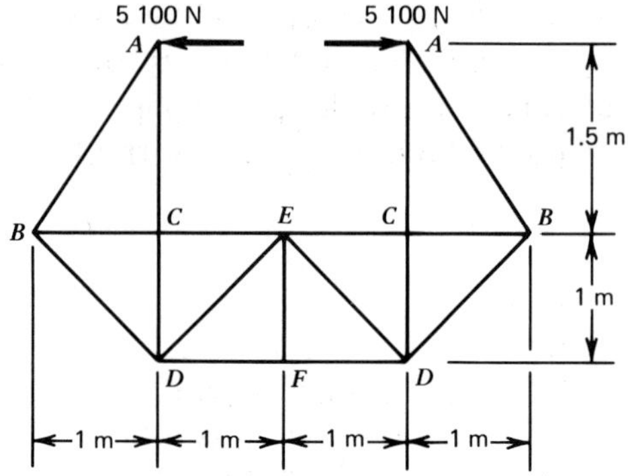

Fig. 2-94. Truss for Problem 2.

18. How are the belt tensions on a pulley related when the belt is about to slip on the pulley?
19. In what units do you express the angle of contact of a belt on a pulley?

REVIEW PROBLEMS Let's finish the chapter with these review problems.

1. Calculate the force in each member of the cantilever truss in Fig. 2-93.
2. Calculate the force in each member of the special truss in Fig. 2-94.
3. A steel cable weighs 1.5 lb/ft. The span is 430 ft and the sag is 20 ft. What is the maximum tension in the cable?
4. An aluminum alloy cable weighs 16 N/m. The span is 280 m and the sag is 18 m. What is the maximum tension in the cable?
5. What magnitude of horizontal force is needed to push a 200-lb box up a 15° inclined plane? The coefficient of friction between the box and plane is 0.2.
6. What magnitude of horizontal force is needed to push a 600 N crate down a 10° inclined plane if the coefficient of friction between the crate and plane is 0.25?
7. Figure 2-95 shows a uniform ladder that weighs 60 lb. A man who weighs 150 lb stands on the ladder at *A*. The wall is smooth. What coefficient of friction is needed between the ladder and floor to keep the ladder from slipping?

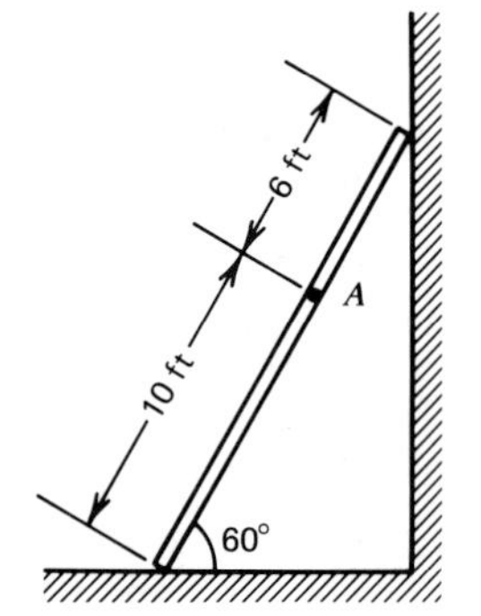

Fig. 2-95. Ladder for Problem 7.

8. Figure 2-96 shows a uniform ladder that weighs 250 N. A man who weighs 700 N stands on the ladder at point *A*. The wall is smooth.

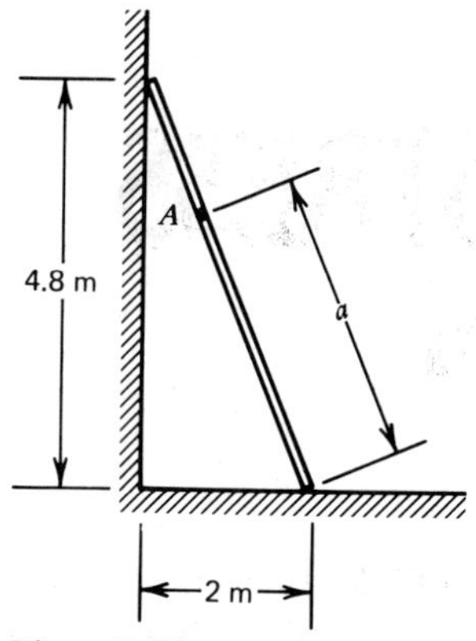

Fig. 2-96. Ladder for Problem 8.

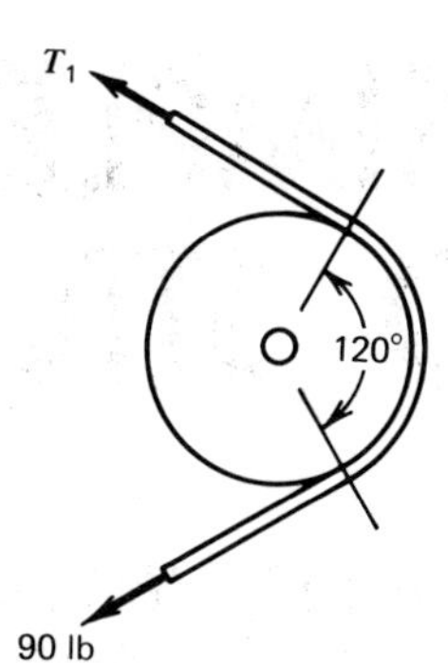

Fig. 2-97. Belt and pulley for Problem 9.

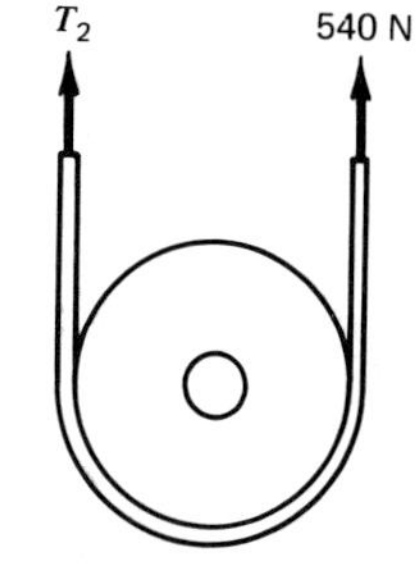

Fig. 2-98. Belt and pulley for Problem 10.

The coefficient of friction between the ladder and floor is 0.3. The ladder is on the point of slipping. Calculate the distance a.

9. In Fig. 2-97, the coefficient of friction between the belt and pulley is 0.45. The belt is at the point of sliding counterclockwise relative to the pulley. Calculate the belt tension T_1.
10. In Fig. 2-98, the coefficient of friction between the belt and pulley is 0.55. The belt is about to slide counterclockwise relative to the pulley. Calculate the belt tension T_2.

3 SIMPLE STRESS AND STRAIN

PURPOSE OF THIS CHAPTER. We will make real progress in strength of materials in this chapter. First, we will become acquainted with some of the technical terms of the subject and learn how to use them—terms such as *stress* (force divided by area) and *strain* (deformation divided by length). These are important terms, because the amount of stress or strain in a machine or structural part is an indication of how severely the part is loaded and is a factor in determining whether the forces applied are reasonable, enabling the part to stand up, or whether the forces are so great that the part will break. Designers use their knowledge of stress to make sure that a stress is reasonable and that each part of the machine or structure is strong enough. We will see how this is done and try our hand at designing a few simple members.

Also in this chapter we will see what happens to a piece of steel or other material when force is applied to it, how it deforms, and how strong it is. Tables in the chapter give the properties of most of the common materials used in machines and structures, and you will get an idea of how to express the strength of a material and, from the point of view of a designer, learn the significant properties of that material. You will be interested to find how much stronger some materials are than others and to see just how strong the different materials are. Designers need to know the properties of different materials so that they can select the most suitable material for each part of a machine or structure.

3-1. WHAT STRESS IS

Stress is force per unit of area or force divided by area. In Chapter 1 you learned what a force is; and in Chapter 2 you learned to calculate forces acting on a body in equilibrium. Now we will go one step farther. We will divide the force acting on the body by the area of the cross section of the body. This will be force divided by area, and we will call it *stress* (some books use the term *unit stress*, but here we will simply say *stress* for force divided by area).

UNITS

If we are using English gravitational units it is most likely that the units for stress will be in pounds per square inch. This results from dividing a force in pounds by an area in square inches. It is commonly abbreviated as *psi*. A steel bar might have a stress of 20 000 psi, for example.

There is some use of the kip as a unit of force in structural engineering. When this is done the stress is expressed in kips per square inch, abbreviated as ksi. A typical value of stress in a steel bar might be 20 ksi.

In the metric system the force is expressed in Newtons, and the area could be expressed in square meters, so the stress could be expressed in newtons per square meter. This combination of units is called a pascal and is abbreviated as Pa. However, stresses in pascals usually turn out to be in very large numbers, so it is convenient to express these in megapascals. Mega is a prefix meaning *million*, so a megapascal is a million pascals and is abbreviated as MPa. A typical value of stress in a steel bar might be 140 MPa.

The word *stress* is probably the most important word in the subject of strength of materials, because the amount of the stress in a member of a machine or structure is the real indication of how severely the member is loaded; if the stress is above a certain figure, the part is loaded too heavily and will probably fail. The amount of the force alone may not mean much. A column in a tall building may be able to withstand a force of several million pounds without serious damage, because the area of its cross section is so great that the stress (force per unit of area) is reasonable in amount. On the other hand, a camera mechanism may fail under a load of 5 lb because the area of the cross section of the member is so small that the load causes a very high stress.

Our first job in this chapter is to get acquainted with the three simple stresses. The simple stresses are: (1) *tension*, in which the material stretches, (2) *compression*, in which the material shortens, and (3) *shear*, in which one part of the material slides over another part. We are going to take these stresses one at a time and study them in detail.

3-2. TENSION AND TENSILE STRESS

A member is in tension when the forces acting on it tend to stretch it or increase its length. Figure 3-1*a* shows a bar in tension. If the bar is in equilibrium, the two forces P must be equal for the sum of the horizontal forces acting on the bar to be zero. Now suppose we consider the portion of the bar to the right of the section B-B. Let's draw the free-body diagram of this part of the bar (Fig. 3-1*b*) so that we can see how the forces act on the cross section. First, we show the force P (directed to the right) at the right end of the bar. Then we draw the force on the

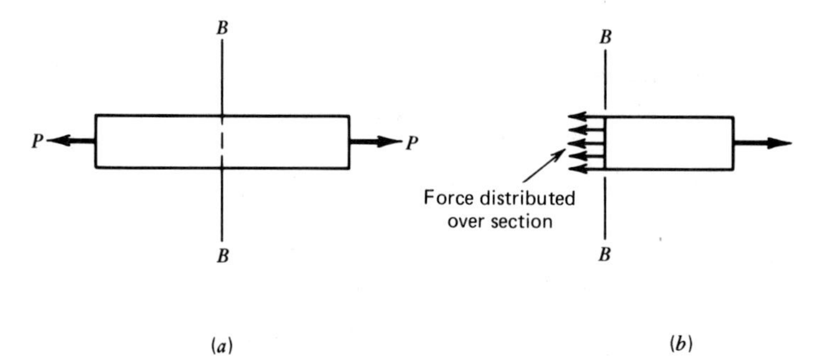

Fig. 3-1. Bar in tension. (*a*) Free-body diagram of bar. (*b*) Free-body diagram of part of bar.

cross section, and this is distributed over the area of the cross section. The force on the cross section must be directed to the left, and the total amount of it must be P for this part of the bar to be in equilibrium. The total force P is distributed over the cross section, and the magnitude of the stress (force per unit of area) is equal to the force P divided by the area A of the cross section. Thus

$$S = \frac{P}{A}$$

where S is the stress and where the units of stress are pounds per square inch if we are using gravitational units or megapascals if we are using the metric system. So we have reached the conclusion that the stress in a bar is equal to the total force divided by the area. We say that the bar is in *tension*, and we call the stress *tensile stress* when the member is stretched.

There are certain conditions that must be met for the tensile stress to be given by the simple formula $S = P/A$. These are: (1) the force must be perpendicular to the cross section of the bar, and (2) the force must be centrally applied (that is, it must go through the center of the cross section of the bar). If these conditions are not met, the stress condition is more complicated, and a different formula must be used. (More information is given on this in Chapter 7.) In Fig. 3-2*a* the force P passes

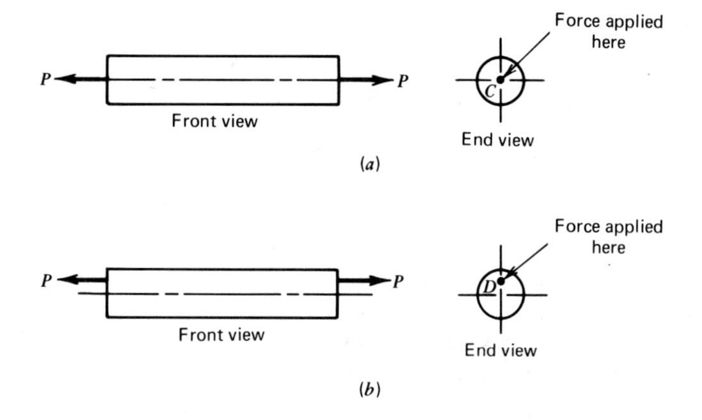

Fig. 3-2. Force applied to bar. (*a*) Force applied at center of bar. (*b*) Force applied off center of bar.

through the center C of the circular cross section; but in Fig. 3-2*b* the force P passes through the point D of the cross section, and point D is not at the center. Therefore, the formula $S = P/A$ can be used to calculate the stress in the bar in Fig. 3-2*a* but cannot be used to calculate the stress in the bar in Fig. 3-2*b*. We will look at this type of problem later. For the present it is enough to keep in mind the two conditions already given. Now let's try the formula $S = P/A$ in some problems.

Illustrative Example 1. Figure 3-3 shows two views of a bar in tension. Find the tensile stress.

Solution:

1. The force P is 12 000 lb.
2. The cross section is a rectangle, and its area is equal to the product of the width and the height. Thus

$$A = 0.375 \times 2 = 0.75 \text{ in.}^2$$

3. The stress in the bar is

$$S = \frac{P}{A} = \frac{12\,000}{0.75} = 16\,000 \text{ psi}$$

A FEW WORDS ABOUT PASCALS AND MEGAPASCALS

The dimensions of cross sections are usually expressed in millimeters in the metric system. This leads to calculating areas in square millimeters (mm^2). So, if the force is in newtons, the stress is easily obtained in N/mm^2, by dividing the force by the area. There, since $1\ mm = 1 \times 10^{-3}\ m$, it follows that $1\ mm^2 = 1 \times 10^{-6}\ m^2$, and

$$1\,\frac{\text{N}}{\text{mm}^2} = 1\,\frac{\text{N}}{10^{-6}\ \text{m}^2} = 1 \times 10^6 \text{ Pa} = 1 \text{ MPa}$$

Let's try this in an example.

Illustrative Example 2. Calculate the stress in a wire 4 mm in diameter when it is subjected to an axial tensile force of 1 970 N.

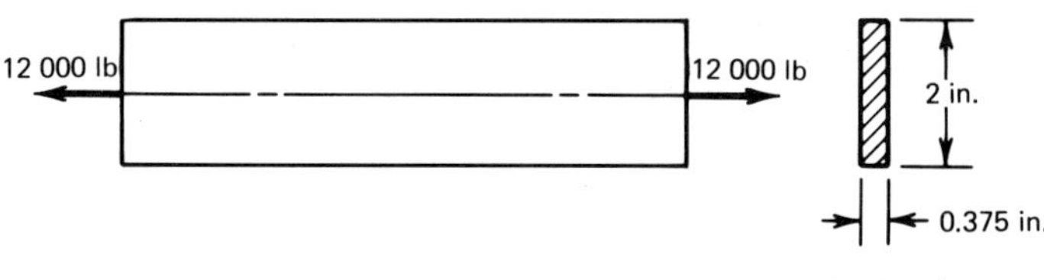

Fig. 3-3. Bar in tension.

Solution:

1. The magnitude of the force P is 1 970 N.
2. The cross section is circular, for which the area is $\pi d^2/4$.

$$A = \pi \frac{(4)^2}{4} = 12.56 \text{ mm}^2$$

3. The stress in the wire is

$$S = \frac{P}{A} = \frac{1\,970}{12.56} = 156.8 \text{ N/mm}^2 \qquad \text{or} \qquad 156.8 \text{ MPa}$$

SMALLEST CROSS SECTION

In these two examples each member had the same cross section at all points along its length. This isn't always the case; often the member has holes drilled in it or has changes in its lateral dimensions so that the area of the cross section is not the same at all points. The stress changes, too, and naturally it is greatest where the area of the cross section is smallest. We are usually interested in the maximum stress, and so ordinarily we select the smallest area of cross section.

Illustrative Example 3. The bar in Fig. 3-4*a* is loaded by pins through the holes at the ends. Find the maximum tensile stress.

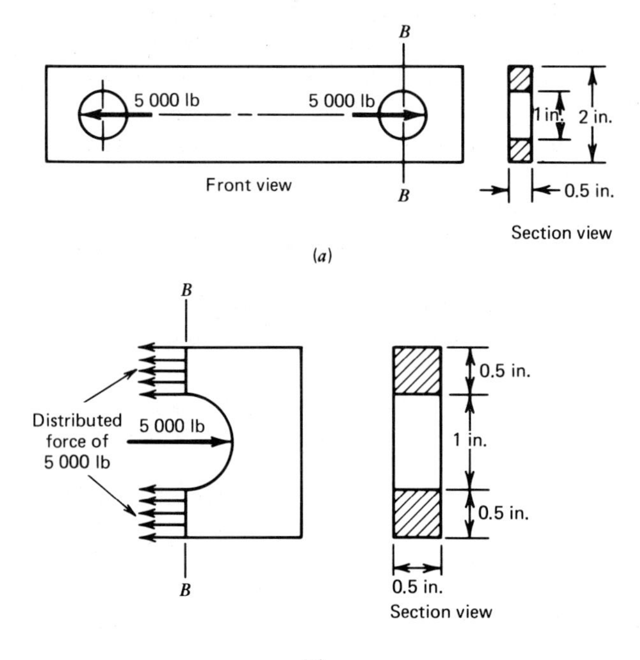

Fig. 3-4. Bar loaded by pins. (*a*) Bar in tension. (*b*) Free-body diagram of part of bar.

Solution:

1. The force on the bar is 5 000 lb.
2. The smallest area is at a cross section through the center of the holes; such a cross section is *B-B* in Fig. 3-4*a*. Figure 3-4*b* shows the free-body diagram of the part of the bar to the right of the section *B-B*. We start this by drawing the force of 5 000 lb to the right; then we show a force of 5 000 lb to the left, and this force is distributed over the area of the cross section. The cross section through the center of the hole consists of two small rectangles, each 0.5 in. by 0.5 in., and the total force is distributed over this area. Then the area of the cross section is

$$A = 2 \times 0.5 \times 0.5 = 0.5 \text{ in.}^2$$

3. The maximum stress in the bar is

$$S = \frac{P}{A} = \frac{5\,000}{0.5} = 10\,000 \text{ psi}$$

Illustrative Example 4. In Fig. 3-5*a* the two small rods at the ends are

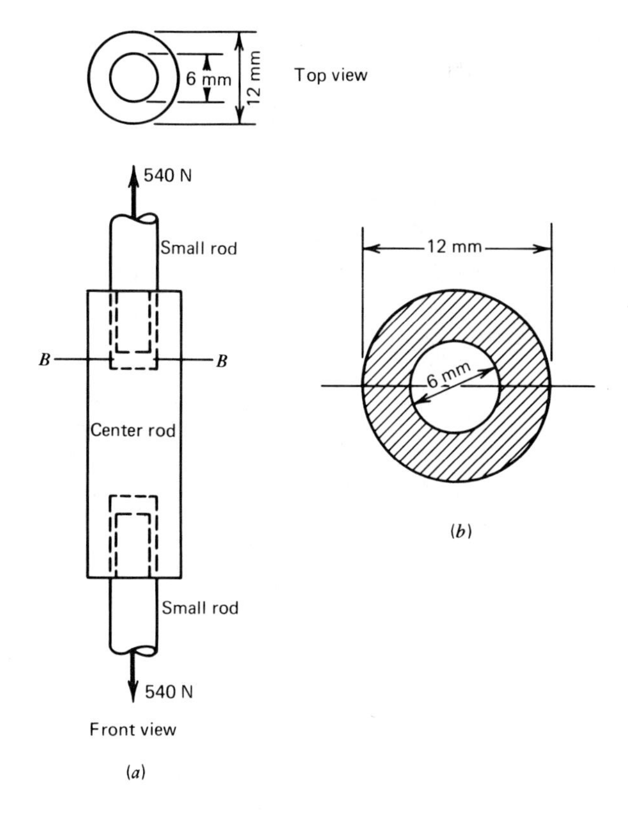

Fig. 3-5. Bar in tension. (*a*) Bar in tension. (*b*) Section view.

6 mm in diameter and are threaded into tapped holes in the center rod, which is 12 mm in diameter. Find the tensile stress on the cross section *B-B*.

Solution:

1. The total force P is 540 N.
2. Figure 3-5*b* shows the cross section *B-B*. This is a hollow circle, and the area of the cross section is equal to the area of the large circle minus the area of the small circle. The area of the large circle is

$$\frac{\pi d^2}{4} = \frac{\pi(12)^2}{4} = 113.1 \text{ mm}^2$$

and the area of the small circle is

$$\frac{\pi d^2}{4} = \frac{\pi(6)^2}{4} = 28.3 \text{ mm}^2$$

The area A is the difference between these values, so it is

$$A = 113.1 - 28.3 = 84.8 \text{ mm}^2$$

3. The stress on the section *B-B* is

$$S = \frac{P}{A} = \frac{540}{84.8} = 6.37 \text{ MPa}$$

CALCULATING THE FORCE

The equation $S = P/A$ can also be written as $P = AS$. In this form it is used to calculate the force that a member can carry without exceeding a given stress.

Illustrative Example 5. What force in tension can be applied to a rectangular steel bar 1 in. by 3 in. in cross section without exceeding a stress of 9 000 psi?

Solution:

1. The area of the cross section is

$$A = 1 \times 3 = 3 \text{ in.}^2$$

2. The stress is $S = 9\,000$ psi.
3. The force is $P = AS = 3 \times 9\,000 = 27\,000$ lb. The total force here is 27 000 lb, but the stress (force per unit area) is only 9 000 psi.

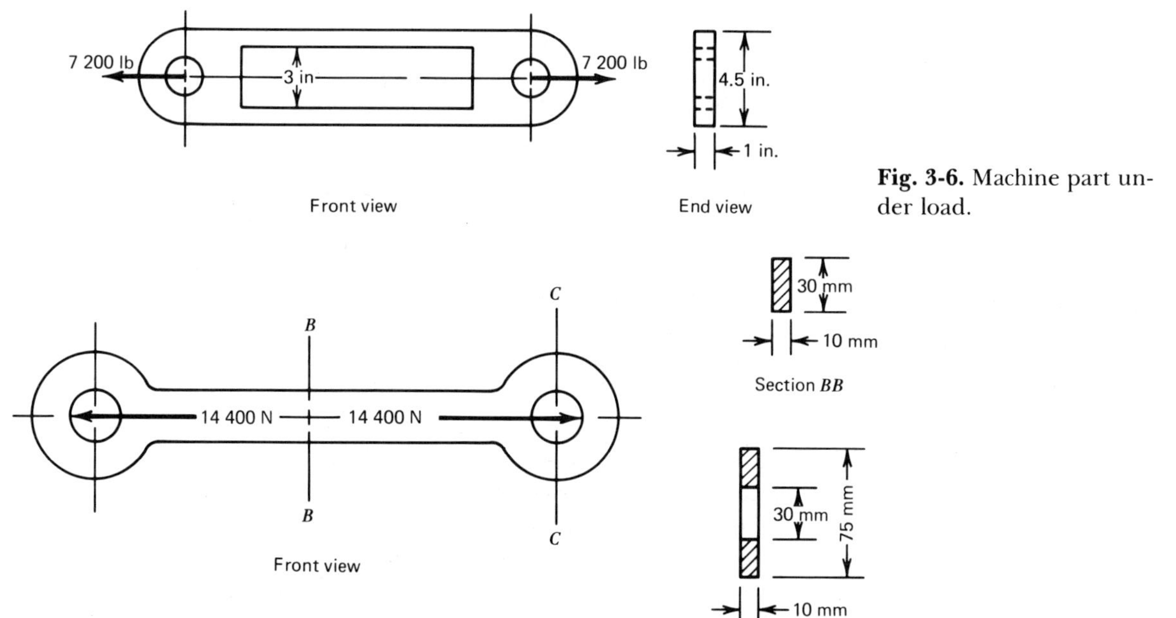

Fig. 3-6. Machine part under load.

Fig. 3-7. Bar in tension.

Practice Problems (Section 3-2). Here are practice problems on tension and tensile stress.

1. Find the tensile stress in a steel rod, 0.6 in. in diameter, under a force of 3 200 lb.
2. A metal strap that has a rectangular cross section, 0.25 in. by 1.5 in. carries a tensile load of 1 700 lb. Find the tensile stress.
3. The machine part, in Fig. 3-6, is loaded as shown. Find the maximum tensile stress.
4. Find the tensile stress on section *B-B* of the bar in Fig. 3-7.
5. Find the tensile stress on section *C-C* of the bar in Fig. 3-7.
6. What tensile force can a steel wire 4 mm in diameter carry without exceeding a stress of 560 MPa?
7. The tensile stress in a square bar, 1.2 in. on a side, is 8 000 psi. Find the tensile force.

3-3. COMPRESSION AND COMPRESSIVE STRESS

A member is in compression when the forces acting on it tend to shorten it. The forces tend to squeeze the material together, and this tendency is resisted by internal forces or stresses; one part of the body pushes

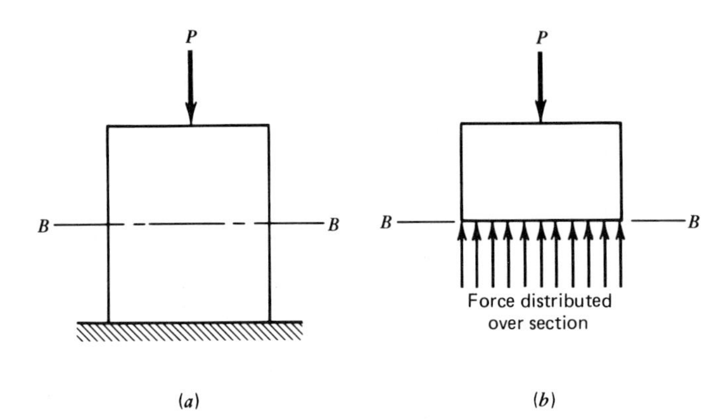

Fig. 3-8. Block in compression. (*a*) Block. (*b*) Free-body diagram of part of block.

against other parts to resist the deformation. For example, the force P in Fig. 3-8*a* tends to shorten the body, so it is in compression. The free-body diagram of the part of the body above section B-B is shown in Fig. 3-8*b*. Here the downward force P is drawn first and then the force on the cross section. The total force on the cross section must be equal to P (so the sum of the vertical forces will be zero; remember, from Chapter 2, $\Sigma F_y = 0$), and this force is distributed over the area of the cross section. The stress is equal to the force divided by the area, or

$$S = \frac{P}{A}$$

The stress is called *compressive stress*, and the member is said to be in *compression*.

The same two conditions must be met as in the case of tension if the stress is to be given by the formula $S = P/A$. These are: (1) the force must be perpendicular to the cross section of the member, and (2) the force must pass through the center of the cross section.

Illustrative Example 1. Figure 3-9 shows a link in the mechanism of a shaker conveyor. Find the compressive stress on section C-C.

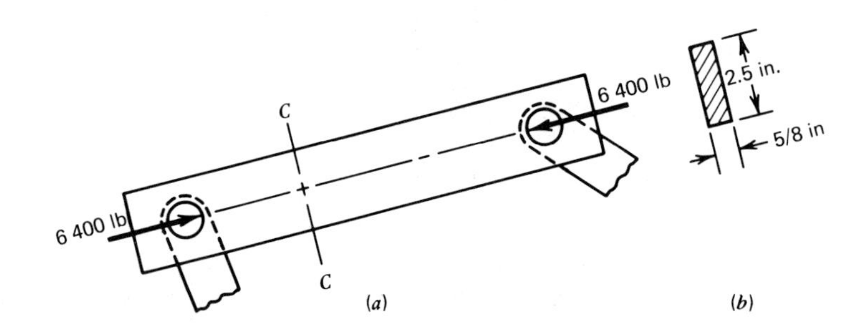

Fig. 3-9. Bar in compression. (*a*) Loading. (*b*) Section view.

Solution:

1. The force P is 6 400 lb. You can see that the forces tend to shorten the bar from the direction of the arrows; from this tendency to shorten the bar we know that the stress is compressive stress.
2. The cross section is a rectangle, $\frac{5}{8}$ in. by 2.5 in. Its area is

$$A = \frac{5}{8} \times 2.5 = 1.56 \text{ in.}^2$$

3. The stress is equal to the force divided by the area,

$$S = \frac{P}{A} = \frac{6\,400}{1.56} = 4\,100 \text{ psi}$$

Illustrative Example 2. Figure 3-10 shows two views of a rubber washer of a type used for motor mounts. Find the compressive stress.

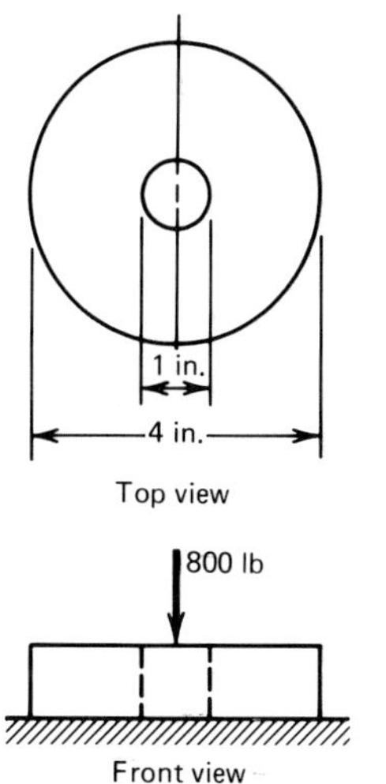

Fig. 3-10. Washer in compression.

Solution:

1. The force P is 800 lb.
2. The cross section is a hollow circle, with outer diameter of 4 in. and inner diameter of 1 in. The area is equal to the area of the large circle minus that of the small circle. The area of the large circle is

$$\frac{\pi}{4} d^2 = \frac{\pi}{4} (4)^2 = 12.563 \text{ in.}^2$$

and the area of the small circle is

$$\frac{\pi}{4} d^2 = \frac{\pi}{4} (1)^2 = 0.785 \text{ in.}^2$$

so the area of the cross section is

$$A = 12.563 - 0.785 = 11.778 \text{ in.}^2$$

3. The compressive stress is equal to the force divided by the area,

$$S = \frac{P}{A} = \frac{800}{11.778} = 67.92 \text{ psi}$$

CALCULATING THE FORCE

Just as we did in studying tension, we can multiply the equation $S = P/A$ by A and write it,

$$P = AS$$

This form of the equation is useful in calculating the total force when the area and stress are known.

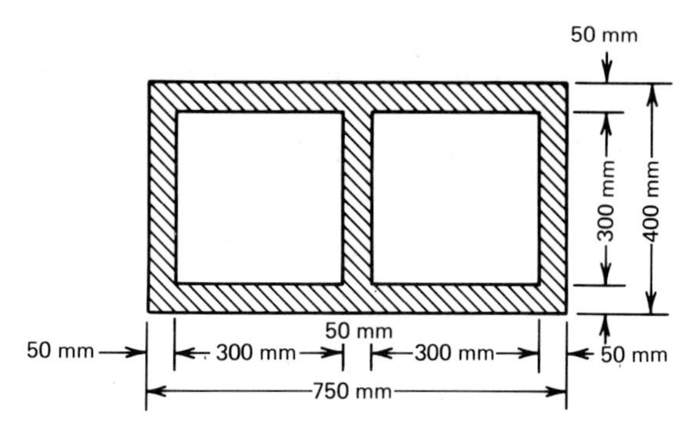

Fig. 3-11. Cross section of compression member.

Illustrative Example 3. Figure 3-11 shows a type of cross section used in the frames of large presses. What force will cause a compressive stress of 60 MPa?

Solution:

1. The cross section can be regarded as a large rectangle, 400 mm × 750 mm, with two squares 300 mm on a side taken away. The area of the cross section is equal to the area of the large rectangle minus the area of the two squares. The area of the large rectangle is

$$400 \times 750 = 300\,000 \text{ mm}^2$$

and the area of the two squares is

$$2 \times 300 \times 300 = 180\,000 \text{ mm}^2$$

Then the area of the cross section is

$$A = 300\,000 - 180\,000 = 120\,000 \text{ mm}^2$$

2. The stress is 60 MPa, or 60 N/mm^2.
3. The total force on the cross section is

$$P = 120\,000 \times 60 = 7\,200\,000 \text{ N}$$

This is a large force, but it is distributed over an area of 120 000 mm^2, so the stress is only 60 MPa.

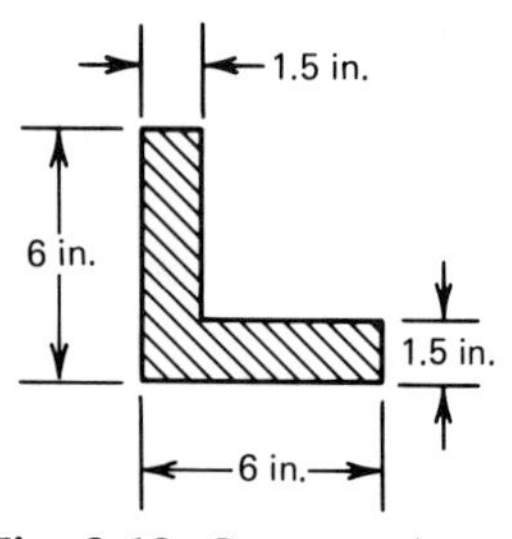

Fig. 3-12. Cross section of machine frame member.

Practice Problems (Section 3-3). Now that you know something about compression, you should work this set of practice problems. It will help you to fix in mind what you have learned.

1. A circular bar 3 in. in diameter is subjected to a compressive force of 62 000 lb. Find the stress.
2. A member of a machine frame has the cross section shown in Fig. 3-12. It is subjected to a force of 80 000 lb in compression. Calculate the stress.

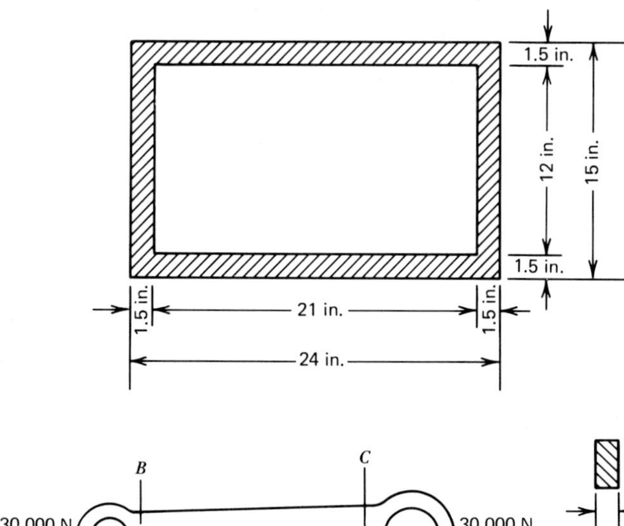

Fig. 3-13. Machine base.

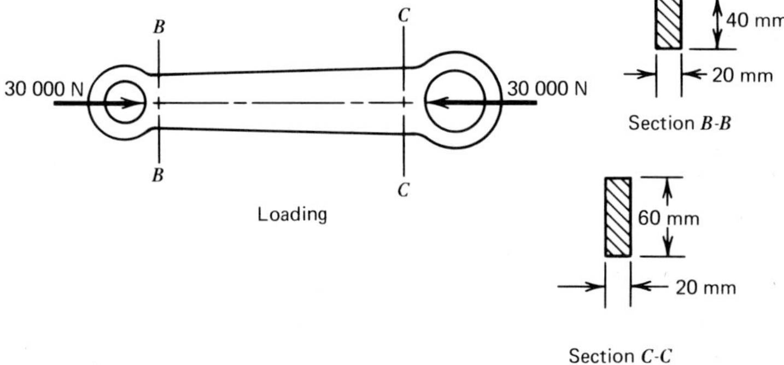

Fig. 3-14. Connecting rod.

3. The area of contact between the base of a certain machine and the floor is the hollow rectangle in Fig. 3-13; the machine weighs 7 200 lb. Find the stress between the machine and the floor.
4. Find the compressive stress on section *B-B* in Fig. 3-14.
5. Find the compressive stress on section *C-C* in Fig. 3-14.
6. A cross section of a rubber motor mounting is a hollow square. The outer square is 4 in. on a side, and the inner square is 2 in. on a side. The compressive stress is 180 psi. Find the total force.
7. The compressive stress in a hollow circular bar of 150-mm outer diameter and 100-mm inner diameter is 42 MPa. What is the total force?

3-4. SHEAR AND SHEARING STRESS

A member is in shear when the forces acting on it tend to slide one part of the member with respect to another part. One common example of shear is the stud or projection used to transmit force from one member to another. Figure 3-15*a* shows one view of a stud and Fig. 3-15*b* shows

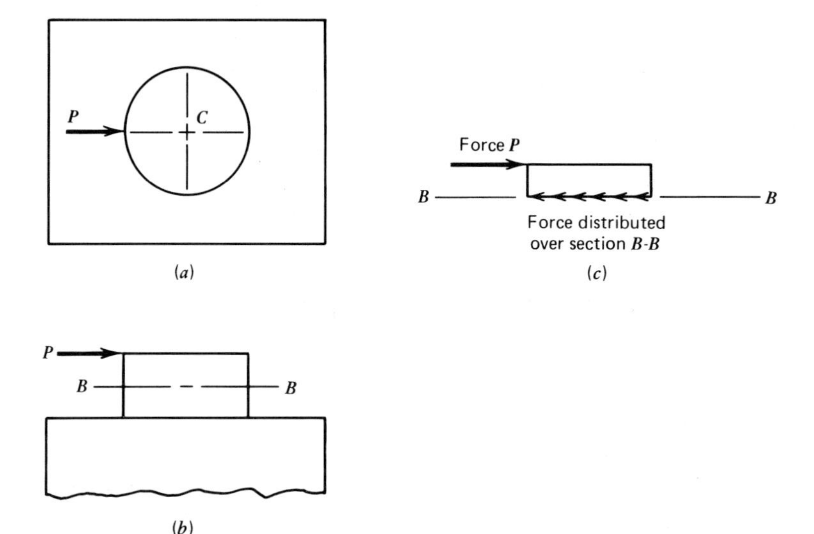

Fig. 3-15. Stud in shear. (*a*) Top view. (*b*) Front view. (*c*) Free-body diagram of part of stud.

another view. The force P tends to slide the upper part of the stud across the lower part on any cross section such as B-B. This tendency is resisted by shearing stresses in the plane of the cross section. Figure 3-15*c* shows the free-body diagram of the part of the stud above the section B-B. The force P directed to the right is balanced by the shearing stresses in the plane of the section. (They must act to the left so that the sum of the horizontal components of forces can be zero, because the body is in equilibrium. Remember $\Sigma F_x = 0$ from Chapter 2. A force directed to the right must be balanced by a force directed to the left.) The total force directed to the left is equal to P and is distributed over the area of the cross section, which in this case is the circle in Fig. 3-15*a*. The magnitude of the stress is

$$S_s = \frac{P}{A_s}$$

where the subscript s for S and A stands for shear.

It is to be noticed especially that shearing stress is in the plane of the cross section and is parallel to the plane. This is in contrast to tensile or compressive stress, which is perpendicular to the plane of the cross section. Shearing stress is similar to tensile and compressive stresses in that all are stresses and all resist the tendency of forces to deform the body. Also, when certain conditions are met, all can be calculated from the formula $S = P/A$. These conditions have already been stated for tension and compression. For shear they are: (1) the applied force P must be parallel to the plane of the cross section on which the stress is calculated; and (2) the projection of the force in the plane of the cross section must go through the center of the cross section. (This projection is shown in Fig. 3-15*a* where point *C* in the top view is the center of the cross section.) The units in which shearing stress is expressed are the same as for tensile and compressive stress, namely, pounds per square inch (psi) in gravitational units or megapascals (MPa) in the metric system.

The hardest part about shear problems is finding the area correctly. Once this is done, the shearing stress is just

$$S_s = \frac{P}{A_s}$$

Illustrative Example 1. Figure 3-16*a* shows two views of a tension member that is loaded by pins through the holes. Section *B-B* is taken through the center of the holes. The force at the right end tends to shear out the shaded part, that is, to make it slide out of the body. What shearing stress is developed?

Solution:

1. The force *P* is 6 000 lb.
2. Figure 3-16*b* shows the free-body diagram of the shaded part. We see here the force *P*, which tends to push the part to the right, and this action is balanced by shearing stresses along the upper and lower faces of the part. The areas in shear are rectangles, each 1 in. by 0.5 in. (1 in. is the horizontal dimension in the plane of the paper, and 0.5 in. is the thickness of the member). The area of each rectangle is

$$1 \times 0.5 = 0.5 \text{ in.}^2$$

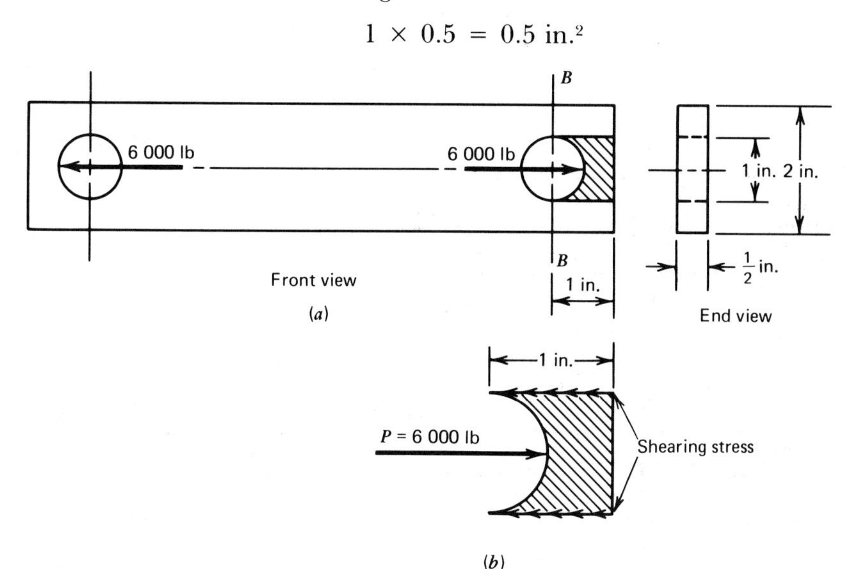

Fig. 3-16. Shear Problem. (*a*) Tension member. (*b*) Free-body diagram.

There are two rectangular areas in shear, so

$$A_s = 2 \times 0.5 = 1 \text{ in.}^2$$

3. The shearing stress is

$$S_s = \frac{P}{A_s} = \frac{6\,000}{1} = 6\,000 \text{ psi}$$

Illustrative Example 2. Figure 3-17*a* shows two views of a pair of plates fastened together by a pin 20 mm in diameter. What is the shearing stress in the pin?

Solution:

1. The force P is 14 000 N.
2. The applied forces tend to make one plate slide over the other and shear the pin in two on the plane of contact of the plates. Figure 3-17*b* shows the free-body diagram of the lower plate with the lower half of the pin. The tendency to slide is resisted by the shearing force exerted on the lower half of the pin by the upper half. The area in shear, then, is the area of the cross section of the pin. This is

$$A_s = \frac{\pi}{4} d^2 = \frac{\pi}{4} (20)^2 = 314 \text{ mm}^2$$

3. The shearing stress is

$$S_s = \frac{P}{A_s} = \frac{14\,000}{314} = 44.6 \text{ MPa}$$

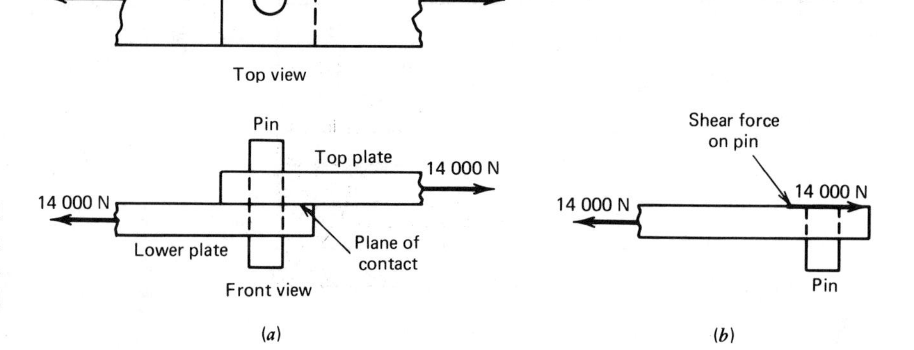

Fig. 3-17. Plates fastened together by pin. (*a*) Pin in shear. (*b*) Free-body diagram of lower plate and lower half of pin.

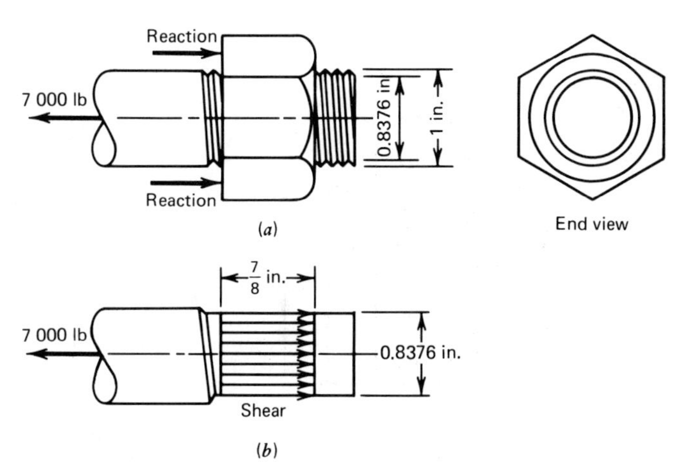

Fig. 3-18. Shear in bolt threads. (*a*) Load on bolt. (*b*) Free-body diagram.

Illustrative Example 3. Figure 3-18*a* shows two views of a 1-in. bolt and nut under load. There is a tendency for the threads to strip from the bolt. Find the shearing stress at the root of the threads. The root diameter is 0.8376 in., as shown.

Solution:

1. The force P is 7 000 lb.
2. The threads tend to shear off at the root. Figure 3-18*b* shows the free-body diagram of the bolt with the threads removed to show the shearing stresses exerted by the threads on the bolt. The shearing stresses are developed on a cylindrical surface 0.8376 in. in diameter and ⅞ in. long. The area of a cylindrical surface is equal to the product of the circumference of the cylinder and the length. This is

$$A_s = \pi dh = \pi \times 0.8376 \times 7/8 = 2.30 \text{ in.}^2$$

3. The shearing stress is

$$S_s = \frac{P}{A_s} = \frac{7\ 000}{2.30} = 3\ 040 \text{ psi}$$

Illustrative Example 4. Figure 3-19*a* shows two views of a sandwich-type engine mounting. The shaded parts are made of a soft rubber compound, and the other parts are made of steel to which the rubber is bonded. Find the shearing stress in the rubber. (Mountings of this type are used to support engines. The engine is fastened to the center piece of steel, and the steel pieces, B, are fastened to the frame of the vehicle. The function of the mounting is to keep vibrations of the engine from being transmitted to the rest of the vehicle.)

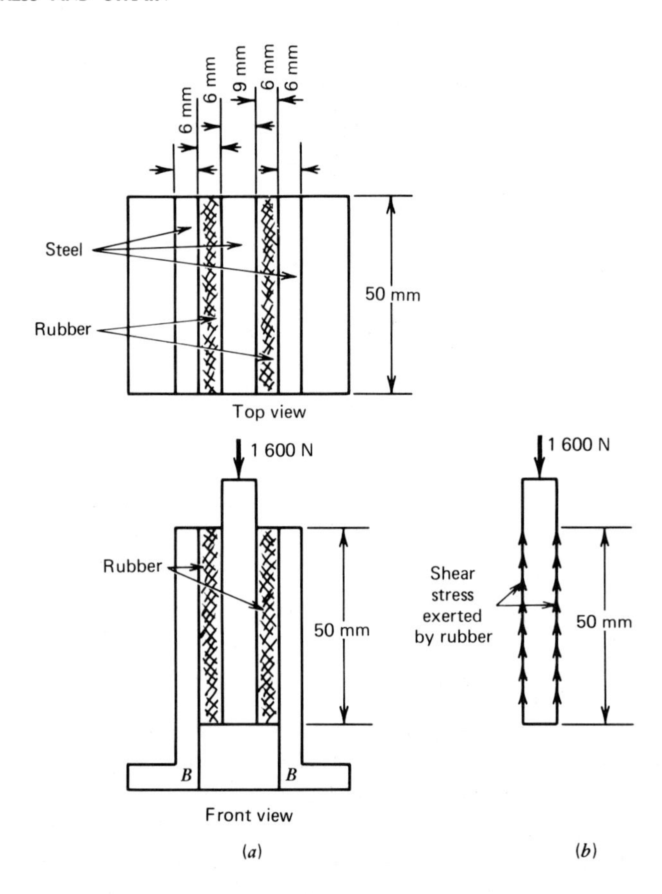

Fig. 3-19. Shear in motor mounting. (*a*) Motor mounting. (*b*) Free-body diagram of center part.

Solution:

1. The force P is 1 600 N.
2. Figure 3-19*b* shows the free-body diagram of the center piece of steel. The rubber exerts shearing stresses on each face of the steel to resist the tendency of the steel to slide out from between the two pieces of rubber. The area in shear consists of two rectangles, each 50 mm by 50 mm. The total area is

$$A_s = 50 \times 50 \times 2 = 5\ 000 \text{ mm}^2$$

3. The shearing stress is

$$S_s = \frac{P}{A_s} = \frac{1\ 600}{5\ 000} = 0.32 \text{ N/mm}^2 = 0.32 \text{ MPa}$$

CALCULATING THE FORCE

In many cases the area in shear can be calculated, and the stress is known; then the problem is to find the total force. This can be done by rewriting

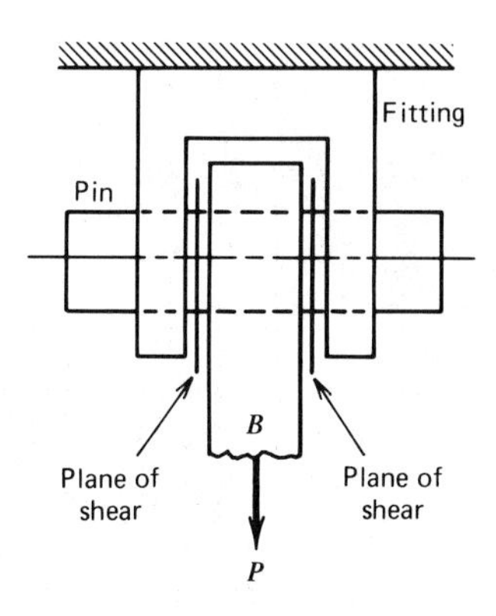

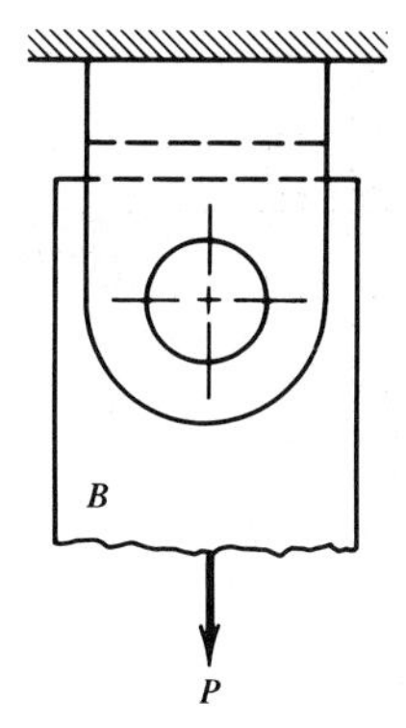

Fig. 3-20. Pin in shear.

the formula $S_s = P/A_s$. When both sides of the equation are multiplied by A_s, the result is

$$P = A_sS_s$$

This can be used to calculate P if A_s and S_s are known.

Illustrative Example 5. Figure 3-20 shows two views of the upper end of a bar B that is fastened to a fitting by means of a pin 2 in. in diameter. The shearing stress in the pin is 10 000 psi. Find the force P.

Solution:

1. It would be necessary to shear the pin in two planes to pull the bar away from the fitting. Each plane furnishes a circular area 2 in. in diameter, in shear, so the area is

$$A_s = \frac{2\pi d^2}{4} = \frac{2\pi(2)^2}{4} = 6.28 \text{ in.}^2$$

2. The stress is 10 000 psi.
3. The force P is

$$P = A_sS_s = 6.28 \times 10\,000 = 62\,800 \text{ lb}$$

Practice Problems (Section 3–4). Try your knowledge of shear on these practice problems.

1. A force of 27 000 lb is distributed over an area of 2.84 in.^{2}, in shear. What is the shearing stress?
2. The pin in Fig. 3-21 is ⅞ in. in diameter. Find the shearing stress in it.

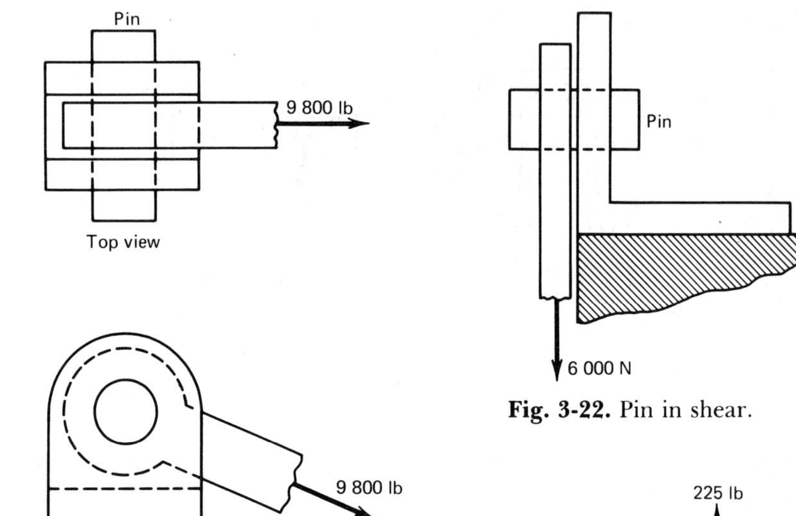

Fig. 3-21. Pin in shear.

Fig. 3-22. Pin in shear.

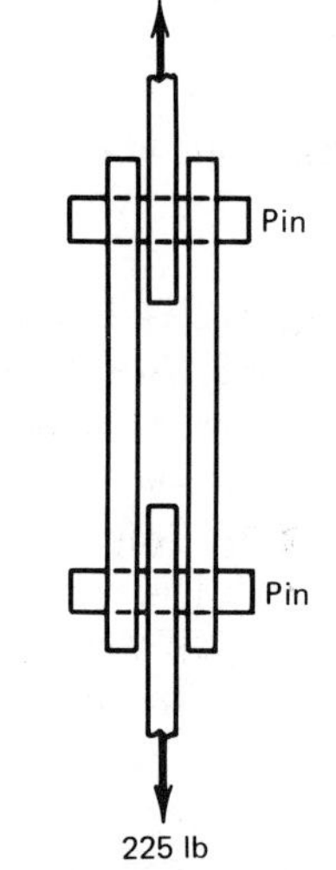

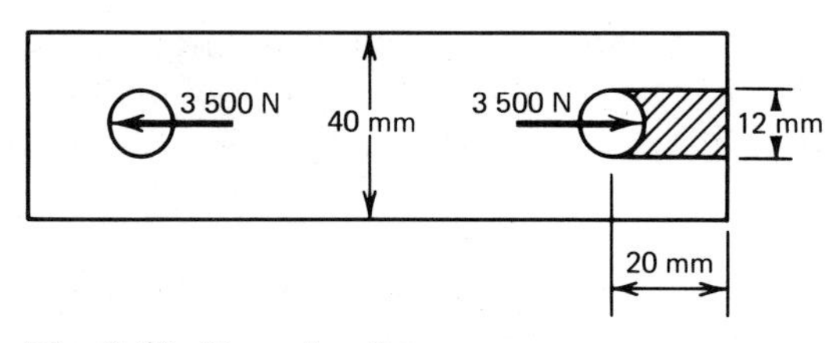

Fig. 3-23. Shear in plate.

Fig. 3-24. Pins in shear.

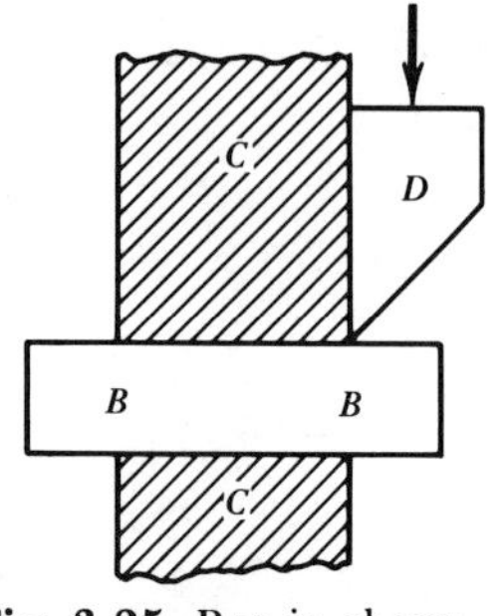

Fig. 3-25. Bar in shear.

3. In Fig. 3-22 the pin is 12 mm in diameter. What is the shearing stress?
4. The plate in Fig. 3-23 is 6 mm thick. What is the shearing stress along the edge of the shaded part?
5. The pins in Fig. 3-24 are ¼ in. in diameter. Calculate the shearing stress.
6. An area of 1.93 in.^{2} is subjected to a shearing stress of 9 200 psi. Find the total force.
7. In Fig. 3-25 the bar *B-B* is 24 mm square. It is held by the jaws *C* and sheared by the blade *D*. The shearing stress in the bar is 270 MPa. Find the total force.

3-5. STRAIN IN TENSION AND COMPRESSION

All materials deform when subjected to stress, and it is necessary to be able to calculate the *deformation* of a body under load, because, in many load-carrying members, the deformation is more significant than the stress. This is especially true in metal-cutting machines, where the machine parts must be made heavy, to limit the deformation and avoid chatter. When the members of the machine are made large enough to keep the deformation within reasonable limits, they are quite likely to be several times as strong as they need to be, and there is no danger of breaking.

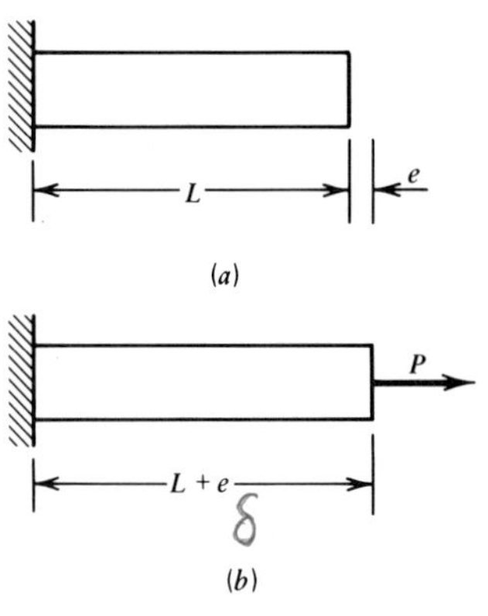

Fig. 3-26. Bar in tension. (*a*) Bar with no load. (*b*) Bar under load.

In metal machine parts the deformation is usually so small that it cannot be seen with the eye, and special instruments are required to measure it; in the case of a soft-rubber motor mounting, however, the deformation may be as much as ½ in. or more and is easily apparent. Also, in such processes as forging, extruding, drawing, and rolling of metals, the deformation is often great and easily noticed. Our next job is to study the deformation of materials and learn to calculate it.

Deformation is simply change of dimension or shape. In tension and compression it is just a change in length. A bar in tension stretches, and a bar in compression shortens. The bar in Fig. 3-26*a* has the original length L, and when a tensile force P (Fig. 3-26*b*) is applied it stretches by the amount e, as shown in Fig. 3-26*b*. There e is the deformation, because e is the change in length. In Fig. 3-27*a* the bar has the original length L, and when a compressive force P (Fig. 3-27*b*) is applied it shortens by the amount e, as shown in Fig. 3-27*b*. Here e is the deformation.

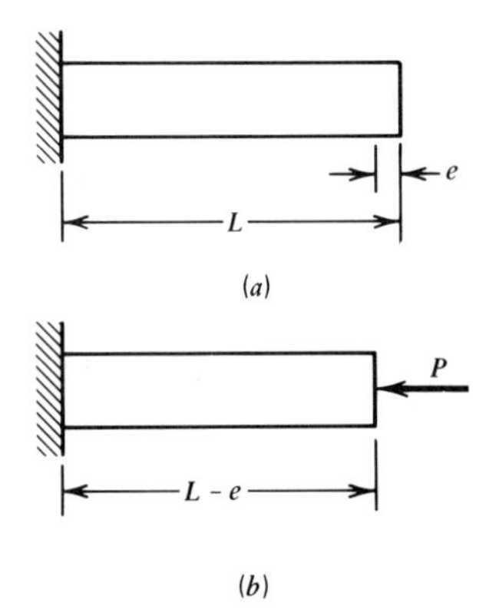

Fig. 3-27. Bar in compression. (*a*) Bar with no load. (*b*) Bar under load.

The deformation is something we need to know about a part of a machine or structure, but the deformation alone is not enough to indicate the severity of loading or how near the member is to failure. A deformation of ½ in. is very large for a connecting rod 10 in. long, but it is trivial in a mine cable 1 000 ft long. A more significant quantity than deformation is strain. Strain is unit deformation or unit change of length; that is, it is the deformation divided by the length over which the deformation occurs. For the bar in Fig. 3-26 or the bar in Fig. 3-27, the deformation is e, and the length over which it occurs is L. Then the strain is e divided by L. Strain is usually represented by ϵ (Greek letter epsilon) and is, then,

$$\epsilon = \frac{e}{L}$$

The deformation e is expressed in some unit of length and the length L in the same unit of length, so, when we calculate strain, we are dividing one number by another number in the same unit. The result is a pure number or ratio that has no dimension, so the strain would be the same whether we are working with gravitational units or metric units.

Values of the strain may vary from 0.0002 for a metal part in service to 10 for a piece of rubber under test.

Illustrative Example 1. A steel rod 30 in. long when not under load is stretched to a length of 30.03 in. Calculate the strain.

Solution:

1. The original length L is 30 in.
2. The deformation e is equal to the difference between the final length and the original length. This is

$$e = 30.03 - 30 = 0.03 \text{ in.}$$

3. The strain is

$$\epsilon = \frac{e}{L} = \frac{0.03}{30} = 0.001$$

Illustrative Example 2. A rubber motor mounting has an original height of 40 mm, as shown in Fig. 3-28*a* and is compressed to a final height of 32 mm, as shown in Fig. 3-28*b*. Find the strain.

Solution:

1. The original length L is 40 mm.
2. The deformation e is equal to the difference between the original and final lengths or heights. Then,

$$e = 40 - 32 = 8 \text{ mm}$$

3. The strain is

$$\epsilon = \frac{e}{L} = \frac{8}{40} = 0.2$$

Illustrative Example 3. A hoisting cable 1 200 ft long is stretched 4 in. What is the strain?

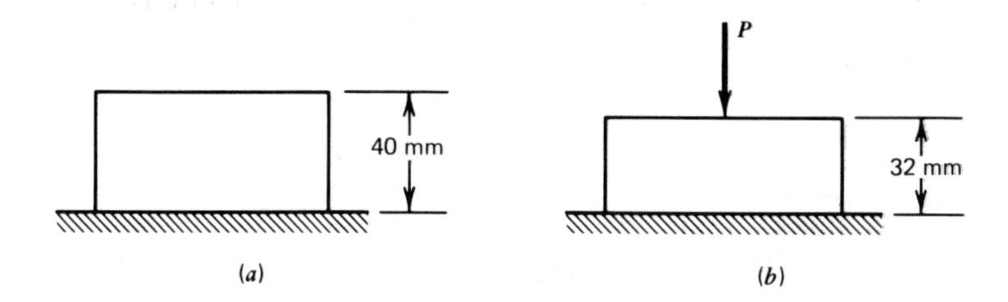

Fig. 3-28. Compression of motor mounting. (*a*) Free. (*b*) Under load.

Solution:

1. The original length is 1 200 ft. This should be converted to inches by multiplying by 12, since there are 12 inches in 1 foot. Thus

$$L = 1\,200 \times 12 = 14\,400 \text{ in.}$$

2. The deformation e is the change in length and is 4 in.
3. The strain is

$$\epsilon = \frac{e}{L} = \frac{4}{14\,400} = 0.000278$$

Practice Problems (Section 3-5). Strain is simple and easily understood, but a few practice problems will help you to remember it.

1. A rod 48 in. long is stretched 0.018 in. Find the strain.
2. A post 8 ft long is compressed 1⁄32 in. What is the strain?
3. A rubber sample with an original length of 3 in. is stretched to a final length of 15 in. Calculate the strain.
4. A steel plate with an original thickness of 16 mm is compressed to a final thickness of 15.99 mm. How much is the strain?
5. A steel bar 50 mm long is stretched 0.041 mm. Find the strain.
6. A steel tension specimen 2 in. long is stretched by an amount of 0.0035 in. What is the strain?

3-6. STRAIN IN SHEAR

Deformation in shear looks quite different from deformation in tension and compression, but the formula for strain is the same in each case. The stud or projection in Fig. 3-29*a* deforms to the shape in Fig. 3-29*b* when the force P is applied. This deformation is due to small amounts of sliding on a great many horizontal planes, and the picture in Fig. 3-29*c* represents this. The picture shows sliding on only a few planes, although it actually occurs on a great many, but it presents the correct

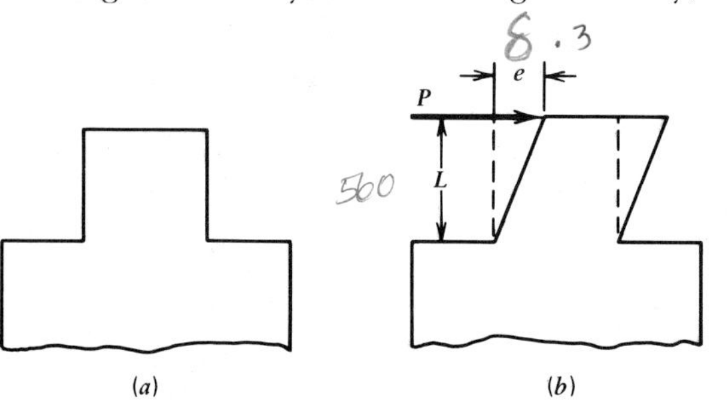

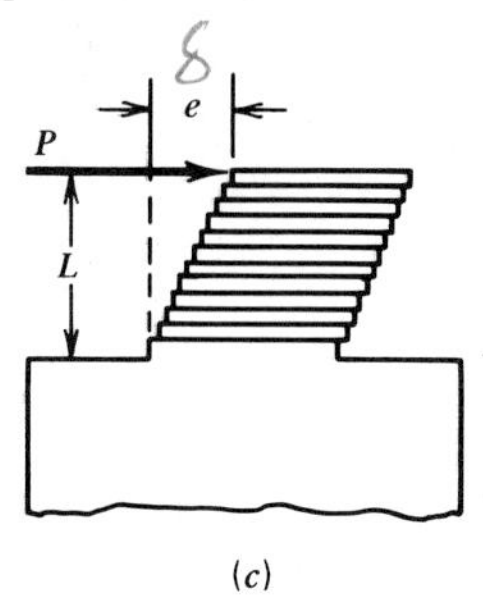

Fig. 3-29. Deformation in shear. (*a*) Before loading. (*b*) After loading. (*c*) Sliding.

idea of deformation in shear, which is that of sliding or slipping. There are so many planes on which sliding occurs and they are so close together that the view to the observer is as shown in Fig. 3-29*b*. Here the distance e is the deformation, and L is the length in which it occurs. (Notice that the deformation is perpendicular to the length in which it occurs in shear; it is parallel in tension and compression.) The strain in shear is equal to the deformation divided by the length in which it occurs, so we have the same formula for it.

$$\epsilon_s = \frac{e}{L}$$

where the subscript s designates the strain as shearing strain.

Illustrative Example 1. Figure 3-30*a* shows a sandwich-type motor mounting. The shaded parts are rubber, and the other parts are steel to which the rubber is bonded. Under a load P the center steel plate is pushed down 0.15 in. as shown in Fig. 3-30*b*. What is the shearing strain in the rubber?

Solution:

1. The length L is ¼ in. or 0.25 in.
2. The deformation e is 0.15 in.
3. The shearing strain is

$$\epsilon_s = \frac{e}{L} = \frac{0.15}{0.25} = 0.6$$

Illustrative Example 2. A steel bar is given a shearing deformation of 0.300 mm in a length of 560 mm. Find the shearing strain.

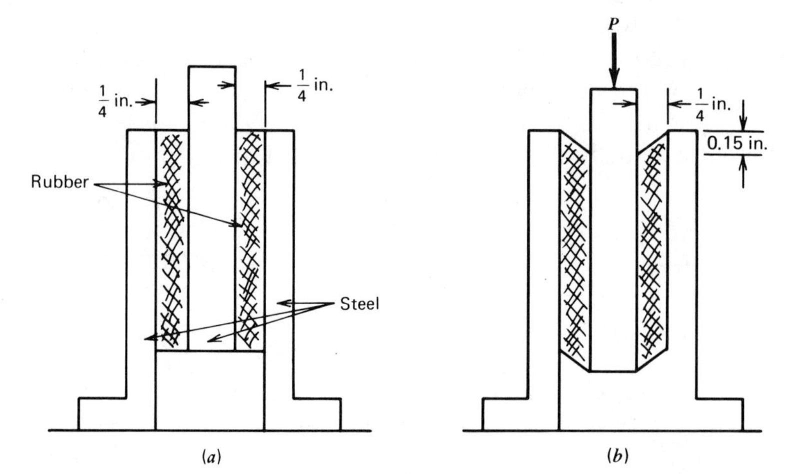

Fig. 3-30. Shear deformation in motor mounting. (*a*) Before loading. (*b*) After loading.

Solution:

1. The length L is 560 mm.
2. The deformation e is 0.300 mm.
3. The shearing strain is

$$\epsilon_s = \frac{e}{L} = \frac{0.300}{560} = 0.000536$$

Practice Problems (Section 3-6). It will help you to remember shearing strain if you work these practice problems.

1. What strain results from a shearing deformation of 0.0568 in. in a length of 72 in.?
2. A straight pin is deformed to the shape in Fig. 3-31. Calculate the shearing strain.
3. Figure 3-32*a* shows a layer of cement between two blocks of wood. The layer of cement is deformed to the shape in Fig. 3-32*b*. Find the shearing strain.
4. The steel bar in Fig. 3-33 is twisted so that the free end rotates through an angle of 2°. Find the shearing strain at the surface of the bar.

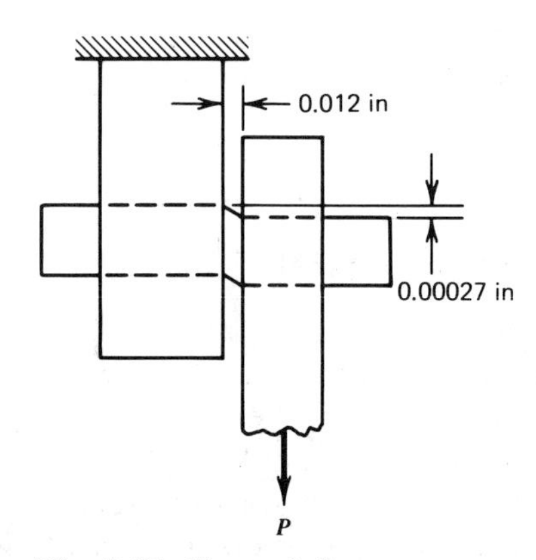

Fig. 3-31. Shear deformation in pin.

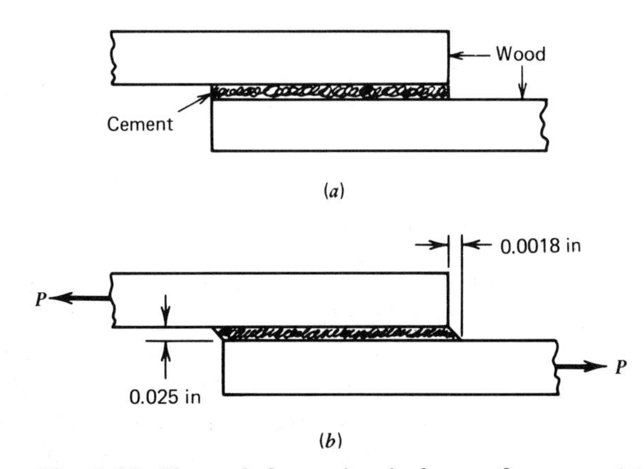

Fig. 3-32. Shear deformation in layer of cement. (*a*) Before loading. (*b*) After loading.

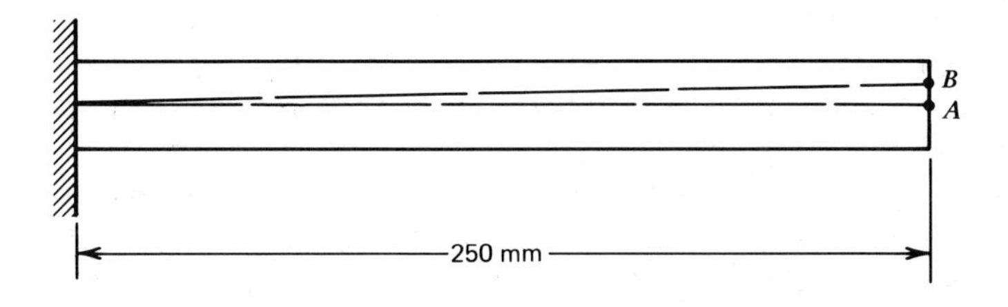

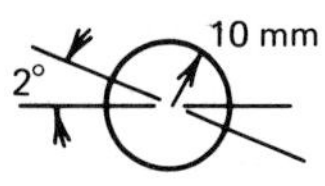

Fig. 3-33. Twisted bar.

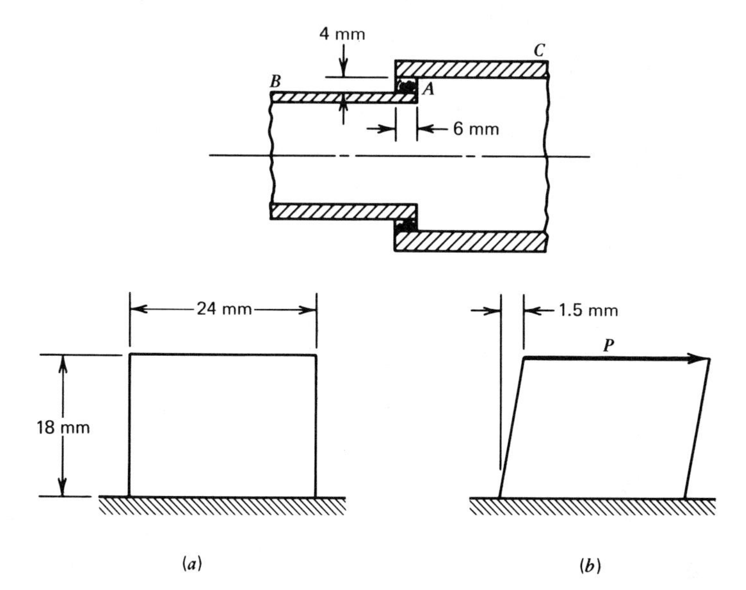

Fig. 3-34. Pipe seal.

Fig. 3-35. Shear deformation in block. (*a*) Before loading. (*b*) After loading.

5. Figure 3-34 shows a rubber seal *A* between a pipe *B* and a coupling *C*. The coupling moves 2 mm longitudinally with respect to the pipe. Find the shearing strain in the seal.
6. The plastic block in Fig. 3-35*a* is deformed to the shape in Fig. 3-35*b*. Find the shearing strain.

3-7. RELATION BETWEEN STRESS AND STRAIN

Stress and strain nearly always occur together. Whenever a material is subjected to stress, it deforms, and whenever a material is deformed, there must be strain. Fortunately, there is a relation between the stress and strain for any given material, and when this relation has been determined, the stress can be found if the strain is known, or the strain can be found if the stress is known. The relation between stress and strain is not the same for all materials and is found by experiment. (Enough experiments have been performed on the common materials, such as steel and its alloys, cast iron, and aluminum alloys, so that their properties are well known.) The most common type of test is a tension test in which a bar is pulled apart in tension. The load is increased by steps from zero until the bar breaks, and the deformation is measured at each step or different load. Values of stress and strain are calculated for each load, and then the stress and strain are plotted in a stress-strain curve. Stress is plotted vertically and strain horizontally, as in Fig. 3-36.

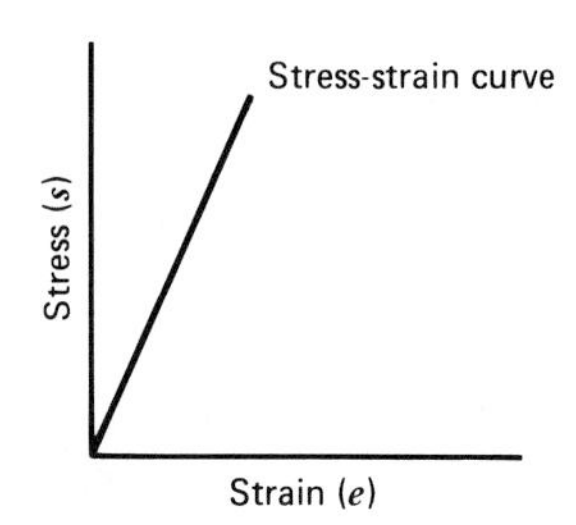

Fig. 3-36. Part of a stress-strain curve.

Then a line is drawn through the plotted points to give the stress-strain curve. As long as the stress is not too great, the stress-strain curve can be taken as a straight line for most materials. (It isn't always a straight line, but is straight enough for practical purposes and is almost always taken this way.) This straight-line relation means that the stress is proportional to the strain, and the statement of this relation is called *Hooke's law*. The mathematical relation between stress and strain is

$$S = E\epsilon$$

Here S is stress, ϵ is strain, and E is a constant called *Young's modulus* or the *modulus of elasticity*. (These are just names; don't worry about them.) Another way of writing the same equation is

$$\frac{S}{E} = \epsilon$$

The modulus of elasticity E is an important property of a material, since it relates the stress and strain so that one can be calculated when the other is known. It is usually expressed in pounds per square inch (psi). In the metric system, the modulus of elasticity is expressed in gigapascals (abbreviated as GPa). The prefix *giga* means 10^9, so a gigapascal is 10^9 pascals. Average values of E are given in Table 3.1 for a number of the common materials. (You need not attempt to memorize them, but you should remember where they can be found for future use.) You should notice that one value of E is given for tension and compression and a separate value for shear.

Let's look at the units of the quantities in Table 3.1. The modulus of elasticity for steel in tension or compression is expressed in units of 10^6 psi and the numerical value is 30. So, for steel in tension or compression, $E = 30 \times 10^6$ psi, which could also be written as $E = 30\ 000\ 000$ psi.

When this property of steel is expressed in the metric system, the value is $E = 20$ GPa, that is, 20 gigapascals. A gigapascal is equivalent to 10^9 pascals.

Illustrative Example 1. A steel rod is stretched to a strain of 0.0008. What is the tensile stress in psi?

Solution:

1. From Table 3.1, $E = 30 \times 10^6$ psi for steel in tension.
2. The strain ϵ is 0.0008.
3. The tensile stress is

$$S = E\epsilon = 30 \times 10^6 \times 0.0008 = 24\ 000 \text{ psi}$$

TABLE 3.1 AVERAGE VALUES OF MODULUS OF ELASTICITY FOR COMMON ENGINEERING MATERIALS

	Tension and Compression, E		*Shear, E_s*	
Material	*10^6 psi*	*GPa*	*10^6 psi*	*GPa*
Steel	30	207	12	82.7
Wrought iron	27	186	10	69.0
Malleable cast iron	24	165	10	69.0
Gray cast iron	15	103	6	41.4
Copper, cast	13	89.6	6	41.4
Copper, hard drawn	17	117	6	41.4
Brass (60% copper/40% zinc) cast	13	89.6	5	34.5
Bronze (90% copper/10% tin) cast	12	82.7	5	34.5
Aluminum and its alloys	10	69.0	4	27.6
Magnesium and its alloys	6.5	44.8	2.4	16.6
Titanium	16.5	114	6.6	45.5
Lead, rolled	1	6.90		
Yellow pine	1.65[a]	11.4		
Douglas fir	1.7	11.7		
Concrete	2–3[b]	13.8–20.7		
Rubber compounds	$(1.8–3.0) \times 10^{-4}$	$(1.24–2.04) \times 10^{-3}$	$(0.5–2) \times 10^{-4}$	$(0.345–1.38) \times 10^{-3}$
Epoxy cast resins, no filler	0.35	2.41		
Epoxy resins, glass fiber filled	3	20.7		
Acrylic	0.35–0.45	2.41–3.10		
Polystyrene, 20–30% glass fiber filled	0.84–1.29	5.79–8.90		
Phenolic, unfilled	0.75–1	5.17–6.90		
Polyethylene, molding grade	0.05–5	0.344–3.44		
Nylon, molding and extusion compounds	0.38	2.62		

[a]For a stress parallel to the grain.

[b]For compression only. Concrete is not ordinarily considered to have tensile strength.

Illustrative Example 2. A piece of magnesium is subjected to a shearing strain of 0.00105. Find the shearing stress in megapascals.

Solution:

1. From Table 3.1, $E_s = 16.6$ GPa for magnesium in shear.
2. The strain is $\epsilon_s = 0.00105$.
3. The shearing stress is

$$S_s = E_s\epsilon_s = 16.6 \times 0.00105$$
$$= 0.01743 \text{ GPa}$$
$$= 17.43 \text{ MPa}$$

Illustrative Example 3. What is the strain in an aluminum bar that is subjected to a compressive stress of 9 000 psi?

Solution:

1. From Table 3.1, $E = 10 \times 10^6$ psi for aluminum in compression.
2. The stress is $S = 9\,000$ psi.
3. The equation $S = E\epsilon$ can be solved for ϵ to give

$$\epsilon = \frac{S}{E} = \frac{9\,000}{10 \times 10^6} = 0.0009$$

RELATION BETWEEN FORCE AND DEFORMATION

Most problems are more comprehensive. For instance, we may know the force applied to a member and its dimensions and want to know the deformation. Or we may know the deformation and want to find the force. We can derive a formula for such problems by combining formulas used previously. We have used $\epsilon = e/L$, or $e = \epsilon L$. Then, we can use $\epsilon = S/E$ to get $e = SL/E$, but since $S = P/A$ we can arrive at

$$e = \frac{PL}{AE}$$

If we use the English gravitational system of units, we will have P in pounds, L in inches, A in square inches, and E in psi, so we will get the deformation e in inches.

Illustrative Example 4. A round steel bar 2 in. in diameter and 40 in. long is subjected to a tensile force of 33 000 lb. How much does it stretch?

Solution:

1. We want to find the deformation e, so we use the formula as,

$$e = \frac{PL}{AE}$$

2. The force P is 33 000 lb.
3. The length L is 40 in.
4. The area is circular, so

$$A = \frac{\pi d^2}{4} = \frac{\pi(2)^2}{4} = 3.14 \text{ in.}^2$$

5. From Table 3.1, $E = 30 \times 10^6$ psi.
6. The deformation is

$$e = \frac{PL}{AE} = \frac{33\,000 \times 40}{3.14 \times 30 \times 10^6} = 0.014 \text{ in.}$$

METRIC UNITS

In the metric system we will have P in newtons, L in millimeters, A in square millimeters, and E in gigapascals (10^9 pascals). If we multiply E by 10^3 to convert it from gigapascals to megapascals, we get e in millimeters. Let's try it.

Illustrative Example 5. A rectangular aluminum alloy bar is 720 mm long, 60 mm wide, and 20 mm thick. How much does it stretch when subjected to a force of 60 000 N?

Solution:

1. The force P is 60 000 N.
2. The length L is 720 mm.
3. The area is $A = 60 \times 20 = 1\,200$ mm^2.
4. From Table 3.1, $E = 69$ GPa, or 69×10^3 MPa.
5. Then,

$$e = \frac{PL}{AE} = \frac{60\,000 \times 720}{1\,200 \times 69 \times 10^3} = 0.522 \text{ mm}$$

CALCULATING THE FORCE

The formula

$$e = \frac{PL}{AE}$$

can be written in several different ways so that it can be solved for different factors. For instance, we can write it as

$$P = \frac{AEe}{L}$$

a formula we can use to find P when the other quantities are known.

Illustrative Example 6. Calculate the magnitude of the compressive force that must be applied to a rectangular, gray, cast-iron block, 150 mm by 200 mm and 400 mm long, to shorten it 0.0625 mm.

Solution:

1. Since we want to calculate the force, we use the formula

$$P = \frac{AEe}{L}$$

2. The deformation e is 0.0625 mm.
3. The area of the rectangular cross section is

$$A = 150 \times 200 = 30\,000 \text{ mm}^2$$

4. From Table 3.1, the modulus of elasticity E is 103 GPa for gray cast iron.
5. The length L is 400 mm.
6. Then, the compressive force P is

$$P = \frac{AEe}{L} = \frac{30\,000 \times 103 \times 10^3 \times 0.0625}{400}$$
$$= 482\,800 \text{ N}$$

FINDING THE LENGTH

Let's try another version of the formula

$$e = \frac{PL}{AE}$$

We can also write the formula as

$$L = \frac{AEe}{P}$$

We can use this to find the length (L) if we know the area (A), the modulus of elasticity (E), the deformation (e), and the load (P).

Illustrative Example 7. How long must a rectangular magnesium bar 50 mm × 10 mm be to stretch 0.55 mm under a tensile force of 28 000 N?

Solution:

1. Here we need the length, so we write the formula as

$$L = \frac{AEe}{P}$$

2. The deformation e is 0.55 mm.
3. The area is a rectangle, 50 mm × 10 mm

$$A = 50 \times 10 = 500 \text{ mm}^2$$

4. From Table 3.1, $E = 44.8$ GPa.
5. The force is $P = 28\ 000$ N.
6. The necessary length of the bar is

$$L = \frac{AEe}{P} = \frac{500 \times 44.8 \times 10^3 \times 0.55}{28\ 000} = 440 \text{ mm}$$

OTHER WAYS

There are two other ways in which we can write this formula. One of them is

$$A = \frac{PL}{Ee}$$

which we can use to find the area (A) when we know the other quantities; the other is

$$E = \frac{PL}{Ae}$$

which we can use to find the modulus of elasticity (E) when we know P, L, A, and e. You can satisfy yourself that they are correct by deriving them from

$$e = \frac{PL}{AE}$$

Practice Problems (Section 3-7). Practice on the following problems. When you need a modulus of elasticity, take it from Table 3.1.

1. A circular brass rod 7/8 in. in diameter and 4 in. long is subjected to a tensile force of 4 700 lb. How much does it stretch?
2. A circular copper tube (hard drawn), 36 mm outer diameter, 30 mm inner diameter, and 260 mm long, is subjected to a compressive force of 4 300 N. How much does it shorten?

3. What tensile force must be applied to a rectangular aluminum strip 2 in. by ⅛ in. and 12 in. long to stretch it 0.001 in.?
4. The shearing stress in a magnesium casting is 26.5 MPa. Find the shearing strain.
5. A steel rod 420 mm long is stretched 0.032 mm by a tensile force of 9 000 N. What is the area of the cross section of the rod?
6. The shearing strain in a malleable cast-iron machine part is 0.00077. Find the shearing stress in psi.
7. What compressive force must be applied to a steel cylinder 6 in. in diameter and 14 in. long to shorten it 0.0175 in.?
8. A circular steel rod 12 mm in diameter is stretched 4.8 mm by a tensile force of 16 200 N. How long is the rod?

3-8. PROPORTIONAL LIMIT AND ULTIMATE STRENGTH

We have already studied the behavior of materials under load for low values of stress, and we have seen that it is reasonable to say that stress is proportional to strain as long as the stress is not too high. Now let's see how a material behaves under larger stresses.

The most common type of test is a tension test in which a tensile force is applied to a bar and is increased in steps until the bar breaks. The stress and strain are calculated for each load, and then the values of stress are plotted against the values of strain to give a stress-strain curve; stress is plotted vertically and strain horizontally. The stress-strain curve that results provides quite a bit of information about the behavior of the material under load.

We saw part of a stress-strain curve in Fig. 3-36, and now Fig. 3-37 shows the complete stress-strain curve for a bar of low-carbon steel in tension. For low values of stress, the curve is the straight line AB. Stress is proportional to strain in this region, and the relation between them is

$$S = E\epsilon$$

At point B there is a transition from the straight line AB to the curve BC. Stress is not proportional to strain for any point on the curve BC, so the stress at point B is the greatest stress for which stress is proportional to strain. It marks the limit of proportionality of stress to strain, and the stress at this point is called the *proportional limit.* (The value is about 35 000 psi for low-carbon steel.)

At point C the stress-strain curve becomes horizontal. This point is called the yield point. Then, as the stress is increased, there is a great increase in deformation. The high point, D, of the curve represents the greatest stress that can be applied to the material, a stress sufficient to

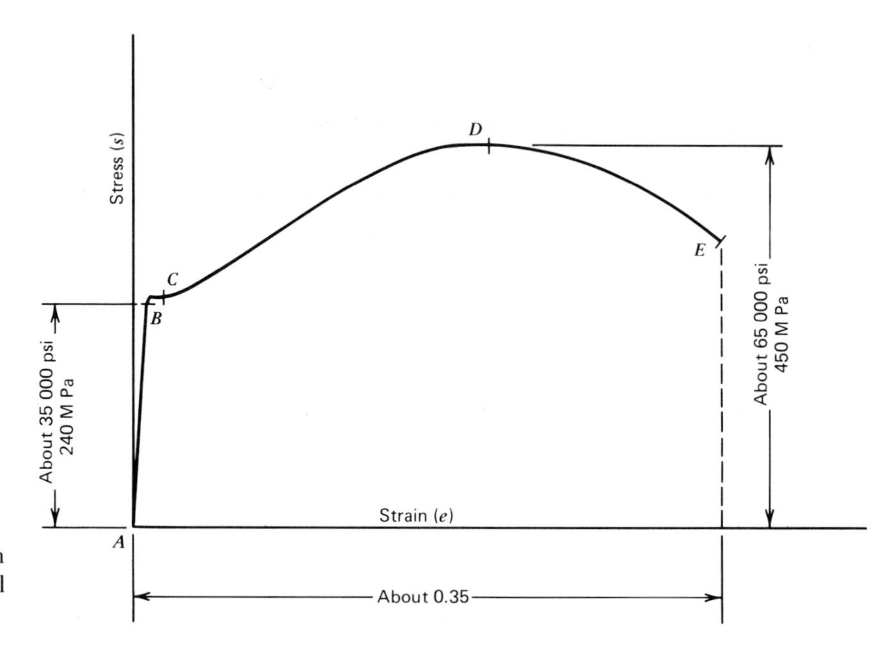

Fig. 3-37. Stress-strain curve for low-carbon steel in tension.

break it. This value of the stress is called the *ultimate strength*. (The value is about 60 000 psi for low-carbon steel.)

As the test continues, the stress drops off a bit while the bar continues to stretch, until it breaks at the point E. (The value of the strain at E is about 0.35 for low-carbon steel.) The value of the strain at which the bar breaks is an important characteristic of a material; a material that stretches quite a bit before breaking is called a *ductile material*, and one that breaks with practically no stretching is called a *brittle material*. Soft steel and copper are often cited as examples of ductile materials, whereas cast iron and chalk are often cited as examples of brittle materials. The numerical figure usually given as a measure of ductility is the elongation (same as deformation, e) in a 2 in. length ($L = 2$ in.) at failure.

Figure 3-38 shows stress-strain curves for a number of the common materials. It is interesting here to compare strengths and to see how much stronger high-carbon steel is than low-carbon steel or magnesium and also to see how much more ductile low-carbon steel is than high-carbon steel or cast iron. It is to be noticed that for some materials there is no straight-line part of the curve. Stress is never *exactly* proportional to strain for such materials, but the proportion is near enough so that it is satisfactory for practical purposes to assume that it is. Even so, no value can be obtained for the proportional limit, because the stress is never proportional to the strain. In any such case a *yield strength* is found instead. The way to find the yield strength is to draw a straight line AB (as in Fig. 3-39) tangent to the stress-strain curve at the origin, then to offset 0.002 along the strain axis to point C, and next draw the line CD

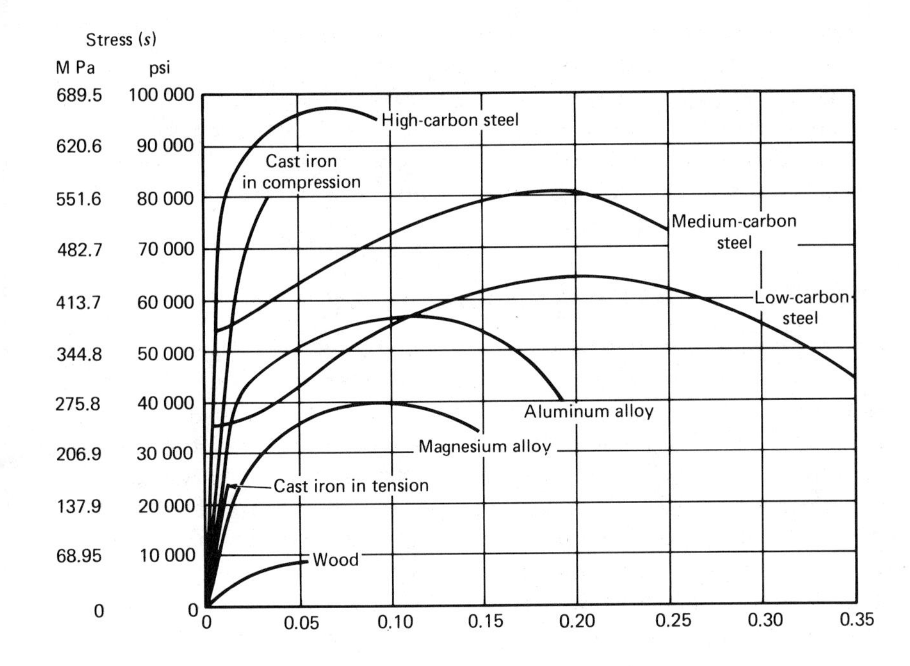

Fig. 3-38. Several stress-strain curves.

parallel to the line *AB*. The value of the stress at the point where the line *CD* intersects the stress-strain curve is the yield strength.

You are not expected to take numerical values of stress and strain from Fig. 3-38. It is included here to give you an idea of how the different

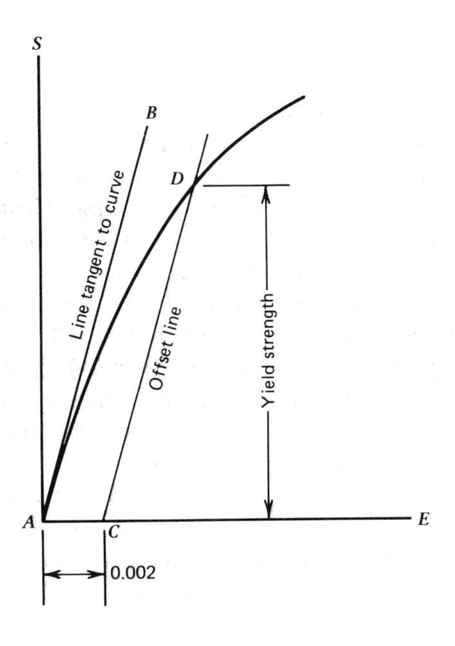

Fig. 3-39. Offset method.

materials compare with one another. Numerical values of yield point and ultimate strength should be taken from tables.

Table 3.2 gives values of the proportional limit, ultimate strength, and elongation for a number of the common engineering materials. (This table need not be memorized, but you should study it enough to know what it contains and remember where it is for future reference.)

3-9. WORKING STRESS AND FACTOR OF SAFETY

One of the most important questions in machine and structural design is: to what stress should the material be subjected in service? The stress the material is expected to bear in service, and which is used in design calculations, is called the *working stress*. A few general facts about working stress are obvious. Certainly the working stress is going to be less than the ultimate strength of the material, because the member is going to break if the stress reaches the ultimate. Also, it is pretty sure to be well below the proportional limit of the material, because most materials undergo more deformation than would be permissible in a member when the stress goes beyond the proportional limit. (Why hold dimensions of machine parts to 0.001 in. in the shop and then stress them enough to change a dimension by 0.01 in.?)

It isn't difficult for us to agree that the working stress must be well below the proportional limit of the material. Now let us be sure that we know what we mean by *working stress* so we can go a little farther. The *working stress* is the stress to which we expect to subject the material in service. If everything works out exactly as expected (it probably won't), the stress in the machine part will be equal to the working stress. But there are always a number of uncertainties in a design problem. The first, and one of the greatest, is the load. It is difficult to decide exactly what load is going to be applied to a member in service. Ordinarily, the figure used in design is just a good guess. A second uncertainty is in regard to the formulas used in calculating stress. Most of the formulas we use are not exactly correct, only good approximations. Sometimes the approximation is especially good and sometimes not too good. At any rate, there is some chance for error in the formulas we use. Finally, there is the material itself. The figures in Table 3.2 represent average values from a large number of tests, as the heading shows. Some individual members may be stronger than these figures indicate, but some are going to be weaker; to be safe, we must assume always that the member in question is weaker than the average.

In spite of the uncertainties previously mentioned, the designer must make sure that the stress in the member in service remains below the proportional limit and well below the ultimate strength. To do this, he or she uses a *factor of safety* and divides the ultimate strength by this

TABLE 3.2 AVERAGE VALUES OF STRENGTH AND ELONGATION FOR COMMON ENGINEERING MATERIALS

	Proportional Limit						*Ultimate Strength*						
	Tension		*Compression*		*Shear*		*Tension*		*Compression*		*Shear*		
Material	*10^3 psi*	*MPa*	*10^3 psi*	*MPa*	*10^3 psi*	*MPa*	*10^3 psi*	*MPa*	*10^3 psi*	*MPa*	*10^3 psi*	*MPa*	*Elongation %*
Steel, 0.2% carbon													
hot rolled	35	241	35	241	21	145	60	414	90	621	45	310	35
Annealed casting	30	207	30	207	18	124	60	414	[a]	[a]	45	310	30
Steel, 0.6% carbon													
hot rolled	60	414	60	414	36	248	100	690	[a]	[a]	80	552	15
Steel, 1% carbon													
hot rolled	80	552	80	552	48	331	135	931	[a]	[a]	115	793	10
oil quenched	135	931	135	931	80	552	320	2206	[a]	[a]	185		1
Wrought iron	30	207	30	207	18	124	50	345	60	414	40	276	30
Gray cast iron ASTM A48 grade 20	—	—	—	—	—	—	20	138	80	552	32	221	—
Copper													
annealed	3.2	22	3.2	22	1.9	13	32	221	[a]	[a]	—	—	56
hard-drawn	38	262	38	262	23	159	55	379	[a]	[a]	—	—	4
Aluminum alloy 6061-T6	40	276	—	—	—	—	45	310	—	—	30	207	17
Magnesium alloy ASTM AZ63A-T6	19	131	—	—	—	—	40	276	—	—	—	—	5
Acrylic, cast	—	—	—	—	—	—	8–11	55–76	11–19	76–131	—	—	2–7
Epoxy resin, glass fiber filled	—	—	—	—	—	—	10–20	7–14	25–40	172–276	—	—	
Polyethelene molding grades	—	—	—	—	—	—	1.6–4.8	11–33	2–5.5	14–38	—	—	10–440
Polyerethane cast unsaturated	—	—	—	—	—	—			—	—	—	—	3–6

[a]For practical purposes, this can be taken as equal to the proportional limit in tension.

factor to obtain the working stress. The factor of safety can be represented by f, so the relation is

$$\text{working stress} = \frac{\text{ultimate strength}}{f}$$

or

$$f = \frac{\text{ultimate strength}}{\text{working stress}}$$

The factor of safety varies from about 1.5 to about 10, depending on the material, the type of loading, the judgment of the individual designer, and the industry involved. The factor of safety is supposed to take care of all the uncertainties in stress calculations so that even though the actual stress in the material turns out to be greater than the working stress, it is still below the ultimate strength of the material. It is not uncommon for the factor of safety to be about 4 for static loading where the load is applied gradually and remains constant. When the load is repeated a great many times or is applied with shock, the factor of safety may be as high as 8 or 10. Factors of safety as low as 1.2 are sometimes used for aeronautical structures. Of course, the higher the factor of safety, the lower the working stress, and vice versa.

The wide variety of uses of engineering materials and the vast changes that can be made in their properties by heat treatment and cold working make it impossible to give anything like a complete list of working stresses.

Table 3.3 gives a few values, but this table is not to be regarded as complete. The logical thing for the designer to do is to get in touch with the manufacturer of the material, who will be glad to furnish complete information of the properties of the material and to make recommendations for the working stress. Many manufacturers issue handbooks containing this information.

3-10. HOW TO DETERMINE THE SIZE OF MEMBERS

The most important use of the equation for simple stress is in determining the size of the member or machine part. We have written the equation as,

$$S = \frac{P}{A}$$

where S is the stress, P is the load, and A is the area of the cross section. Usually P and S are known and the problem is to find the size of the member. To do this we write the equation in the form,

TABLE 3.3 A FEW VALUES OF WORKING STRESSES

A. Static Loading

Material	*Tension*		*Compression*		*Shear*	
	psi	*MPa*	*psi*	*MPa*	*psi*	*MPa*
Low-carbon steel	12 000–24 000	83–166	12 000–24 000	83–166	8 000–16 000	55–110
Medium-carbon steel	16 000–30 000	110–207	16 000–30 000	110–207	12 000–20 000	83–138
Cast steel	8 000–15 000	55–103	8 000–15 000	55–103	6 000–12 000	41– 83
Cast iron	3 000– 4 000	21– 28	10 000–16 000	70–110	3 000– 4 000	21– 28

B. Repeated or Shock Loading

Material	*Tension*		*Compression*		*Shear*	
	psi	*MPa*	*psi*	*MPa*	*psi*	*MPa*
Low-carbon steel	6 000–12 000	42– 84	6 000–12 000	42– 84	4 000– 8 000	28–56
Medium-carbon steel	8 000–15 000	55–103	8 000–15 000	55–103	6 000–12 000	42–84
Cast steel	4 000– 7 500	28– 52	4 000– 7 500	28– 52	3 000– 6 000	21–42
Cast iron	1 500– 2 000	10– 14	5 000– 8 000	35– 55	1 500– 2 000	10–14

$$A = \frac{P}{S}$$

We then solve the equation for A. There are usually requirements that determine the shape (a bolt is circular, for instance), and we can solve for the dimensions. In another case, we may decide that we want to cut the member out of a plate that is ¼ in. thick. Then we know the thickness, and we have to find the width.

Illustrative Example 1. A square tension member is to carry a load of 48 000 lb, and the working stress is 18 000 psi. Find the size required for the member.

Solution:

1. The load P is 48 000 lb.
2. The stress S is 18 000 psi.
3. The area required is

$$A = \frac{P}{S} = \frac{48\,000}{18\,000} = 2.67 \text{ in.}^2$$

4. Let one side of the square cross section be represented by d. The area of the cross section is d^2, so

$$d^2 = A = 2.67$$

$$d = \sqrt{2.67} = 1.63 \text{ in.}$$

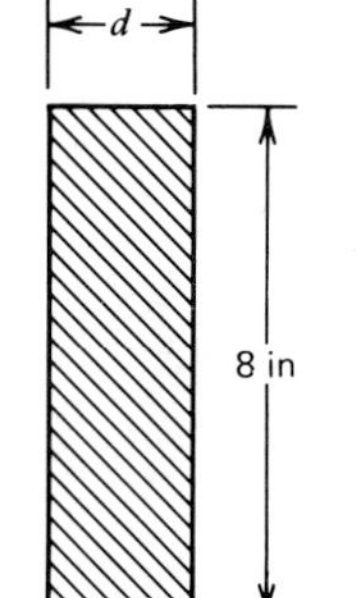

Fig. 3-40. Cross section of machine frame member.

Illustrative Example 2. A cast-iron compression member in a machine frame is to have the rectangular cross section shown in Fig. 3-40; here d is an unknown quantity and is to be found. The load is 160 000 lb, and the working stress is 9 000 psi. Find d.

Solution:

1. The load P is 160 000 lb.
2. The stress S is 9 000 psi.
3. The area required is

$$A = \frac{P}{S} = \frac{160\,000}{9\,000} = 17.78 \text{ in.}^2$$

4. The area of the rectangular cross section is $8d$, so

$$8d = A = 17.78 \qquad d = 2.22 \text{ in.}$$

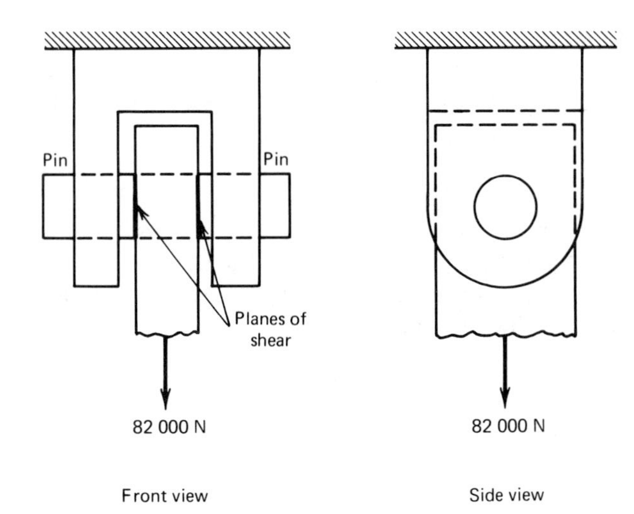

Fig. 3-41. Pin in shear.

Illustrative Example 3. Figure 3-41 shows two views of a bar that is fastened to a bracket by means of a pin. The working stress for the pin in shear is 52 MPa. What diameter pin should be used?

Solution:

1. The load P is 82 000 N.
2. The stress S is 52 MPa = 52 N/mm².
3. Next we find the area needed in shear. This is

$$A = \frac{P}{S} = \frac{82\,000}{52} = 1\,577 \text{ mm}^2$$

4. The pin is in shear on two planes as shown in the front view of Fig. 3-41, so there are two cross sections of the pin in shear. Each cross section provides half of the area, so each must be

$$\frac{1\,577}{2} = 788.5 \text{ mm}^2$$

Let the diameter of the pin be d. The area of the circle is $\pi d^2/4$, so

$$\frac{\pi d^2}{4} = 788.5$$

and

$$d^2 = \frac{4}{\pi} \times 788.5 = 1004$$

$$d = \sqrt{1004} = 31.7 \text{ mm}$$

NET SECTION

When the cross section is not the same throughout the length of the member, the smallest cross section must be used. This is called the *net section*.

Illustrative Example 4. Figure 3-42 shows a tension member that is loaded by pins through holes in the ends. The bar is 12 mm thick, and the working stress in tension is 130 MPa; the holes are 25 mm in diameter. Find the dimension d.

Solution:

1. The load P is 76 000 N.
2. The stress S is 130 MPa = 130 N/mm².
3. The area required is

$$A = \frac{P}{S} = \frac{76\ 000}{130} = 584.6 \text{ mm}^2$$

4. The smallest cross section is the one through the center of the hole. This is shown in Fig. 3-42, where the shaded areas carry stress. The net area is regarded as a positive rectangle d inches high and a negative rectangle 25 mm high. The area of the large rectangle is $12d$, and the area of the negative rectangle is 25×12, so

$$A = 12d - 25 \times 12 = (d - 25)12$$

This expression for A is equated to the 584.6 that we found we needed in part 3.

$$(d - 25)12 = 584.6$$

Now let's divide by 12:

$$d - 25 = \frac{584.6}{12} = 48.72$$

and

$$d = 25 + 48.72 = 73.72 \text{ mm}$$

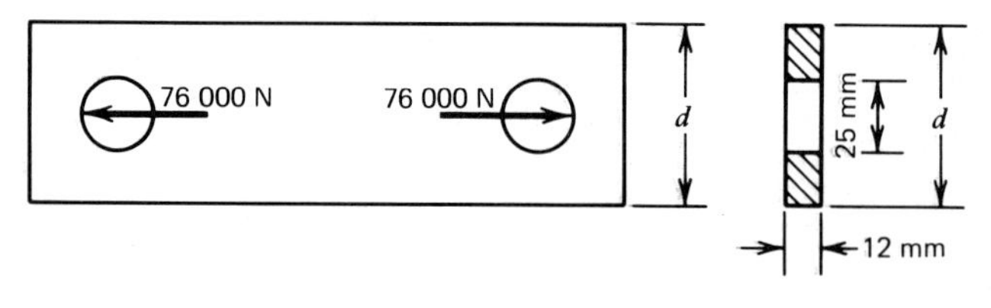

Fig. 3-42. Bar in tension.

HOLLOW MEMBERS

Heavy machine frames are usually hollow for greater economy and better qualities of the casting. This creates interesting problems in designing sections.

Illustrative Example 5. The cross section in Fig. 3-43*a* is to carry a load of 820 000 lb in compression. The wall thickness is 1.5 in. as shown, and the working stress is 12 000 psi. Find the dimension d.

Solution:

1. The load P is 820 000 lb.
2. The stress S is 12 000 psi.
3. The area required is

$$A = \frac{P}{S} = \frac{820\ 000}{12\ 000} = 68.33 \text{ in.}^2$$

4. The area can be considered as a large rectangle 10 in. wide and d inches high, minus a small rectangle 7 in. wide and $(d - 3)$ inches high. See Fig. 3-43*b*. The area of the large rectangle is $10d$, and the area of the small rectangle is (product of width and height)

$$7(d - 3) = 7d - 21$$

Then the area A is

$$A = 10d - (7d - 21) = 10d - 7d + 21$$
$$= 3d + 21$$

The area was calculated as 68.33 in.^{2} in part 3, so

$$3d + 21 = 68.33$$

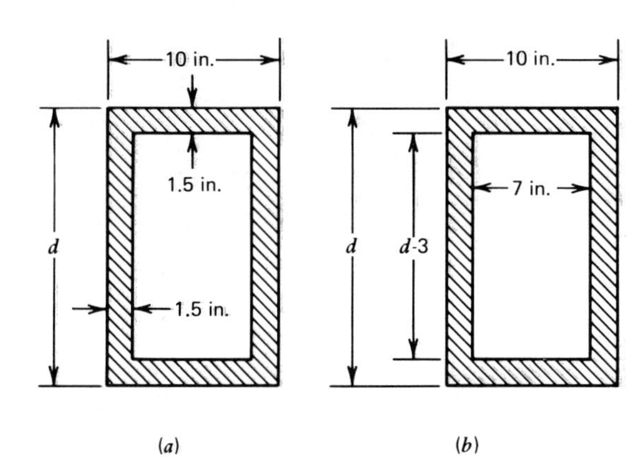

Fig. 3-43. Cross section of member in compression.

Let's transpose the 21. Then,

$$3d = 68.33 - 21 = 47.33$$

Then we divide by 3,

$$d = 15.78 \text{ in.}$$

Practice Problems (Section 3-9 and 3-10). Working the following problems will help you remember how to determine the size of members.

1. A circular area is to carry a load of 11 000 lb in shear with a working stress of 12 000 psi. What diameter is required?
2. A rectangular area 48 mm long is to carry a load of 13 600 N in shear with a working stress of 44 MPa. Find the width of the rectangle.
3. The tension member in Fig. 3-44 is ½ in. thick and the working stress is 16 000 psi. Find the maximum width of the slot in the center.
4. A steel wire must carry a load of 6 300 N in tension with a working stress of 180 MPa. What diameter is required?
5. A square compression member is to carry a load of 380 000 lb with a working stress of 8 600 psi. Find the size of the square.
6. Figure 3-45 shows the cross section of a cast-iron compression member. The load is 900 000 N, and the working stress is 70 MPa. Find the dimension t.

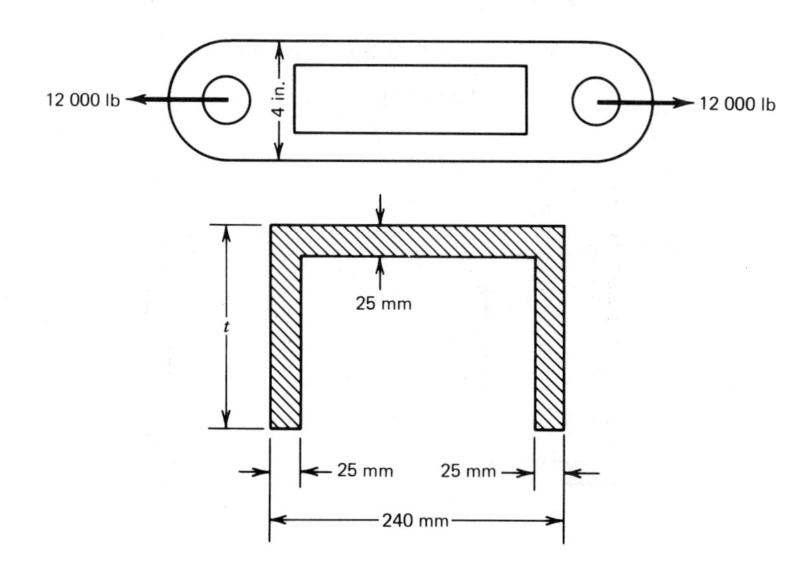

Fig. 3-44. Slotted bar in tension.

Fig. 3-45. Cross section of a compression member.

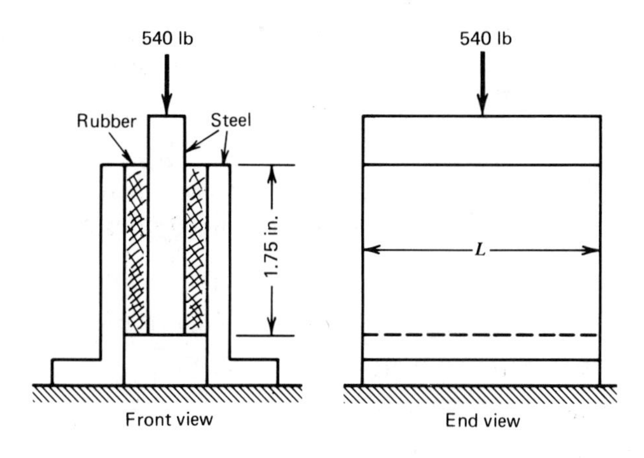

Fig. 3-46. Motor mounting.

7. Figure 3-46 shows two views of a motor mounting in which rubber is used in shear. The shaded parts are of rubber, and the other parts are of steel to which the rubber is bonded. The working stress in shear in the rubber is 60 psi. Find the dimension L.
8. A circular steel tube in compression is to carry a load of 39 000 N with a working stress of 26 MPa. The inner diameter is to be 50 mm. Find the outer diameter.

3-11. THERMAL DEFORMATION AND STRESS

It is one of the facts of nature that most materials expand when heated and contract when cooled. The amount of the expansion or contraction is expressed by the *coefficient of expansion* α (Greek letter alpha), which is the unit deformation per degree of temperature change. The total unit deformation is the product of α, and the temperature change ΔT(Δ is the Greek letter delta), thus $\alpha\Delta T$. Table 3.4 gives values of α for some of the common materials in both systems of units.

TABLE 3.4 COEFFICIENTS OF EXPANSION

Material	$\alpha \times 10^6$ *in./in./°F*	$\alpha \times 10^6$ *mm/mm/°C*
Aluminum alloys	12.5	22.5
Low-carbon steel	6.6	11.9
Stainless steel	6.1	11.0
Magnesium alloys	14.5	26.1
Copper-base alloys	10.0	18.0
Concrete	6.2	11.2
Gray cast iron	6.0	10.8

The change in length of a member is calculated by multiplying the unit deformation by the length of the member. It is $\alpha \Delta T L$. Now, let's try a couple of examples.

Illustrative Example 1. A steel rod 30 in. long is subjected to a temperature rise of 120°F. What is the change in length?

Solution:

1. From Table 3.4, $\alpha = 6.6 \times 10^{-6}$ in./in./°F.
2. The temperature change is $\Delta T = 120$°F.
3. The length L is 30 in.
4. Then the change in length is

$$\begin{aligned} \Delta L = \alpha \Delta T L &= 6.6 \times 10^{-6} \times 120 \times 30 \\ &= 23\,760 \times 10^{-6} \text{ in.} \\ &= 0.02376 \text{ in.} \end{aligned}$$

Illustrative Example 2. A copper wire 2 m long is subjected to a drop in temperature of 60°C. Find the change in length.

Solution:

1. From Table 3.4, $\alpha = 18 \times 10^{-6}$ mm/mm/°C.
2. The temperature change is $\Delta T = -60$°C.
3. The length is 2 m = 2 000 mm.
4. Then, the change in length is

$$\begin{aligned} \Delta L &= 18 \times 10^{-6} \times (-60) \times 2\,000 \\ &= -2\,160\,000 \times 10^{-6} \text{ mm} \\ &= -2.16 \text{ mm} \end{aligned}$$

STRESS

A member of a machine or structure is fastened to other members. These other members prevent free expansion because of temperature change, so that stresses occur when the temperature changes. The easiest way to calculate the stress is to regard the deformation due to temperature change as a strain; thus ϵ_t, where the subscript t indicates that this strain is due to temperature change. Then, the total strain is the sum of ϵ_t and the strain caused by stress. Let's try an example or two.

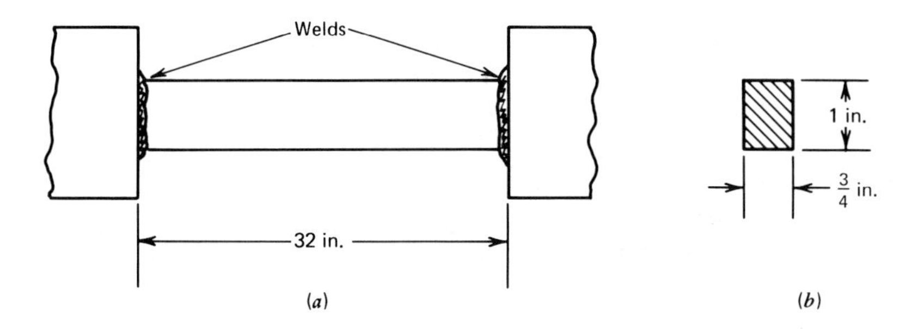

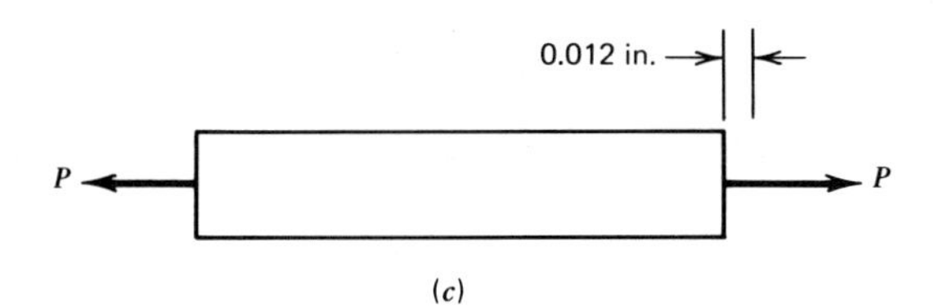

Fig. 3-47. Welded bar. (*a*) Welded bar. (*b*) Section. (*c*) Shortened bar.

Illustrative Example 3. Figure 3-47*a* shows a steel bar which is welded to other parts of a machine so that the bar cannot contract freely when the temperature drops. The decrease in length is limited to 0.012 in. when the temperature drops by 90°F, as shown in Fig. 3-47*c*. Find the force P.

Solution:

1. From Table 3.4, $\alpha = 6.6 \times 10^{-6}$ in./in./°F.
2. The temperature change is $\Delta T = -90$°F.
3. The length L is 32 in.
4. The change in length because of the temperature change is

$$\Delta L = \alpha \Delta T L = 6.6 \times 10^{-6} \times (-90) \times 32$$
$$= -19\,000 \times 10^{-6} \text{ in.}$$
$$= -0.0190 \text{ in.}$$

5. The force P must stretch the bar enough to limit the decrease in length to 0.012 in. So the deformation due to the force P is

$$e = 0.0190 - 0.0120 = 0.0070 \text{ in.}$$

6. The area of the bar is $\frac{3}{4} \times 1 = 0.75$ in.^{2}
7. The modulus of elasticity is 30×10^6 psi for steel.
8. Now, using a formula we already know,

$$P = \frac{AEe}{L} = \frac{0.75 \times 30 \times 10^6 \times 0.0070}{32} = 4920 \text{ lb}$$

Illustrative Example 4. An aluminum alloy tube is subjected to a temperature rise of 110°C. The length cannot change because of restrictions at the ends of the tube. Find the stress in the tube.

Solution:

1. From Table 3.4, $\alpha = 22.5 \times 10^{-6}$ mm/mm/°C.
2. The temperature change is $\Delta T = 110$°C.
3. The strain due to temperature change is

$$\epsilon_t = \alpha \Delta T = 22.5 \times 10^{-6} \times 110 = 2\,475 \times 10^{-6} \text{ mm/mm}$$

4. There must be a compressive strain of the same amount as ϵ_t. So $\epsilon = 2\,475 \times 10^{-6}$ mm/mm.
5. The modulus of elasticity is 68.95 GPa or 68.95×10^3 MPA or 68.95×10^3 N/mm^2.
6. Then the stress is

$$S = E\epsilon = 68.95 \times 10^3 \times 2\,475 \times 10^{-6}$$
$$= 170\,700 \times 10^{-3} \text{ MPa}$$
$$= 170.7 \text{ MPa}$$

Practice Problems (Section 3-11). Here are a few problems of thermal deformation for you to work.

1. What will be the change in length of a stainless steel rod 80 in. long when there is a temperature drop of 70°F?
2. A cast-iron block has a length of 31 in. How much will the length change with a temperature rise of 106°F?
3. A magnesium alloy bar is 180 mm long. Find the change in length with a temperature drop of 48°C.
4. A circular concrete column has a diameter of 540 mm. How much will the diameter increase if the temperature rises 84°C?
5. Figure 3-48 shows a low-carbon steel bar that is welded to a body *A*. There is a gap between the bar and a second body *B*. Both *A* and *B* may be considered rigid. Find the stress in the bar when the temperature rises 210°F.
6. A magnesium alloy bar is fastened to supports at the ends. These supports can be considered to be rigid. There is no stress in the

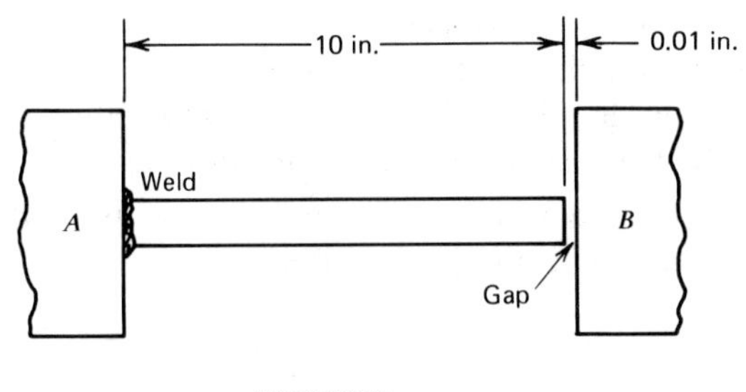

Fig. 3-48. Bar with gap.

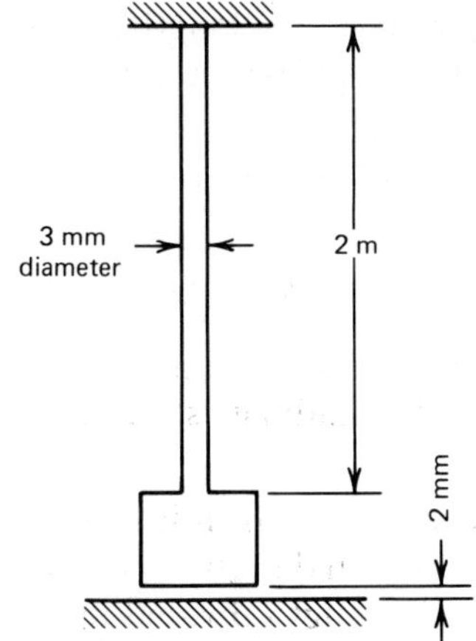

Fig. 3-49. Body supported by wire.

bar at a temperature of 60°F. What is the stress when the temperature is 130°F?

7. Figure 3-49 shows a low-carbon steel wire that supports a body. The weight of the body is 800 N. How much must the temperature rise for the stress in the bar to be 20 MPa?
8. An aluminum alloy bar is subjected to a temperature rise of 90°C. What stress must develop to keep the bar from changing length?

SUMMARY

Here is a brief summary of the material of Chapter 3.

1. Stress is force divided by area and is usually expressed in pounds per square inch, which is abbreviated as psi, or in megapascals, which is abbreviated as MPa.

   (a) The stress is $S = \dfrac{P}{A}$

   (b) The load is $P = AS$

   (c) The area is $A = \dfrac{P}{S}$

2. There are three simple stresses.
   (a) Tension, in which the material is stretched.
   (b) Compression, in which the material is shortened or compressed.
   (c) Shear, which resists a sliding deformation.
3. Strain is unit deformation and is equal to the deformation divided by the length in which it occurs. Thus

$$\epsilon = \frac{e}{L}$$

   The deformation is parallel to the length in which it occurs in tension and compression, but it is perpendicular to the length in which it occurs in shear.
4. For low values of stress it is satisfactory to say that stress is proportional to strain. Thus

$$S = E\epsilon$$

5. The proportional limit is the maximum stress for which stress is proportional to strain. The ultimate strength is the breaking strength of the material.
6. Working stress is the stress to which it is planned to subject the material and is the stress used in design calculations. The factor of safety is equal to the ultimate strength divided by the working stress.
7. The unit deformation due to a temperature change is expressed by the coefficient of thermal expansion α. The total deformation due to temperature change is

$$\Delta L = \alpha \Delta T L$$

REVIEW QUESTIONS Answer the following questions from memory. Try not to refer back to the chapter for your answers.

1. What is stress?
2. What are the units in which stress is expressed?
3. Give the meaning of each term in the formula,

$$S = \frac{P}{A}$$

4. How would you write the formula to solve for load?
5. How would you write the formula to solve for the area?
6. What is tension?
7. What is compression?
8. What is shear?

9. What is strain?
10. How does strain in tension and compression differ from strain in shear?
11. What is the relation between stress and strain?
12. What is the modulus of elasticity?
13. What is the proportional limit?
14. What is the yield point?
15. What is the ultimate strength?
16. What is the difference between a ductile material and a brittle material?
17. What is working stress?
18. What is factor of safety?
19. What are the units in which strain is expressed?
20. Give the meaning of each term in the formula

$$e = \frac{PL}{AE}$$

21. What is a coefficient of thermal expansion?
22. In what units is the coefficient of expansion expressed in the gravitational system of units?
23. In what units is the coefficient of expansion expressed in the metric system?

REVIEW PROBLEMS Some of the review problems are more comprehensive than the practice problems.

1. A rectangular steel bar ⅞ in. × ⅜ in. and 18 in. long is subjected to a tensile force of 4 800 lb. Find the stress, the strain, and the change in length.
2. The working stress for the bar in Fig. 3-50 is 48 MPa. Find the dimensions d_1 and d_2. The hole is 30 mm in diameter.

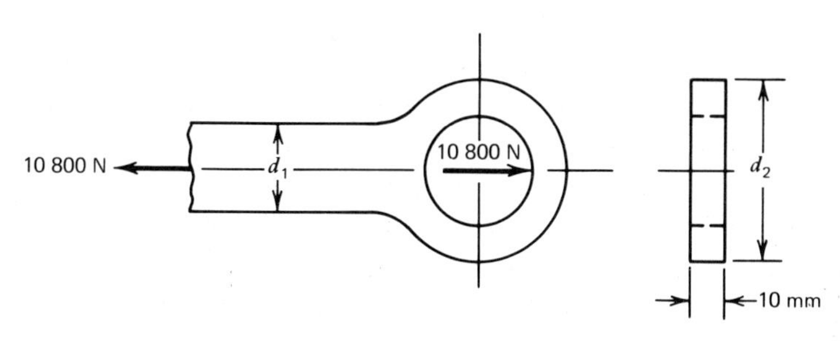

Fig. 3-50. End of tension member.

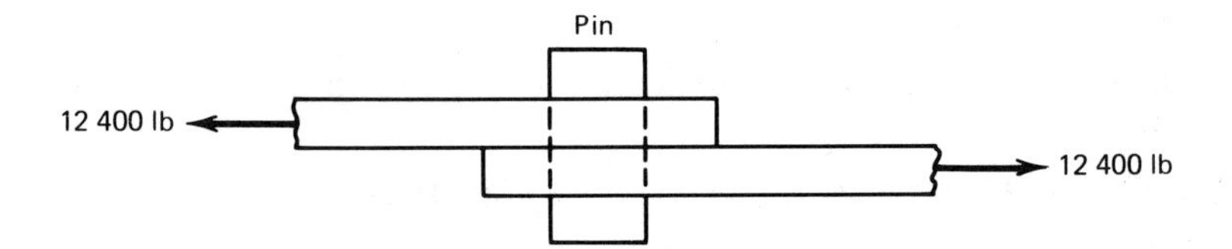

Fig. 3-51. Pin in shear.

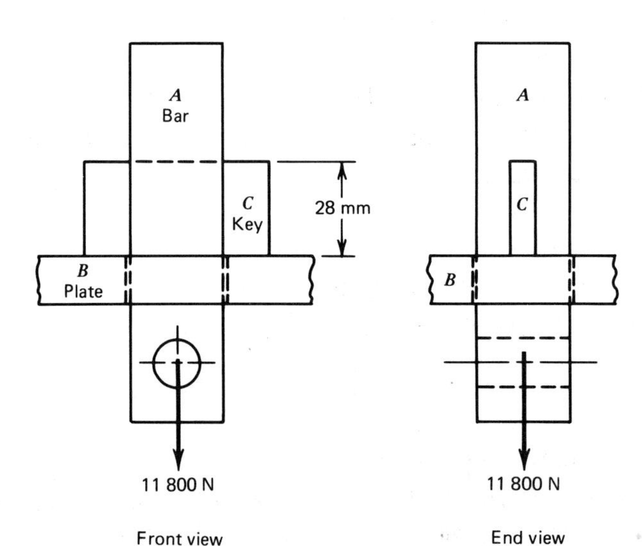

Fig. 3-52. Key in shear.

3. The two plates in Fig. 3-51 are fastened together by the circular pin for which the working stress in shear is 8 000 psi. What diameter is required for the pin?
4. An aluminum alloy tube 50 mm outer diameter, 40 mm inner diameter, and 300 mm long is subjected to a compressive force of 42 000 N. Find the stress, the strain, and the change in length.

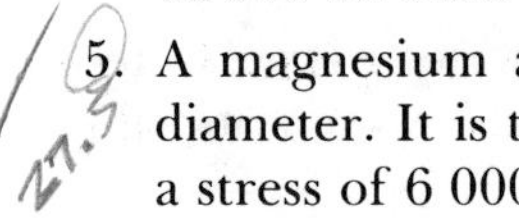

5. A magnesium alloy bar is circular in cross section and ¾ in. in diameter. It is to be stretched 0.01 in. under a force that develops a stress of 6 000 psi. Find its length.
6. In Fig. 3-52 the circular bar *A* passes through the plate *B* and is held by the key *C*. The key is 7 mm thick. Find the shearing stress in the key.
7. A gray cast-iron frame has the cross section shown in Fig. 3-53 and is 3 ft long. It is subjected to a compressive force of 360 000 lb. Find the stress, strain, and the change in length.
8. How wide should a strip of steel 3 mm thick be to carry a tensile force of 3 500 N with a working stress of 70 MPa?
9. Find the proper diameter for a circular compression member to carry a load of 106 000 lb with a working stress of 8 000 psi.

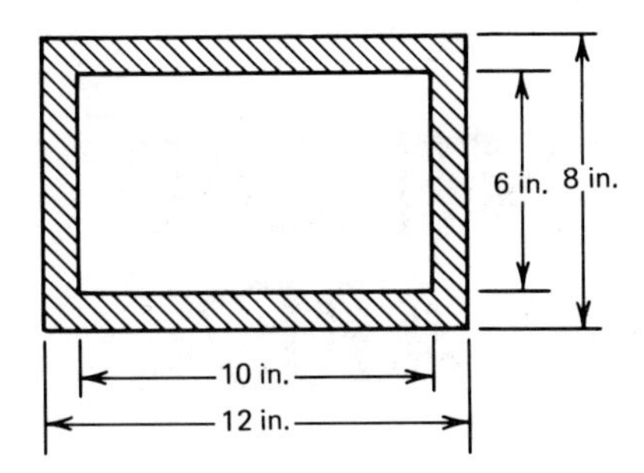

Fig. 3-53. Cross section of frame.

10. Figure 3-54 shows a circular shaft forged in one piece with the collar bearing B. The working stress in shear between A and B is 26 MPa. Find the required thickness t.
11. A low carbon steel bar is 20 in. long and is 0.035 in. longer than the space it must fit into. How much must the bar be cooled to slip into the space?
12. A circular magnesium alloy rod is 33 mm in diameter. It is to be inserted into a circular hole 32.85 mm in diameter in a magnesium-alloy plate. How much must the plate be heated for the rod to be inserted into the hole?

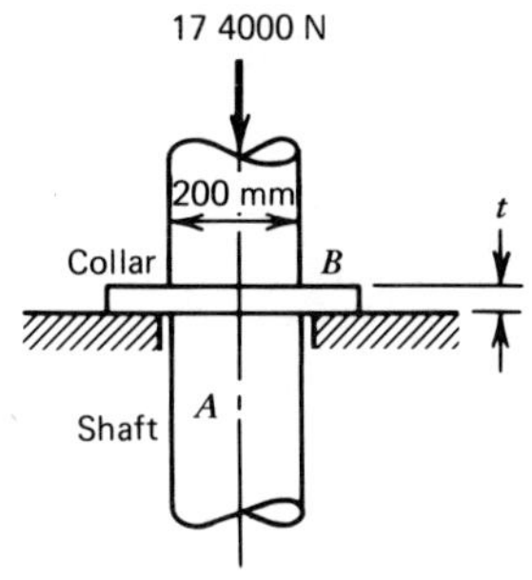

Fig. 3-54. Collar bearing.

4 CENTROID OF AN AREA

PURPOSE OF THIS CHAPTER. We are going to study areas in this chapter and investigate something probably unfamiliar to you before: the *centroid of an area*. So far we have used the following formula to good advantage

$$S = \frac{P}{A}$$

(S is stress, P is force, and A is area) to calculate stress. This formula gives us the stress in simple cases of tension, compression, and shear. But we must go on and learn how to calculate stress in a machine or structural part when it is bent or twisted, and we must learn how to design members that are bent or twisted. We have to know how to find the centroid of an area in order to do this. Therefore, the purpose of this chapter is to learn what the centroid of an area is and how to find it.

4-1. WHAT THE CENTROID IS

The *centroid* of an area is a certain point in the area. The distance of the centroid from any axis, such as the x or y axis, is equal to the average distance of the area from the axis. We will learn how to locate the centroid in this chapter. Then we will use it in later chapters when we study beams (Chapter 6) and columns (Chapter 12).

Let's show each area with a set of x and y axes when we work with centroids, because this will make it easier to describe the area and its location. For instance, we show the circular area in Fig. 4-1 with the x and y axes. The centroid of an area is a point whose distance from any axis is equal to the average distance of the area from that axis. For the circle in Fig. 4-1, the average distance of the area from the y axis is just the distance of the center of the circle from the y axis. This is 6 in. We designate the centroid by G and the distance of the centroid from the y axis as $\bar{x}$ (pronounced *x bar*; $\bar{x}$ is just the x coordinate of G). We refer to the distance of the centroid from the x axis as $\bar{y}$ (pronounced *y bar*;

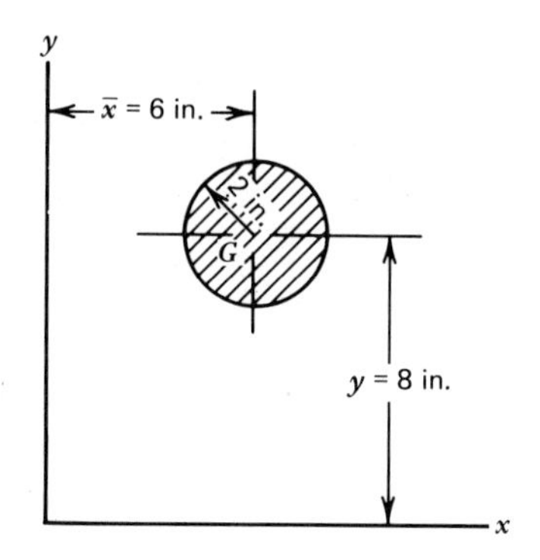

Fig. 4-1. Circular area.

$\bar{y}$ is the y coordinate of G). In Fig. 4-1, $\bar{y}$ equals 8 in. Now you see how it goes. We letter the centroid G, and its distance from any axis is the average distance of the area from the axis. It takes two coordinates to locate the centroid, and we call these two coordinates $\bar{x}$ and $\bar{y}$.

Illustrative Example 1. Figure 4-2 shows a rectangular area 4 in. wide and 6 in. high. Locate the centroid of the rectangle.

Solution:

1. To find $\bar{x}$: the distance from the y axis to the left side of the rectangle is 5 in., and then the distance from the left side of the rectangle to the centroid is 2 in. (the width is 4 in. and half of 4 in. is 2 in.). So,

$$\bar{x} = 5 + 2 = 7 \text{ in.}$$

2. To find $\bar{y}$: the distance from the x axis to the bottom of the rectangle is 2 in., and the distance from the bottom of the rectangle to the center is 3 in. more: this gives us

$$\bar{y} = 2 + 3 = 5 \text{ in.}$$

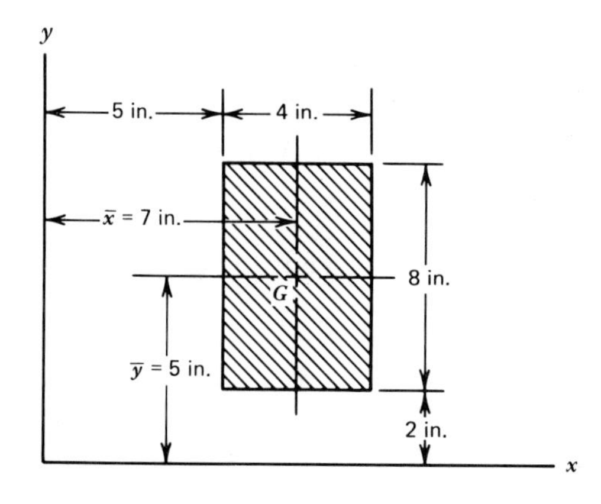

Fig. 4-2. Rectangular area.

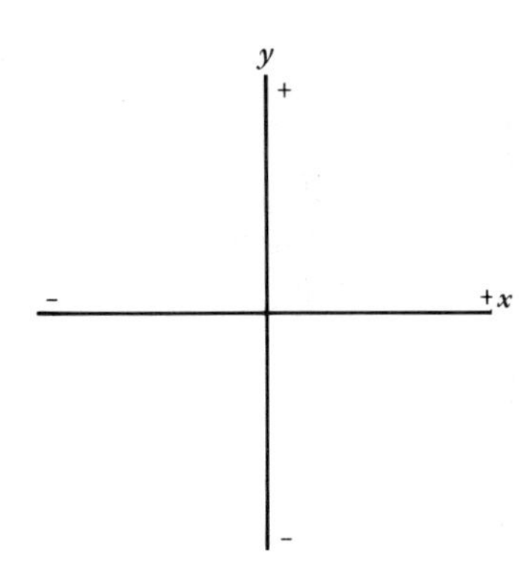

Fig. 4-3. Positive and negative coordinates.

ALGEBRAIC SIGNS

Sometimes an area is placed to the left of the y axis or below the x axis. We need a system of coordinates that can indicate this, and we use the usual convention, which is shown in Fig. 4-3. We say that an x coordinate is *positive* if it is measured to the right from the y axis and *negative* if measured to the left; a y coordinate is *positive* if it is measured upward from the x axis and *negative* if measured downward. Just follow these rules in designating $\bar{x}$ and $\bar{y}$ as positive or negative.

Illustrative Example 2. Locate the centroid of the square A in Fig. 4-4. Here is a chance to use the metric system.

Solution:

1. The square is to the left of the y axis, so $\bar{x}$ is negative. The distance from the y axis to the right side of the square is 270 mm, and it is 180 mm more to the center of the square. Write it this way,

$$\bar{x} = -(270 + 180) = -450 \text{ mm}$$

2. The coordinate $\bar{y}$ is positive, because the square is located above the x axis. The distance from the x axis to the lower side of the square is 210 mm, and the distance from the lower side of the square to the centroid is 180 mm, so

$$\bar{y} = 210 + 180 = 390 \text{ mm}$$

Illustrative Example 3. Locate the centroid of the circular area B in Fig. 4-4.

Solution:

1. In this case $\bar{x}$ is negative, because the area is to the left of the y axis:

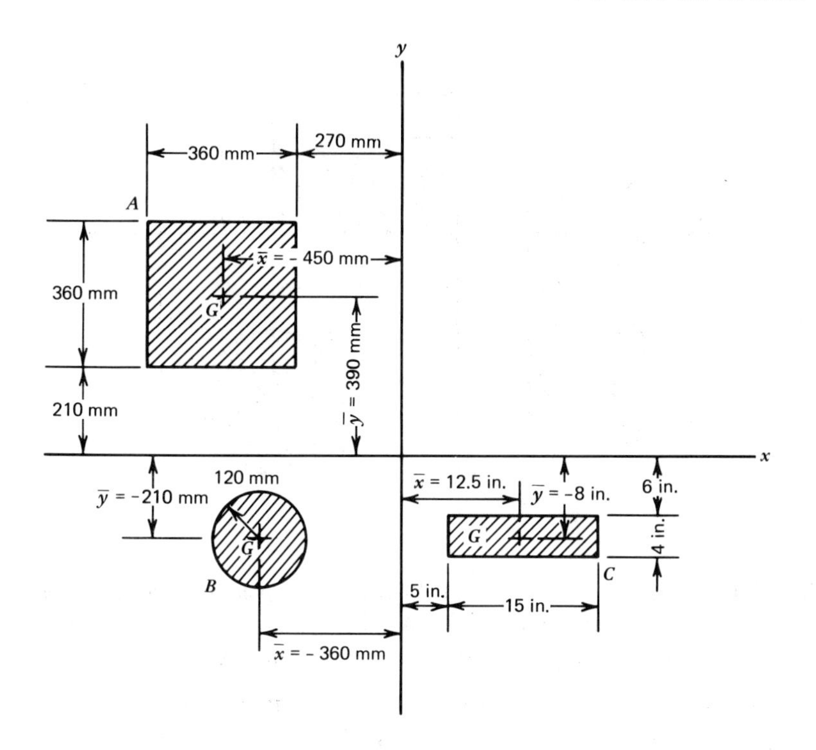

Fig. 4-4. Assorted areas.

$$\bar{x} = -360 \text{ mm}$$

2. Here y is negative, because the area is below the x axis:

$$\bar{y} = -210 \text{ mm}$$

Illustrative Example 4. Find $\bar{x}$ and $\bar{y}$ for the rectangular area C in Fig. 4-4.

Solution:

1. The area is to the right of the y axis, so $\bar{x}$ is positive. It is 5 in. from the y axis to the left side of the rectangle and 7.5 in. more from the left side of the rectangle to the center (7.5 in. is half of the width of 15 in.), so

$$\bar{x} = 5 + 7.5 = 12.5 \text{ in.}$$

2. For $\bar{y}$ it is 6 in. from the x axis to the top of the rectangle and 2 in. more to the centroid. The rectangle is below the x axis, so $\bar{y}$ is negative. Then,

$$\bar{y} = -(6 + 2) = -8 \text{ in.}$$

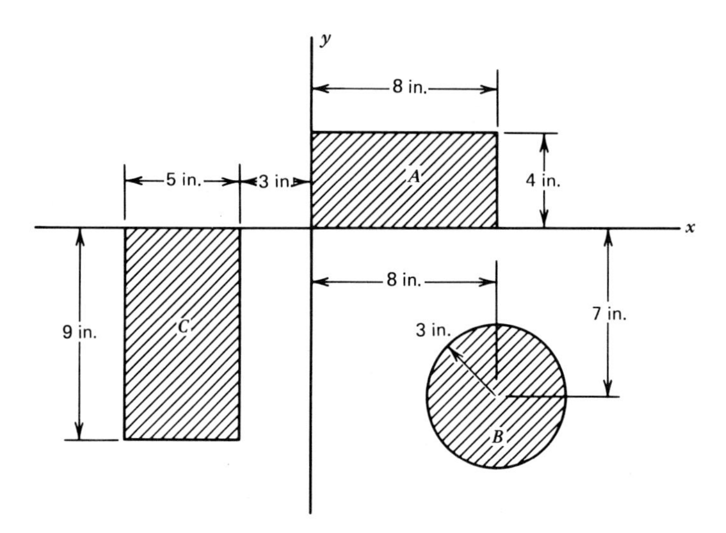

Fig. 4-5. More areas.

Practice Problems (Section 4-1). Work these practice problems before you go on to the next article. Working the problems will be easy if you learn each step thoroughly as you go along.

1. Locate the centroid of the rectangle A in Fig. 4-5.
2. Find $\bar{x}$ and $\bar{y}$ for the circle B in Fig. 4-5.
3. Locate the centroid of the rectangle C in Fig. 4-5.
4. Locate the centroid of the rectangle A in Fig. 4-6.
5. Find $\bar{x}$ and $\bar{y}$ for the rectangle B in Fig. 4-6.
6. Locate the centroid of the circle C in Fig. 4-6.

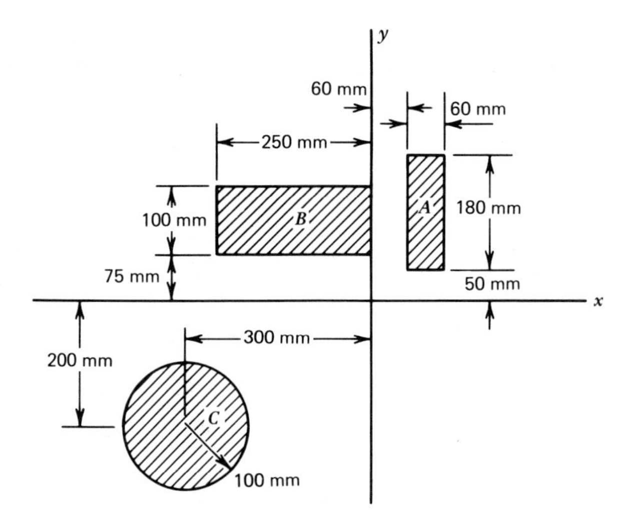

Fig. 4-6. Still more areas.

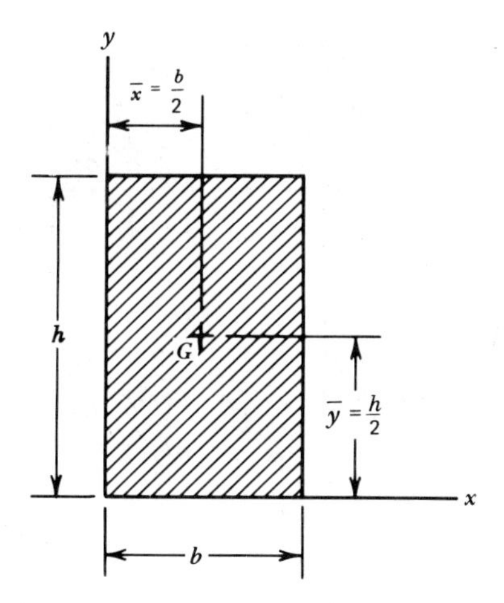

Fig. 4-7. Centroid of a rectangular area.

4-2. CENTROIDS OF SIMPLE AREAS

It is time now to provide you with the properties of a few simple areas. These properties are the amount of the area and the location of the centroid. Some of them you already know, and some are new. (Some of the new formulas would have to be derived by using calculus, but we are more interested here in learning to use them than we are in proving them, so they are given without proof.)

We have already used the properties of a rectangle. In Fig. 4-7, where the width of the rectangle is designated by b and the height by h, the area is the product of b and h. Thus $A = bh$. Also, $\bar{x} = b/2$ and $\bar{y} = h/2$ ($\bar{x}$ and $\bar{y}$ are the coordinates of the centroid).

Figure 4-8 shows a right triangle of width b and altitude h. The area is one-half the product of the width and altitude, so $A = bh/2$. The

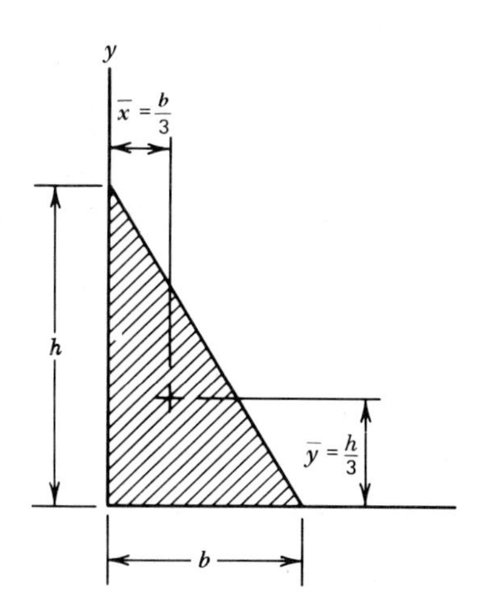

Fig. 4-8. Centroid of a triangular area.

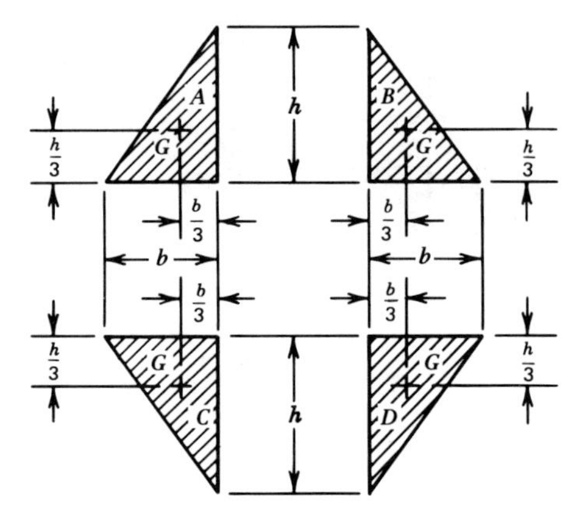

Fig. 4-9. Different orientations of a right triangle.

centroidal distances are $\bar{x} = b/3$ and $\bar{y} = h/3$. (We could prove that $\bar{x} = b/3$ and $\bar{y} = h/3$ by means of calculus.) Notice that these distances are measured from the corner where the right angle is. This is a convenient point from which to measure, and it is worth while to state a general rule about the centroid of a right triangle, so we won't get into trouble when the triangle is turned differently. To assure agreement, let's remember that the sides of the triangle adjacent to the right angle are called *legs*. Then we can say that the distance from either leg to the centroid is one-third of the other leg. See how this works for the triangle in Fig. 4-9*a*. Here b is the horizontal leg, and the distance from it to the centroid of the triangle is one-third of the vertical leg. The vertical leg is h, and one-third of it is $h/3$. Also, h is the vertical leg, and the distance from it to the centroid is one-third of the horizontal leg. The horizontal leg is b, and one-third of it is $b/3$. Now look at Fig. 4-9*b* and see how the rule gives the location of the centroid G. The distance of the centroid from the vertical leg is one-third of the horizontal leg, and the distance of the centroid from the horizontal leg is one-third of the vertical leg. Figures 4-9*c* and 4-9*d* show other ways in which a right triangle may be turned. Look them over and be sure that you understand how the rule applies. Then you will be ready for some examples.

Illustrative Example 1. Locate the centroid of the right triangle in Fig. 4-10.

Solution:

1. The distance from the y axis to the vertical leg of the triangle is 2 in. The horizontal leg is 6 in. and so the centroid is 2 in. (one-third of 6 in.) from the vertical leg. Then,

$$\bar{x} = 2 + 2 = 4 \text{ in.}$$

2. The distance from the x axis to the horizontal leg of the triangle

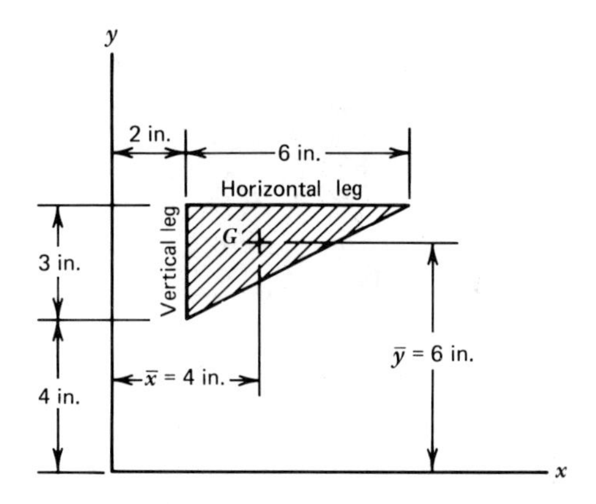

Fig. 4-10. Right triangle for Example 1.

is 7 in. The vertical leg is 3 in., so the centroid is 1 in. (one-third of 3 in.) below the horizontal leg. Then the distance from the x axis to the centroid G is the difference between 7 in. and 1 in., or

$$\bar{y} = 7 - 1 = 6 \text{ in.}$$

Illustrative Example 2. Locate the centroid of the right triangle in Fig. 4-11.

Solution:

1. The distance from the y axis to the vertical leg of the triangle is 100 mm, but the centroid is to the left of the vertical leg; the distance from the vertical leg to the centroid is one-third of the horizontal leg or one-third of 50 mm, which is 16.7 mm. Then,

$$\bar{x} = 100 - 16.7 = 83.3 \text{ mm}$$

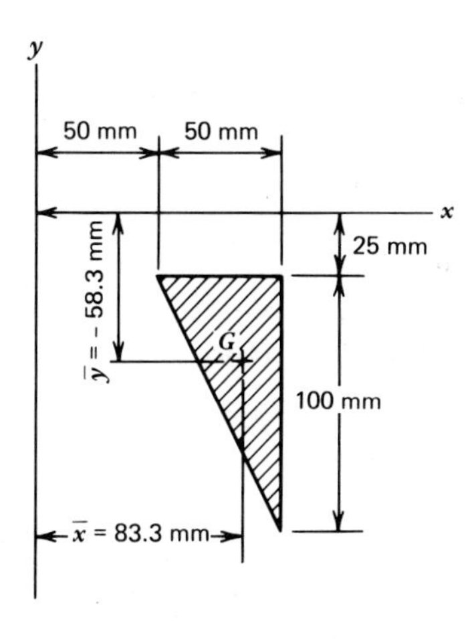

Fig. 4-11. Right triangle for Example 2.

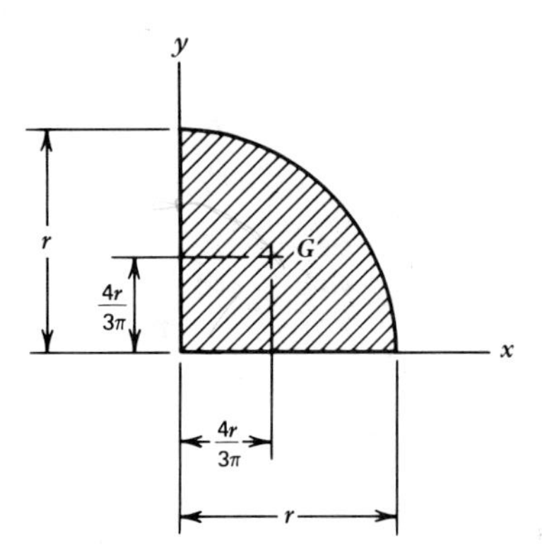

Fig. 4-12. Centroid of a quadrant.

2. The horizontal leg of the triangle is 25 mm below the x axis, and the centroid is one-third of the vertical leg below the horizontal leg, or one-third of 100, which is 33.3 mm. Remember that $\bar{y}$ is negative when the centroid is below the x axis. So,

$$\bar{y} = -(25 + 33.3) = -58.3 \text{ mm}$$

QUADRANT

Figure 4-12 shows the quadrant of a circle that has the radius r. The centroid is at G, and its distances from the axes are $\bar{x} = 4r/3\pi$ and $\bar{y} = 4r/3\pi$. (It is necessary to use calculus to derive this formula, so it is given here without proof.) Here the x axis coincides with a radius and is a boundary for the area, so the x axis could be called a *bounding radius*. So could the y axis. Then we have the following:

Rule 1. *The centroid of a quadrant is at a distance of $4r/3\pi$ from a bounding radius.* The area of a quadrant is just one-fourth of a circular area, so it is $(\pi/4)/r^2$.

Illustrative Example 3. Locate the centroid of the quadrant in Fig. 4-13.

Solution:

1. The vertical side of the quadrant is 2 in. from the y axis, and since the vertical side is a bounding radius, the centroid G is $4r/3\pi$ from the vertical side. This is

$$4 \times \frac{4}{3\pi} = 1.70 \text{ in., so}$$

$$\bar{x} = 2 + 1.70 = 3.70 \text{ in.}$$

2. The horizontal side of the quadrant is 3 in. from the x axis, and

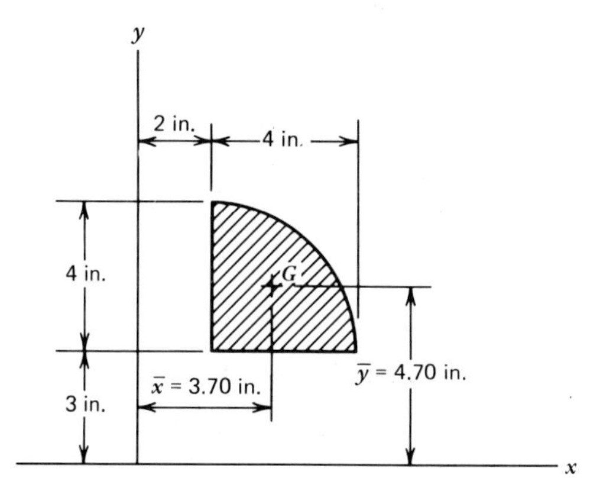

Fig. 4-13. A quadrant.

then the distance from the horizontal side to the centroid is $4r/3\pi$. We found the value of $4r/3\pi$ in Solution 1 of this example, and it is 1.70 in. Then,

$$\bar{y} = 3 + 1.70 = 4.70 \text{ in.}$$

OTHER POSITIONS

Quadrants may be turned in other positions, but the same rule holds: *The distance of the centroid from a bounding radius is* $4r/3\pi$. Figure 4-14 shows the four different positions that are going to be of interest to us. In Fig. 4-14*a* the centroid of the quadrant is above and to the left of the center of the circle, but the distance of the centroid from each bounding radius is $4r/3\pi$. The distance is always $4r/3\pi$ no matter whether the centroid is above or below the bounding radius or whether it is to the

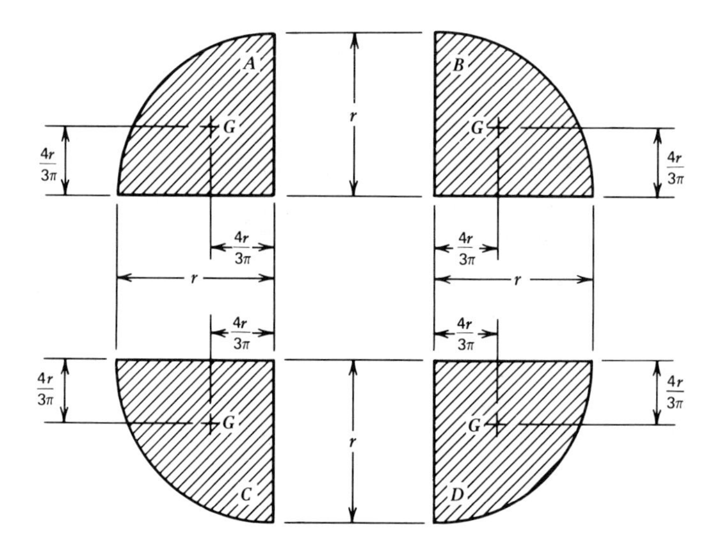

Fig. 4-14. Different orientations of a quadrant.

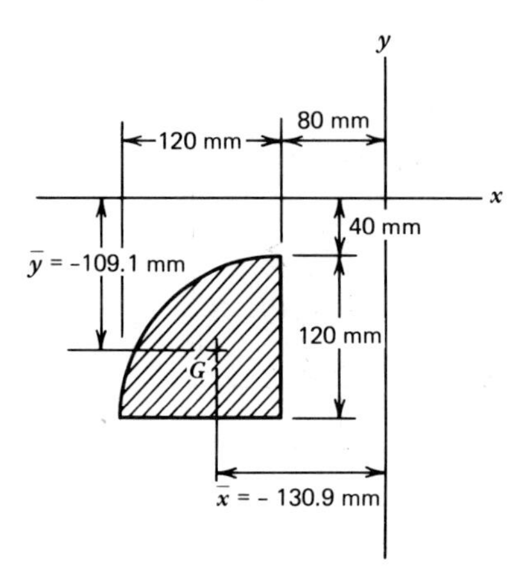

Fig. 4-15. Quadrant for Example 4.

right or to the left. It will pay you to study Figures 4-14*b*, 4-14*c*, and 4-14*d* carefully, to be sure you realize that the distance from the centroid to a bounding radius is always $4r/3\pi$.

Illustrative Example 4. Locate the centroid of the quadrant in Fig. 4-15.

Solution:

1. The distance from the y axis to the vertical side of the quadrant is 80 mm, and the additional distance to the centroid G is $4r/3\pi$. The radius of the quadrant is 120 mm, and $4r/3\pi = 4 \times 120/3\pi = 50.9$ mm. In this case $\bar{x}$ is negative, because the centroid is to the left of the y axis, and

$$\bar{x} = -(80 + 50.9) = -130.9 \text{ mm}$$

2. The distance from the x axis to the horizontal boundary of the quadrant is 160 mm, but the centroid is above this boundary by the distance $4r/3\pi$, which we found in Solution 1 to be 50.9 mm. We must subtract 50.9 from 160 to get $\bar{y}$, but we must remember that $\bar{y}$ is negative, because the centroid of the quadrant is below the x axis. So let's write it this way,

$$\bar{y} = -(160 - 50.9) = -109.1 \text{ mm}$$

INTERSECTING AXES

It looks a little more complicated when the x axis or the y axis (or both) passes through the area, but it really isn't. We can still find the location of the centroid. Then $\bar{x}$ is positive when the centroid is to the right of the y axis and negative when it is to the left. Also, $\bar{y}$ is positive when the

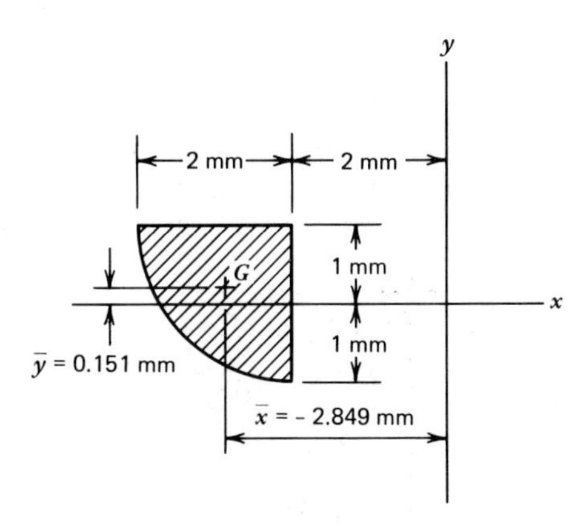

Fig. 4-16. Quadrant for Example 5.

centroid is above the x axis and negative when it is below. A pair of examples will illustrate.

Illustrative Example 5. Locate the centroid of the quadrant in Fig. 4-16.

Solution:

1. The distance from the y axis to the vertical side of the quadrant is 2 mm, and the distance from the vertical side to the centroid is $4 \times 2/3\pi = 0.849$ mm. Then,

$$\bar{x} = -(2 + 0.849) = -2.849 \text{ mm}$$

$\bar{x}$ is negative, because the centroid is to the left of the y axis.

2. The distance from the x axis to the horizontal side of the quadrant is 1 mm, and the centroid is below this horizontal side by the distance $4r/3\pi$. We found in Solution 1 that $4r/3\pi = 0.849$ mm, so

$$\bar{y} = 1 - 0.849 = 0.151 \text{ mm}$$

Here $\bar{y}$ is positive because the centroid is above the x axis.

Illustrative Example 6. Locate the centroid of the quadrant in Fig. 4-17.

Solution:

1. The distance from the y axis to the vertical side of the quadrant is 50 mm, measured to the left. The distance from the vertical side of the quadrant to the centroid is $4r/3\pi = 4 \times 300/3\pi = 127.3$ mm. Then,

$$\bar{x} = 127.3 - 50 = 77.3 \text{ mm}$$

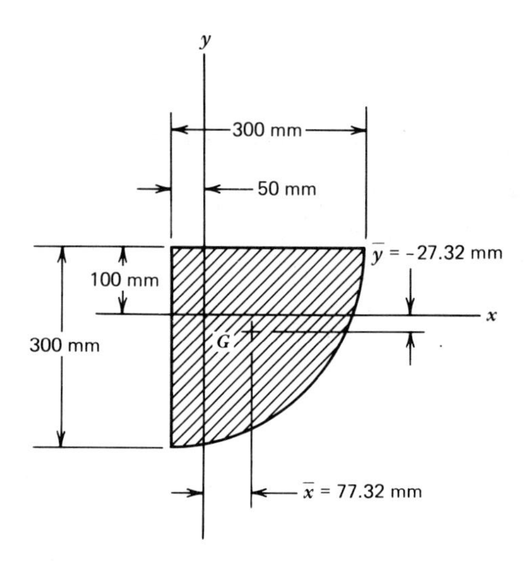

Fig. 4-17. Quadrant for Example 6.

2. The distance from the x axis to the horizontal side of the quadrant is 100 mm, measured upward. The distance from this horizontal side to the centroid is $4r/3\pi = 4 \times 300/3\pi = 127.3$ mm. Then,

$$\bar{y} = 100 - 127.3 = -27.3 \text{ mm}$$

Practice Problems (Section 4-2). These practice problems will help you to remember the locations of the centroids of the simple areas. Find $\bar{x}$ and $\bar{y}$ for each area described below.

1. The triangle A in Fig. 4-18.

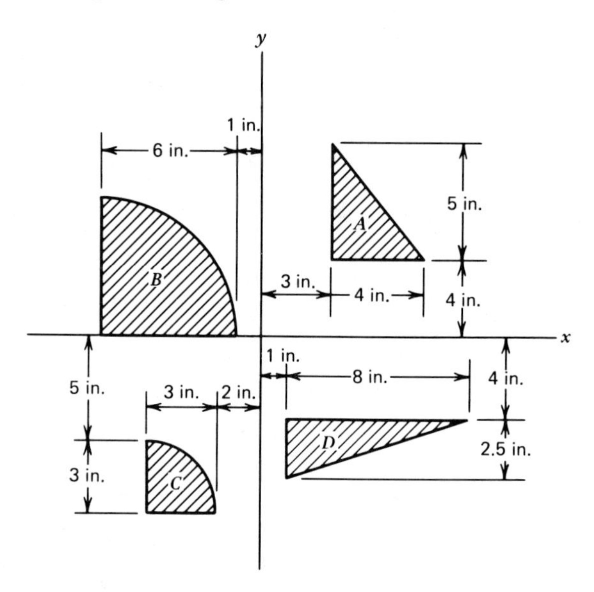

Fig. 4-18. Assorted areas.

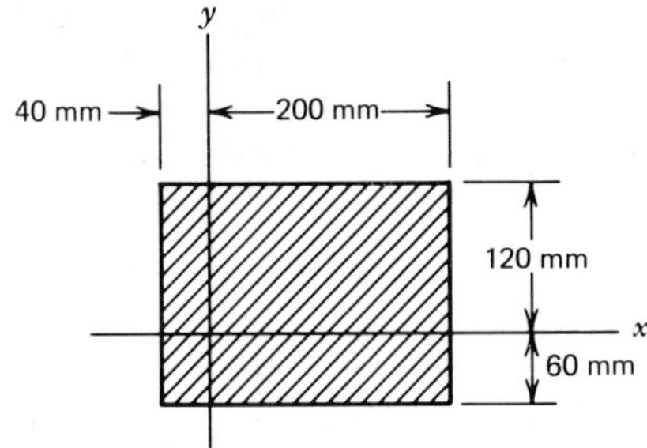

Fig. 4-19. Rectangle for Problem 5.

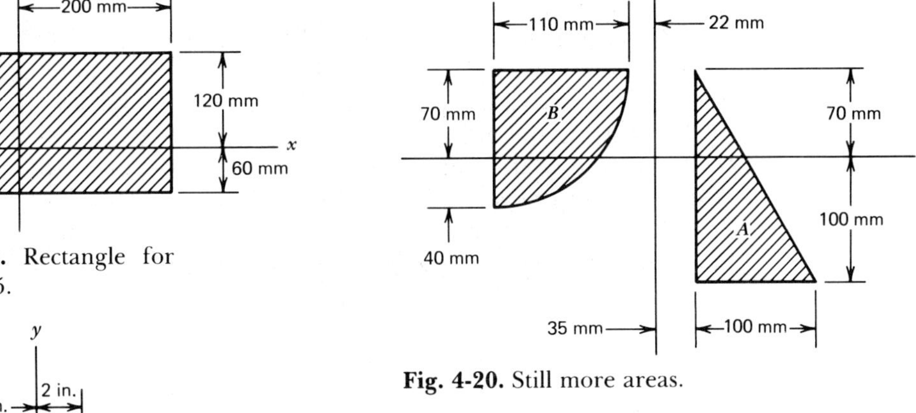

Fig. 4-20. Still more areas.

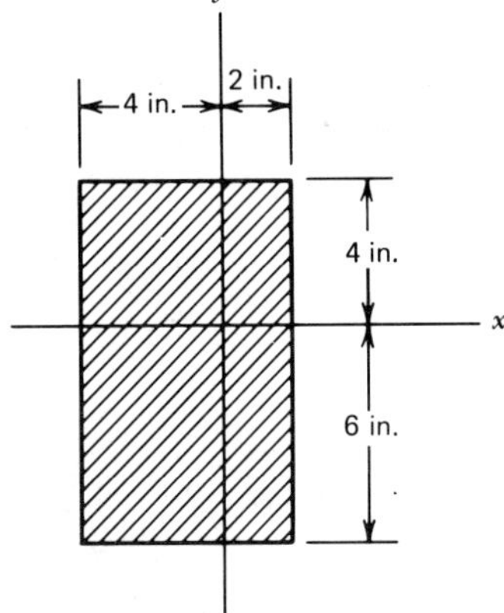

Fig. 4-21. Rectangle for Problem 8.

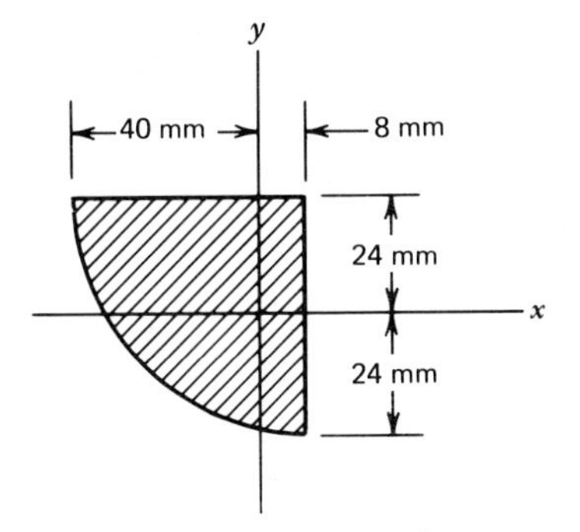

Fig. 4-22. Quadrant for Problem 9.

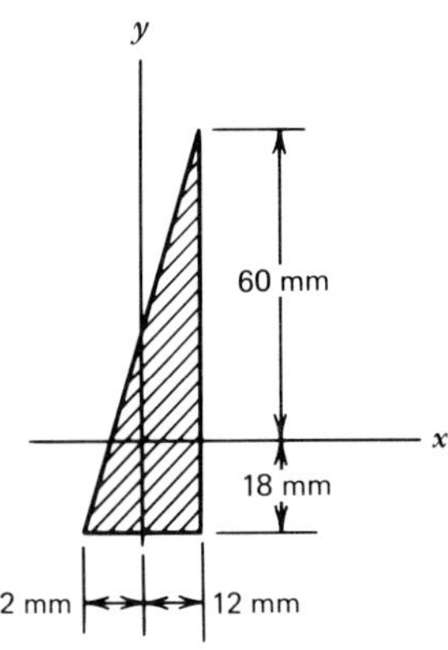

Fig. 4-23. Triangle for Problem 10.

2. The quadrant B in Fig. 4-18.
3. The quadrant C in Fig. 4-18.
4. The triangle D in Fig. 4-18.
5. The rectangle in Fig. 4-19.
6. The triangle A in Fig. 4-20.
7. The quadrant B in Fig. 4-20.
8. The rectangle in Fig. 4-21.
9. The quadrant in Fig. 4-22.
10. The triangle in Fig. 4-23.

4-3. MOMENT OF A SIMPLE AREA

Moment of an area is a name given to the product of the amount of the area and the distance of the centroid of the area from an axis. We must learn how to calculate it, because it is part of the process of locating the centroid of a complicated area of the type that we will study in the last part of this chapter.

We can calculate the moment of an area with respect to either the x axis or the y axis, and usually we do both. The moment of an area with respect to the x axis is equal to the product of the area and the distance of the centroid of the area from the x axis. We can represent the area by A and the centroidal distance by $\bar{y}$. The moment of the area with respect to the x axis is represented by M_x where M stands for moment and the subscript x indicates that it is with respect to the x axis. The equation for the moment of the area with respect to the x axis is

$$M_x = A\bar{y}$$

The moment of the area with respect to the y axis is equal to the product of the area and the distance of the centroid of the area from the y axis. The area is A, the distance of the centroid from the y axis is $\bar{x}$, and the moment of the area with respect to the y axis is M_y. Then the equation is

$$M_y = A\bar{x}$$

UNITS

Moment of area is expressed in a new sort of dimensional unit. We are multiplying area, which is expressed as the square of a unit of length, by a unit of length. So, we are going to get

$$\text{length}^2 \times \text{length} = \text{length}^3$$

that is, length to the third power. This is the same sort of multiplication you learned in algebra when you multiplied powers of x. You said that x^2 multiplied by x gave you x^3.

If we are working with gravitational units, we usually express area in square inches and length in inches and have moment of area in inches to the third power, thus in^3.

If we are working in the metric system, we usually express area in square millimeters and length in millimeters, so we have moment of area in millimeters to the third power, thus mm^3.

It is easy to find the area of a rectangle, right triangle, or quadrant. You know how to find $\bar{x}$ and $\bar{y}$ for one of these simple areas. Now let's multiply the area by a centroidal distance to get the moment of the area.

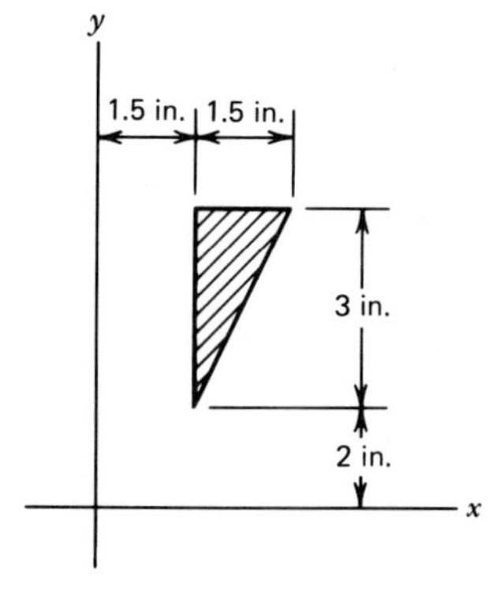

Fig. 4-24. Triangle for Example 1.

Illustrative Example 1. Find the moment of the triangle in Fig. 4-24 with respect to the x axis and the y axis.

Solution:

1. The vertical leg of the triangle is 1.5 in. from the y axis and the centroid is 0.5 in. (0.5 in. is one-third of 1.5 in.) to the right of the vertical leg. So,

$$\bar{x} = 1.5 + 0.5 = 2 \text{ in.}$$

2. The horizontal leg of the triangle is 5 in. $(2 + 3 = 5)$ above the x axis, and the centroid is 1 in. (1 in. is one-third of the vertical leg of 3 in.) below the horizontal leg. Then,

$$\bar{y} = 5 - 1 = 4 \text{ in.}$$

3. The area of the triangle is one-half of the product of the base and the altitude.

$$A = \tfrac{1}{2} \times 1.5 \times 3 = 2.25 \text{ in.}^2$$

4. The moment of the area with respect to the x axis is the product of A and $\bar{y}$. Thus

$$M_x = A\bar{y} = 2.25 \times 4 = 9 \text{ in.}^3$$

5. The moment of the area with respect to the y axis is the product of A and $\bar{x}$. So,

$$M_y = A\bar{x} = 2.25 \times 2 = 4.5 \text{ in.}^3$$

Illustrative Example 2. Calculate the moment of the area of the quadrant in Fig. 4-25 with respect to the x axis and the y axis.

Solution:

1. The distance from the y axis to the vertical side of the quadrant is 4 in., and the distance from the vertical side to the centroid of the area is $4r/3\pi = 4 \times 4/3\pi = 1.70$ in. Then,

$$\bar{x} = 4 + 1.70 = 5.70 \text{ in.}$$

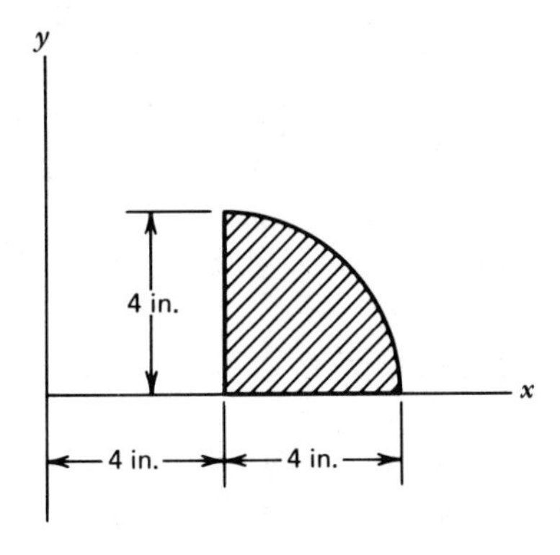

Fig. 4-25. Quadrant for Example 2.

2. The horizontal side of the quadrant is on the x axis, so $\bar{y}$ is just the distance from the horizontal side to the centroid. This is $4r/3\pi$, and in part 1 we found that $4r/3\pi$ is equal to 1.70 in., so

$$\bar{y} = 1.70 \text{ in.}$$

3. The area of a quadrant is $\pi r^2/4$. Thus

$$A = \frac{\pi r^2}{4} = \frac{\pi(4)^2}{4} = 12.56 \text{ in.}^2$$

4. The amount of the area with respect to the x axis is

$$M_x = A\bar{y} = 12.56 \times 1.70 = 21.4 \text{ in.}^3$$

5. The moment of the area with respect to the y axis is

$$M_y = A\bar{x} = 12.56 \times 5.70 = 71.6 \text{ in.}^3$$

POSITIVE AND NEGATIVE

We have seen that the distances $\bar{x}$ and $\bar{y}$ can be positive or negative ($\bar{x}$ is positive if measured to the right and negative if measured to the left; $\bar{y}$ is positive if measured upward and negative if measured downward). If the area A is positive and $\bar{x}$ is negative, the product of A and $\bar{x}$ is going to be negative. Also, if $\bar{y}$ is negative, the product of A and $\bar{y}$ is negative. This is just according to the rules of algebra, in which the product of one positive quantity and one negative quantity is negative.

Illustrative Example 3. Calculate the moment of the rectanglar area in Fig. 4-26 with respect to the x axis and the y axis.

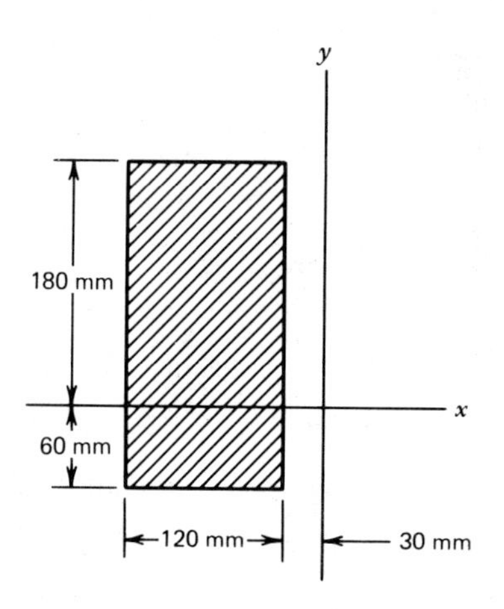

Fig. 4-26. Rectangle for Example 3.

Solution:

1. The distance from the y axis to the right side of the rectangle is 30 mm, and it is 60 mm farther (60 mm is one-half of the width of 120 mm) to the centroid. Hence the centroid is 30 + 60 = 90 mm from the y axis, and because the centroid is to the left of the y axis, $\bar{x}$ is negative. So

$$\bar{x} = -90 \text{ mm}$$

2. The bottom of the rectangle is 60 mm below the x axis, but the centroid is 120 mm above the bottom of the rectangle (120 mm is one-half of the altitude of 240 mm). So

$$\bar{y} = 120 - 60 = 60 \text{ mm}$$

3. The area of a rectangle is equal to the product of the width and the altitude.

$$A = 120 \times 240 = 28\,800 \text{ mm}^2$$

4. The moment of the area with respect to the x axis is

$$M_x = A\bar{y} = 28\,800 \times 60 = 1\,728\,000 \text{ mm}^3$$

5. The moment of the area with respect to the y axis is

$$M_y = A\bar{x} = 28\,800 \times (-90) = -2\,592\,000 \text{ mm}^3$$

The answer is negative, because it is the product of one positive quantity and one negative quantity.

Illustrative Example 4. Calculate the moment of the triangular area in Fig. 4-27 with respect to the x axis and the y axis.

Solution:

1. The vertical leg of the triangle is 6 mm to the left of the y axis, and the centroid of the triangular area is 4 mm (4 mm is one-

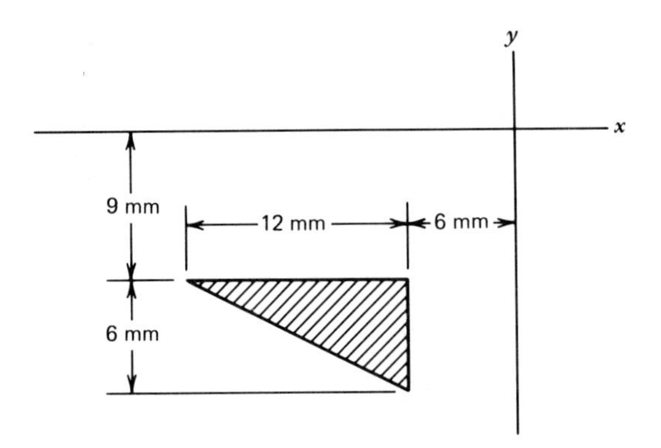

Fig. 4-27. Triangle for Example 4.

third of 12 mm) further to the left. In this case $\bar{x}$ is negative because the centroid is to the left of the y axis. Then,

$$\bar{x} = -(6 + 4) = -10 \text{ mm}$$

2. The horizontal leg of the triangle is 9 mm below the x axis, and the centroid of the area is 2 mm (one-third of 6 mm) below the horizontal leg. Since the centroid of the area is below the x axis, $\bar{y}$ is negative. Thus

$$\bar{y} = -(9 + 2) = -11 \text{ mm}$$

3. The area of the triangle is one-half the product of the base and altitude. So

$$A = \tfrac{1}{2} \times 12 \times 6 = 36 \text{ mm}^2$$

4. The moment of the area with respect to the x axis is

$$M_x = A\bar{y} = 36 \times (-11) = -396 \text{ mm}^3$$

5. The moment of the area with respect to the y axis is

$$M_y = A\bar{x} = 36 \times (-10) = -360 \text{ mm}^3$$

Practice Problems (Section 4-3). Now try it for yourself in this set of practice problems. In each problem, calculate the moment of the area with respect to the x axis and the y axis.

1. The rectangle A in Fig. 4-28.
2. The circle B in Fig. 4-28.

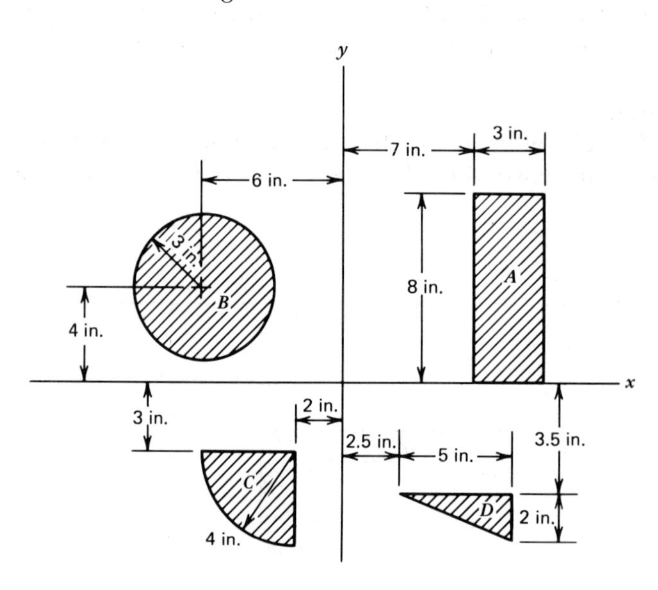

Fig. 4-28. Assorted areas.

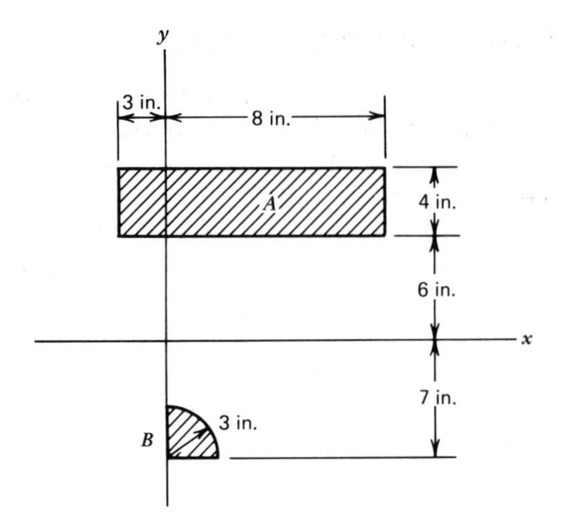

Fig. 4-29. Areas for Problems 5 and 6.

3. The quadrant C in Fig. 4-28.
4. The triangle D in Fig. 4-28.
5. The rectangle A in Fig. 4-29.
6. The quadrant B in Fig. 4-29.
7. The rectangle A in Fig. 4-30.
8. The quadrant B in Fig. 4-30.
9. The triangle C in Fig. 4-30.
10. The rectangle D in Fig. 4-30.

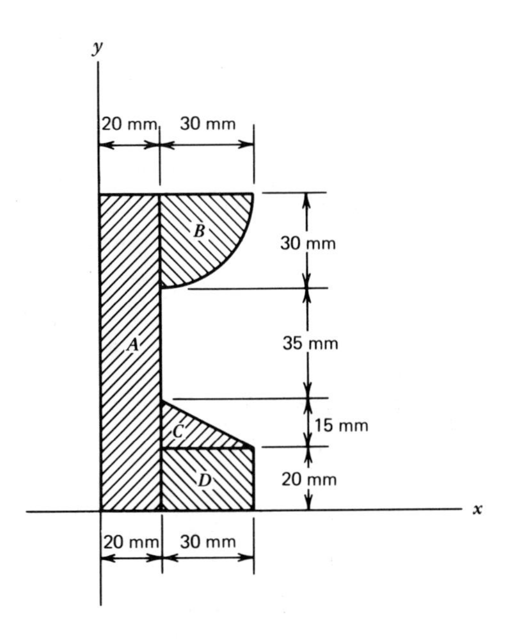

Fig. 4-30. More areas.

Fig. 4-31. Beam section.

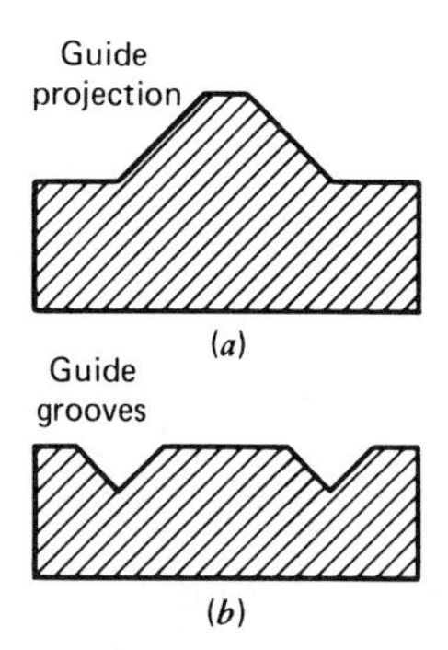

Fig. 4-32. Beam sections that serve as guides.

4-4. MOMENT OF A COMPOSITE AREA

Many of the areas that we encounter in studying beams (Chapter 6) are more irregular and more complicated than rectangles and triangles. There are several reasons for this. For one thing, the more complicated areas lead to economy in the use of material. This is the reason why many beams are of an I shape, as shown in Fig. 4-31. We'll see why this type of section is economical when we study beams. Another reason why beam sections are not always simple is that a machine part may have to serve as a guide for another part of the machine, in addition to carrying a load. Figure 4-32 shows two beam sections that could perform this function. The projection in Fig. 4-32*a* or the grooves in Fig. 4-32*b* could serve as guides for the moving part of a planer or for the carriage of a lathe.

We can usually divide the complicated areas into simple parts such as rectangles, right triangles, and quadrants. For this reason we call them *composite areas* (composite means made up of distinct elements or parts). Figure 4-33 shows how the beam section of Fig. 4-31 can be divided into the three rectangles A, B, and C. Such a division of the composite area into simple parts makes it easier to calculate its moment with respect to an axis, because we can calculate the moment of each simple part and then add the moments of the parts to find the moment of the whole area. This is a use of the general axiom that the whole is equal to the sum of all of its parts. (No doubt you have used this axiom in geometry and algebra; we will have more use for it later.) For this particular application we could state it this way: The moment of an area with respect to any axis is equal to the sum of the moments of the different parts of the area with respect to the axis. You have learned to calculate the moment of a simple area with respect to an axis in the first part of this chapter and, of course, you know how to add. That's all there is to it.

Let's think of a complicated area as divided into simple parts, as A, B, and C. We can find the area of each part and refer to their areas, as A_a, A_b, and A_c, where the capital A stands for area and the subscript shows

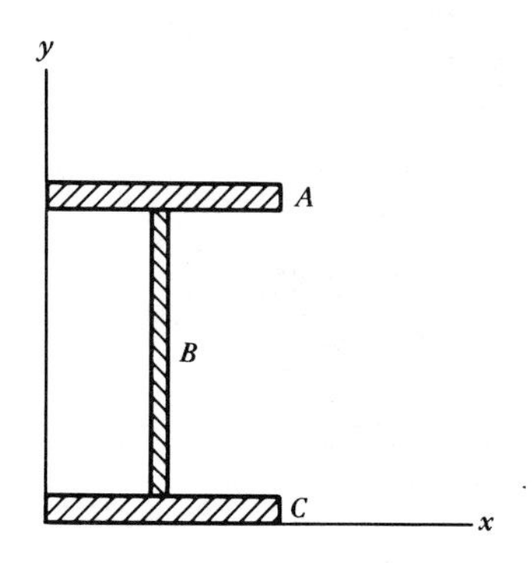

Fig. 4-33. Beam section again.

to which part it refers. Also, we can find the distance from the y axis to the centroid of each part and designate these distances, as $\bar{x}_a$, $\bar{x}_b$, and $\bar{x}_c$ where $\bar{x}$ stands for the distance of a centroid from the y axis and the subscript a, b, or c indicates to which part it belongs. The moment of area A with respect to the y axis is $A_a\bar{x}_a$, the moment of area B is $A_b\bar{x}_b$, the moment of area C is $A_c\bar{x}_c$, and the moment of the entire area with respect to the y axis is equal to the sum of the moments of its parts with respect to the y axis. Thus

$$M_y = A_a\bar{x}_a + A_b\bar{x}_b + A_c\bar{x}_c$$

We know how to find the distance from the x axis to the centroid of each part of the area, and we can designate these distances by $\bar{y}_a$, $\bar{y}_b$, and $\bar{y}_c$. The moment of area A with respect to the x axis is $A_a\bar{y}_a$, the moment of area B is $A_b\bar{y}_b$, and so on, and the moment of the whole area with respect to the x axis is equal to the sum of the moments of its parts with respect to the x axis. So

$$M_x = A_a\bar{y}_a + A_b\bar{y}_b + A_c\bar{y}_c$$

Now let's try an example or two.

Illustrative Example 1. Calculate the moment of the area in Fig. 4-34*a* with respect to the x axis and the y axis.

Solution:

1. Figure 4-34*b* shows the area divided into the two rectangles A and B
2. For the rectangle A
   (a) The distance $\bar{x}_a$ is just one-half of the width of the rectangle A, so it is

$$\bar{x}_a = \tfrac{1}{2} \times 2 = 1 \text{ in.}$$

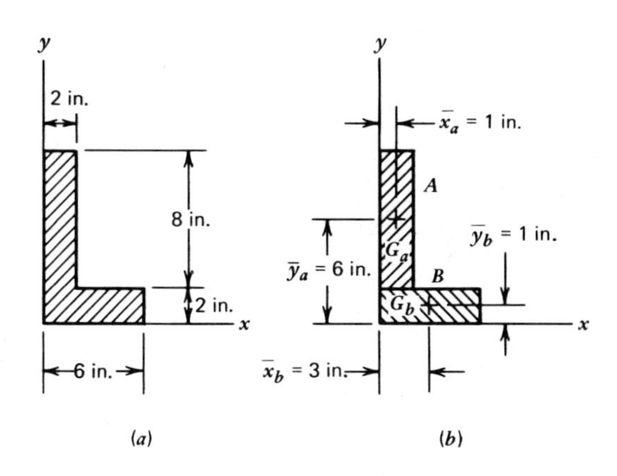

Fig. 4-34. Division of an area into parts.

(b) The distance $\bar{y}_a$ is equal to the distance from the x axis to the bottom of rectangle A plus one-half the height. So

$$\bar{y}_a = 2 + \tfrac{1}{2} \times 8 = 2 + 4 = 6 \text{ in.}$$

(c) The area of rectangle A is

$$A_a = 2 \times 8 = 16 \text{ in.}^2$$

3. For the rectangle B

(a) The distance $\bar{x}_b$ is one-half the width of the rectangle B

$$\bar{x}_b = \tfrac{1}{2} \times 6 = 3 \text{ in.}$$

(b) The distance $\bar{y}_b$ is one-half the height of the rectangle B, so

$$\bar{y}_b = \tfrac{1}{2} \times 2 = 1 \text{ in.}$$

(c) The area of rectangle B is

$$A_b = 6 \times 2 = 12 \text{ in.}^2$$

4. For the entire area

(a) The moment of the area with respect to the x axis is equal to the sum of the moments of the parts of the area with respect to the x axis. Thus

$$M_x = A_a\bar{y}_a + A_b\bar{y}_b = 16 \times 6 + 12 \times 1 = 108 \text{ in.}^3$$

(b) The moment of the entire area with respect to the y axis is equal to the sum of the moments of the parts of the area with respect to the y axis. Like this:

$$M_y = A_a\bar{x}_a + A_b\bar{x}_b = 16 \times 1 + 12 \times 3 = 52 \text{ in.}^3$$

Illustrative Example 2. Calculate the moment of the area of the trapezoid in Fig. 4-35*a* with respect to the x axis and with respect to the y axis.

Solution:

1. Figure 4-35*b* shows the area divided into the rectangle A and the triangle B
2. For the rectangle A

(a) The distance $\bar{x}_a$ is one-half the width of the rectangle A

$$\bar{x}_a = \tfrac{1}{2} \times 75 = 37.5 \text{ mm}$$

(b) The distance $\bar{y}_a$ is equal to one-half of the height of the rectangle A

$$\bar{y}_a = \tfrac{1}{2} \times 225 = 112.5 \text{ mm}$$

(c) The area of rectangle A is

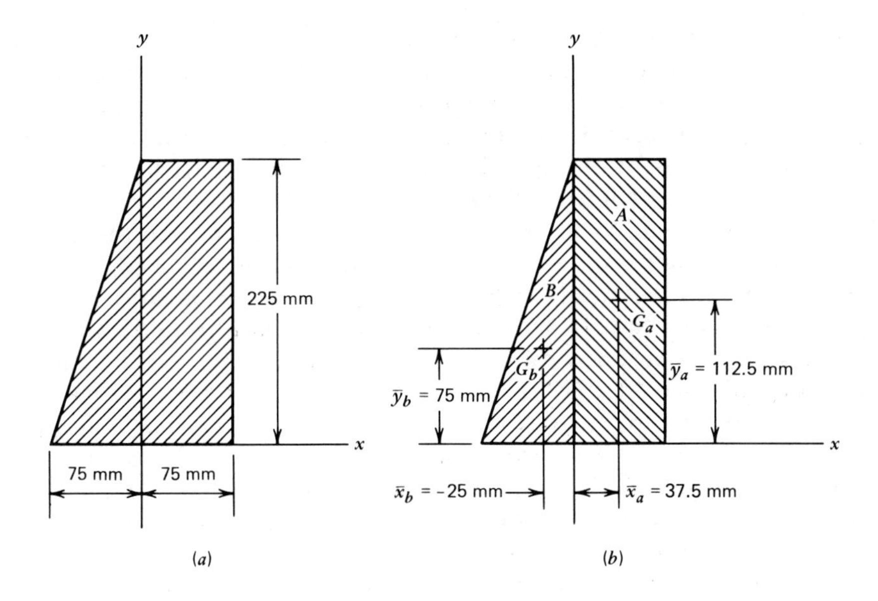

Fig. 4-35. Division of a composite area into parts.

$$A_a = 75 \times 225 = 16\,900 \text{ mm}^2$$

3. For the triangle B,
   (a) The distance $\bar{x}_b$ is one-third the width of the triangle B and is negative because it is measured to the left from the y axis. So

$$\bar{x}_b = -\tfrac{1}{3} \times 75 = -25 \text{ mm}$$

   (b) The distance $\bar{y}_b$ is one-third the altitude of the triangle B

$$\bar{y}_b = \tfrac{1}{3} \times 225 = 75 \text{ mm}$$

   (c) The area of the triangle B is

$$A_b = \tfrac{1}{2} \times 75 \times 225 = 8\,440 \text{ mm}^2$$

4. For the entire area,
   (a) The moment of the entire area with respect to the x axis is

$$M_x = A_a\bar{y}_a + A_b\bar{y}_b = 16\,900 \times 112.5 + 8\,440 \times 75$$

$$= 1\,901\,000 + 633\,000 = 2\,534\,000 \text{ mm}^3$$

$$= 2.534 \times 10^6 \text{ mm}^3$$

   (b) The moment of the entire area with respect to the y axis is

$$M_y = A_a\bar{x}_a + A_b\bar{x}_b = 16\,900 \times 37.5 + 8\,440\,(-25)$$

$$= 634\,000 - 211\,000 = 423\,000 \text{ mm}^3$$

$$= 0.423 \times 10^6 \text{ mm}^3$$

AREA SYMMETRICAL WITH RESPECT TO AN AXIS

An area is symmetrical with respect to an axis if, for each part on one side of the axis, there is a matching part on the other side of the axis. For example, the area in Fig. 4-36 is symmetrical with respect to the y axis. The rectangle A on the right side of the y axis is matched by the rectangle A on the left side of the y axis; the two rectangles are of the same size and same shape and are at the same distance from the axis. Also, the quadrant B on the right side of the y axis is matched by the quadrant B on the left side of the axis, because the two quadrants are of the same size, same shape, and at the same distance from the y axis. When an area is symmetrical with respect to an axis, the axis is called an *axis of symmetry*.

It is important to recognize an axis of symmetry because the moment of an area is zero with respect to an axis of symmetry. (You can just record the moment as zero and avoid having to calculate it.) In Fig. 4-36, for instance, the moment of the rectangle A on the right side of the y axis is positive with respect to the y axis, because the distance from the y axis to its centroid is positive, but the moment of the rectangle A on the left side of the y axis is negative with respect to the y axis, because the distance from the y axis to its centroid is negative. The two moments are of the same amount, and since one is positive and the other negative, they cancel. Also, the moment of the quadrant B on the right side of the y axis is positive with respect to the y axis; the moment of the quadrant B on the left side of the y axis is negative with respect to the y axis and of the same amount as the other, so they cancel. If an area is symmetrical with respect to an axis, its moment with respect to that axis is zero, because there is an equal negative moment to match and cancel every positive moment. Hereafter we will just put down the *moment is zero* and give symmetry as the reason for doing this.

Illustrative Example 3. Find the moment of the area in Fig. 4-36 with respect to the x axis and the y axis.

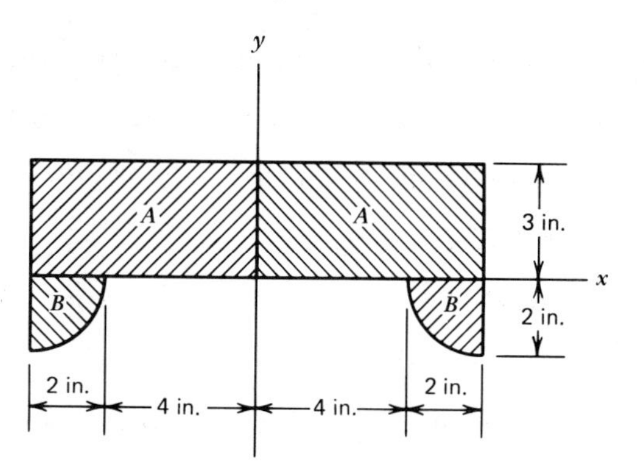

Fig. 4-36. Area symmetrical with respect to the Y axis.

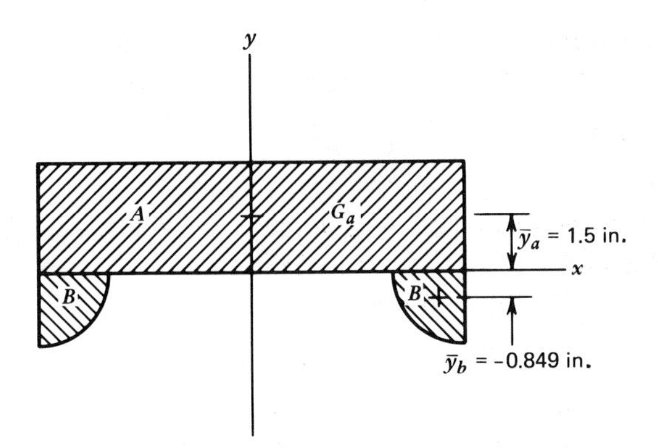

Fig. 4-37. Area divided into a rectangle and two quadrants.

Solution:

1. Figure 4-37 shows the area divided into the rectangle A and the two quadrants B. The y axis is an axis of symmetry, so

$$M_y = 0$$

2. For the rectangle A
   (a) The distance $\bar{y}_a$ from the x axis to the centroid of rectangle A is one-half the height of the rectangle

$$\bar{y}_a = \tfrac{1}{2} \times 3 = 1.5 \text{ in.}$$

   (b) The area of rectangle A is

$$A_a = 12 \times 3 = 36 \text{ in.}^2$$

3. For the quadrants B
   (a) The centroid of each quadrant B is below the x axis, so $\bar{y}_b$ is negative. It is $4r/3\pi$ in amount and $r = 2$ in. So

$$\bar{y}_b = -4 \times \tfrac{2}{3\pi} = -0.849 \text{ in.}$$

   (b) The area of one quadrant B is $\pi r^2/4$

$$A_b = \frac{\pi}{4}(2)^2 = \pi = 3.14 \text{ in.}^2$$

4. The moment of the entire area with respect to the x axis is equal to the sum of the moments of the parts of the area with respect to the x axis. There are two quadrants B, so

$$M_x = A_a\bar{y}_a + A_b\bar{y}_b + A_b\bar{y}_b,$$

or

$$M_x = A_a\bar{y}_a + 2A_b\bar{y}_b = 36 \times 1.5 + 2 \times 3.14\,(-0.849)$$

$$= 54 - 5.33 = 48.67 \text{ in.}^3$$

Illustrative Example 4. Calculate the moment of the area of the beam section in Fig. 4-38*a* with respect to the x axis and with respect to the y axis.

Solution:

1. Figure 4-38*b* shows the area divided into the rectangle A, the two rectangles B and the two triangles C. The area is symmetrical with respect to the x axis, so

$$M_x = 0$$

2. For the rectangle A,
   (a) The distance $\bar{x}_a$ is one-half the width of the rectangle A,

$$\bar{x}_a = \tfrac{1}{2} \times 8 = 4 \text{ mm}$$

   (b) The area of the rectangle A is

$$A_a = 8 \times 80 = 640 \text{ mm}^2$$

3. For one of the rectangles B,
   (a) The distance $\bar{x}_b$ is equal to the distance from the y axis to the left side of the rectangle B plus one-half the width of B. Thus

$$\bar{x}_b = 8 + \tfrac{1}{2} \times 12 = 8 + 6 = 14 \text{ mm}$$

   (b) The area of one of the rectangles B is

$$A_b = 12 \times 8 = 96 \text{ mm}^2$$

4. For one of the triangles C,

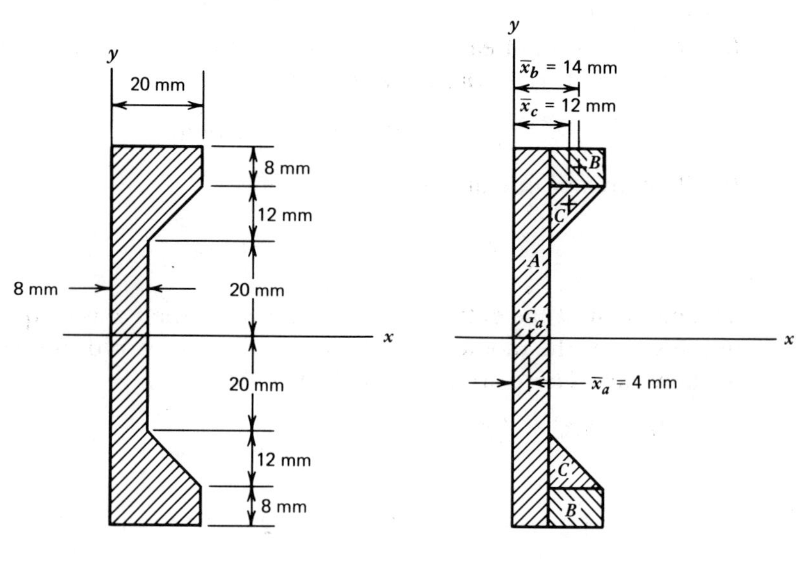

Fig. 4-38. Beam section divided into parts.

(a) The distance $\bar{x}_c$ is equal to the distance from the y axis to the vertical leg of the triangle plus one-third of the width of the triangle,

$$\bar{x}_c = 8 + \tfrac{1}{3} \times 12 = 8 + 4 = 12 \text{ mm}$$

(b) The area of one of the triangles C is

$$A_c = \tfrac{1}{2} \times 12 \times 12 = 72 \text{ mm}^2$$

5. For the whole area, remember that there are two rectangles B and two triangles C,

$$M_y = A_a\bar{x}_a + 2A_b\bar{x}_b + 2A_c\bar{x}_c$$

$$= 640 \times 4 + 2 \times 96 \times 14 + 2 \times 72 \times 12$$

$$= 2\,560 + 2\,688 + 1\,728 = 6\,976 \text{ mm}^3$$

NEGATIVE AREAS

Sometimes the easiest way to divide an area into simple parts is to regard one part of it as a negative area, that is, a part subtracted or cut out of the area. Look at this idea in a pair of examples.

Illustrative Example 5. Find the moment of the area in Fig. 4-39*a* with respect to the x axis and y axis.

Solution:

1. Figure 4-39*b* shows the division of the area into the rectangle A, 7 in. wide and 5 in. high, and the negative quadrant B. (Just think of this as a rectangle with a quadrant cut out at the corner.)

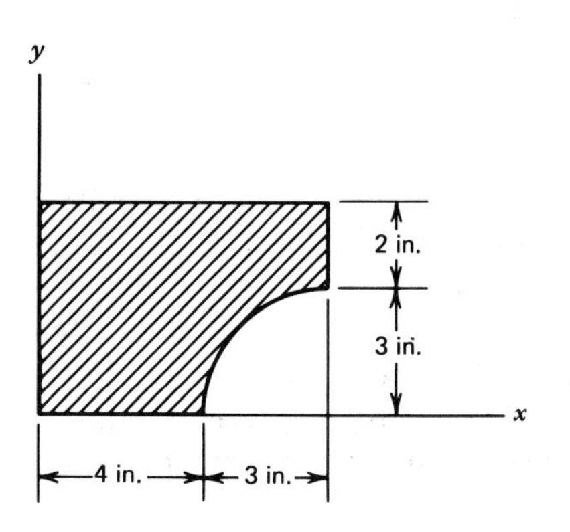

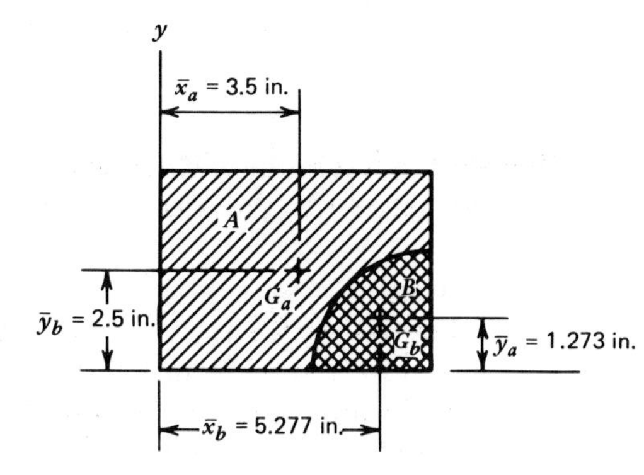

Fig. 4-39. Composite area divided into parts.

2. For the rectangle A,
   (a) The distance $\bar{x}_a$ is one-half the width of the rectangle A,

$$\bar{x}_a = \tfrac{1}{2} \times 7 = 3.5 \text{ in.}$$

   (b) The distance $\bar{y}_a$ is one-half the height of the rectangle A,

$$\bar{y}_a = \tfrac{1}{2} \times 5 = 2.5 \text{ in.}$$

   (c) The area of the rectangle A is

$$A_a = 7 \times 5 = 35 \text{ in.}^2$$

3. For the quadrant B,
   (a) The distance $\bar{x}_b$ is equal to the distance from the y axis to the vertical side of the quadrant minus the distance from the vertical side of the quadrant to the centroid of the quadrant. So

$$\bar{x}_b = 7 - \frac{4r}{3\pi} = 7 - 4 \times \frac{3}{3\pi} = 7 - 1.273 = 5.727 \text{ in.}$$

   (b) The distance $\bar{y}_b$ is just $\dfrac{4r}{3\pi}$, so $\bar{y}_b = 4 \times \dfrac{3}{3\pi} = 1.273$ in.

   (c) The area of the quadrant is negative:

$$A_b = -\frac{\pi r^2}{4} = -\frac{\pi}{4}(3)^2 = -7.07 \text{ in.}^2$$

4. For the entire area,
   (a) $M_x = A_a\bar{y}_a + A_b\bar{y}_b$, but remember that A_b is negative:

   $M_x = 35 \times 2.5 + (-7.07) \times 1.273 = 87.5 - 9.00 = 78.5$ in.^{3}

   (b) $M_y = A_a\bar{x}_a + A_b\bar{x}_b = 35 \times 3.5 + (-7.07) \times 5.727$

   $= 122.5 - 40.5 = 82$ in.^{3}

Illustrative Example 6. Calculate the moment of the area in Fig. 4-40*a* with respect to the x axis and with respect to the y axis.

Solution:

1. Figure 4-40*b* shows the area divided into the positive rectangle A, 150 mm wide and 120 mm high, and the negative rectangle B, 75 mm wide and 90 mm high (we think of the rectangle B as cut out of the rectangle A).
2. For the rectangle A,
   (a) The distance $\bar{x}_a$ is one-half the width of the rectangle A, so

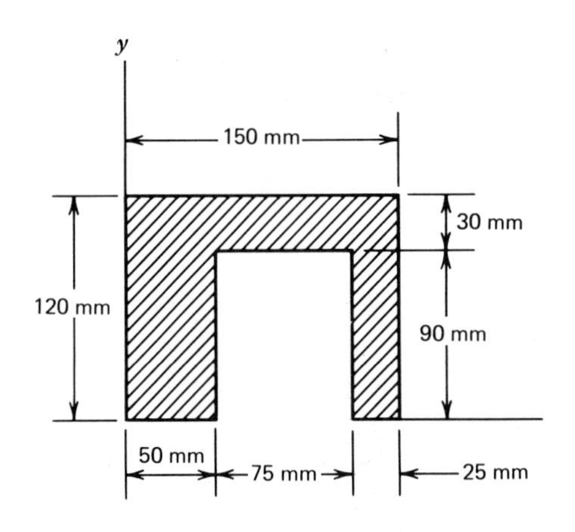

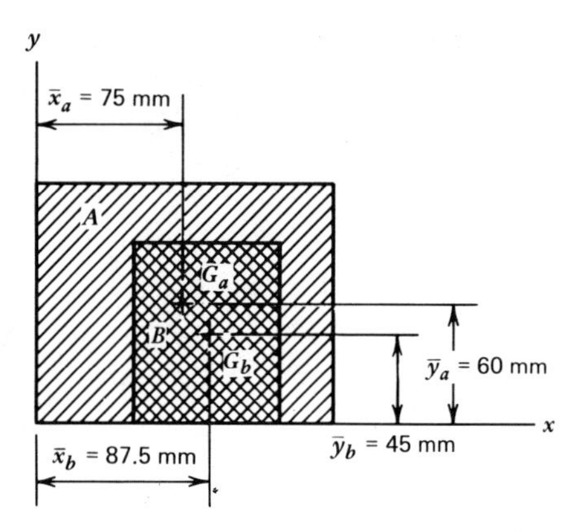

Fig. 4-40. Composite area divided into parts.

$$\bar{x}_a = \tfrac{1}{2} \times 150 = 75 \text{ mm}$$

(b) The distance $\bar{y}_a$ from the x axis to the centroid of the rectangle A is one-half the height of the rectangle,

$$\bar{y}_a = \tfrac{1}{2} \times 120 = 60 \text{ mm}$$

(c) The area of the rectangle A is

$$A_a = 150 \times 120 = 18\,000 \text{ mm}^2$$

3. For the rectangle B,
   (a) The distance $\bar{x}_b$ from the y axis to the centroid of the rectangle B is equal to the distance from the y axis to the left side of the rectangle plus one-half the width of the rectangle. Then,

$$\bar{x}_b = 50 + \tfrac{1}{2} \times 75 = 50 + 37.5 = 87.5 \text{ mm}$$

   (b) The distance $\bar{y}_b$ is one-half the height of the rectangle B. This is

$$\bar{y}_b = \tfrac{1}{2} \times 90 = 45 \text{ mm}$$

   (c) The area of the rectangle B is negative, because the area is considered to be cut away, so

$$A_b = -75 \times 90 = -6\,750 \text{ mm}^2$$

4. Now for the composite area (remember that A_b is negative)
   (a) $M_x = A_a\bar{y}_a + A_b\bar{y}_b = 18\,000 \times 60 + (-6\,750) \times 45$

$$= 1\,080\,000 - 304\,000 = 776\,000 \text{ mm}^3$$

   (b) $M_y = A_a\bar{x}_a + A_b\bar{x}_b = 18\,000 \times 75 + (-6\,750) \times 87.5$

$$= 1\,350\,000 - 591\,000 = 759\,000 \text{ mm}^3$$

Practice Problems (Section 4-4). In each problem, find the moment of the area with respect to the x axis and with respect to the y axis.

1. The area in Fig. 4-41.
2. The area in Fig. 4-42.
3. The area in Fig. 4-43.
4. The area in Fig. 4-44.
5. The area in Fig. 4-45.
6. The area in Fig. 4-46.

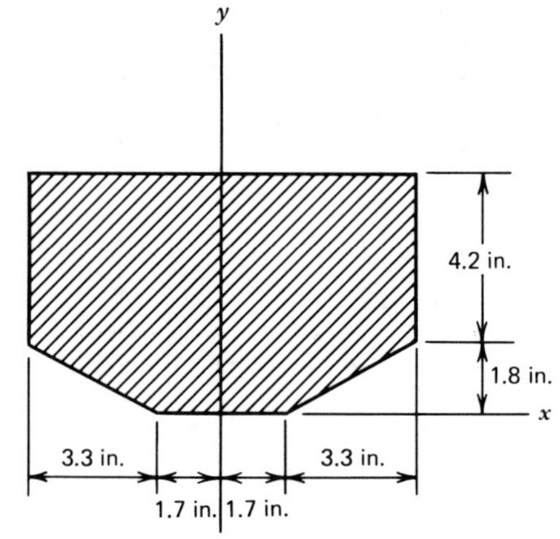

Fig. 4-41. Area for Problem 1.

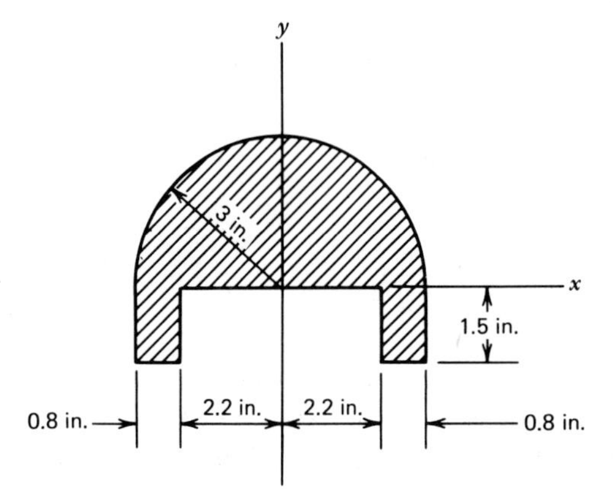

Fig. 4-42. Area for Problem 2.

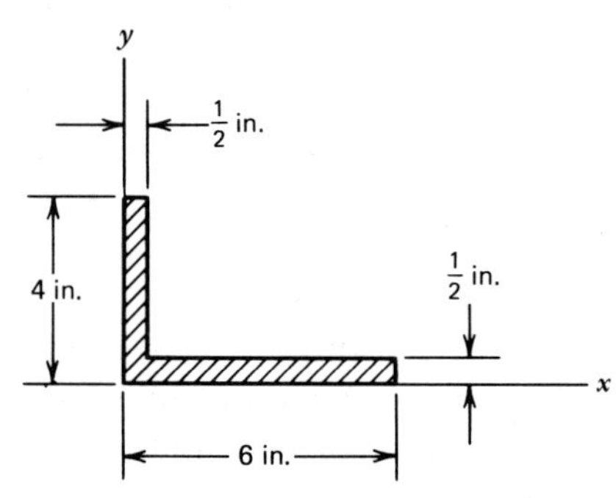

Fig. 4-43. Angle section for Problem 3.

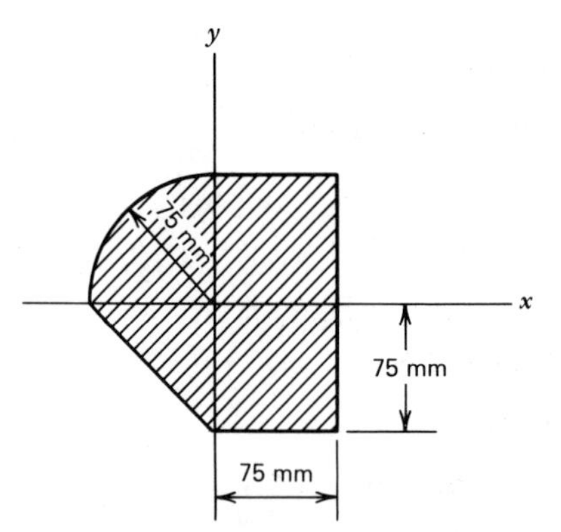

Fig. 4-44. Area for Problem 4.

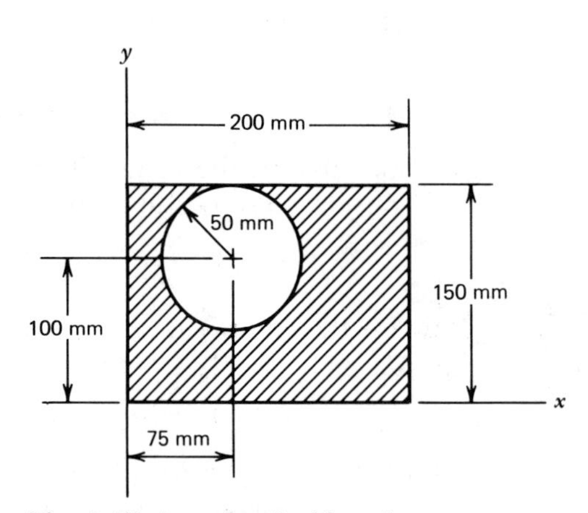

Fig. 4-45. Area for Problem 5.

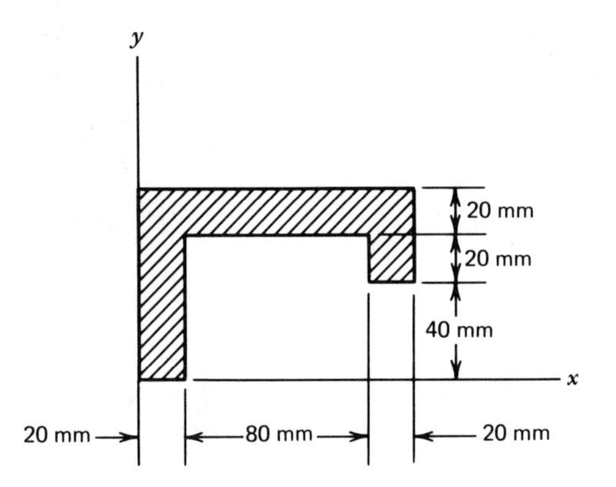

Fig. 4-46. Area for Problem 6.

4-5. CENTROID OF A COMPOSITE AREA

Now we are ready for the real object of this chapter, which is to locate the centroid of a composite area. (If you don't learn to do this, you will get into trouble when you reach Chapter 7). We know how to calculate the moment of a composite area by dividing it into simple areas, such as A, B, and C. We have the moment of the composite area with respect to the x axis, such as

$$M_x = A_a\bar{y}_a + A_b\bar{y}_b + A_c\bar{y}_c$$

and the moment of the area with respect to the y axis, such as

$$M_y = A_a\bar{x}_a + A_b\bar{x}_b + A_c\bar{x}_c$$

The amount of the composite area is the sum of the amounts of the simple areas. So

$$A = A_a + A_b + A_c$$

Here A is the total area.

Let's represent the distance from the y axis to the centroid of the composite area by $\bar{x}$. The moment of the composite area with respect to the y axis is the product of the area and the distance of the centroid of the area from the y axis. This is

$$M_y = A\bar{x}$$

We can solve this equation for $\bar{x}$ and have

$$\bar{x} = \frac{M_y}{A}$$

We know how to calculate M_y and A, so we can find $\bar{x}$. Here $\bar{x}$ is the coordinate of the centroid of the composite area. It is equal to the moment of the area with respect to the y axis divided by the area.

The distance from the x axis to the centroid of the composite area is represented by the symbol $\bar{y}$. The moment of the composite area with respect to the x axis is then

$$M_x = A\bar{y}$$

We want to find $\bar{y}$, so we solve this equation for $\bar{y}$

$$\bar{y} = \frac{M_x}{A}$$

Here $\bar{y}$ is the y coordinate of the centroid of the composite area. It is equal to the moment of the area with respect to the x axis divided by the area.

There isn't much that's new in this type of problem. We have already practiced on calculating M_x and M_y. The only new feature is the adding of the areas of the different parts to find A and then to divide M_y by A to find $\bar{x}$ and to divide M_x by A to find $\bar{y}$. Why not try an example? You should be so well acquainted with the first part of this problem that we can shorten it a bit.

Illustrative Example 1. Locate the centroid of the area in Fig. 4-47*a*.

Solution:

1. Figure 4-47*b* shows the area divided into the rectangle A and the triangle B.
2. For the rectangle A,
   (a) $\bar{x}_a = \frac{1}{2} \times 3 = 1.5$ in.
   (b) $\bar{y}_a = 1.5 + \frac{1}{2} \times 1 = 1.5 + 0.5 = 2$ in.
   (c) $A_a = 3 \times 1 = 3$ in.^{2}
3. For the triangle B,
   (a) $\bar{x}_b = \frac{1}{3} \times 3 = 1$ in.
   (b) $\bar{y}_b = 1.5 - \frac{1}{3} \times 1.5 = 1.5 - 0.5 = 1$ in.
   (c) $A_b = \frac{1}{2} \times 3 \times 1.5 = 2.25$ in.^{2}
4. For the entire area,
   (a) $M_x = A_a\bar{y}_a + A_b\bar{y}_b = 3 \times 2 + 2.25 \times 1$
   $= 6 + 2.25 = 8.25$ in.^{3}
   (b) $M_y = A_a\bar{x}_a + A_b\bar{x}_b = 3 \times 1.5 + 2.25 \times 1$
   $= 4.5 \times 2.25 = 6.75$ in.^{3}
   (c) $A = A_a + A_b = 3 + 2.25 = 5.25$ in.^{2}
   (d) Now, $\bar{x}$ for the composite area is equal to M_y divided by A,

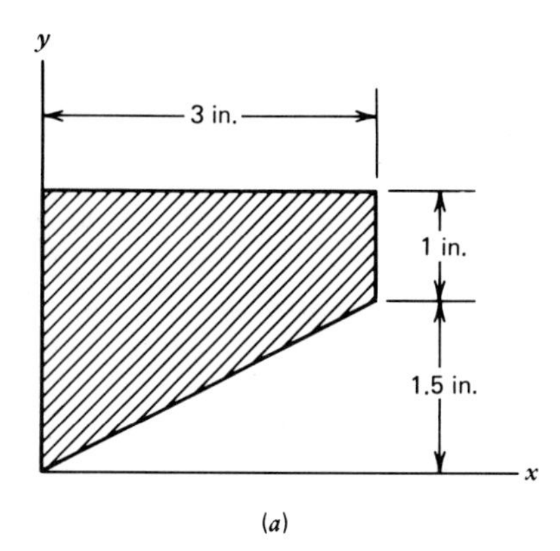

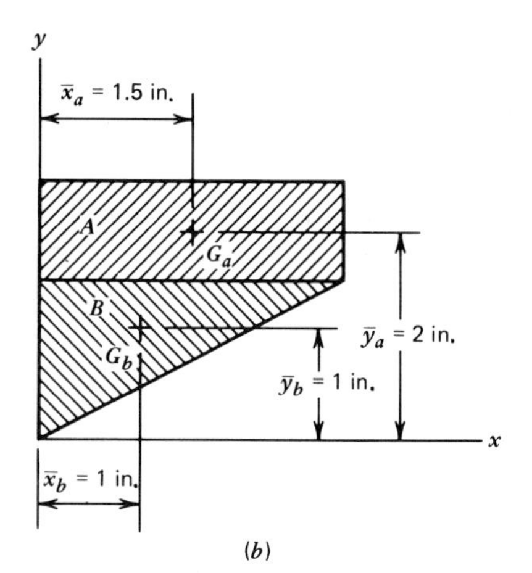

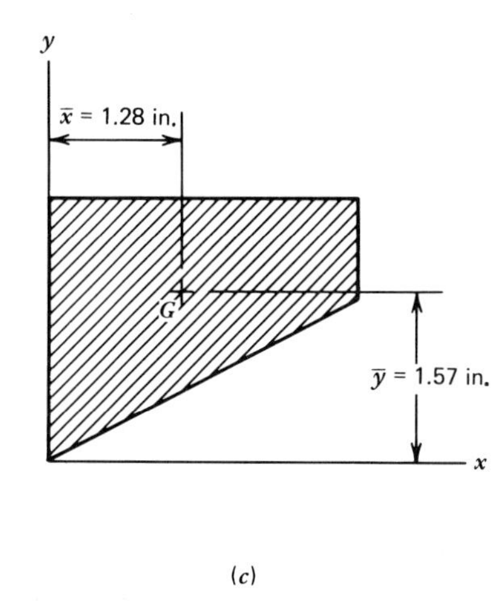

Fig. 4-47. Composite area for Example 1. (*a*) Area. (*b*) Area divided into parts. (*c*) Results.

$$\bar{x} = \frac{M_y}{A} = \frac{6.75}{5.25} = 1.29 \text{ in.}$$

(e) And $\bar{y}$ for the composite area is equal to M_x divided by A,

$$\bar{y} = \frac{M_x}{A} = \frac{8.25}{5.25} = 1.57 \text{ in.}$$

SYMMETRY

When there is an axis of symmetry for a composite area, the centroid lies on this axis. Suppose the y axis is an axis of symmetry. Then the moment of the area with respect to the y axis is zero (remember from Section 4 of this chapter). That is, $M_y = 0$. The distance of the centroid from the y axis is $\bar{x}$, and

$$\bar{x} = \frac{M_y}{A}$$

But if $M_y = 0$, then,

$$\bar{x} = \frac{0}{A} = 0$$

Here $\bar{x}$ is the distance of the centroid from the y axis, and if $\bar{x}$ is zero, it means that the centroid is on the y axis.

If the x axis is an axis of symmetry, the moment of the area with respect to the x axis is zero. The distance of the centroid from the x axis is $\bar{y}$, and

$$\bar{y} = \frac{M_x}{A}$$

When $M_x = 0$,

$$\bar{y} = \frac{0}{A} = 0$$

When $\bar{y}$ is zero, the centroid is on the x axis.

Hereafter, we will just put down that $\bar{y} = 0$ when we see that the x axis is an axis of symmetry and that $\bar{x} = 0$ when we see that the y axis is an axis of symmetry.

Illustrative Example 2. Locate the centroid of the T section in Fig. 4-48*a*.

Solution:

1. Figure 4-48*b* shows the composite area divided into the rectangle A and the rectangle B. The y axis is an axis of symmetry, so

$$\bar{x} = 0$$

2. For the rectangle A,
   (a) $\bar{y}_a = \frac{1}{2} \times 60 = 30$ mm
   (b) $A_a = 20 \times 60 = 1\,200$ mm^2
3. For the rectangle B,
   (a) $\bar{y}_b = -\frac{1}{2} \times 20 = -10$ mm, negative because the centroid of the rectangle B is below the x axis,
   (b) $A_b = 60 \times 20 = 1\,200$ mm^2
4. For the composite area,
   (a) $M_x = A_a\bar{y}_a + A_b\bar{y}_b = 1\,200 \times 30 + 1\,200(-10)$
   $= 36\,000 - 12\,000 = 24\,000$ mm^3
   (b) $A = A_a + A_b = 1\,200 + 1\,200 = 2\,400$ mm^2
   (c) The y coordinate of the centroid of the composite area is

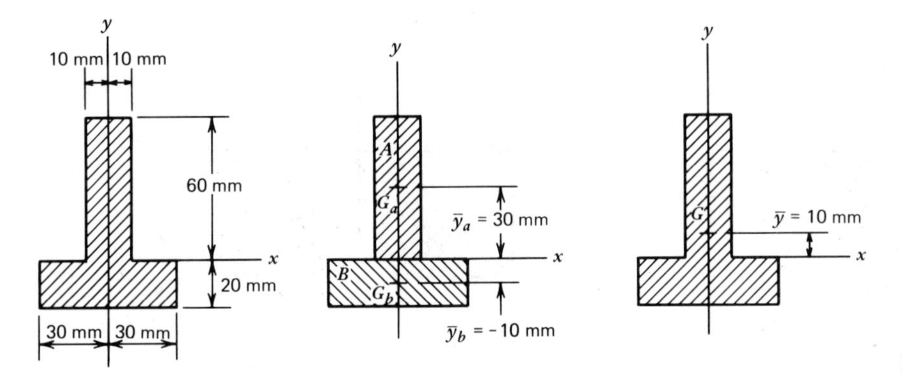

Fig. 4-48. *T* Section. (*a*) Area. (*b*) Area divided into parts. (*c*) Results.

$$\bar{y} = \frac{M_x}{A} = \frac{24\ 000}{2\ 400} = 10 \text{ mm}$$

The location of the centroid of the composite area is shown in Fig. 4-48*c*.

NEGATIVE AREA

Let's try a problem with a negative area. Remember to subtract the negative area when you calculate the total area A.

Illustrative Example 3. Locate the centroid of the area in Fig. 4-49*a*.

Solution:

1. Figure 4-49*b* shows the area divided into the positive rectangle A and the two negative triangles B. The y axis is an axis of symmetry, so

$$\bar{x} = 0$$

2. For the rectangle A,
   (a) $\bar{y}_a = \frac{1}{2} \times 4 = 2$ in.
   (b) $A_a = 10 \times 4 = 40$ in.^{2}
3. For one triangle B,
   (a) $\bar{y}_b = \frac{1}{3} \times 2 = 0.667$ in.
   (b) $A_b = -\frac{1}{2} \times 2 \times 2 = -2$ in.^{2}
4. For the entire area,
   (a) $M_x = A_a\bar{y}_a + 2A_b\bar{y}_b = 40 \times 2 + 2(-2) \times 0.667$
   $= 80 - 2.67 = 77.33$ in.^{3}

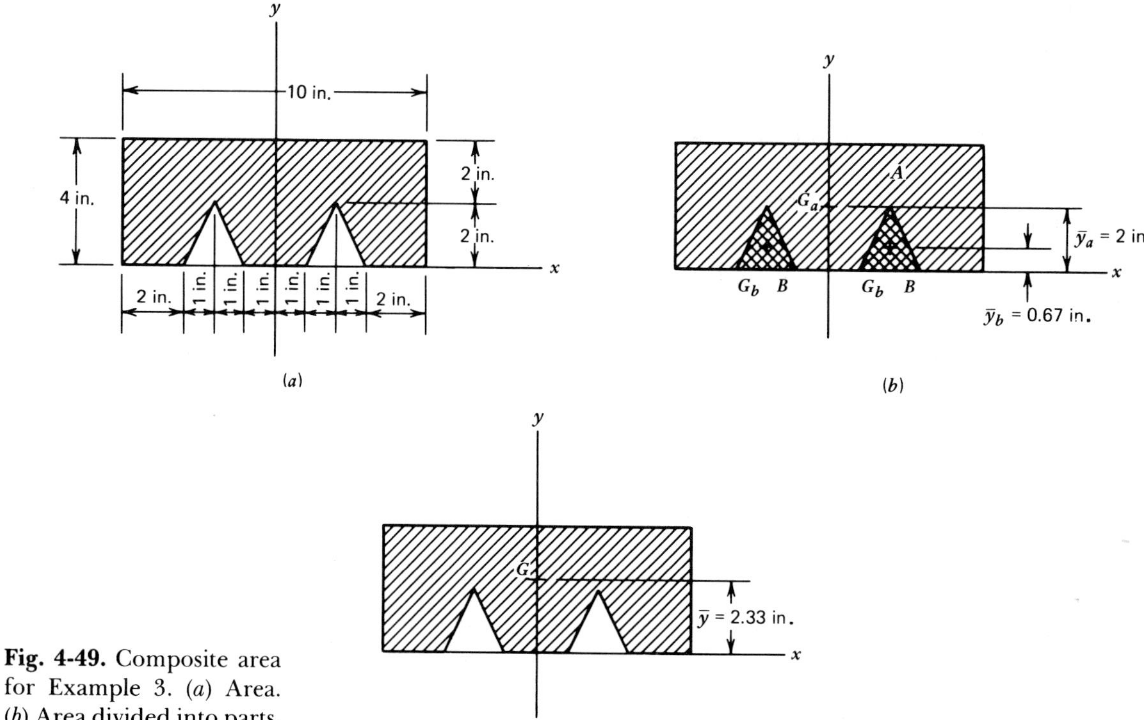

Fig. 4-49. Composite area for Example 3. (*a*) Area. (*b*) Area divided into parts. (*c*) Results.

(b) $A = A_a + 2A_b = 40 + 2(-2) = 40 - 4 = 36 \text{ in.}^2$

(c) Then,

$$\bar{y} = \frac{M_x}{A} = \frac{77.33}{36} = 2.15 \text{ in.}$$

Figure 4-49*c* shows the location of the centroid of the composite area.

SUMMARY Here is a brief statement of the important points in this chapter.

1. The centroid of an area is a point whose distance from any axis is equal to the average distance of the area from that axis. Think of it in terms of an *xy* coordinate system.
   (a) $\bar{x}$ is the x coordinate of the centroid.
   (b) $\bar{y}$ is the y coordinate of the centroid.
2. Remember these properties of the simple areas.

(a) The area of a rectangle is equal to the product of the width and the height. The centroid is at the center of the rectangle.
(b) The area of a triangle is one-half the product of the width and the altitude. The distance of the centroid from either leg is one-third of the other leg.
(c) The area of a quadrant is $\pi r^2/4$ where r is the radius. The distance of the centroid from a radius boundary is $4r/3\pi$.

3. The moment of a part of the area with respect to any axis is equal to the product of the area of the part and the distance of the centroid of the part from the axis.
4. The moment of a composite area with respect to an axis is equal to the sum of the moments of the parts of the area with respect to the axis. Thus
   (a) $M_x = A_a\bar{y}_a + A_b\bar{y}_b + A_c\bar{y}_c$
   (b) $M_y = A_a\bar{x}_a + A_b\bar{x}_b + A_c\bar{x}_c$
5. The total area is equal to the sum of the parts.

$$A = A_a + A_b + A_c$$

6. The two coordinates of the centroid of a composite area are $\bar{x}$ and $\bar{y}$.
   (a) $\bar{x} = \dfrac{M_y}{A}$
   (b) $\bar{y} = \dfrac{M_x}{A}$

REVIEW QUESTIONS These questions are intended to test your knowledge of the chapter and should be answered without looking back at the chapter.

1. What is the centroid of an area?
2. Where is the centroid of a circular area?
3. What is the area of a circle?
4. Where is the centroid of a rectangular area?
5. What is the area of a triangle?
6. Where is the centroid of a triangular area?
7. Where is the centroid of a quadrant?
8. What is the area of a quadrant?
9. What is the moment of an area with respect to an axis?
10. What is a composite area?

11. How do you find the moment of a composite area with respect to an axis?
12. How do you locate the centroid of a composite area?

REVIEW PROBLEMS If you can work these problems, you know the material of this chapter. If you can't, you need to study Section 1-5 again. The problems are all the same. Just locate the centroid of the area. This means to find $\bar{x}$ and $\bar{y}$.

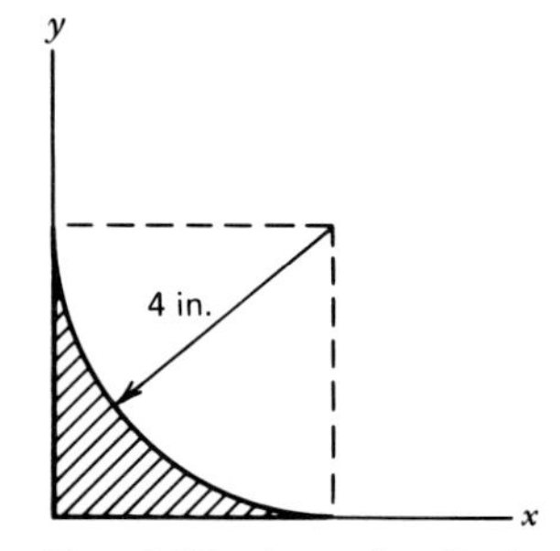

Fig. 4-50. Area for Problem 1.

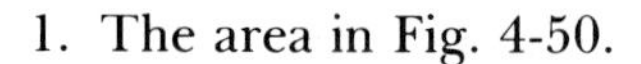

1. The area in Fig. 4-50.
2. The area in Fig. 4-51.
3. The area in Fig. 4-52.
4. The area in Fig. 4-53.
5. The area in Fig. 4-54.
6. The area in Fig. 4-55.
7. The area in Fig. 4-56.
8. The area in Fig. 4-57.

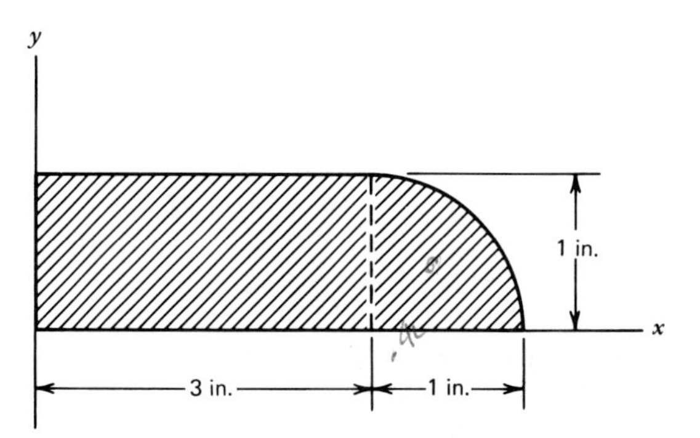

Fig. 4-51. Area for Problem 2.

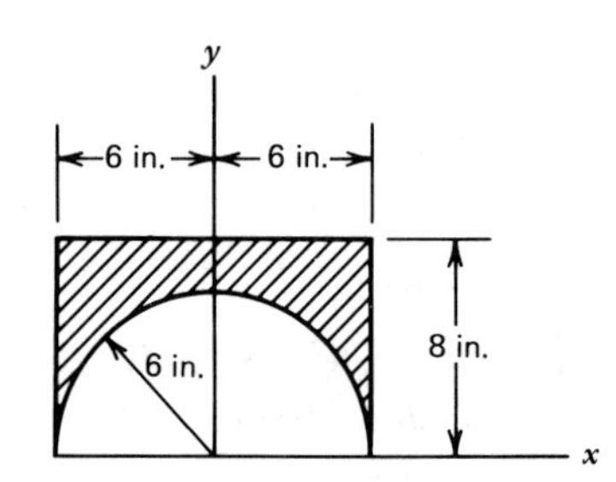

Fig. 4-52. Area for Problem 3.

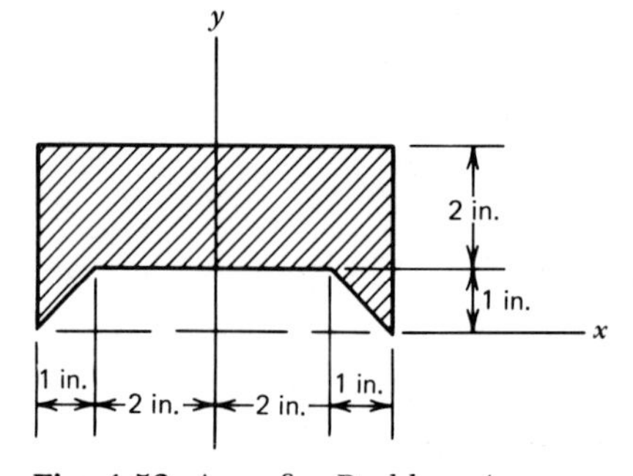

Fig. 4-53. Area for Problem 4.

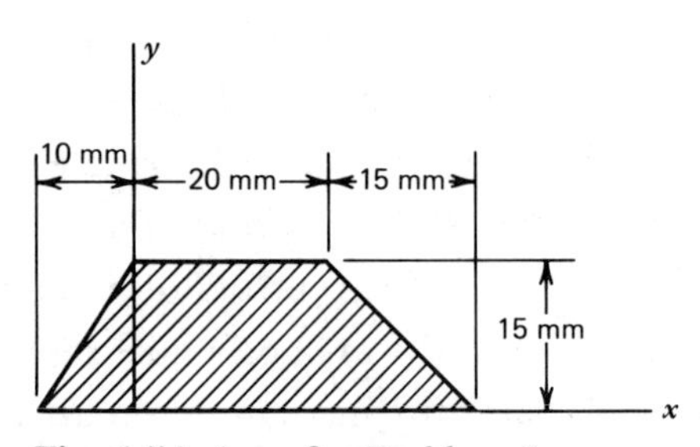

Fig. 4-54. Area for Problem 5.

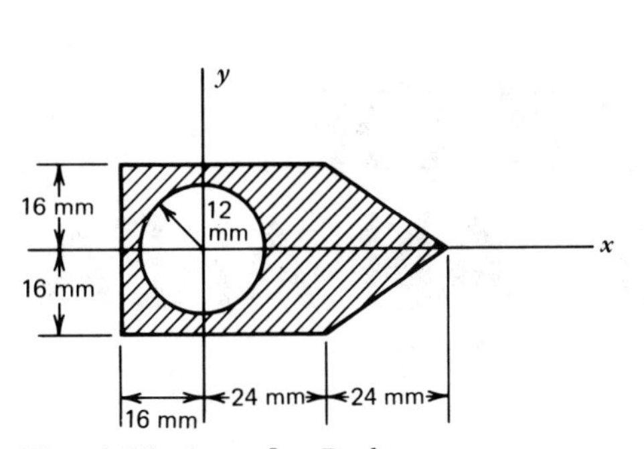

Fig. 4-55. Area for Problem 6.

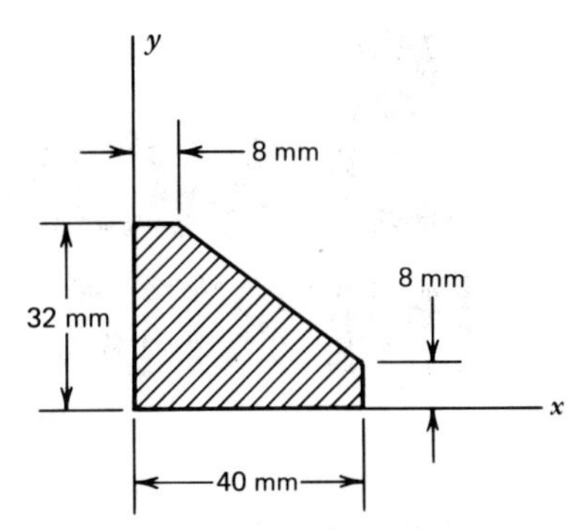

Fig. 4-56. Area for Problem 7.

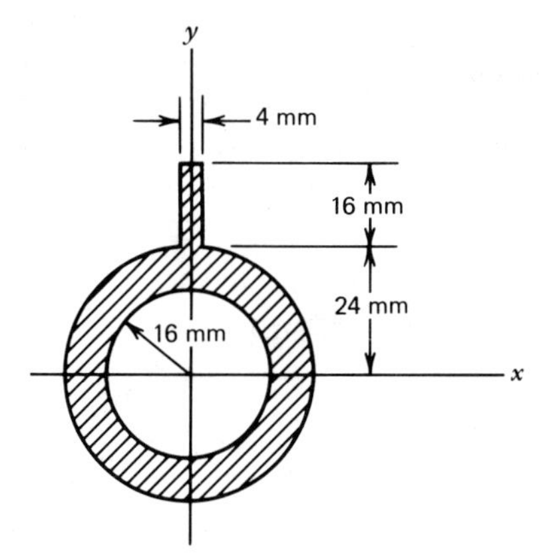

Fig. 4-57. Area for Problem 8.

5 MOMENT OF INERTIA OF AN AREA

PURPOSE OF THIS CHAPTER. In this chapter we will go farther in our study of areas. We will learn how to calculate the *moment of inertia* of an area. This is another one of the properties of an area that a designer must understand in order to design parts that are bent or twisted. The designer must know a lot of facts and methods, and the best way to learn them is to study them one at a time. Right now we will learn how to calculate moment of inertia of an area. Then we will be ready to study beams and to learn how to use moment of inertia.

5-1. WHAT MOMENT OF INERTIA IS

Moment of inertia is a mathematical property of an area that we have to know something about in order to study beams and columns, because the moment of inertia of the area of the cross section of the beam or column is a measure of the stiffness of the beam or column in bending. It is hard to get a full appreciation of what is meant by the term until after you have studied beams (you will in Chapter 6) and have seen how it is used there. Probably it will be best for you to take moment of inertia on faith for the time being and be sure you learn enough about how to calculate the moment of inertia so you can do it when you need to.

Here is a definition of moment of inertia. Let's think of the area in Fig. 5-1 as divided into a great many small areas, each of the amount a. The distance of one small area from the x axis is y. Then the moment of inertia of the area with respect to the x axis is equal to the sum of the quantities obtained by multiplying each small area a by the square of its distance y from the x axis. For each small area we would have ay^2, and for the whole area we would have the sum of the quantities ay^2. This sum could be written

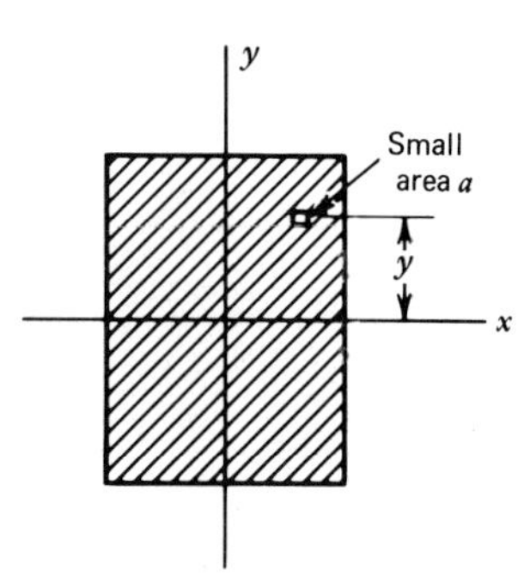

Fig. 5-1. Rectangular area made up of many small parts.

$$\Sigma ay^2$$

(Remember we used Σ for *sum* in Chapter 2.) The symbol used to represent moment of inertia is I, and a subscript is added to designate the axis to which it refers. Thus I_x represents the moment of inertia with respect to the x axis, and we have

$$I_x = \Sigma ay^2$$

We aren't in a position to derive any fundamental formulas for moment of inertia of area. It would be necessary to divide the large area into an infinitely large number of small areas a, each infinitely small, and use calculus to derive the formulas. Instead, we are going to give you the formulas you need and show you how to use them. Then you will be able to calculate the moment of inertia of any practical area you encounter later.

UNITS

Moment of inertia is expressed in unusual dimensional units, and we can see what they are from the definition

$$I_x = \Sigma ay^2$$

If we are using gravitational units, the small area a is expressed in square inches, or inches to the second power (in.^{2}); y is in inches, and y^2 is in inches to the second power (in.^{2}). When we multiply a in in.^{2} by y^2 in in.^{2}, the units are

$$\text{in.}^2 \times \text{in.}^2 = \text{in.}^4$$

just as we would say

$$x^2 \times x^2 = x^4$$

(You should remember from algebra that you add exponents when you multiply.)

If we are using the metric system, the small area is expressed in square millimeters, or millimeters to the second power (mm^2); y is in millimeters, and y^2 is millimeters to the second power (mm^2). When we multiply a in mm^2 by y^2 in mm^2, the units are

$$\text{mm}^2 \times \text{mm}^2 = \text{mm}^4$$

Each numerical answer for moment of inertia will be expressed in a unit of length to the fourth power. Don't waste time trying to get a picture of length to the fourth power, because there isn't any such thing. However, you will be able to see the meaning of this sort of unit when you study beams.

5-2. MOMENTS OF INERTIA OF SIMPLE AREAS

We are now going to give you the formulas for moment of inertia of some of the simple areas. Let's start with a rectangle. The moment of inertia of a rectangular area, with respect to an axis through the centroid and parallel to the base, is $\frac{1}{12}$ times the width times the cube of the

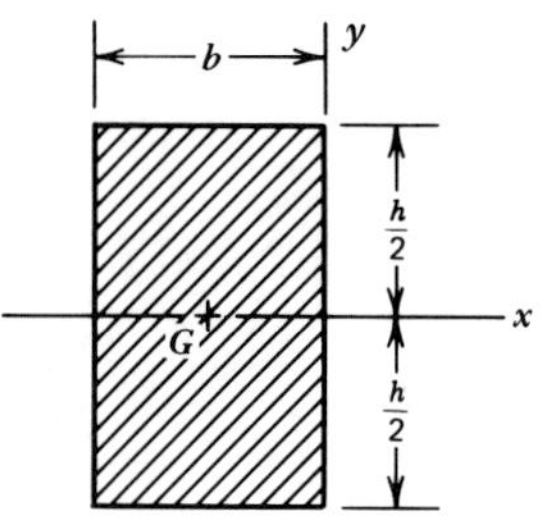

Fig. 5-2. Rectangular area.

depth. Now look at Fig. 5-2 to put this as a formula. The width of the rectangle is b and the depth is h. The centroid of the area is at G, halfway between the top and bottom, so the x axis passes through the centroid; also, the x axis is parallel to the base. The formula is

$$I_x = \frac{1}{12} bh^3$$

The subscript x shows that the moment of inertia is with respect to the x axis. Now let's put some numbers in the formula.

Illustrative Example 1. Find the moment of inertia of a rectangular area 6 in. wide and 8 in. deep with respect to a centroidal axis parallel to the base.

Solution:

1. The width b of the rectangle is 6 in.
2. The depth h of the rectangle is 8 in.
3. The moment of inertia of the area with respect to the x axis is

$$I_x = \frac{1}{12} bh^3 = \frac{1}{12} \times 6 \times (8)^3 = 256 \text{ in.}^4$$

Illustrative Example 2. Calculate the moment of inertia of a rectangular area 120 mm wide and 10 mm deep with respect to a centroidal axis parallel to the base.

Solution:

1. $b = 120$ mm.
2. $h = 10$ mm.
3. $I = 1/12 \times 120 \times (10)^3 = 10\,000 \text{ mm}^4$.

TRIANGLE

The formula for the moment of inertia of a triangular area with respect to a centroidal axis parallel to the base is similar to that for a rectangular area. Just change the fraction from 1⁄12 to 1⁄36 and you have it. It's $\frac{1}{36} bh^3$. In Fig. 5-3 the centroid of the area is at G, so the x axis is a centroidal axis; also, it is parallel to the base of the triangle. The width is b and the depth is h. Then,

$$I_x = \frac{1}{36} bh^3$$

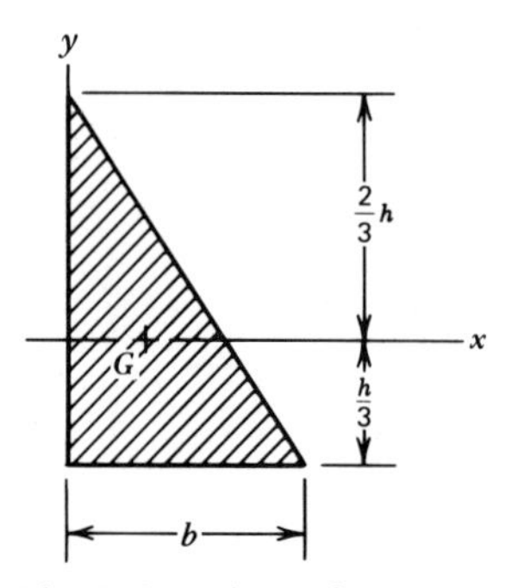

Fig. 5-3. Triangular area.

A right triangle can be turned in any of the four ways shown in Fig.

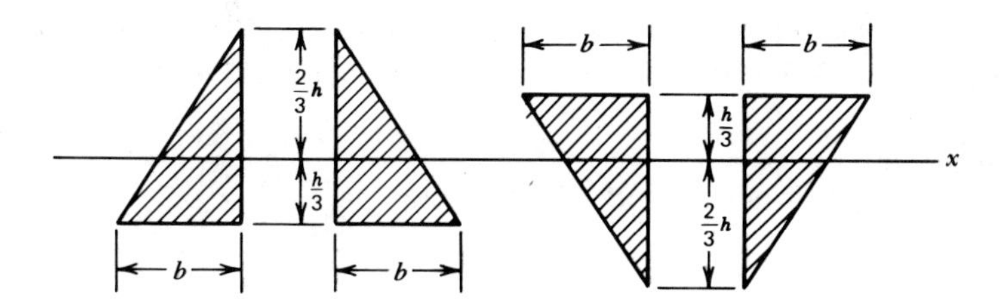

Fig. 5-4. Different orientations of a right triangle.

5-4, but so long as the axis goes through the centroid and is parallel to the base, the moment of inertia is the same. Each of the four triangles in Fig. 5-4 has the same moment of inertia with respect to the x axis. It is

$$I_x = \frac{1}{36} bh^3$$

Illustrative Example 3. Calculate the moment of inertia of the area of a right triangle 3 in. wide and 6 in. deep with respect to a centroidal axis parallel to the base.

Solution:

1. $b = 3$ in.
2. $h = 6$ in.
3. $I = \frac{1}{36} bh^3 = \frac{1}{36} \times 3 \times (6)^3 = 18$ in.^{4}

Illustrative Example 4. Find the moment of inertia of a right triangular area 48 mm wide and 20 mm deep with respect to a centroidal axis parallel to the base.

Solution:

1. $b = 48$ mm.
2. $h = 20$ mm.
3. $I = \frac{1}{36} bh^3 = \frac{1}{36} \times 48 \times (20)^3 = 10\,700$ mm^4.

CIRCLE

The moment of inertia of a circular area with respect to a diametral axis is $\pi r^4/4$, where r is the radius of the circle. Figure 5-5 shows a circular area. The centroid is at G, and the x axis coincides with the diameter, so it is a *diametral axis*. Thus

$$I_x = \frac{\pi}{4} r^4$$

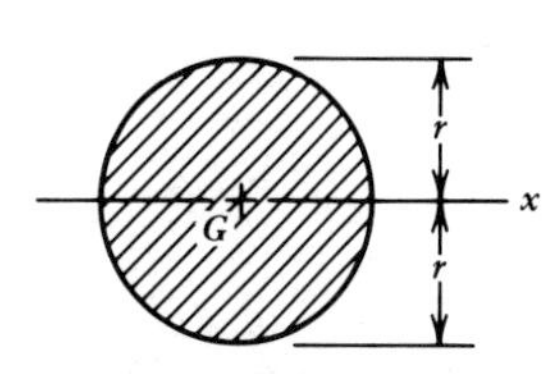

Fig. 5-5. Circular area.

Illustrative Example 5. Calculate the moment of inertia of a circular area 3 in. in diameter with respect to a diametral axis.

Solution:

1. The radius is one-half the diameter, so

$$r = \tfrac{1}{2} \times 3 = 1.5 \text{ in.}$$

2. $I = \dfrac{\pi}{4} r^4 = \dfrac{\pi}{4}(1.5)^4 = 3.976 \text{ in.}^4$

IN TERMS OF DIAMETER

Some people prefer to have the formula for the moment of inertia of a circular area given in terms of the diameter d. We can change the formula easily, since the radius is one-half the diameter; $r = d/2$. Then,

$$I = \frac{\pi}{4} r^4 = \frac{\pi}{4}\left(\frac{d}{2}\right)^4 = \frac{\pi}{64} d^4$$

This formula is just as easy to use.

Illustrative Example 6. Find the moment of inertia of a circular area 12 mm in diameter with respect to a diametral axis.

Solution:

1. $d = 12.$

2. $I = \dfrac{\pi}{64} d^4 = \dfrac{\pi}{64}(12)^4 = 1\,018 \text{ mm}^4.$

SEMICIRCLES

Later we will need a formula for finding the moment of inertia of a semicircular area with respect to a diameter boundary. This is just half of the moment of inertia of a full circular area and is $\pi r^4/8$. The semicircle can be above the x axis like A in Fig. 5-6a or below the axis like B in Fig. 5-6a. The formula is still the same. In fact, it can even be used for an axis that bisects a semicircular area. The moment of inertia of the area

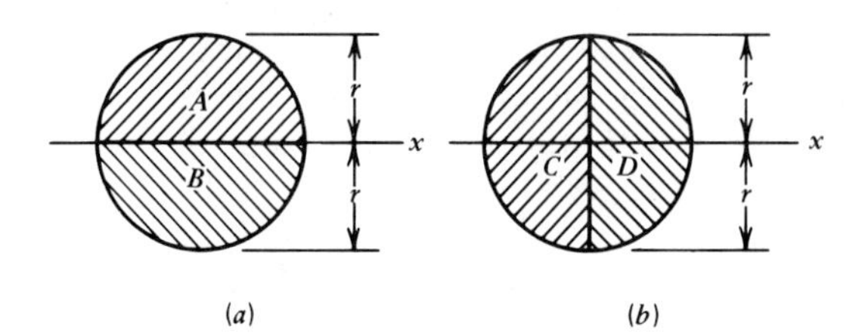

Fig. 5-6. Circular areas divided into semicircles.

C or the area D in Fig. 5-6b is also $\pi r^4/8$. Thus the moment of inertia of each of the areas A, B, C, and D in Fig. 5-6 with respect to the x axis is

$$I_x = \frac{\pi}{8}r^4$$

QUADRANT

The moment of inertia of a quadrant with respect to a radius boundary is one-fourth of the moment of inertia of a whole circular area, no matter which quadrant is taken. The moment of inertia of each of the four quadrants A, B, C, and D in Fig. 5-7 with respect to the x axis is

$$I_x = \frac{\pi}{16}r^4$$

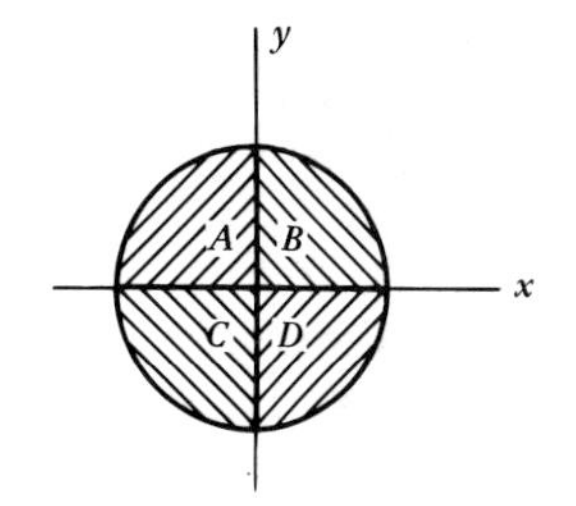

Fig. 5-7. Circular area divided into quadrants.

VERTICAL AXIS

So far we have calculated moment of inertia only with respect to a horizontal axis, but frequently we want to find a moment of inertia with respect to a vertical axis. This makes it necessary for us to expand our ideas of the width of an area and the depth of an area. Hereafter, we will think of the width as the dimension parallel to the axis we are interested in and the depth as the dimension perpendicular to this axis. With respect to the x axis, the width is the horizontal dimension, and the depth is the vertical dimension, but with respect to the y axis, the width is the vertical dimension, and the depth is the horizontal dimension. Let's see how this works in examples. We will still use $\frac{1}{12}\,bh^3$ for a rectangle and $\frac{1}{36}\,bh^3$ for a triangle, but we will have to be careful in choosing b and h.

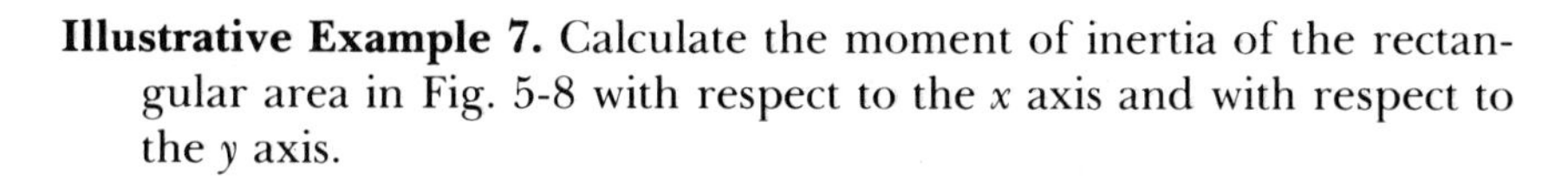

Illustrative Example 7. Calculate the moment of inertia of the rectangular area in Fig. 5-8 with respect to the x axis and with respect to the y axis.

Solution:

1. For I_x,
   (a) $b = 2$ in.
   (b) $h = 4$ in.
   (c) $I_x = \frac{1}{12}\,bh^3 = \frac{1}{12} \times 2 \times (4)^3 = 10.67 \text{ in.}^4$
2. For I_y,
   (a) $b = 4$ in.
   (b) $h = 2$ in.
   (c) $I_y = \frac{1}{12}\,bh^3 = \frac{1}{12} \times 4 \times (2)^3 = 2.67 \text{ in.}^4$

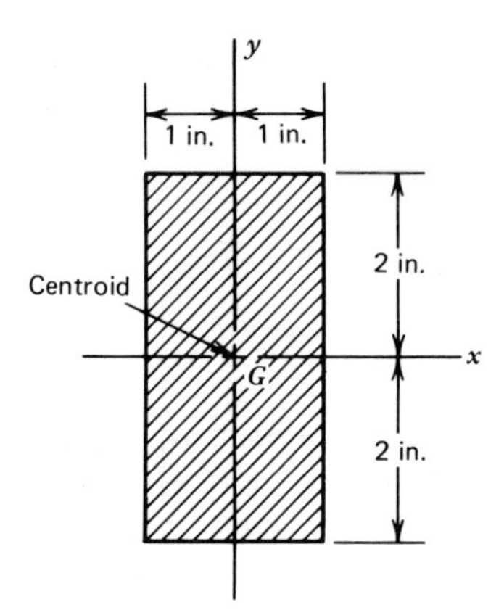

Fig. 5-8. Rectangular area for Example 7.

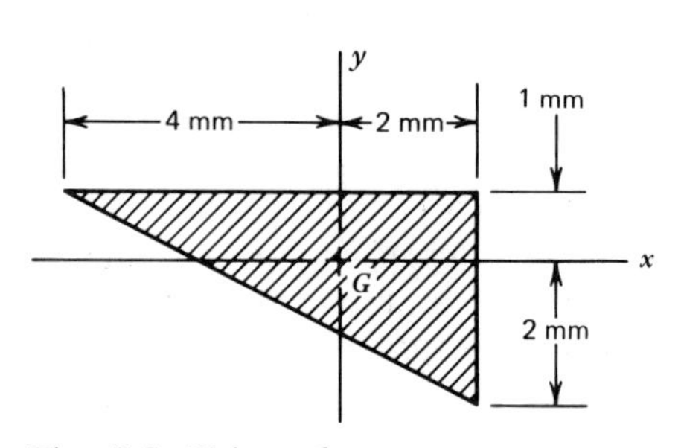

Fig. 5-9. Triangular area for Example 8.

Illustrative Example 8. Find the moment of inertia of the triangular area in Fig. 5-9 with respect to both the x and y axes.

Solution:

1. For I_x,
   (a) $b = 6$ mm. (b) $h = 3$ mm.
   (c) $I_x = \frac{1}{36}bh^3 = \frac{1}{36} \times 6 \times (3)^3 = 4.5 \text{ mm}^4$.
2. For I_y,
   (a) $b = 3$ mm. (b) $h = 6$ mm.
   (c) $I_y = \frac{1}{36}bh^3 = \frac{1}{36} \times 3 \times (6)^3 = 18 \text{ mm}^4$.

Practice Problems (Section 5-2). Work these practice problems so that you will be sure to remember the formulas for moment of inertia of a *rectangle, triangle,* and *circle.* You will need them later. Just find I_x and I_y for each of the areas indicated in the following problems.

1. The rectangular area in Fig. 5-10.
2. The triangular area in Fig. 5-11.
3. The circular area in Fig. 5-12.
4. The rectangular area in Fig. 5-13.
5. The triangular area in Fig. 5-14.
6. The semicircular area in Fig. 5-15.
7. The quadrant in Fig. 5-16.
8. The rectangular area in Fig. 5-17.
9. The triangular area in Fig. 5-18.
10. The semicircular area in Fig. 5-19.

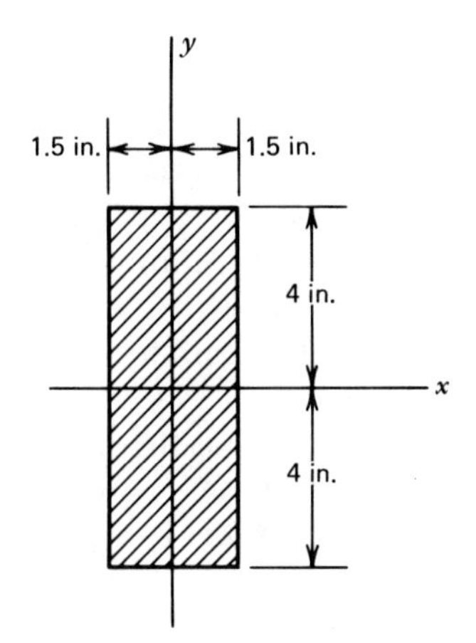

Fig. 5-10. Rectangular area for Problem 1.

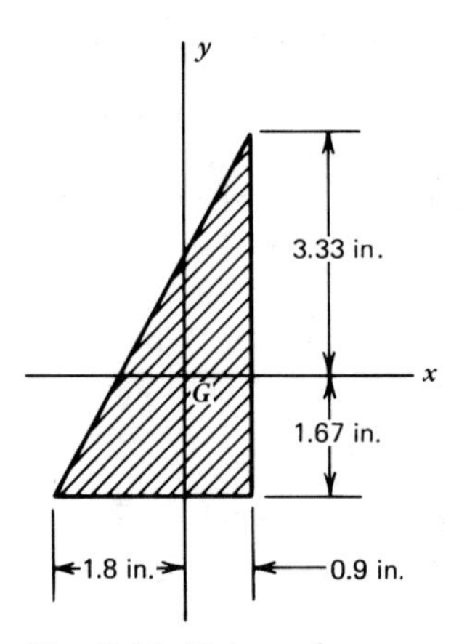

Fig. 5-11. Triangular area for Problem 2.

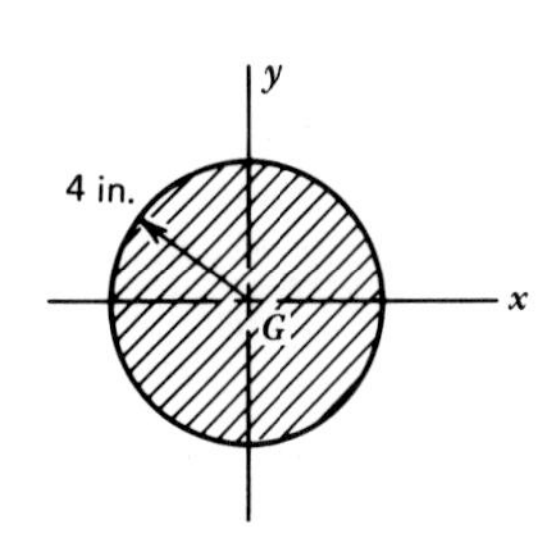

Fig. 5-12. Circular area for Problem 3.

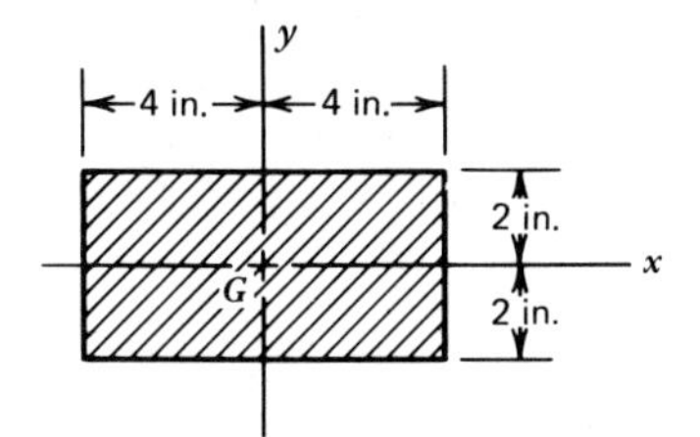

Fig. 5-13. Rectangular area for Problem 4.

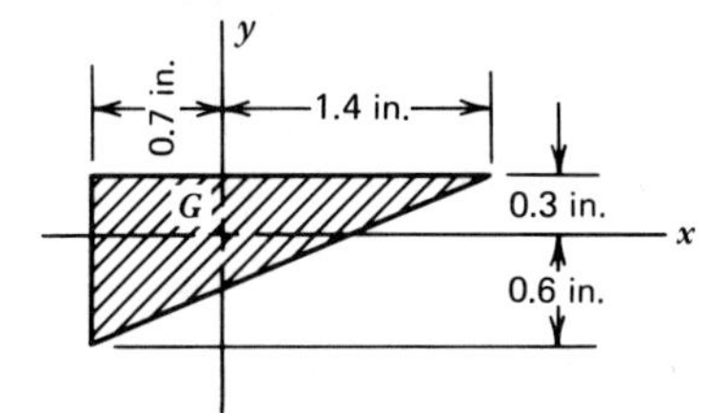

Fig. 5-14. Triangular area for Problem 5.

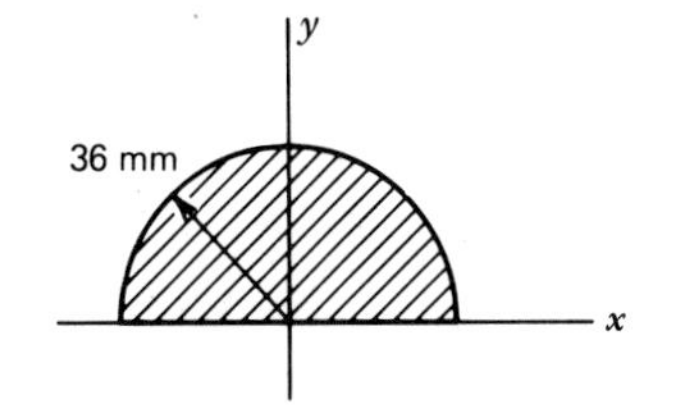

Fig. 5-15. Semicircular area for Problem 6.

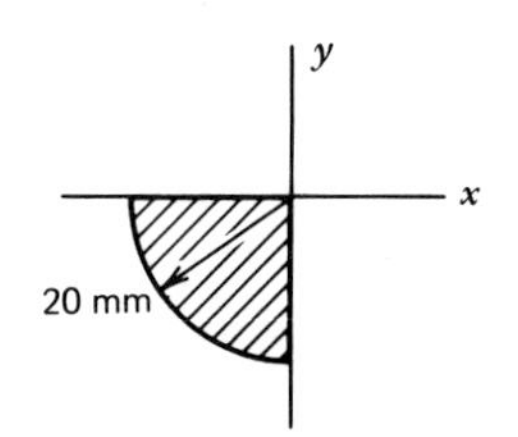

Fig. 5-16. Quadrant for Problem 7.

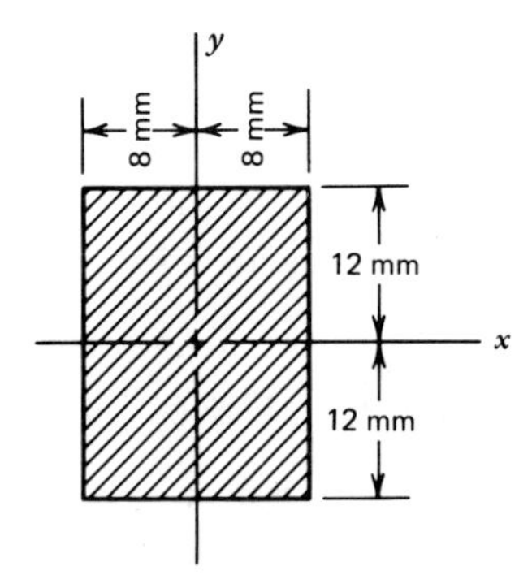

Fig. 5-17. Rectangular area for Problem 8.

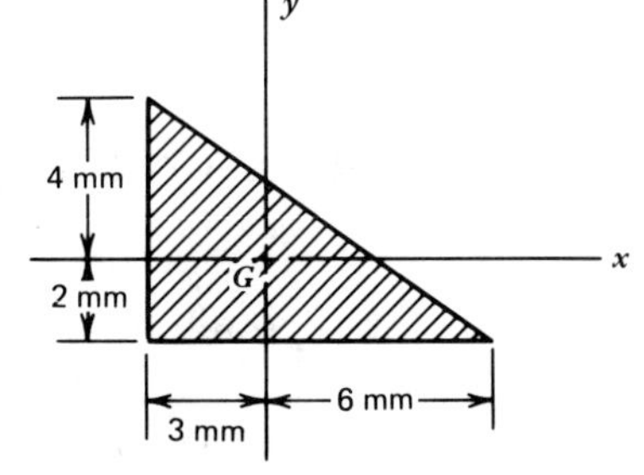

Fig. 5-18. Triangular area for Problem 9.

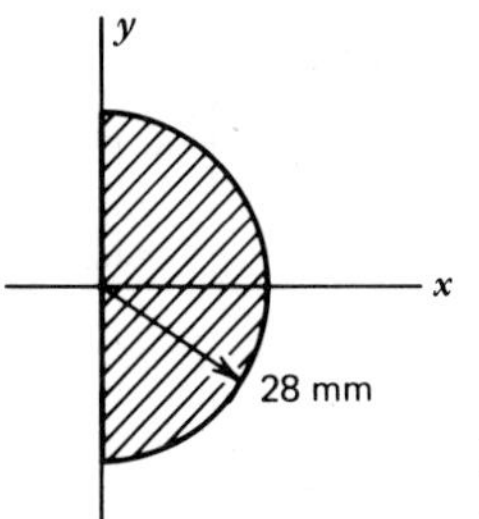

Fig. 5-19. Semicircular area for Problem 10.

5-3. PARALLEL-AXIS THEOREM

The parallel-axis theorem enables us to find the moment of inertia of an area with respect to any axis if we know the moment of inertia with respect to a parallel axis through the centroid of the area. The great advantage of knowing the theorem is that it reduces the number of formulas that have to be remembered. If you know how to use the parallel-axis theorem, you only have to remember a few moments of inertia with respect to centroidal axes.

Let's look at Fig. 5-20 while we think about the parallel-axis theorem. Point G is the centroid of the area, and the axis x_g passes through the centroid (the subscript g indicates this). The x axis is parallel to the x_g axis, and the perpendicular distance between the two axes is c. Now, according to the parallel-axis theorem, the moment of inertia of an area with respect to any axis is equal to the moment of inertia with respect to a parallel axis through the centroid of the area, plus the product of the area and the square of the distance between the two axes. (The proof of this theorem requires a knowledge of calculus, but we are more interested in the use than the proof, so it is given here without proof.) The formula that says the same thing as the theorem is

$$I_x = I_{xg} + Ac^2$$

Here I_x is the moment of inertia of the area with respect to the x axis; I_{xg} is the moment of inertia with respect to the centroidal axis x_g; A is the area, and c is the perpendicular distance between the two axes. The formula is ordinarily used to find I_x when the formula for I_{xg} is known. Now we are ready to try it.

Illustrative Example 1. Calculate the moment of inertia of the rectangular area in Fig. 5-21 with respect to the x axis.

Solution:

1. $I_{xg} = \frac{1}{12} bh^3 = \frac{1}{12} \times 6 \times (2)^3 = 4 \text{ in.}^4$
2. $A = 6 \times 2 = 12 \text{ in.}^2$

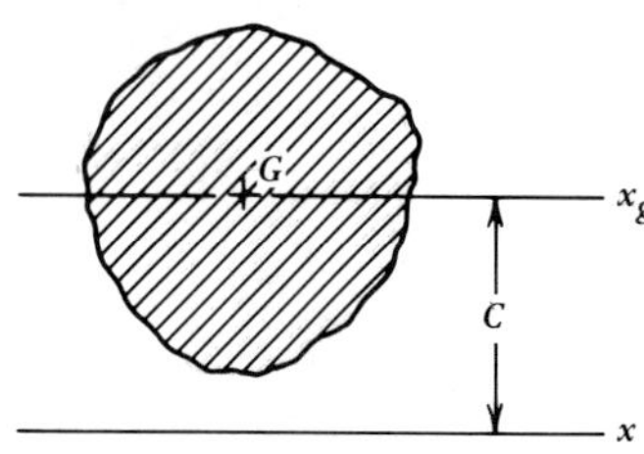

Fig. 5-20. Illustration for parallel-axis theorem.

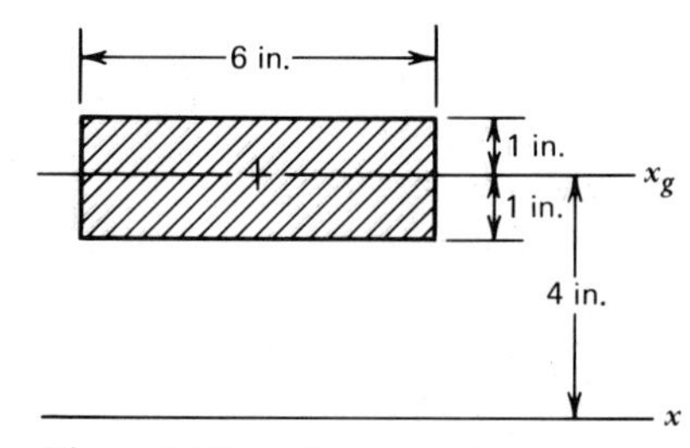

Fig. 5-21. Rectangular area for Example 1.

3. $c = 4$ in.
4. $I_x = I_{xg} + Ac^2 = 4 + 12(4)^2 = 4 + 192 = 196$ in.^{4}

Illustrative Example 2. Find the moment of inertia of the triangular area in Fig. 5-22 with respect to the x axis.

Solution:

1. $I_{xg} = \frac{1}{36}\,bh^3 = \frac{1}{36} \times 50 \times (30)^3 = 37\,500$ mm^4.
2. $A = \frac{1}{2} \times 50 \times 30 = 750$ mm^2.
3. $c = 50$ mm.
4. $I_x = I_{xg} + Ac^2 = 37\,500 + 750(50)^2 = 37\,500 + 1\,875\,000$
   $= 1\,912\,500$ mm^4 or 1.9125×10^6 mm^4

UP OR DOWN
It doesn't matter whether the x axis is above or below the centroidal axis. You still add Ac^2 to I_{xg} to get I_x.

Illustrative Example 3. Calculate the moment of inertia of the circular area in Fig. 5-23 with respect to the x axis.

Solution:

1. $I_{xg} = \dfrac{\pi}{4}r^4 = \dfrac{\pi}{4}(1.25)^4 = 1.917$ in.^{4}
2. $A = \pi r^2 = \pi(1.25)^2 = 4.909$ in.^{2}
3. $c = 1.5$ in.
4. $I_x = I_{xg} + Ac^2 = 1.917 + 4.909(1.5)^2 = 1.917 + 11.043$
   $= 12.960$ in.^{4}

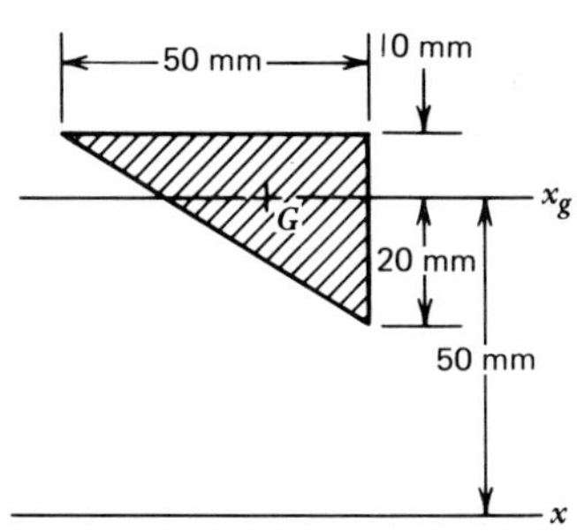

Fig. 5-22. Triangular area for Example 2.

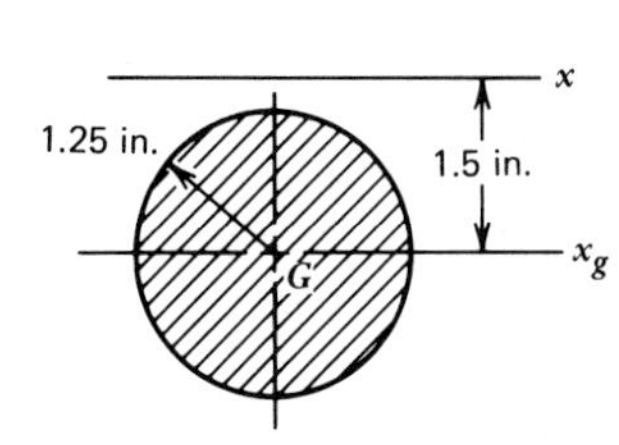

Fig. 5-23. Circular area for Example 3.

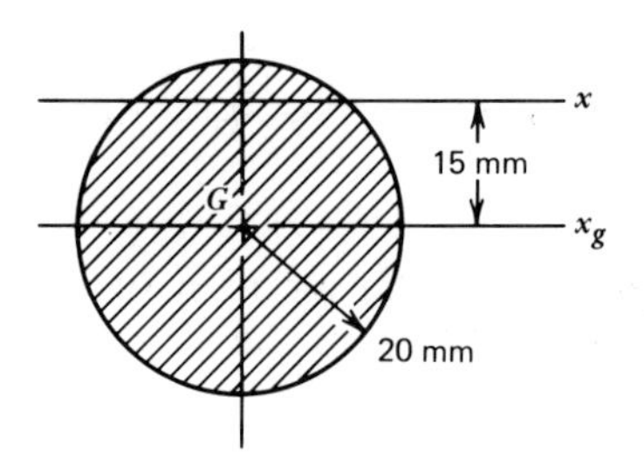

Fig. 5-24. Circular area for Example 4.

INTERSECTING AXIS
It often happens that the x axis passes through the area. This doesn't make any difference. Just go ahead and use the theorem.

Illustrative Example 4. Find the moment of inertia of the circular area in Fig. 5-24 with respect to the x axis.

Solution:

1. $I_{xg} = \frac{\pi}{4}r^4 = \frac{\pi}{4}(20)^4 = 125\ 600\ \text{mm}^4$.
2. $A = \pi r^2 = \pi(20)^2 = 1\ 256\ \text{mm}^2$.
3. $c = 15$ mm.
4. $I_x = I_{xg} + Ac^2 = 125\ 600 + 1\ 256(15)^2 = 125\ 600 + 282\ 600$
   $= 408\ 200\ \text{mm}^4$

VERTICAL AXIS
We must be able to calculate a moment of inertia with respect to a vertical axis, so let's write the parallel-axis formula in a suitable manner for this purpose. The parallel-axis theorem states that the moment of inertia of an area with respect to any axis is equal to the moment of inertia with respect to a parallel axis through the centroid of the area, plus the product of the area and the square of the distance between the two axes. Figure 5-25 shows a rectangular area for which y_g is a centroidal axis and y is an axis parallel to y_g; the distance between the two axes is c. For such vertical axes, the parallel-axis formula is

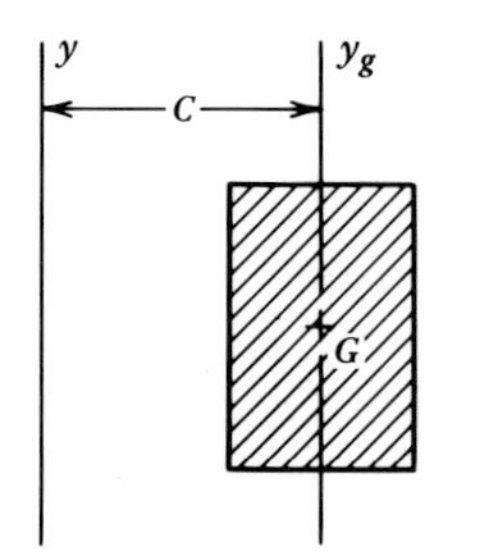

Fig. 5-25. Rectangular area.

$$I_y = I_{yg} + Ac^2$$

Let's use it.

Illustrative Example 5. Calculate the moment of inertia of the triangular area in Fig. 5-26 with respect to the y axis.

Solution:

1. $I_{yg} = 1/36\ bh^3 = 1/36 \times 2 \times (3)^3 = 1.5$ in.^{4}
2. $A = 1/2 \times 2 \times 3 = 3$ in.^{2}
3. $c = 3$ in.
4. $I_y = I_{yg} + Ac^2 = 1.5 + 3 \times (3)^2 = 1.5 + 27 = 28.5$ in.^{4}

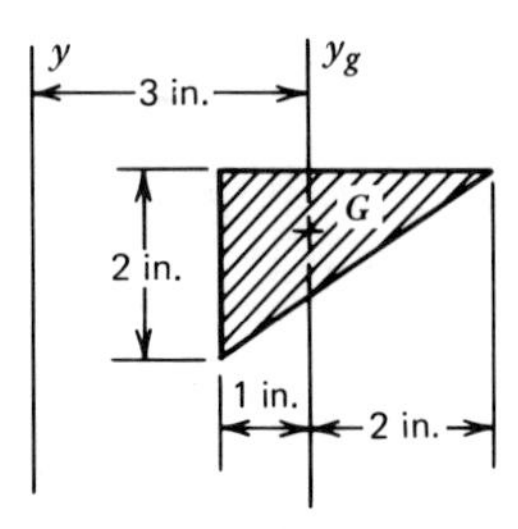

Fig. 5-26. Triangular area for Example 5.

INTERSECTING AXIS

It doesn't make any difference whether the area is to the right or left of the y axis. Neither does it matter whether the axis passes through the area. The formula stays the same.

Illustrative Example 6. Find the moment of inertia of the rectangular area in Fig. 5-27 with respect to the y axis.

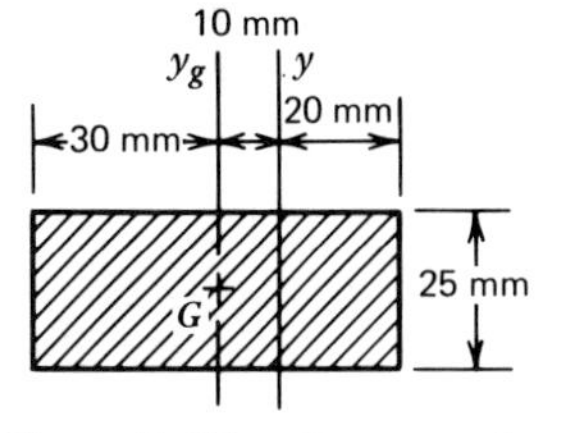

Fig. 5-27. Rectangular area for Example 6.

Solution:

1. $I_{yg} = 1/12\ bh^3 = 1/12 \times 25 \times (60)^3 = 450\ 000$ mm^4
2. $A = 25 \times 60 = 1\ 500$ mm^2
3. $c = 10$ mm
4. $I_y = I_{yg} + Ac^2 = 450\ 000 + 1\ 500 \times (10)^2 = 450\ 000 + 150\ 000$
   $= 600\ 000$ mm^4

Illustrative Example 7. Calculate the moment of inertia of the circular area in Fig. 5-28 with respect to the y axis.

Solution:

1. $I_{yg} = (\pi/4)r^4 = \pi/4(0.8)^4 = 0.322$ in.^{4}

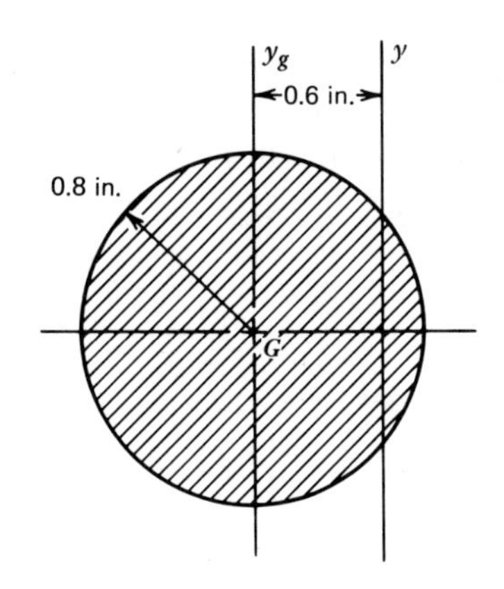

Fig. 5-28. Circular area for Example 7.

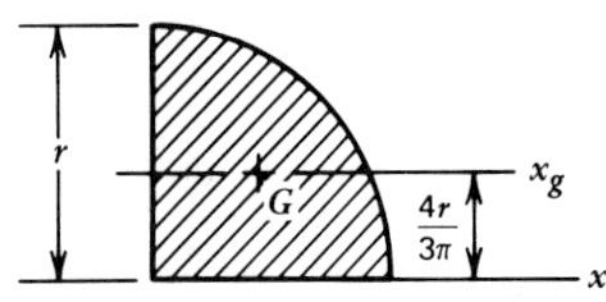

Fig. 5-29. Quadrant.

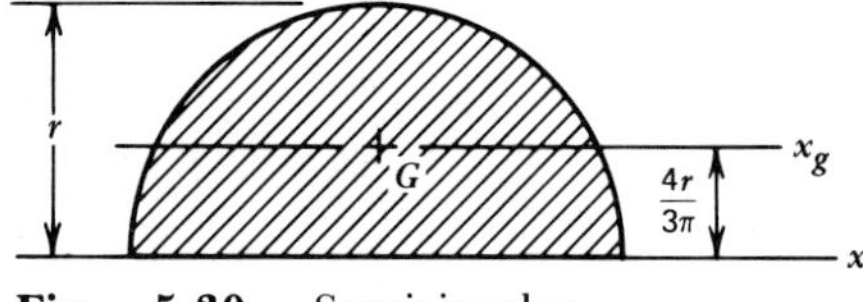

Fig. 5-30. Semicircular area.

2. $A = \pi r^2 = \pi(0.8)^2 = 2.01\text{ in.}^2$
3. $c = 0.6$ in.
4. $I_y = I_{yg} + Ac^2 = 0.322 + 2.01 \times (0.6)^2 = 0.322 + 0.724 = 1.046$ in.^{4}

QUADRANT

Now let's use the parallel-axis theorem to derive a formula. Looking at the quadrant in Fig. 5-29, we can write

$$I_x = I_{xg} + Ac^2$$

We already know I_x, A, and c but we don't know I_{xg}. Let's find it, because we are going to use it in Section 5-4. We know that $I_x = (\pi/16)r^4$ and that $A = (\pi/4)r^2$, and c is shown on the figure as $4r/3\pi$. When we substitute the known quantities, we have

$$I_x = I_{xg} + Ac^2$$

$$\frac{\pi}{16}r^4 = I_{xg} + \frac{\pi}{4}r^2\left(\frac{4r}{3\pi}\right)^2$$

$$0.1963r^4 = I_{xg} + 0.1415r^4$$

Transposing,

$$I_{xg} = 0.1963r^4 - 0.1415r^4 = 0.0548r^4$$

The moment of inertia of the semicircular area in Fig. 5-30 with respect to the x_g axis is just twice the moment of inertia of the quadrant (because the semicircle consists of two quadrants). It is, then,

$$I_{xg} = 2 \times 0.0548r^4 = 0.1096r^4$$

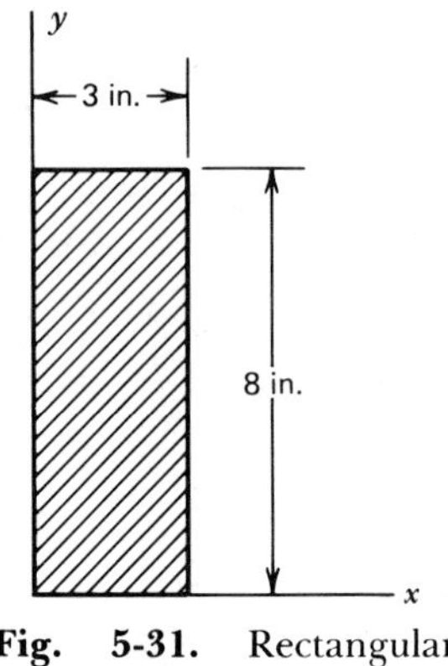

Fig. 5-31. Rectangular area for Problem 1.

Practice Problems (Section 5-3). Now practice using the parallel-axis theorem. Remember that the two axes in the problem must be parallel and that one of them must go through the centroid of the area. Calculate:

1. I_x and I_y for the rectangular area in Fig. 5-31.

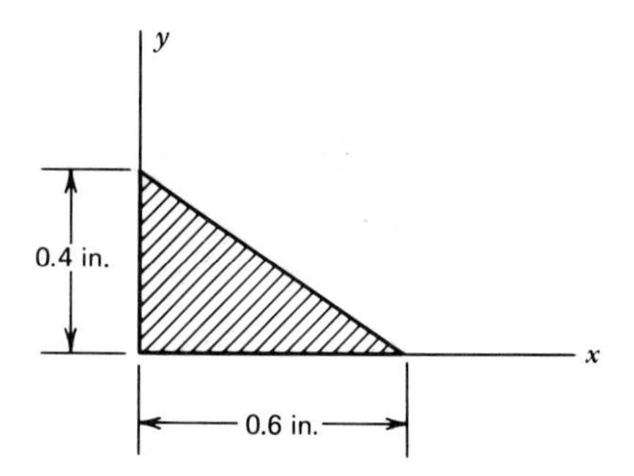

Fig. 5-32. Triangular area for Problem 2.

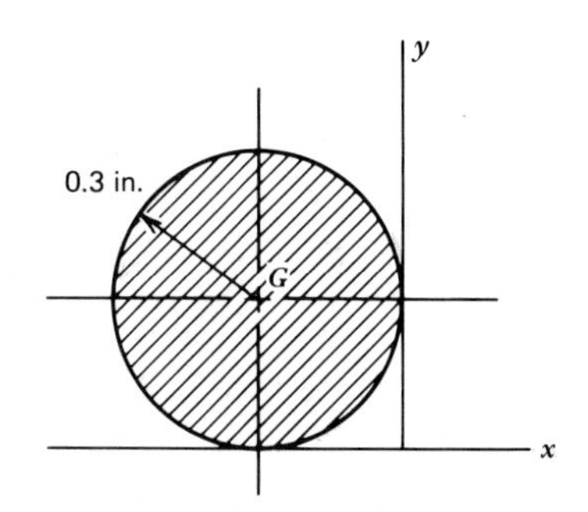

Fig. 5-33. Circular area for Problem 3.

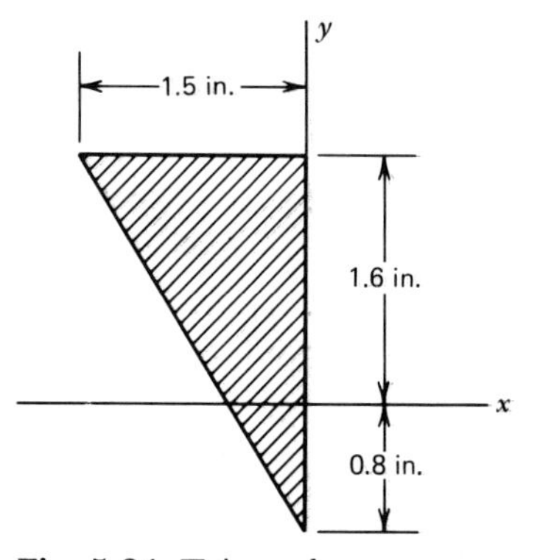

Fig. 5-34. Triangular area for Problem 4.

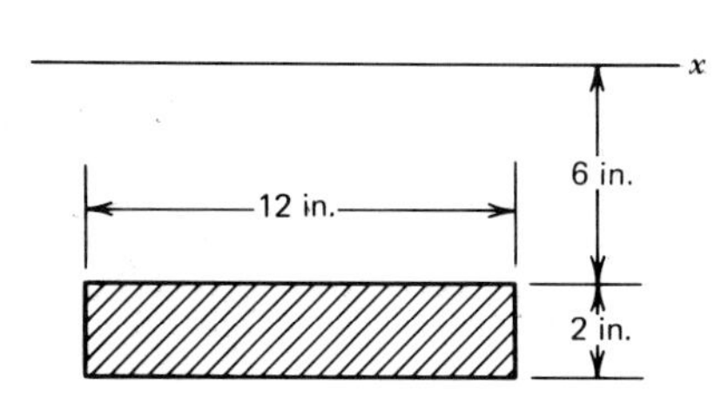

Fig. 5-35. Rectangular area for Problem 5.

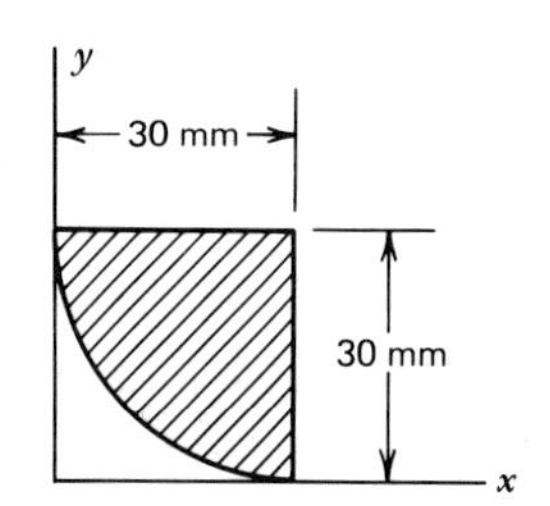

Fig. 5-36. Quadrant for Problem 6.

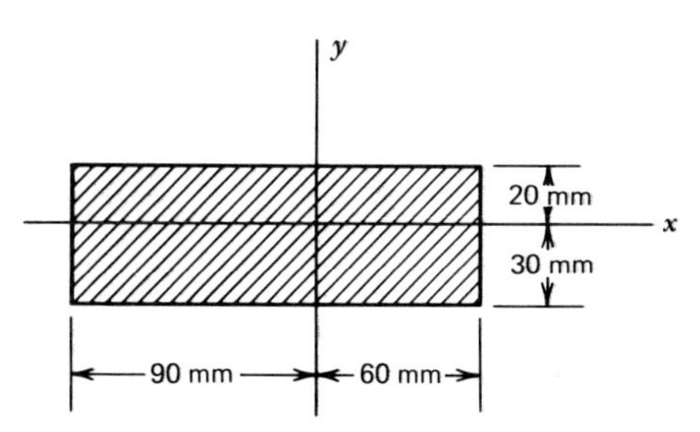

Fig. 5-37. Rectangular area for Problem 7.

2. I_x and I_y for the triangular area in Fig. 5-32.
3. I_x and I_y for the circular area in Fig. 5-33.
4. I_x for the triangular area in Fig. 5-34.
5. I_x for the rectangular area in Fig. 5-35.
6. I_x and I_y for the quadrant in Fig. 5-36.
7. I_x and I_y for the rectangular area in Fig. 5-37.

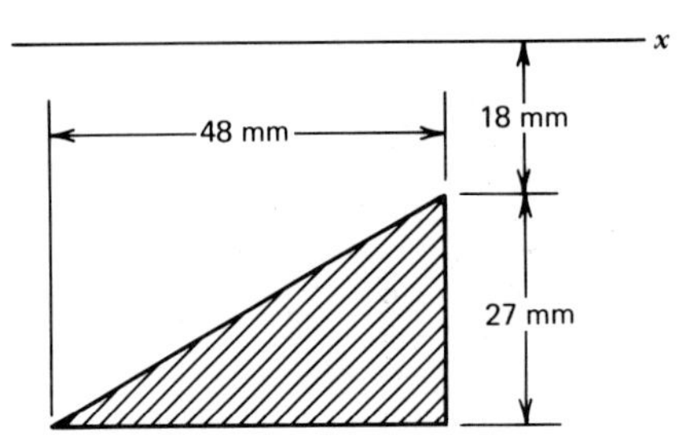

Fig. 5-38. Triangular area for Problem 8.

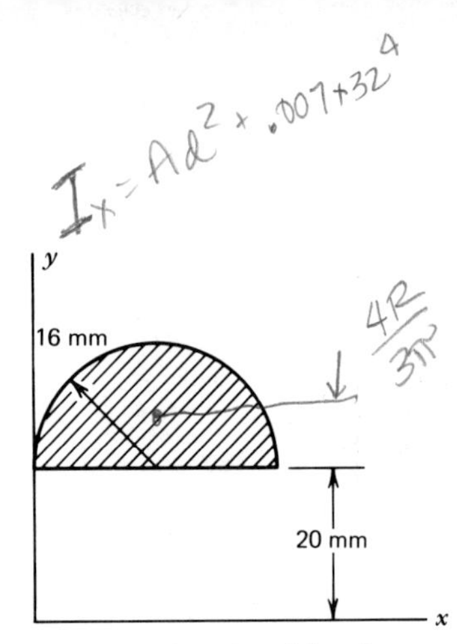

Fig. 5-39. Semicircular area for Problem 9.

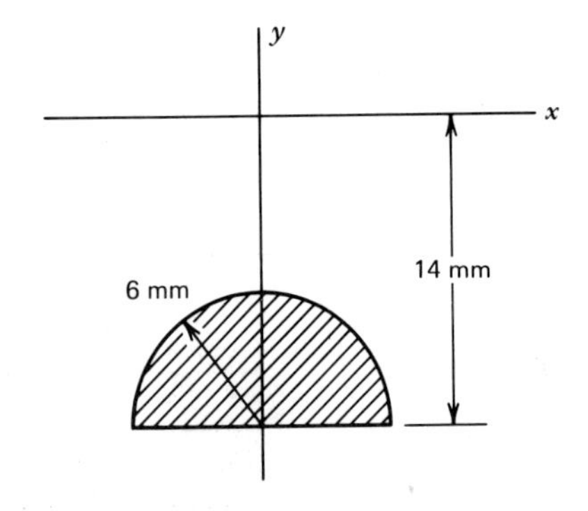

Fig. 5-40. Semicircular area for Problem 10.

8. I_x for the triangular area in Fig. 5-38.
9. I_x and I_y for the semicircular area in Fig. 5-39.
10. I_x and I_y for the semicircular area in Fig. 5-40.

5-4. MOMENT OF INERTIA OF COMPOSITE AREAS

You remember what a composite area is, because you learned how to locate the centroid of a composite area in Chapter 4. A *composite area* is an area that can be divided into simple areas, such as rectangles, triangles, and quadrants. Our previous work in calculating moment of inertia has been leading up to the problem of finding the moment of inertia of a composite area. This type of area is used commonly in machine and structural members, and we must learn how to find the moment of inertia for it.

A composite area can be divided into simple parts, such as rectangles, triangles, circles, and quadrants. From our work thus far in this chapter we know how to calculate the moment of inertia for each part of the area. Then we will say the moment of inertia of the whole area is equal to the sum of the moments of inertia of the parts of the area (this is another version of the old axiom that the whole is equal to the sum of its parts). The only part of this process we haven't done is the addition of the moments of inertia of the parts of the area. Suppose we have a composite area divided into parts A, B, and C. We can calculate the moment of inertia of each part with respect to the x axis, say, and call these moments of inertia of the parts I_{xa}, I_{xb}, and I_{xc}. Then, letting I_x be the moment of inertia of the composite area with respect to the x axis

$$I_x = I_{xa} + I_{xb} + I_{xc}$$

Let's try this in examples. To begin with, we will take such simple areas that we don't even have to use the parallel-axis theorem. We will use the idea of a negative area.

Illustrative Example 1. Calculate the moment of inertia of the hollow circular area of Fig. 5-41 with respect to the x axis.

Solution:

1. The easiest way to divide the area into parts is to take it as,
   (a) A positive circular area A of 1.5 in. radius.
   (b) A negative circular area B of 1.2 in. radius.
   We think of the inner circle as subtracted from the outer one.
2. For area A,

$$I_{xa} = \frac{\pi}{4}r^4 = \frac{\pi}{4}(1.5)^4 = 3.976 \text{ in.}^4$$

3. For area B,

$$I_{xb} = \frac{\pi}{4}r^4 = \frac{\pi}{4}(1.2)^4 = 1.629 \text{ in.}^4$$

4. For the composite area, the moment of inertia is equal to the moment of inertia of A minus the moment of inertia of B. Thus

$$I_x = I_{xa} - I_{xb} = 3.976 - 1.629 = 2.347 \text{ in.}^4$$

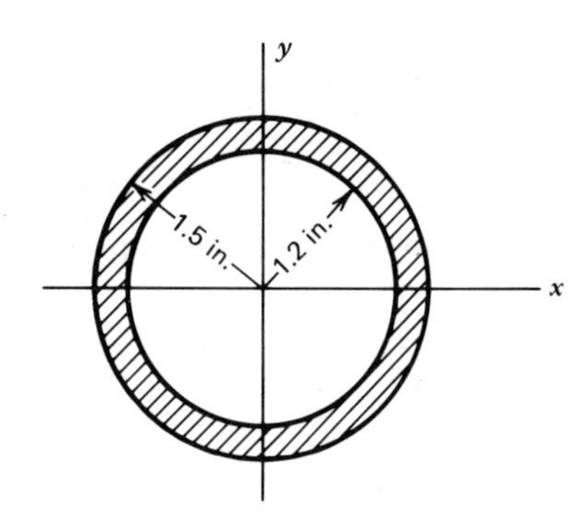

Fig. 5-41. Hollow circular area for Example 1.

Illustrative Example 2. Find the moment of inertia of the area in Fig. 5-42*a* with respect to the *x* axis and the *y* axis.

Solution:

A. For I_x,

1. Figure 5-42*b* shows the area divided into
   (a) A positive rectangle *A*, 40 mm wide and 60 mm deep.
   (b) Two negative rectangles *B*, each 15 mm wide and 40 mm deep.
2. The moment of inertia of the rectangle *A* with respect to the *x* axis is

$$I_{xa} = \frac{1}{12} bh^3 = \frac{1}{12} \times 40 \times (60)^3 = 720\,000 \text{ mm}^4$$

3. The moment of inertia of one rectangle *B* with respect to the *x* axis is

$$I_{xb} = \frac{1}{12} bh^3 = \frac{1}{12} \times 15 \times (40)^3 = 80\,000 \text{ mm}^4$$

4. The moment of inertia of the composite area with respect to the *x* axis is

$$I_x = I_{xa} - 2\,I_{xb} = 720\,000 - 2 \times 80\,000$$
$$= 560\,000 \text{ mm}^4$$

B. For I_y,

1. Figure 5-42*c* shows the area divided into
   (a) Two rectangles *A*, each 10 mm by 40 mm.
   (b) A rectangle *B*, 40 mm by 10 mm.
   Notice that the centroid of each area is on the *y* axis. Also, each area is positive.
2. The moment of inertia of one rectangle *A* with respect to the *y* axis is

$$I_{ya} = \frac{1}{12} bh^3 = \frac{1}{12} \times 10 \times (40)^3$$
$$= 53\,333 \text{ mm}^4$$

3. The moment of inertia of the rectangle *B* with respect to the *y* axis is

$$I_{yb} = \frac{1}{12} bh^3 = \frac{1}{12} \times 40 \times (10)^3$$
$$= 3\,333 \text{ mm}^4$$

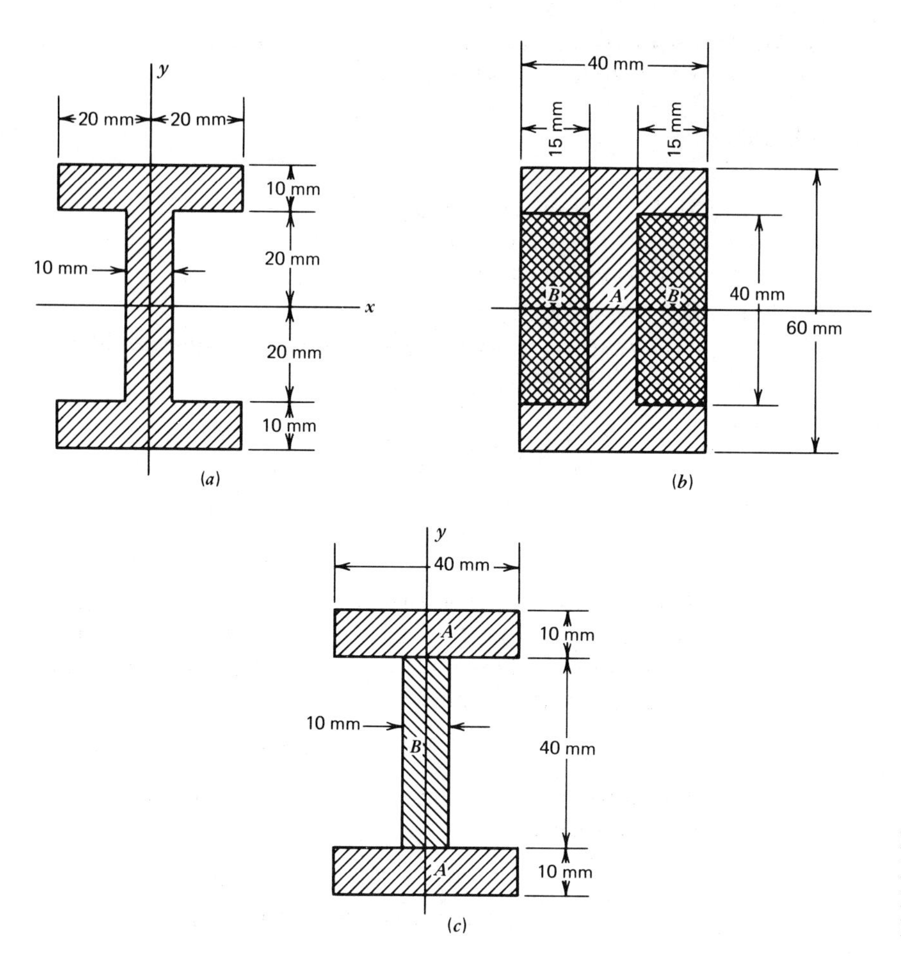

Fig. 5-42. Beam section for Example 2. (*a*) Area. (*b*) Division of area for *Ix*. (*c*) Division of area for *Iy*.

4. The moment of inertia of the composite area with respect to the y axis is the sum of the moments of inertia of the separate parts, so

$$I_y = 2\,I_{ya} + I_{yb} = 2 \times 53\,333 + 3\,333$$
$$= 110\,000 \text{ mm}^4$$

PARALLEL-AXIS THEOREM

Now let's try some harder problems where we have to use the parallel-axis theorem. Remember that the moment of inertia of an area with respect to any axis is equal to the moment of inertia with respect to a parallel axis through the centroid, plus the product of the area and the square of the distance between.

Illustrative Example 3. Calculate the moment of inertia of the area in Fig. 5-43*a* with respect to both the *x* and *y* axes.

Solution:

A. For I_x,

1. Figure 5-43*b* shows the area divided into,
   (a) The rectangular area *A*.
   (b) The triangular area *B*.
2. For the rectangle *A*,
   (a) The moment of inertia with respect to a horizontal axis through the centroid is

$$I_{xg} = \frac{1}{12} bh^3 = \frac{1}{12} \times 4 \times (2)^3 = 2.667 \text{ in.}^4$$

   (b) The area is

$$A_a = 4 \times 2 = 8 \text{ in.}^2$$

   (c) The centroid of the rectangle *A* is 1 in. below the top of the rectangle, and the *x* axis is 1.55 in. below the top of the rectangle. The distance *c* from the centroid of the rectangle to the *x* axis is

$$c = 1.55 - 1 = 0.55 \text{ in.}$$

   (d) The moment of inertia of the rectangle *A* with respect to the *x* axis is

$$I_{xa} = I_{xg} + A_a c^2 = 2.667 + 8(0.55)^2$$
$$= 2.667 + 2.420 = 5.087 \text{ in.}^4$$

3. For the triangle *B*,
   (a) The moment of inertia with respect to a horizontal axis through the centroid is

$$I_{xg} = \frac{1}{36} bh^3 = \frac{1}{36} \times 4(2)^3 = 0.889 \text{ in.}^4$$

   (b) The area is

$$A_b = \tfrac{1}{2} \times 4 \times 2 = 4 \text{ in.}^2$$

   (c) The centroid of the triangle *B* is 0.667 in. (one-third of 2 in.) below the top of the triangle. The *x* axis is 0.45 in. above the top of the triangle, so the distance *c* is

$$c = 0.667 + 0.45 = 1.117 \text{ in.}$$

   (d) The moment of inertia of the triangle *B* with respect to the *x* axis is

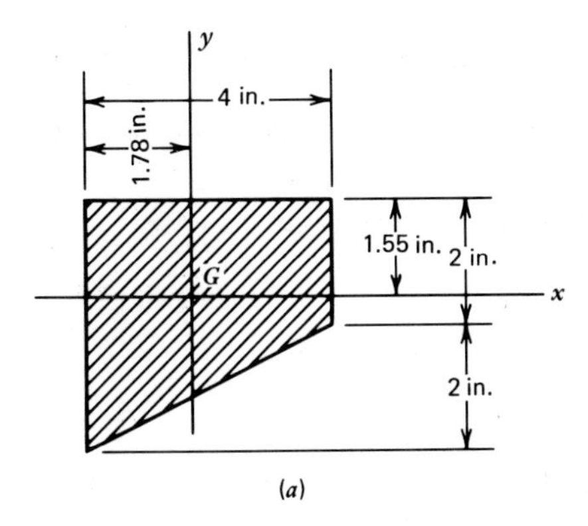

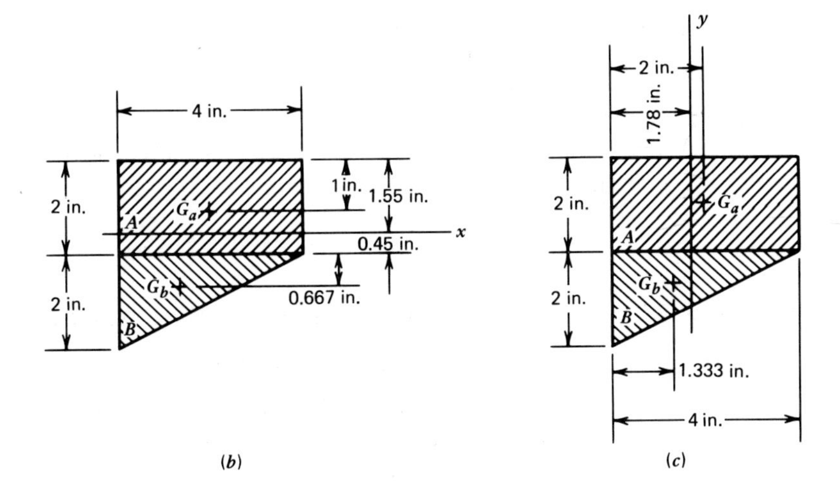

Fig. 5-43. Composite area for Example 3. (*a*) Area. (*b*) Dimensions for calculating *Ix*. (*c*) Dimensions for calculating *Iy*.

$$I_{xb} = I_{xg} + A_b c^2 = 0.889 + 4(1.117)^2$$
$$= 0.889 + 4.991 = 5.880 \text{ in.}^4$$

4. The moment of inertia of the composite area with respect to the x axis is the sum of the moments of inertia of the rectangle A and the triangle B.

$$I_x = I_{xa} + I_{xb} = 5.087 + 5.880 = 10.967 \text{ in.}^4$$

B. For I_y,

1. Figure 5-43*c* shows the area divided the same as in the first part of the problem:
   (a) The rectangle A.
   (b) The triangle B.
2. The rectangle A
   (a) The moment of inertia with respect to a vertical axis

through the centroid of the rectangle is

$$I_{yg} = \frac{1}{12} bh^3 = \frac{1}{12} \times 2\,(4)^3 = 10.667 \text{ in.}^4$$

(b) The area was found before

$$A_a = 8 \text{ in.}^2$$

(c) The distance c is

$$c = 2 - 1.780 = 0.220 \text{ in.}$$

(d) The moment of inertia of the rectangle A with respect to the y axis is

$$I_{ya} = I_{yg} + A_a c^2 = 10.667 + 8(0.220)^2$$
$$= 10.667 + 0.387 = 11.054 \text{ in.}^4$$

3. For the triangle B
   (a) The moment of inertia with respect to a vertical axis through the centroid of the triangle is

$$I_{yg} = \frac{1}{36} bh^3 = \frac{1}{36} \times 2\,(4)^3 = 3.556 \text{ in.}^4$$

   (b) The area was found in Section A of this problem as

$$A_b = \tfrac{1}{2} \times 4 \times 2 = 4 \text{ in.}^2$$

   (c) The centroid of the triangle B is 1.333 in. (one-third of 4 in.) from the vertical leg of the triangle. The distance from the centroid to the y axis is

$$c = 1.780 - 1.333 = 0.447 \text{ in.}$$

   (d) The moment of inertia of the triangle B with respect to the y axis is

$$I_{yb} = I_{yg} + A_b c^2 = 3.556 + 4(0.447)^2$$
$$= 3.556 + 0.799 = 4.355 \text{ in.}^4$$

4. The moment of inertia of the composite area with respect to the y axis is equal to the sum of the moments of inertia of the areas A and B.

$$I_y = I_{ya} + I_{yb} = 11.054 + 4.355 = 15.409 \text{ in.}^4$$

Illustrative Example 4. Calculate the moment of inertia of the tubular section in Fig. 5-44*a* with respect to the y axis.

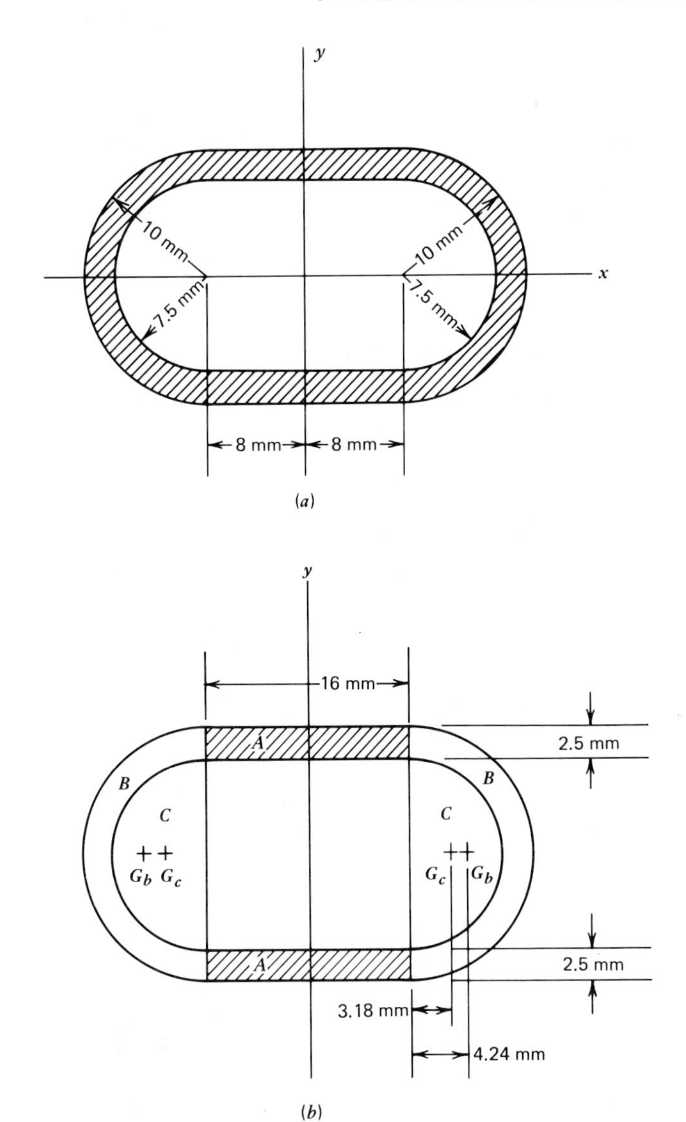

Fig. 5-44. Composite area for Example 4. (*a*) Area. (*b*) Division of area into parts.

Solution:

1. Figure 5-44*b* shows the division of the area into parts.
   (a) Two positive rectangular areas A, each 16 mm horizontal by 2.5 mm vertical.
   (b) The two positive semicircular areas B, 10 mm radius.
   (c) The two negative semicircular areas C, 7.5 mm radius.
2. The centroid of the rectangles A is on the y axis, so the moment of inertia with respect to the y axis is just $2 \times 1/12\ bh^3$

$$I_{ya} = 2 \times \frac{1}{12} bh^3$$

$$= 2 \times \frac{1}{12} \times 2.5\,(16)^3$$

$$= 1\,707 \text{ mm}^4$$

3. For the semicircular areas B, it is necessary to use the parallel-axis formula.
   (a) The moment of inertia of one semicircular area B with respect to a vertical axis through the centroid is

$$I_{yg} = 0.1096r^4 = 0.1096(10)^4 = 1096 \text{ mm}^4$$

   (b) The area of the semicircle B is

$$A_b = \frac{\pi}{2}r^2 = \frac{\pi}{2}(10)^2 = 157 \text{ mm}^2$$

   (c) The distance of the centroid of B from the vertical diameter is $4r/3\pi = 4 \times 10/3\pi = 4.24$ mm. The distance c is equal to this plus 8 mm, so

$$c = 4.24 + 8 = 12.24 \text{ mm}$$

   (d) The moment of inertia of one semicircular area B with respect to the y axis is

$$I_{yb} = I_{yg} + A_b c^2 = 1\,096 + 157\,(12.24)^2$$

$$= 1\,096 + 23\,521 = 24\,617 \text{ mm}^4$$

The moment of inertia of one semicircular area B is the same as the moment of inertia of the other, because they are the same size and shape and are at the same distance from the y axis. We can just multiply the moment of inertia of one semicircular area by two, to get the moment of inertia of both.

4. The moment of inertia of one negative semicircular area C is calculated in the same way as the moment of inertia of a semicircular area B. Only the numbers are different.
   (a) The moment of inertia of one semicircular area C with respect to a vertical axis through the centroid is

$$I_{yg} = 0.1096r^4 = 0.1096(7.5)^4 = 347 \text{ mm}^4$$

   (b) The area is

$$A_c = \frac{\pi}{2}r^2 = \frac{\pi}{2}(7.5)^2 = 88.4 \text{ mm}^2$$

   (c) The distance from the y axis to the vertical diameter of C is 8 mm, and the additional distance to the centroid is $4r/3\pi$, so

$$c = 8 + \frac{4r}{3}\pi = 8 + 4 \times \frac{7.5}{3\pi}$$

$$= 8 + 3.18 = 11.18 \text{ mm}$$

(d) The moment of inertia of one semicircular area C with respect to the y axis is

$$I_{yc} = I_{yg} = A_c c^2 = 347 + 88.4(11.18)^2$$

$$= 347 + 11\,049 = 11\,396 \text{ mm}^4$$

The moment of inertia of the two semicircular areas C is just twice the moment of inertia of one.

5. The moment of inertia of the composite area is found by adding the moments of inertia of the parts of the area. Thus

$$I_y = I_{ya} + 2I_{yb} - 2I_{yc}$$

$$= 1\,707 + 2 \times 24.617 - 2 \times 11\,395$$

$$= 1\,707 + 49.234 - 22\,790 = 28\,151 \text{ mm}^4$$

Practice Problems (Section 5-4). Now you are ready to practice on moment of inertia of composite areas. Look at each carefully, and see how many you can do without having to use the parallel-axis theorem. Calculate:

1. I_x for the area in Fig. 5-45.
2. I_x for the area in Fig. 5-46.

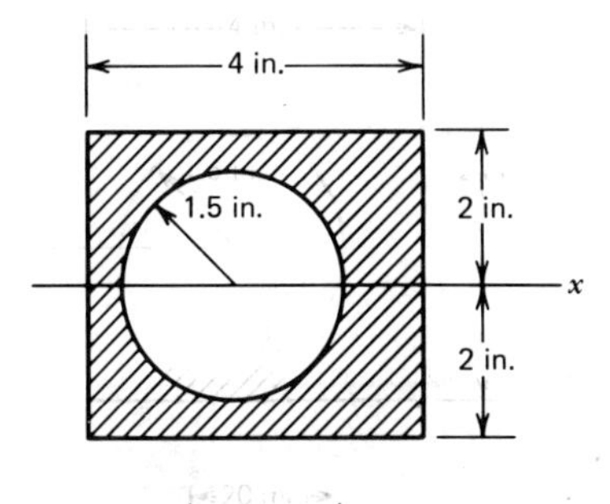

Fig. 5-45. Hollow area for Problem 1.

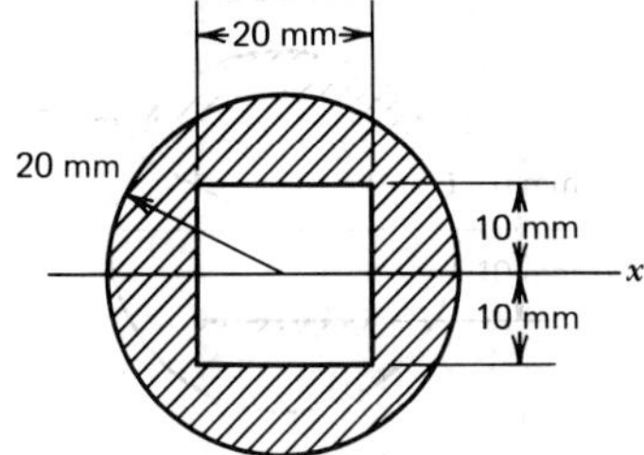

Fig. 5-46. Hollow area for Problem 2.

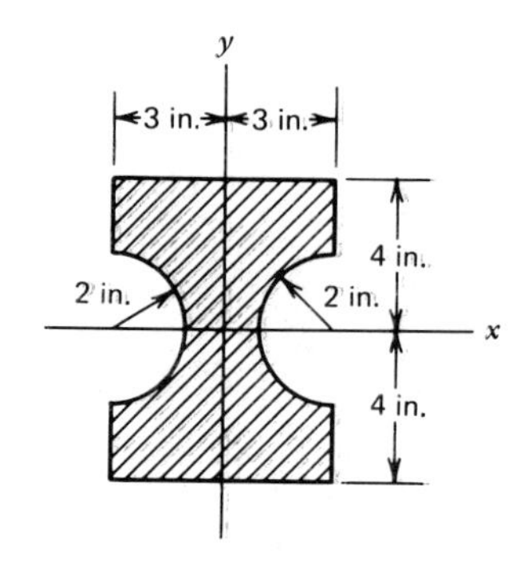

Fig. 5-47. Composite area for Problem 3.

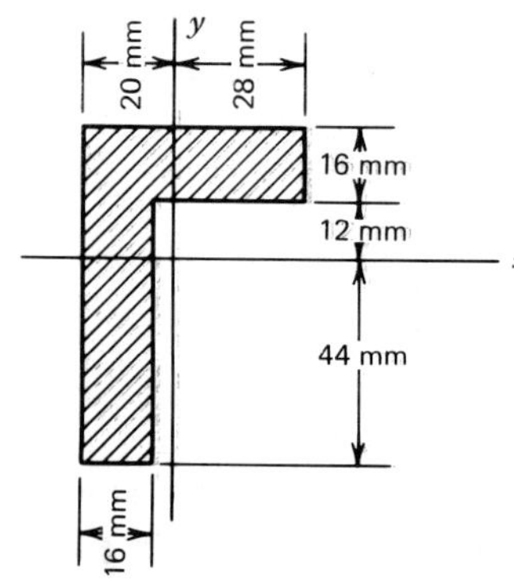

Fig. 5-48. Angle section for Problem 4.

3. I_x and I_y for the area in Fig. 5-47.
4. I_x and I_y for the area in Fig. 5-48.
5. I_x for the area in Fig. 5-49.

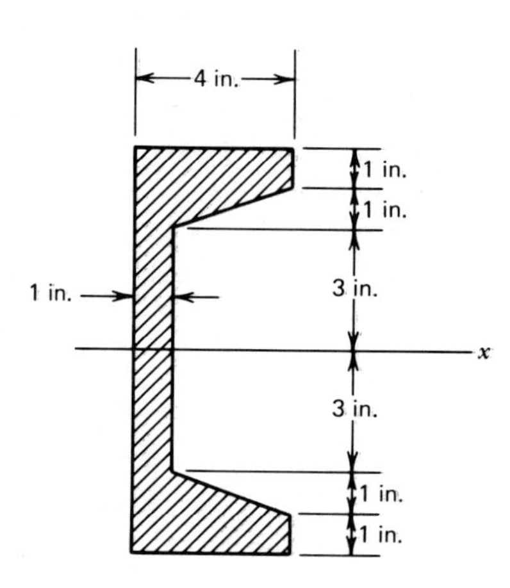

Fig. 5-49. Channel section for Problem 5.

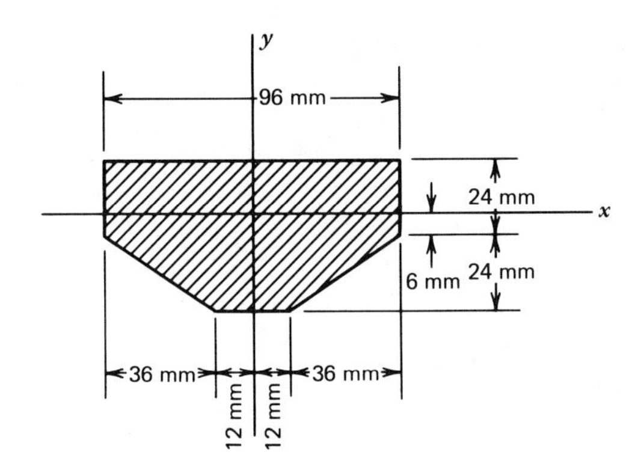

Fig. 5-50. Area for Problem 6.

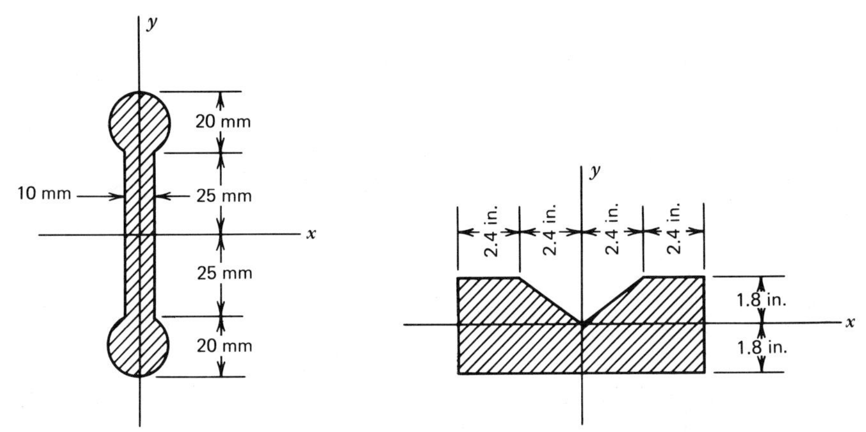

Fig. 5-51. Area for Problem 7.

Fig. 5-52. Area for Problem 8.

6. I_x and I_y for the area in Fig. 5-50.
7. I_x and I_y for the area in Fig. 5-51.
8. I_x and I_y for the area in Fig. 5-52.

SUMMARY The main points of this chapter are:

1. Moment of inertia is a property of an area. If the area is thought of as being divided into extremely small parts, the moment of inertia of the area with respect to an axis is equal to the sum of the quantities obtained by multiplying each small area by the square of its distance from the axis. (This definition can be applied by use of calculus to derive the fundamental formulas.)
2. You should remember the following fundamental formulas for moment of inertia.

(a) A rectangular area, with respect to a centroidal axis parallel to the base

$$\frac{1}{12}bh^3$$

(b) A triangular area, with respect to a centroidal axis parallel to the base

$$\frac{1}{36}bh^3$$

(c) A circular area, with respect to a diametral axis

$$\frac{\pi}{4}r^4$$

(d) a quadrant, with respect to a bounding radius

$$\frac{\pi}{16}r^4$$

(e) A quadrant, with respect to a centroidal axis, parallel to a bounding radius

$$0.0548r^4$$

3. The parallel-axis theorem says that the moment of inertia of an area with respect to any axis is equal to the moment of inertia with respect to a parallel axis through the centroid of the area plus the product of the area and the square of the distance between the two axes. The theorem can be applied to both horizontal and vertical axes. The formulas are
   (a) $I_x = I_{xg} + A_c^2$
   (b) $I_y = I_{yg} + A_c^2$
4. The moment of inertia of a composite area can be found by the following procedure:
   (a) Divide the composite area into simple parts.
   (b) Calculate the moment of inertia of each part, using the parallel-axis theorem when necessary.
   (c) Add the moments of inertia of the parts of the area to get the moment of inertia of the composite area.

REVIEW QUESTIONS These review questions are to be answered without referring to the first part of the chapter.

1. What is the area of a circle?
2. What is the moment of inertia of a circular area with respect to a diametral axis?

3. What is the moment of inertia of a rectangular area with respect to an axis through the centroid and parallel to the base?
4. What is the moment of inertia of a triangular area with respect to an axis through the centroid and parallel to the base?
5. What is the area of a quadrant?
6. What is the moment of inertia of a quadrant with respect to a centroidal axis parallel to a radius boundary?
7. State the parallel-axis theorem.
8. Under what conditions would you think of an area as negative?
9. What is a composite area?
10. Which side of a rectangle do you take as the width for the purpose of calculating its moment of inertia with respect to the y axis?

REVIEW PROBLEMS If you can find the correct moment of inertia for each of these composite areas, you know the material of this chapter. If you can't, you had better review the preceding material. Just find,

1. I_x and I_y for the area in Fig. 5-53.
2. I_x for the area in Fig. 5-54.

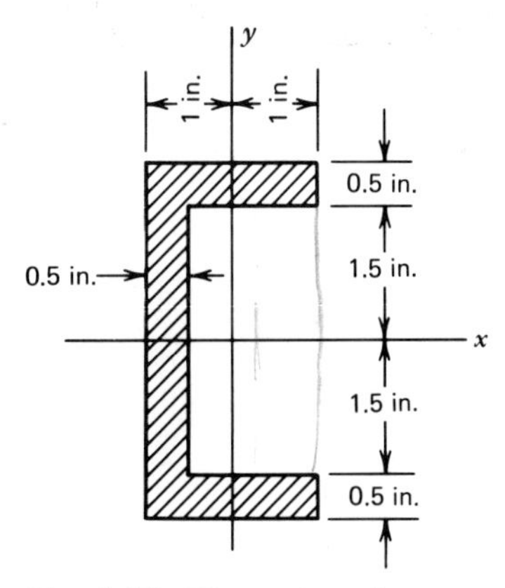

Fig. 5-53. Channel section for Problem 1.

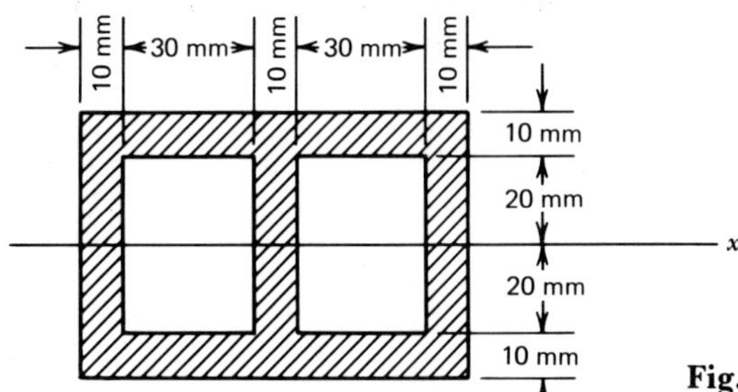

Fig. 5-54. Hollow section for Problem 2.

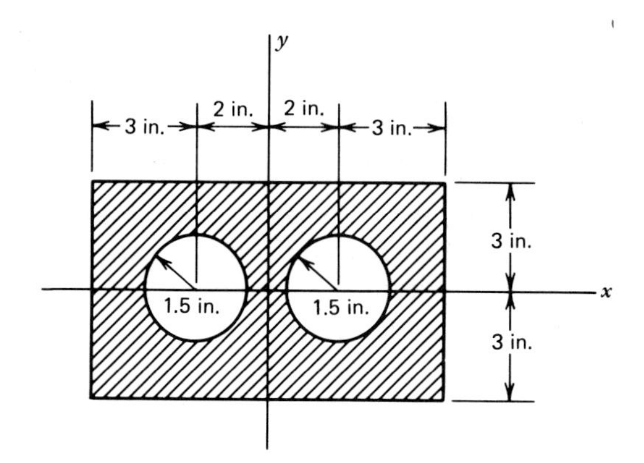

Fig. 5-55. Hollow section for Problem 3.

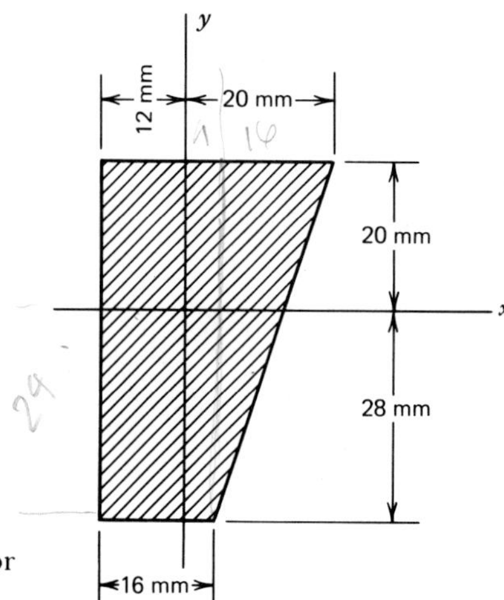

Fig. 5-56. Trapezoid for Problem 4.

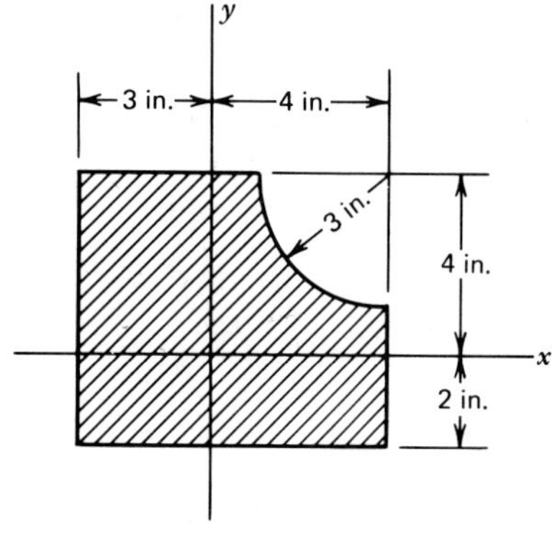

Fig. 5-57. Area for Problem 5.

3. I_x and I_y for the area in Fig. 5-55.
4. I_x and I_y for the area in Fig. 5-56.
5. I_x and I_y for the area in Fig. 5-57.

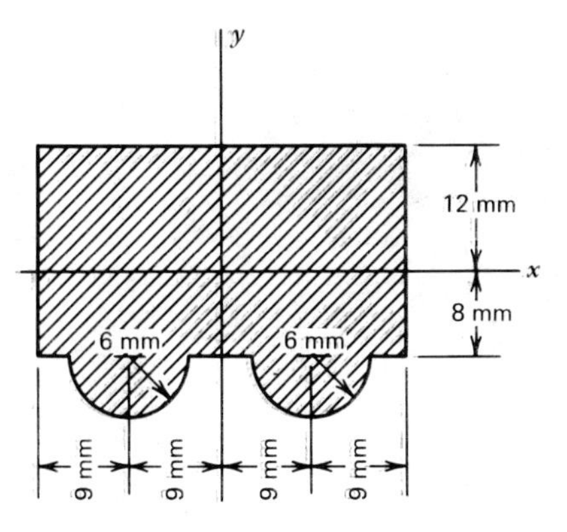

Fig. 5-58. Composite area for Problem 6.

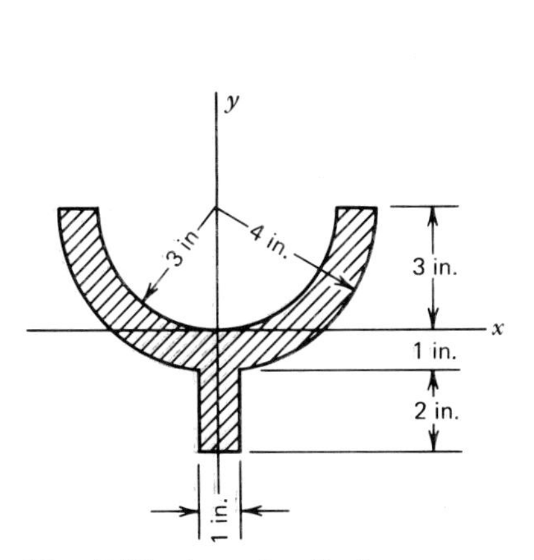

Fig. 5-59. Area for Problem 7.

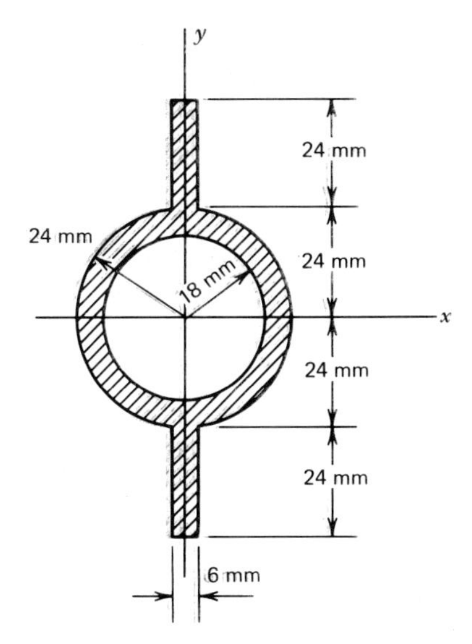

Fig. 5-60. Reinforced pipe area for Problem 8.

6. I_x and I_y for the area in Fig. 5-58.
7. I_x and I_y for the area in Fig. 5-59.
8. I_x and I_y for the area in Fig. 5-60.

6 STRESSES IN BEAMS

PURPOSE OF THIS CHAPTER. We are going to advance considerably in our study of strength of materials in this chapter. We are going to study beams, learn how to find stresses in beams, and how to design them. We will have to use most of the material studied in the first part of the book, because we will solve equilibrium problems (Chapter 2), we will locate the centroid of an area (Chapter 4), and find the moment of inertia of an area (Chapter 5). You will learn several new technical terms in this chapter, but each one will be defined carefully as we come to it.

The study of beams is important to the designer because many parts of machines and structures are beams. Sometimes it is necessary to find the stress in a beam to make certain that it isn't loaded too heavily, and a large part of design consists of designing beams.

6-1. WHAT A BEAM IS

A *beam* is a bar or member that is bent by forces perpendicular to the beam. Figure 6-1 shows a beam that is part of a weighing machine. The beam is subjected to the load P and is supported by knife-edges. Figure 6-2 shows a simple beam that is subjected to the loads P_1 and P_2; it is

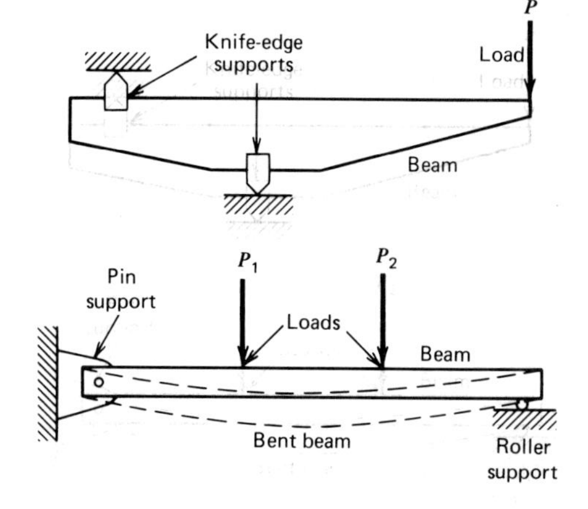

Fig. 6-1. Beam in weighing machine.

Fig. 6-2. Beam supported by pin and roller.

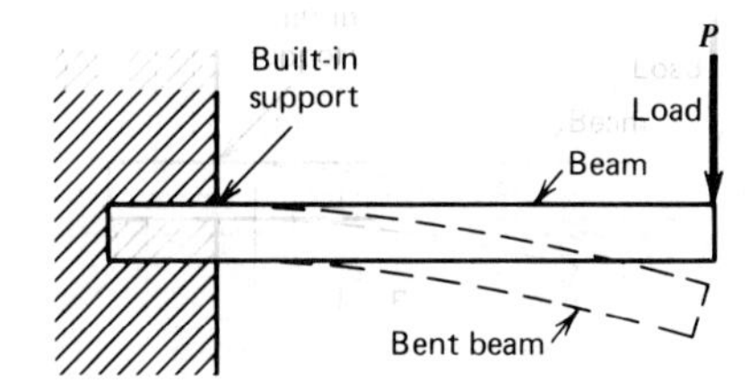

Fig. 6-3. Cantilever beam.

supported by a pin at the left end and a roller at the right end. The dotted lines show the shape of the beam as it is bent under the loads. (The bending is exaggerated in this drawing. Beams don't usually bend this much in service.) Beams that are supported at two points by knife-edges, pins, or rollers are called *simple beams*. The beams in Figs. 6-1 and 6-2 are simple beams.

A beam that is supported at only one end is called a *cantilever beam*. Figure 6-3 shows a cantilever beam that is subjected to a load P. The single support for a cantilever beam must be of a special type to keep the beam from tipping, for instance, a built-in support such as you see in Fig. 6-3. Not only does the support have to hold up the end of the beam, but also the support must hold the end of the beam clamped so that it won't tip. The dotted lines show the bent shape of the beam.

6-2. REACTIONS ON BEAMS

The reactions on a beam are the forces exerted by the supports. Usually we have the loads, which are the known forces on the beam, and the reactions, which are unknown forces. The reactions must be found before we can proceed to the real analysis of the beam. Finding the reactions is just a problem of equilibrium, such as you studied in Chapter 2. First we draw the free-body diagram, which is a picture showing all the forces acting on the beam. Then we write moment equations to find the reactions. (The moment of a force with respect to a point is the product of the force and the perpendicular distance from the point to the force. Clockwise moments are positive, and counterclockwise moments are negative.) Remember that we choose the moment center on one reaction to find the second; next we choose the moment center on the second reaction to find the first one. Then we check the values of the reactions by adding the vertical forces. You already know how to do this, but let's take an example for review.

Illustrative Example 1. The beam in Fig. 6-4*a* is subjected to the loads of 240 lb and 380 lb and is supported by knife-edges at A and B. Find the reactions.

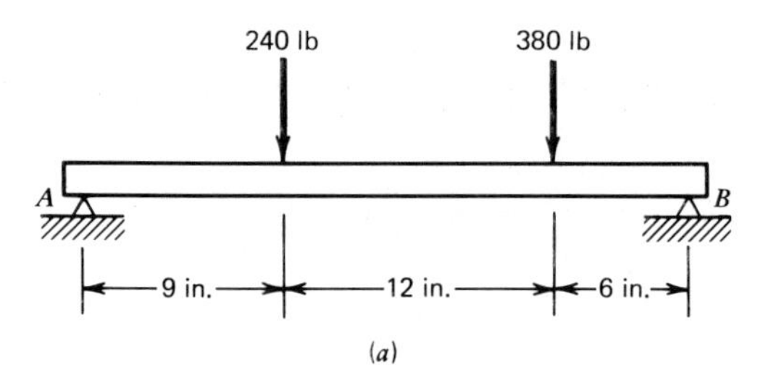

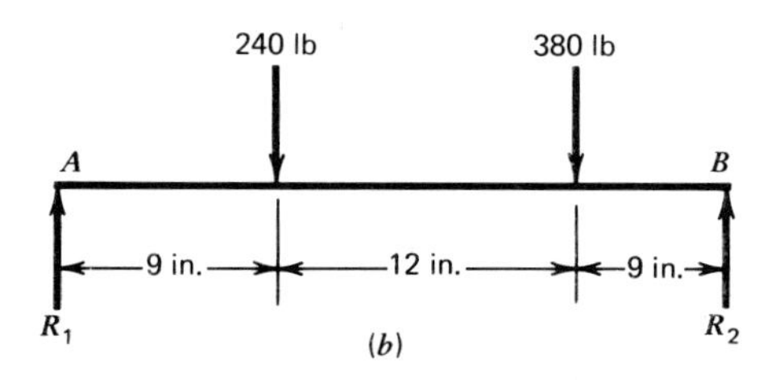

Fig. 6-4. Beam supported on knife-edges. (*a*) Beam. (*b*) Free-body diagram.

Solution:

1. Figure 6-4*b* shows the free-body diagram. The heavy line represents the beam, then the loads are drawn, and finally the reactions R_1 and R_2.
2. We write a moment equation with center at A to find R_2. Taking the forces in order from left to right,

$$\Sigma M_A = 0$$

$$240 \times 9 + 380 \times 21 - 27R_2 = 0$$

$$27R_2 = 240 \times 9 + 380 \times 21$$

$$= 2\,160 + 7\,980$$

$$= 10\,140 \qquad R_2 = 376 \text{ lb}$$

3. We write a moment equation with center at B to find R_1. Taking the forces in order from left to right,

$$\Sigma M_B = 0$$

$$27R_1 - 240 \times 18 - 380 \times 6 = 0$$

Transposing,

$$27R_1 = 240 \times 18 + 380 \times 6 = 4\,320 + 2\,280$$

$$= 6\,600$$

$$R_1 = 244 \text{ lb}$$

4. Now we check the values of the reactions by writing the equilib-

rium equation $\Sigma F_y = 0$ (the sum of the vertical forces must be zero if the beam is in equilibrium). Thus, in order from left to right,

$$\Sigma F_y = 0$$

$$R_1 - 240 - 380 + R_2 = 0$$

We found $R_1 = 244$ lb and $R_2 = 376$ lb, so let's substitute them:

$$244 - 240 - 380 + 376 = 0$$

$$620 - 620 = 0$$

$$0 = 0$$

Good, it checks, and the values of the reactions are correct.

6-3. SHEAR IN BEAMS

There is a shearing force at each cross section of a beam, a force that tends to shear the beam in two. (Remember, from Chapter 3, that shear is sliding.) Let's take a close look at a beam and see how the shearing forces act. Figure 6-5*a* shows the free-body diagram of a beam with loads P_1 and P_2 and reactions R_1 and R_2. (It is easier to visualize the shearing forces if we show the beam as it is, instead of representing it by a line.) The entire beam is in equilibrium, and each part of it is in equilibrium; if it weren't in equilibrium it would move. Now let's imagine that we cut the beam on the cross section *C-C*; Fig. 6-5*b* shows the free-body diagram of the part of the beam to the right of this cross section. We see the load P_2 and the reaction R_2, as they act on this part of the beam, and also we see a shearing force at the left end of this part. (There must be an upward force at the left end of this part to hold it up.) We usually represent a shearing force by V, and we designate this shearing force at C as V_c. The shearing force is exerted on the right-hand part of the beam by the left-hand part. To complete the picture, let's look at the free-body diagram of the left-hand part of the beam, Fig. 6-5*c*. Here we see the load P_1 and the reaction R_1; also we see the shearing force V_c, which is exerted on the left-hand part of the beam by the right-hand part. The force V_c in Fig. 6-5*c* is equal and opposite to the force V_c in Fig. 6-5*b*.

Figure 6-5*c* can be used to establish a general rule for the value of the shearing force V_c. The sum of the vertical forces must be zero, because this part of the beam is in equilibrium. Then,

$$\Sigma F_y = 0$$

$$R_1 - P_1 - V_c = 0$$

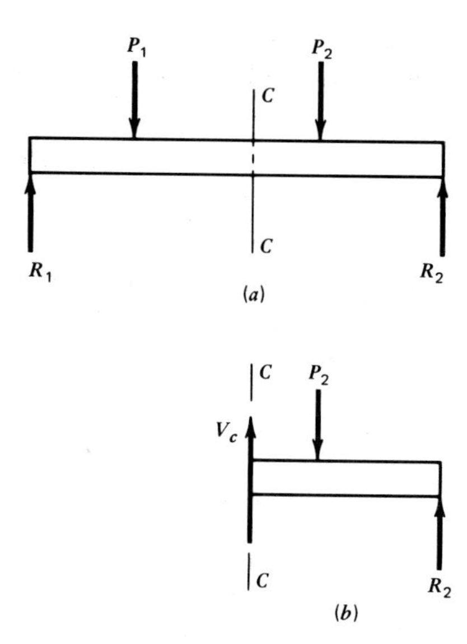

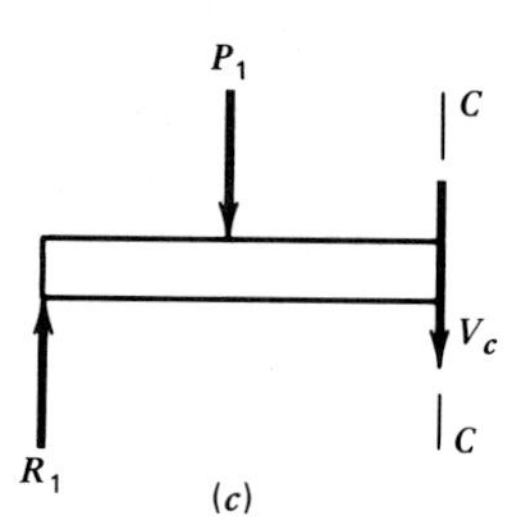

Fig. 6-5. Shear forces in a beam. (*a*) Free-body diagram of beam. (*b*) Free-body diagram of right part of beam. (*c*) Free-body diagram of left part of beam.

Then we transpose V_c, and we have

$$V_c = R_1 - P_1$$

Now we know the value of V_c, but let's look carefully at the result and see what its general meaning is. The forces R_1 and P_1 are the only forces to the left of the cross section *C-C*; R_1 is directed upward, and P_1 is directed downward. If we were to sum up the forces to the left of the cross section *C-C* we would get $R_1 - P_1$, the same result we got for V_c. So the shearing force at this cross section *C-C* is equal to the sum of the vertical forces to the left of the cross section. It works out the same way, no matter what cross section we choose and no matter how the beam is loaded. *The shearing force at any cross section is equal to the sum of the vertical forces to the left of the cross section.* Upward forces are positive, and downward forces are negative when we sum them up. A positive result means that the shearing force exerted by the left part of the beam on the right is upward; a negative result means that this force is downward. Let's try this in an example.

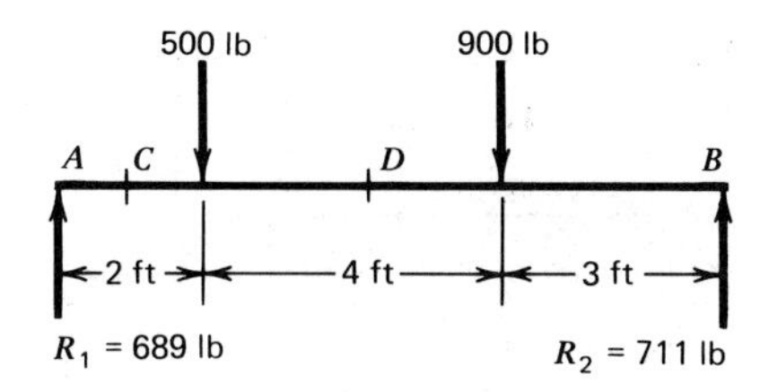

Fig. 6-6. Beam with concentrated loads.

Illustrative Example 1. Figure 6-6 shows the free-body diagram of a beam. The reactions are given. Find the shearing force at a point *C* between the left end and the 500-lb load.

Solution:

At any point between the left end of the beam and the 500-lb load there is only one force to the left and that is R_1. So the sum of the forces to the left of any cross section in this part of the beam is

$$V_c = 689 \text{ lb}$$

where the subscript *c* is to show that this is the shear at point *C*.

Illustrative Example 2. Find the shearing force at a point *D* between the 500-lb and 900-lb loads in Fig. 6-6.

Solution:

1. The sum of the forces to the left of any point between the loads is

$$V_d = 689 - 500 = 189 \text{ lb}$$

DISTRIBUTED LOAD

Simple, wasn't it? Now let's try it for a beam with a uniformly distributed load. The total of the distributed load in a certain length is equal to the product of (1) intensity of the load in units of force per unit of length and (2) the length. We will find that the shear changes from point to point, and we usually have to specify a definite location for the point in which we are interested.

Illustrative Example 3. Find the shearing force at point *C* in the beam in Fig. 6-7.

Solution:

The point *C* is 1.6 m from the left end of the beam. The only force to the left of *C* is part of the uniformly distributed load. So

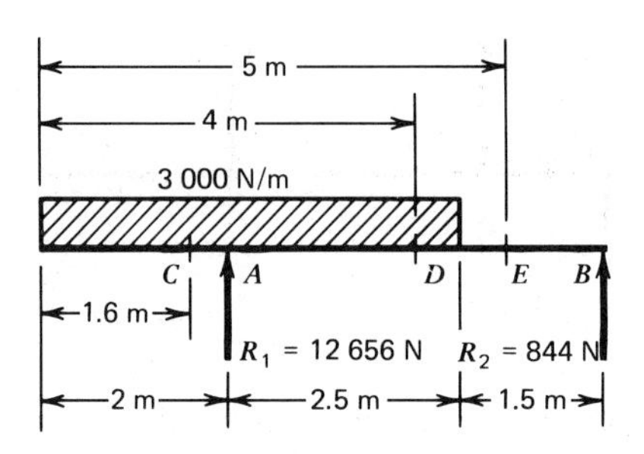

Fig. 6-7. Beam with uniformly distributed load.

$$V_c = -3\ 000 \times 1.6 = -4\ 800 \text{ N}$$

The negative result means that the force exerted by the part of the beam to the left of this point on the part of the beam to the right is downward.

Illustrative Example 4. Find the shearing force at point D in the beam in Fig. 6-7.

Solution:
The point D is 4 m from the left end of the beam. The reaction R_1 is to the left of D and so is 4 m of the uniformly distributed load. So

$$V_d = 12\ 656 - 3\ 000 \times 4 = 12\ 656 - 12\ 000 = 656 \text{ N}$$

Illustrative Example 5. Find the shearing force at the point E in the beam in Fig. 6-7.

Solution:
The point E is 5 m from the left end of the beam. To the left of point E there is the reaction R_1 and all of the uniformly distributed load. Consequently,

$$V_e = 12\ 656 - 3\ 000 \times 4.5 = 12\ 656 - 13\ 500 = -844 \text{ N}$$

Practice Problems (Section 6-3). It's easy to find the shearing force in a beam, but you will learn it better if you practice it. You will have to find the reactions yourself in these problems.

1. Find the shear at point C in Fig. 6-8.
2. Find the shear at point D in Fig. 6-8.
3. Find the shear at point E in Fig. 6-8.
4. Find the shear at point F in Fig. 6-8.

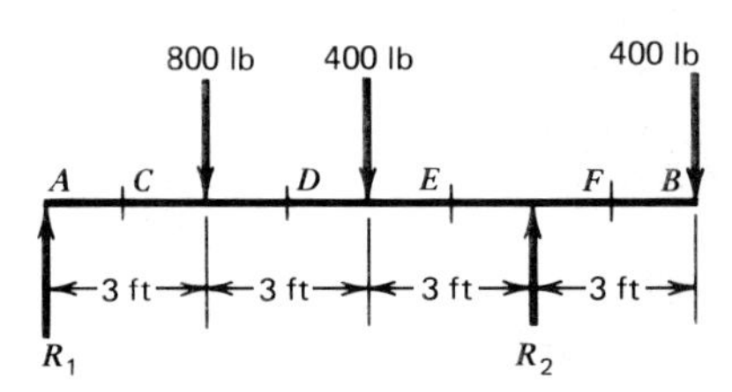

Fig. 6-8. Beam for Problems 1, 2, 3, and 4.

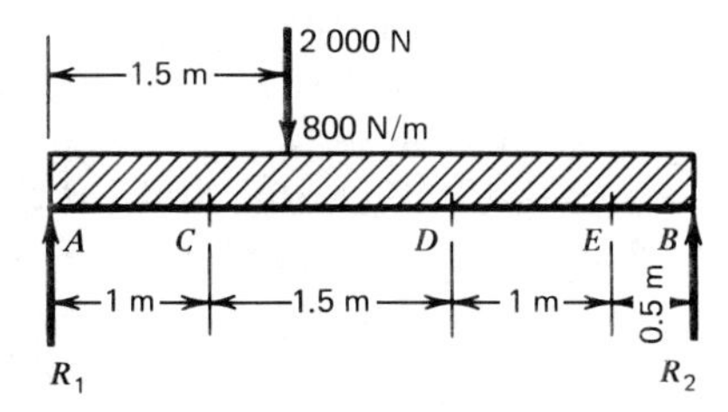

Fig. 6-9. Beam for Problems 5, 6, and 7.

5. Find the shear at point C in Fig. 6-9.
6. Find the shear at point D in Fig. 6-9.
7. Find the shear at point E in Fig. 6-9.

6-4. SHEAR DIAGRAMS

A shear diagram is a diagram that shows the amount of the shearing force at each point along the beam. The shear diagram is useful in finding the maximum stress in a beam and in the problem of designing a beam.

Shear diagrams are probably new to you, so let's look at a completed shear diagram to see what it looks like, and then study how to draw shear diagrams later. Figure 6-10*a* shows the free-body diagram of a simple beam, and Fig. 6-10*b* shows the shear diagram. The beam is supported at points A and B and has a load of 400 lb applied at C. The reactions are 250 lb for R_1 and 150 lb for R_2. (The values of the reactions

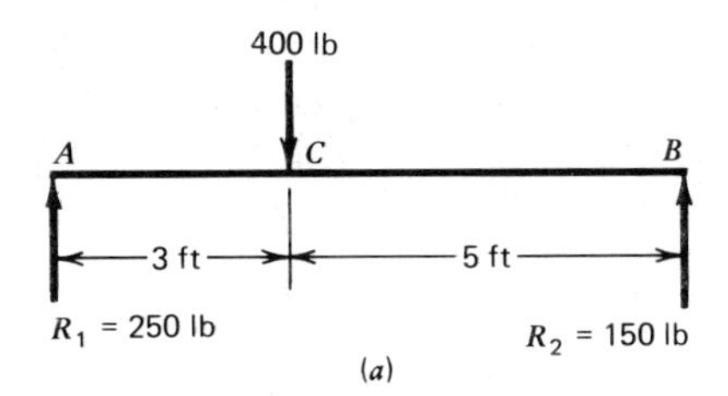

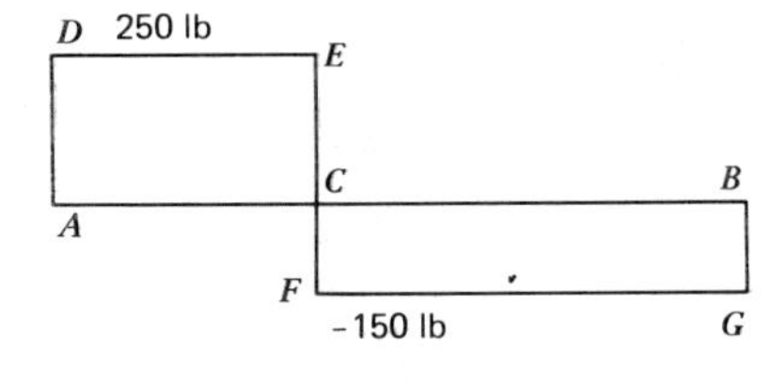

Fig. 6-10. Beam and shear diagram.

are given to save your time.) In Fig. 6-10*b* the line *ACB* is a base line from which the values of shear are measured (positive upward and negative downward). The distance from the base line *ACB* to the line *DE* represents the value of the shear in the length of beam *AC*; the figure of 250 lb written at *D* indicates that the shear in this length is 250 lb, and the line *DE* is above the base line *ACB* because the shear is positive. The distance from the base line *ACB* to the line *FG* represents the shear in the length of beam *CB*; the figure of −150 lb is written at *F* to indicate the value of the shear, and the line *FG* is below the base line *ACB* because the shear is negative. That's the way it goes. The value of the shear is represented by the distance from the base line to the shear diagram. The shear diagram is above the base line when the shear is positive and below the base line when the shear is negative. Shear diagrams need not be drawn to scale. You can make a neat free-hand drawing and write the values of the shear on the diagram.

The easiest way to draw a shear diagram is to start at the left end of the beam and plot the forces as they occur while you move across the beam from left to right. Plot loads and reactions both; plot a force upward if it's directed upward and plot a force downward if it's directed downward. Draw a horizontal line between loads. If you followed this procedure to draw the shear diagram in Fig. 6-10*b*, you would start at *A* and plot the reaction R_1 (250 lb) upward. This would give you point *D*, and you would draw a horizontal line from *D* to the next force, which is at *E*. From *E* you would plot the downward force of 400 lb and reach point *F*. You would draw a horizontal line from *F* to the 150-lb force. This would give you point *G*, and you would finish the shear diagram by plotting the reaction R_2 (150 lb) upward from *G*.

The process of plotting the forces gives you automatically the value of the shear at each point along the beam. Forces that are directed upward are plotted upward, and forces that are directed downward are plotted downward. This means that you are making a graphical sum of the forces to the left of each point, and the distance from the base line to the shear diagram gives the value of the shear.

Illustrative Example 1. Figure 6-11*a* shows the free-body diagram of a beam. Draw the shear diagram.

Solution:

1. The first step is to draw the base line *ABCD* in Fig. 6-11*b*.
2. Start at the left end *A*. Plot the 2 800-lb load downward to point *E* and write the figure −2 800 lb at *E* to show the value of the shear.
3. Draw the horizontal line *EF* from *E* to the line of action of R_1.

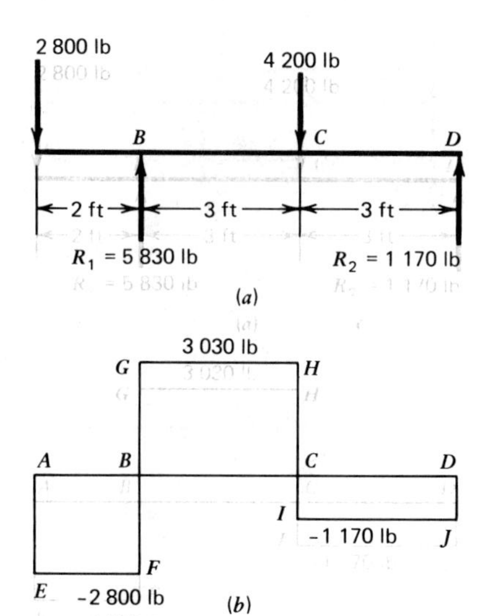

Fig. 6-11. Beam for Illustrative Example 1.

(A horizontal line in a shear diagram always stops at the line of action of the next force.)

4. Plot the reaction R_1 (5 830 lb) upward from point *F* to *G*. You start at 2 800 lb below the base line and move upward 5 830 lb. This places point *G* at

$$5\,830 - 2\,800 = 3\,030 \text{ lb}$$

The shear at *G* is 3 030 lb. Write the figure 3 030 lb at *G* to show the value of the shear.

5. Draw the horizontal line from *G* to *H*, where the 4 200-lb load is applied.
6. Plot the 4 200-lb load downward from *H* to *I*. You start at 3 030 lb above the base line and drop down 4 200 lb. This places point *I* at

$$3\,030 - 4\,200 = -1\,170 \text{ lb}$$

The negative sign shows that point *I* is below the base line. Write the value $-1\,170$ lb at *I* to show the value of the shear.

7. Draw the horizontal line *IJ* from *I* to the line of action of R_2.
8. Plot the reaction R_2 (1 170 lb) upward from *J*. This finishes the shear diagram at *D*.

UNIFORM LOAD

The shear diagram is a little different when there is a uniformly distributed load on a beam. Figure 6-12*a* shows the free-body diagram of a beam with a uniformly distributed load, and Fig. 6-12*b* shows the shear

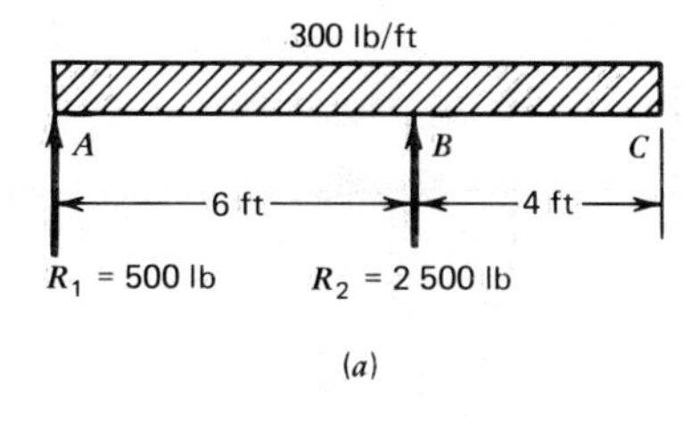

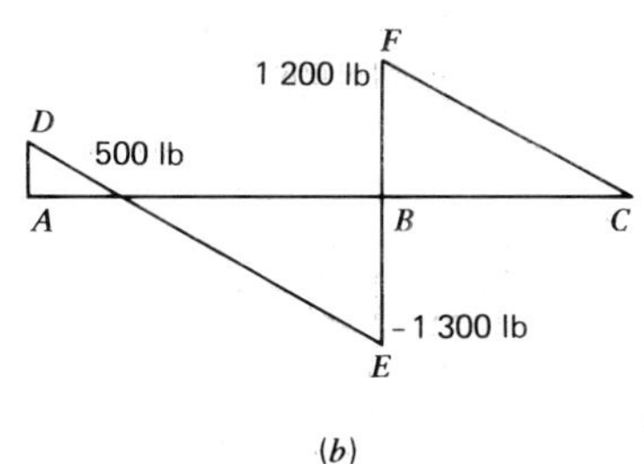

Fig. 6-12. Beam with a uniformly distributed load. (*a*) Free-body diagram. (*b*) Shear diagram.

diagram. Notice that the shear diagram is a straight line that slopes downward to the right between forces. When you plot a uniformly distributed load, you plot just so many pounds per foot downward as you move across the beam. The value of the shear drops twice as much in a length of 2 ft as in a length of 1 ft, six times as much in a length of 6 ft as in a length of 1 ft, and the result is a straight line for the shear diagram. In Fig. 6-12*b*, you would plot R_1 (500 lb) upward from point A to locate point D. Then from D you would drop down and across in a straight line to E. The total amount of the uniformly distributed load in the 6 ft length between D and E is $300 \times 6 = 1\ 800$ lb. The uniformly distributed load is directed downward, so you would plot it downward. You would start then with a value of 500 lb at D and drop 1 800 lb to E. This would give, for the shear at E,

$$500 - 1\ 800 = -1\ 300 \text{ lb}$$

Illustrative Example 2. Figure 6-13*a* shows the free-body diagram of a beam with a concentrated load and a uniformly distributed load. Draw the shear diagram.

Solution:

1. The first step is to draw the base line ABC in Fig. 6-13*b*.
2. Start at A and plot R_1 (6 400 N) upward to D. Write the value of the shear (6 400 N) at D.
3. From D, plot the distributed load between the left end of the beam and the 6 000-N load. The magnitude of the distributed

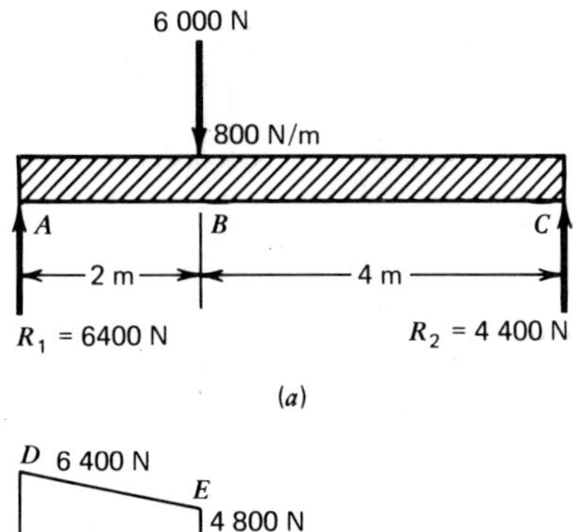

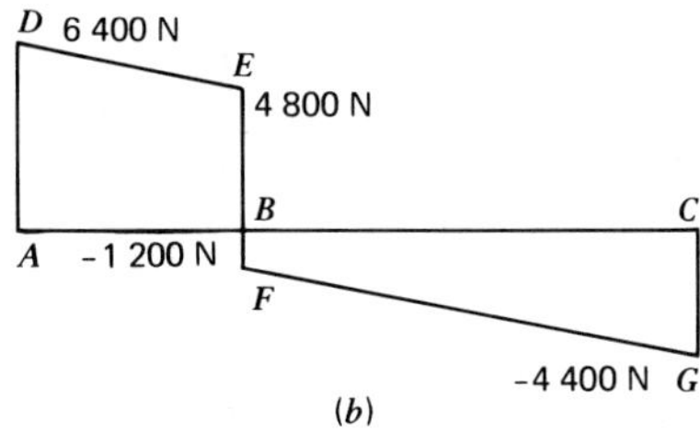

Fig. 6-13. Beam for Illustrative Example 2. (*a*) Free-body diagram. (*b*) Shear diagram.

load is 800 N per m and the length is 2 m, so the total amount is $800 \times 2 = 1\,600$ N. Take this away from 6 400 N which is the shear at *D*. This gives

$$6\,400 - 1\,600 = 4\,800 \text{ N}$$

for the shear at *E*. Write it on the shear diagram.

4. Plot the 6 000-N load downward from *E*. The shear at *E* is 4 800 N, and you drop downward 6 000 N, so the shear at *F* is

$$4\,800 - 6\,000 = -1\,200 \text{ N}$$

Write this figure ($-1\,200$ N) on the diagram.

5. From *F*, plot the uniform load in the 4-m length between the 6 000-N load and the right end of the beam. You start this with the shear equal to $-1\,200$ N, so the shear at *G* is

$$-1\,200 - 800 \times 4 = -1\,200 - 3\,200 = -4\,400 \text{ N}$$

Write the value ($-4\,400$ N) on the shear diagram.

6. Plot the reaction R_2 (4 400 N) upward from *G*. This ends the shear diagram at *C*.

Practice Problems (Section 6-4). Now you can draw shear diagrams for yourself. You can find the reactions on the beams, too. The free-body diagrams are given and you can start right in. In each case you are to draw the shear diagram.

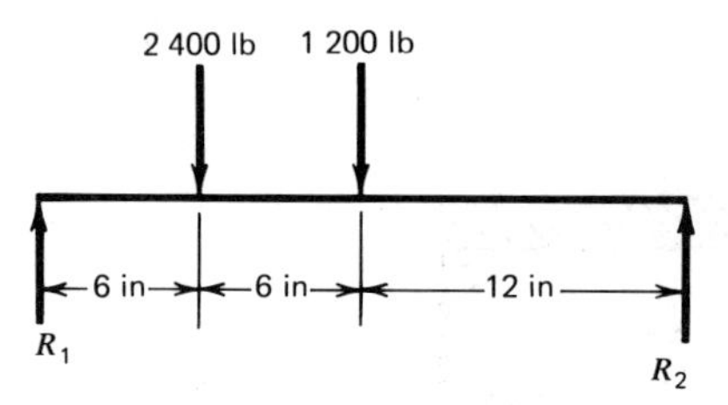

Fig. 6-14. Beam for Problem 1.

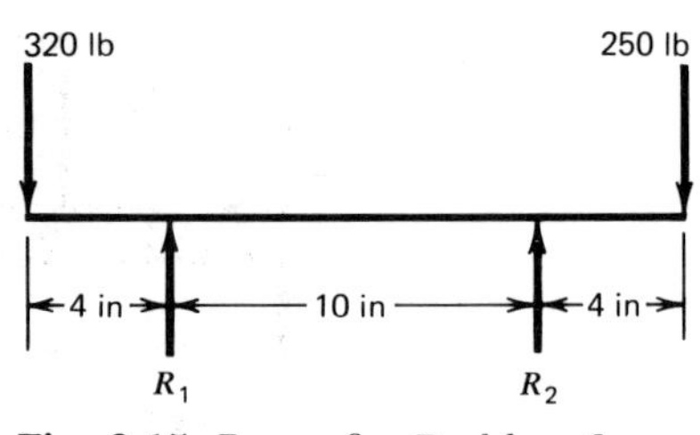

Fig. 6-15. Beam for Problem 2.

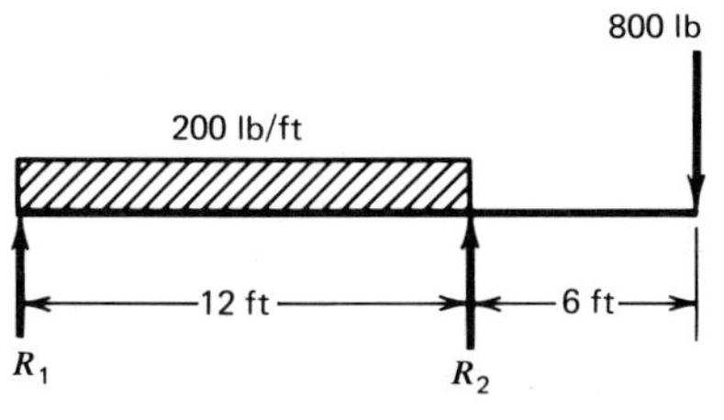

Fig. 6-16. Beam for Problem 3.

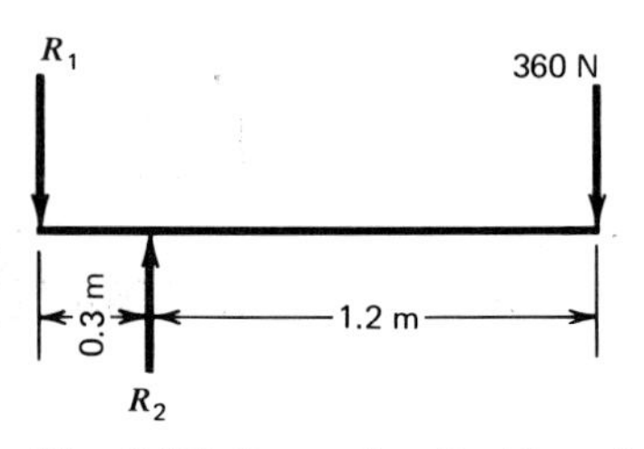

Fig. 6-17. Beam for Problem 4.

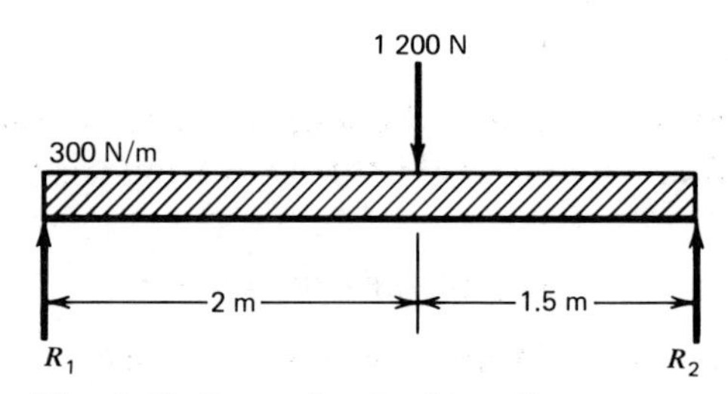

Fig. 6-18. Beam for Problem 5.

1. Figure 6-14.
2. Figure 6-15.
3. Figure 6-16.
4. Figure 6-17.
5. Figure 6-18.

6-5. BENDING MOMENT

The next thing to learn about beams is how to calculate bending moment. The *bending moment* at any point in a beam is the moment, with respect to that point, of all of the forces to the left of the point. For example, let's look at point A in Fig. 6-19, which shows the free-body diagram of a beam. The beam carries the loads P_1 and P_2 and has the reactions R_1 and R_2. The bending moment at point A is the moment, with respect to A as a center, of the forces to the left of A. The forces to the left are the reaction R_1 and the load P_1, so we would have here just the moment of

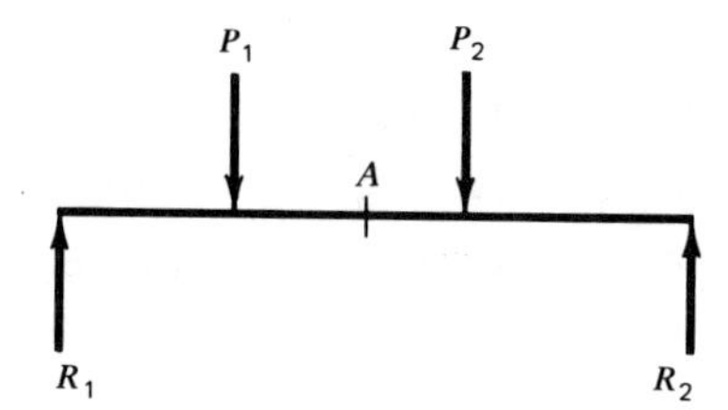

Fig. 6-19. Beam with concentrated loads.

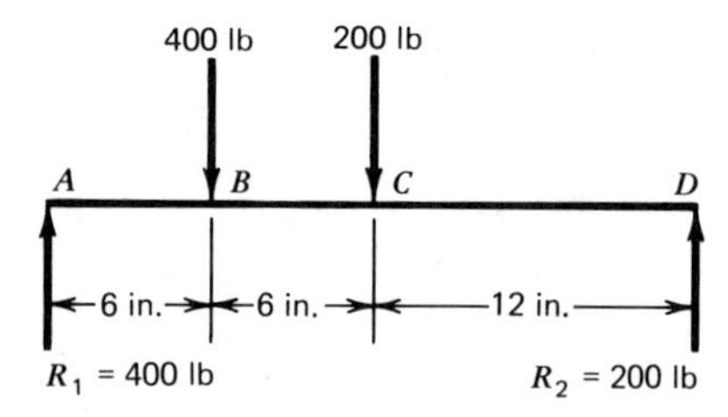

Fig. 6-20. Beam for Illustrative Examples 1 and 2.

R_1 plus the moment of P_1. Clockwise moments are positive, and counterclockwise moments are negative in calculating bending moments. (This is the same sign convention that you have used before.)

Let's learn to calculate bending moment first and then study the stresses caused by the bending moment and see why bending moment is important. The whole picture is too big to grasp all at once, so we will take it a step at a time. It's enough to say now that the stress due to the bending moment is usually the most important stress in a beam.

Now for bending moment. We are going to take the moment of all forces to one side of the point. We represent the bending moment by M and use a subscript to designate the point. Thus M_a designates the bending moment at point A.

UNITS

If we are using the gravitational system of units, the bending moment is usually stated in pound feet (lb ft) or pound inches (lb in.). We have the force in pounds. Then, if the lengths are stated in feet, the bending moment is in pound feet. If the lengths are stated in inches, the bending moment is in pound inches.

Forces are often expressed in kips in structural work. Then, if the lengths are stated in feet, the bending moment is in kip ft. If the lengths are stated in inches, the bending moment is in kip in.

In the metric system the forces are usually stated in newtons. Then, if the lengths are stated in meters, the bending moment is in newton meters (N·m).

Illustrative Example 1. Figure 6-20 shows the free-body diagram of a beam that is supported at the ends and has two concentrated loads. Calculate the bending moment at point B.

Solution:

The bending moment at B is the moment, with respect to point B, of all forces to the left of B. The only force to the left of B is the reaction R_1, which is equal to 400 lb. The moment arm is 6 in., so

$$M_b = 400 \times 6 = 2\,400 \text{ lb in.}$$

The force is in pounds and the distance in inches, so the bending moment is in pound inches. The moment is clockwise, so it is positive.

Illustrative Example 2. Calculate the bending moment at point C in Fig. 6-20.

Solution:
There are two forces to the left of point C. They are the force R_1, which has a moment arm of 12 in. with respect to C, and the force of 400 lb, which has a moment arm of 6 in. The moment of R_1 is clockwise, so it is positive. The moment of the 400-lb force is counterclockwise, and therefore, according to our rule, it is negative. Then the bending moment at C is

$$M_c = 400 \times 12 - 400 \times 6 = 4\,800 - 2\,400 = 2\,400 \text{ lb in.}$$

UNIFORM LOAD

When there is a uniformly distributed load on the beam, we proceed in the same way, but we take only the part of the load to the left of the point. You know how to calculate the moment of a uniform load. You multiply the amount of the load in pounds per foot by the length in feet to find the total load in this length. Then the *moment arm* is the distance from the center of the load to the point where you are calculating the bending moment. If the distributed load is expressed in pounds per foot, and the distances in feet, the bending moment is expressed in pound feet.

If the distributed load is expressed in newtons per meter and the distance in meters, the bending moment is expressed in newton meters.

Illustrative Example 3. Figure 6-21 shows the free-body diagram of a beam with a uniformly distributed load and a concentrated load. Calculate the bending moment at point B.

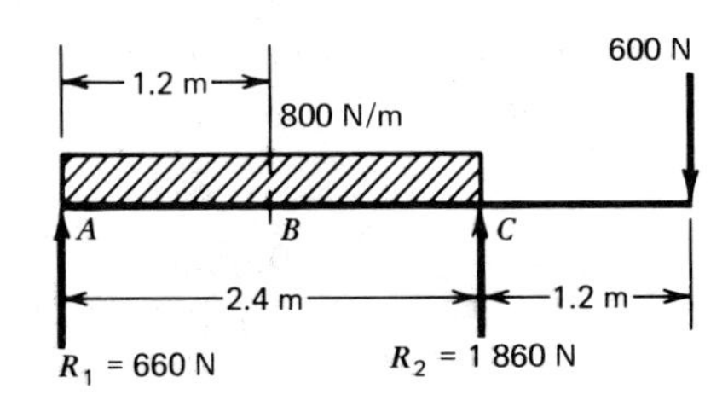

Fig. 6-21. Beam for Illustrative Examples 3 and 4.

Solution:
The forces to the left of point B are the reaction R_1 (660 N) and a 1.2-m length of the uniform load. The uniform load is directed downward, so its moment is counterclockwise with respect to point B. The total amount of the uniform load in the 1.2-m length is 800 × 1.2, and the distance from its center to B is ½ of 1.2 or 0.6 m. Then the bending moment at point B is

$$M_b = 660 \times 1.2 - 800 \times 1.2 \times 0.6 = 792 - 576$$
$$= 216 \text{ N·m}$$

Illustrative Example 4. Calculate the bending moment at point C in Fig. 6-21.

Solution:
The forces to the left of point C are the reaction R_1 and the 2.4-m length of the uniform load. Thus

$$M_c = 660 \times 2.4 - 800 \times 2.4 \times 1.2 = 1\,584 - 2\,304$$
$$= -720 \text{ N·m}$$

The negative sign for the answer shows that the bending moment at point C is counterclockwise.

Practice Problems (Section 6-5). Nothing like practice. We will give you the free-body diagrams, and you do the rest.

1. Calculate the bending moment at point B in Fig. 6-22.
2. Calculate the bending moment at point C in Fig. 6-22.
3. Calculate the bending moment at point B in Fig. 6-23.
4. Calculate the bending moment at point C in Fig. 6-23.
5. Calculate the bending moment at point B in Fig. 6-24.
6. Calculate the bending moment at point C in Fig. 6-24.

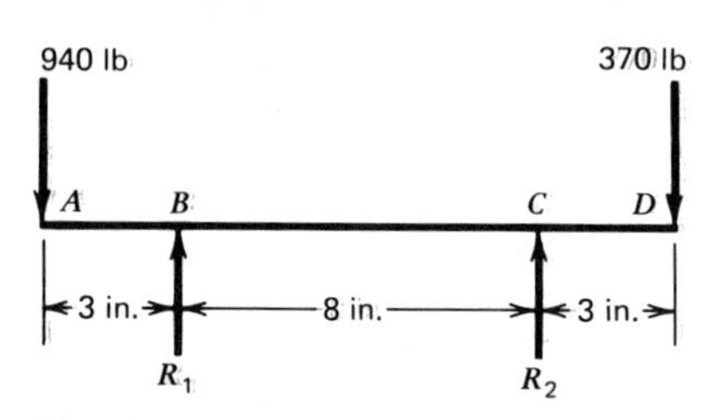

Fig. 6-22. Beam for Problems 1 and 2.

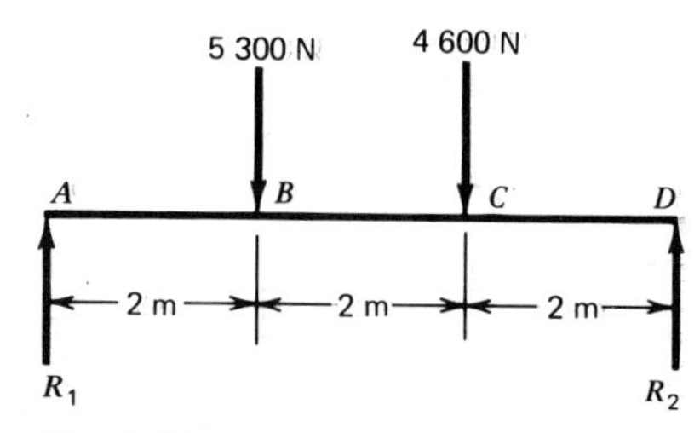

Fig. 6-23. Beam for Problems 3 and 4.

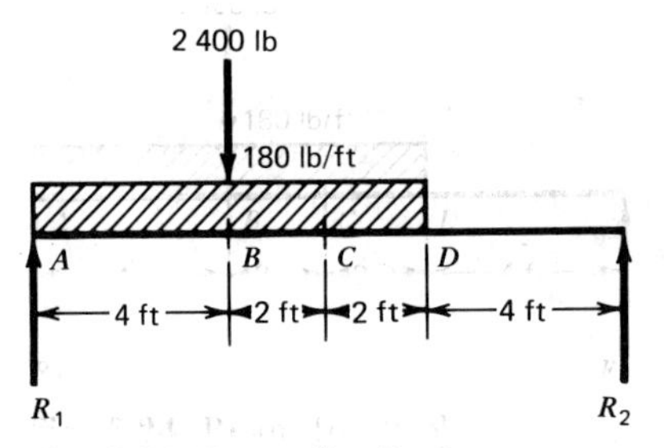

Fig. 6-24. Beam for Problems 5, 6, and 7.

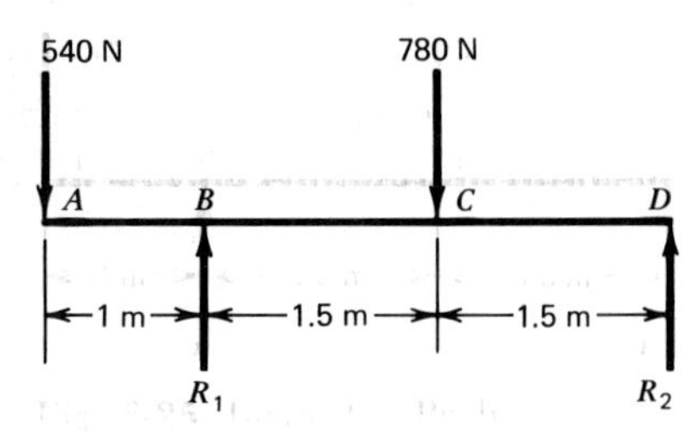

Fig. 6-25. Beam for Problems 8 and 9.

7. Calculate the bending moment at point D in Fig. 6-24.
8. Calculate the bending moment at point B in Fig. 6-25.
9. Calculate the bending moment at point C in Fig. 6-25.

6-6. MOMENT DIAGRAMS

A *moment diagram* is a diagram that shows the value of the bending moment at each point along a beam. A moment diagram is a useful thing to have, because you can look at it and see how the bending moment varies across the beam, where it is maximum, and so on.

The way to draw a moment diagram is to calculate the bending moment at a number of points along the beam and plot each of these bending moments. Then just connect the plotted points with lines, and you have the moment diagram. Plot positive bending moments upward and negative bending moments downward. You don't have to draw the moment diagram to scale. Just make a neat free-hand drawing and write the values of the bending moments on it.

There is one fact that saves time in drawing moment diagrams. In any length of beam between forces, the moment diagram is a straight line. (We won't bother to prove this here, but you can rely on its being true.) This means that if there isn't any uniformly distributed load, the moment diagram is a straight line in each length between forces. All you have to do is to calculate the bending moment at each point where a force is applied, then plot these moments and connect the plotted points by straight lines.

Illustrative Example 1. Figure 6-26*a* shows the free-body diagram of a beam. Draw the moment diagram.

Solution:

1. The base line *ABCD* is drawn in Fig. 6-26*b* to start the moment diagram.

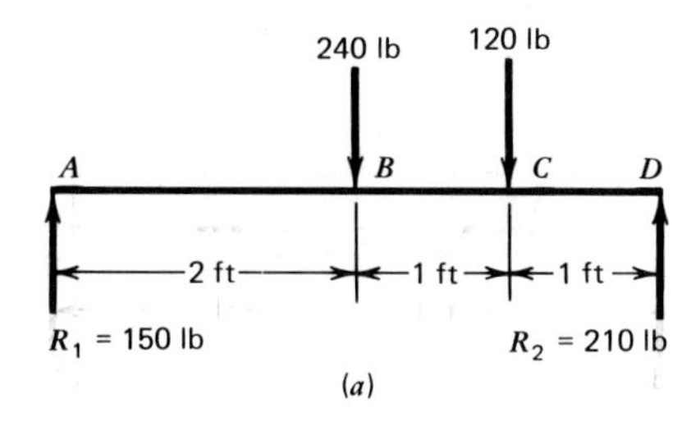

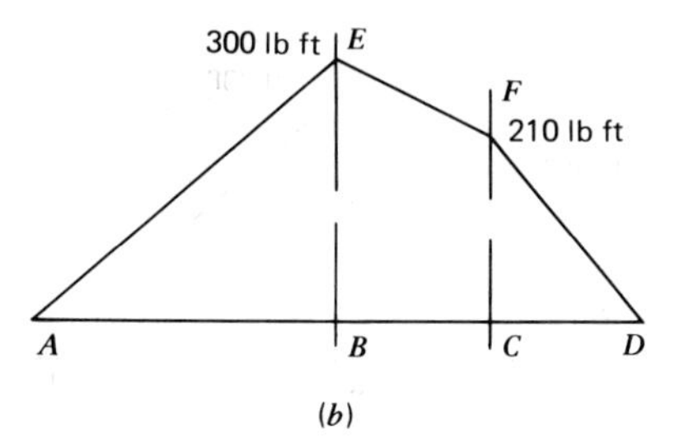

Fig. 6-26. Beam for Illustrative Example 1. (*a*) Free-body diagram. (*b*) Moment diagram.

2. The bending moment at A is zero, because there are no forces to the left of point A.
3. The bending moment at point B is

$$M_b = 150 \times 2 = 300 \text{ lb ft}$$

This positive bending moment is plotted upward from point B, and so point E in the moment diagram is located. The straight line AE is drawn.

4. The bending moment at point C is

$$M_c = 150 \times 3 - 240 \times 1 = 450 - 240 = 210 \text{ lb ft}$$

The bending moment at C is plotted upward (because it's positive) from C to locate point F in the moment diagram. A straight line is drawn from E to F.

5. The bending moment at point D is zero because the moment of the forces to the left of D is the moment of all the forces on the beam. The moment of all the forces must be zero because the beam is in equilibrium (remember $\Sigma M_d = 0$). The straight line FD is drawn to complete the moment diagram.

Illustrative Example 2. Figure 6-27*a* shows the free-body diagram of a beam that carries two loads and is supported at two points. Draw the moment diagram.

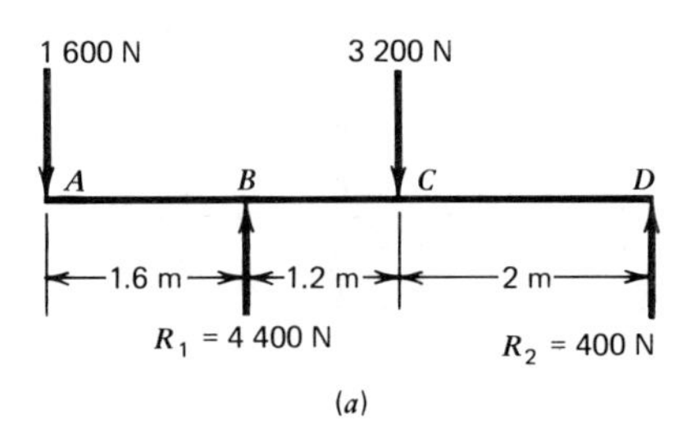

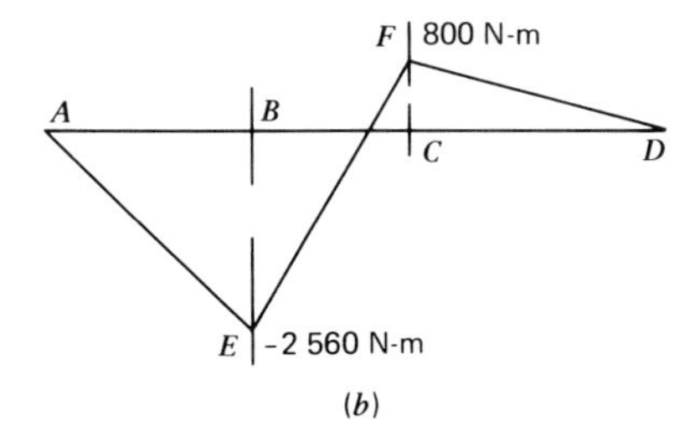

Fig. 6-27. Beam for Illustrative Example 2. (*a*) Free-body diagram. (*b*) Moment diagram.

Solution:

1. The base line *ABCD* is drawn as in Fig. 6-27*b*.
2. The bending moment at *A* is zero, because there are no forces to the left of *A*.
3. The bending moment at *B* is

$$M_b = -1\,600 \times 1.6 = -2\,560 \text{ N·m}$$

This bending moment is plotted downward (downward because it's negative) from *B* to locate point *E* in the moment diagram. Then a straight line is drawn from *A* to *E*.

4. The bending moment at *C* is

$$M_c = -1\,600 \times 2.8 + 4\,400 \times 1.2$$

$$= -4\,480 + 5\,280$$

$$= 800 \text{ N·m}$$

The value of the bending moment at *C* is plotted upward from *C* (upward because it's positive) to locate point *F*. The straight line *EF* is drawn.

5. The bending moment at *D* is zero, because the moment of the forces to the left of *D* is the moment of all the forces. The straight line *FD* is drawn to complete the moment diagram.

UNIFORM LOAD

The *moment diagram* is a curve for any length of a beam where there is a uniformly distributed load. This makes more work for us. We have to calculate the bending moment at enough points to determine the shape of the curve. You may wonder how many points to take. Usually, 3 to 10 points in a length of distributed load will be enough.

You can plot the bending moments as you calculate them and finally draw a curve through the plotted points.

Of course, if there is part of the beam that doesn't have a uniform load, you can just calculate the bending moment at points where forces are applied to this part.

Remember that the bending moment at a point is equal to the moment, with respect to the point, of all of the forces to the left of the point. The moment of a uniformly distributed load is equal to the amount of the load in pounds per foot *times*, the length of the load in feet, *times* the distance from the center of the load to the moment center.

Illustrative Example 3. Figure 6-28*a* shows the free-body diagram of a beam that has a uniformly distributed load over part of its length. Draw the moment diagram.

Solution:

1. Figure 6-28*b* shows the free-body diagram of the beam with the length of the distributed load divided into 1-ft lengths.
2. The base line is drawn in Fig. 6-28*c* to start the moment diagram.
3. The bending moment at A is zero, because there are no forces to the left of A.
4. The bending moment at B is

$$M_b = -900 \times 2 = -1\,800 \text{ lb ft}$$

This is plotted downward (downward because it's negative) from B to locate point I in the moment diagram. A straight line is drawn from A to I.

5. The calculations for the bending moment at C include the 900-lb load, the reaction R_1, and 1 ft of the uniformly distributed load. Consequently,

$$M_c = -900 \times 3 + 2\,100 \times 1 - 300 \times 1 \times \tfrac{1}{2}$$
$$= -2\,700 + 2\,100 - 150 = -750 \text{ lb ft}$$

This value is plotted downward from B to locate point J in the moment diagram, and the value is written on the diagram.

6. The bending moment at D is

$$M_d = -900 \times 4 + 2\,100 \times 2 - 300 \times 2 \times 1$$
$$= -3\,600 + 4\,200 - 600 = 0$$

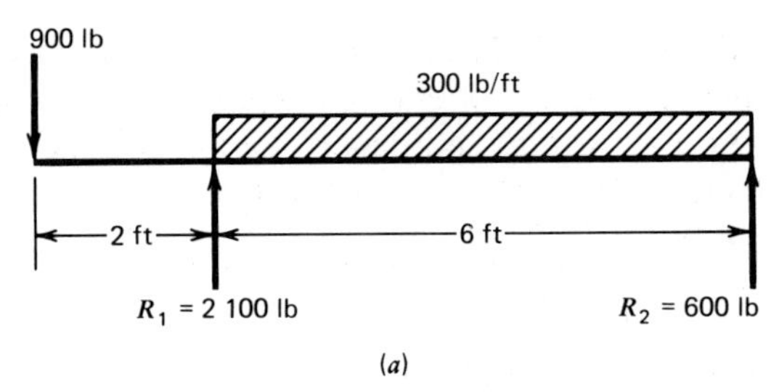

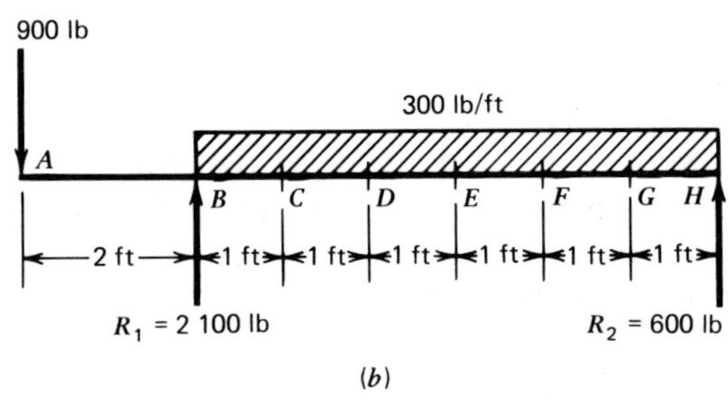

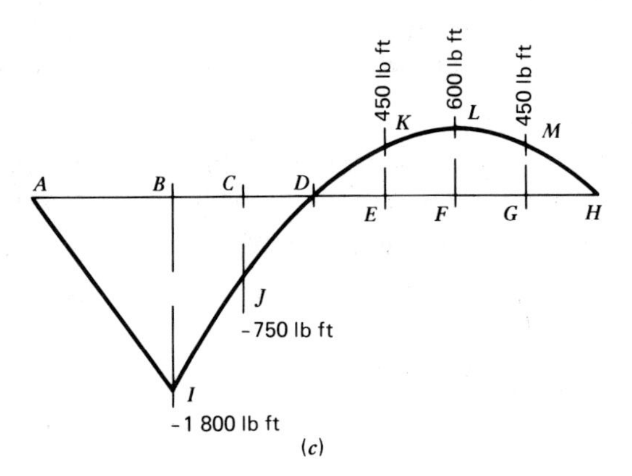

Fig. 6-28. Beam for Illustrative Example 3. (*a*) Free-body diagram. (*b*) Free-body diagram with 1-ft segments under uniformly distributed load. (*c*) Moment diagram.

The value of zero means that D is a point on the moment diagram.

7. The bending moment at point E is

$$M_e = -900 \times 5 + 2\,100 \times 3 - 300 \times 3 \times 1.5$$
$$= -4\,500 + 6\,300 - 1\,350 = 450 \text{ lb ft}$$

This value is plotted upward (upward because it's positive) from E to locate point K in the moment diagram. The value of 450 lb ft is written on the diagram.

8. The bending moment at point F is

$$M_f = -900 \times 6 + 2\,100 \times 4 - 300 \times 4 \times 2$$

$$= -5\,400 + 8\,400 - 2\,400 = 600 \text{ lb ft}$$

The value of 600 lb ft is plotted upward from F to locate point L in the moment diagram and the value is written on the diagram.

9. The bending moment at point G is

$$M_g = -900 \times 7 + 2\,100 \times 5 - 300 \times 5 \times 2.5$$
$$= -6\,300 + 10\,500 - 3\,750 = 450 \text{ lb ft}$$

This value of 450 lb ft is plotted upward from G to locate point M in the moment diagram. The value is written on the diagram.

10. The bending moment at H is zero, because the moment of the forces to the left of H is the moment of all the forces on the beam. ($\Sigma M_h = 0$ for equilibrium.)
11. The curve is drawn through points I, J, D, K, L, M, and H to complete the moment diagram.

Practice Problems (Section 6-6). Now it's time to practice drawing moment diagrams. We will give you the free-body diagrams, and you can proceed from there. Draw the moment diagram in each case.

1. Figure 6-29.
2. Figure 6-30.
3. Figure 6-31.
4. Figure 6-32.
5. Figure 6-33.

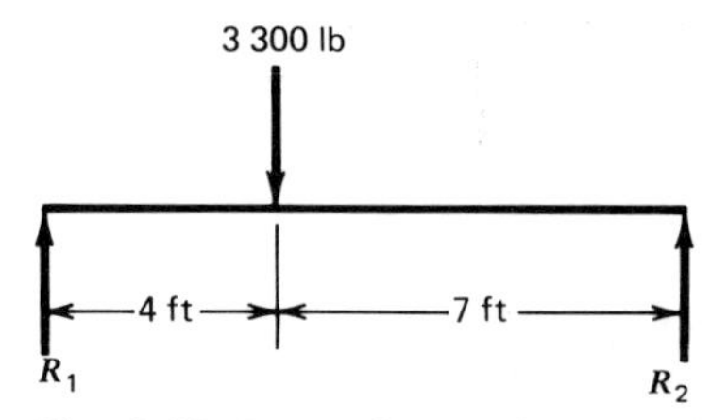

Fig. 6-29. Beam for Problem 1.

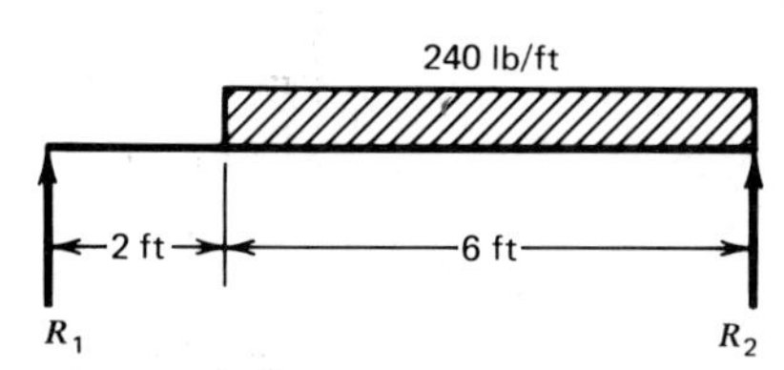

Fig. 6-30. Beam for Problem 2.

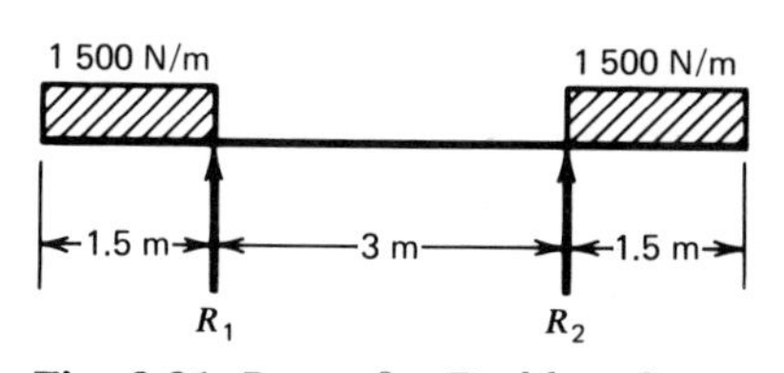

Fig. 6-31. Beam for Problem 3.

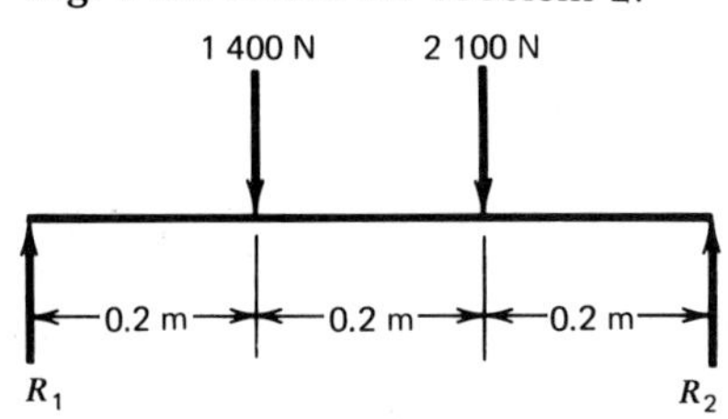

Fig. 6-32. Beam for Problem 4.

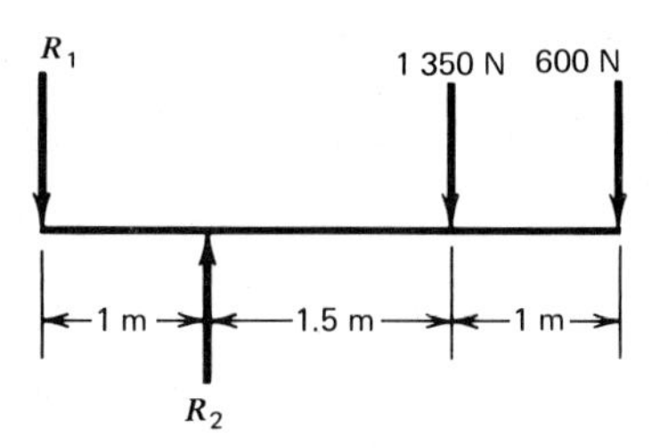

Fig. 6-33. Beam for Problem 5.

6-7. MAXIMUM BENDING MOMENT

The next question is, where does the maximum bending moment occur? The answer is that the maximum bending moment occurs at the point at which the shear is zero. (This statement can be proved by means of calculus.) As we will see later, the maximum stress occurs where the moment is maximum, so it is important to be able to calculate the maximum moment.

Here's the way to proceed. We will draw the shear diagram and note carefully where the shear is zero. The bending moment will have its maximum value at that point, and we will calculate it.

Illustrative Example 1. Figure 6-34*a* shows the free-body diagram of a beam that is supported at the ends and carries two concentrated loads. Find the maximum bending moment.

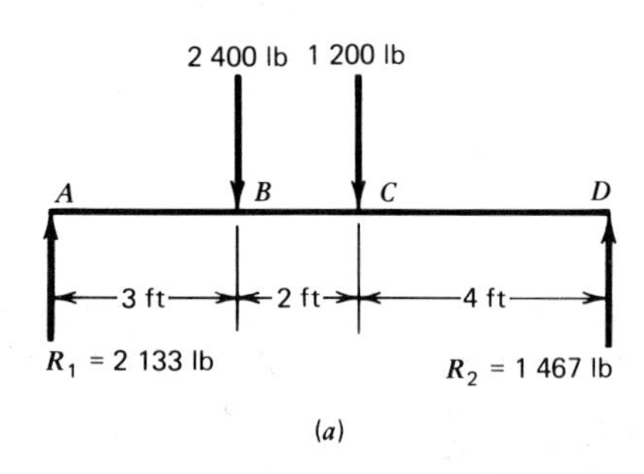

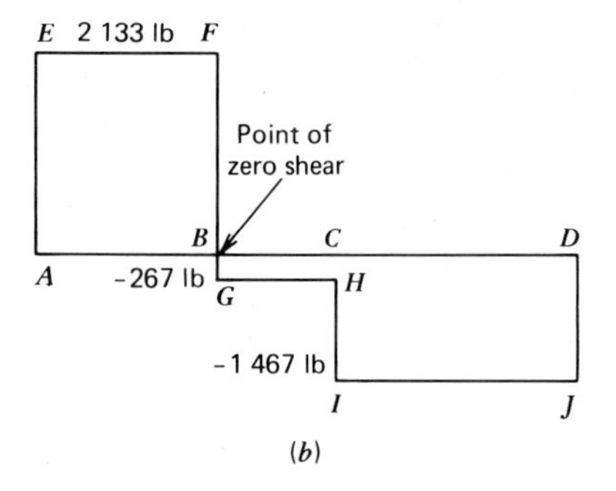

Fig. 6-34. Free-body diagram and shear diagram. (*a*) Free-body diagram. (*b*) Shear diagram.

Solution:

A. We draw the shear diagram as in Fig. 6-34*b*. Remember that we just plot the forces upward or downward, as they are directed, while we move across the beam from left to right.

1. We draw the base line *ABCD* in Fig. 6-34*b*.
2. We start at *A* and plot the reaction R_1 (2 133 lb) upward to locate point *E* in the shear diagram.
3. We draw the horizontal line from *E* to *F*, where it reaches the line of action of the 2 400-lb load.
4. We plot the 2 400-lb load downward from *F* to *G*. We start at 2 133 lb above the base line and drop 2 400 lb. This places point *G* below the base line by an amount of

$$2\,133 - 2\,400 = -267 \text{ lb}$$

5. From *G* we draw the horizontal line *GH* to reach the line of action of the 1 200-lb load.
6. We plot the 1 200-lb load downward from *H* to *I*.

$$-267 - 1\,200 = -1\,467 \text{ lb}$$

for the value of the shear at *I*.

7. We draw the horizontal line *IJ* from *I* to the right end of the beam.
8. We plot the reaction R_2 (1 467 lb) upward from *J* to complete the shear diagram at *D*.

B. We calculate the maximum bending moment.

1. The shear is zero at point *B*, so the maximum bending moment occurs at *B*.
2. The maximum bending moment is

$$M_b = 2\,133 \times 3 = 6\,400 \text{ lb ft}$$

UNIFORM LOAD

When there is a uniformly distributed load on a beam, we may have to solve an equation to locate the point of zero shear. This point may be where a sloping line intersects the base line in the shear diagram. Let's try it.

Illustrative Example 2. Figure 6-35*a* shows the free-body diagram of a beam that has a uniformly distributed load over part of its length. Find the maximum bending moment.

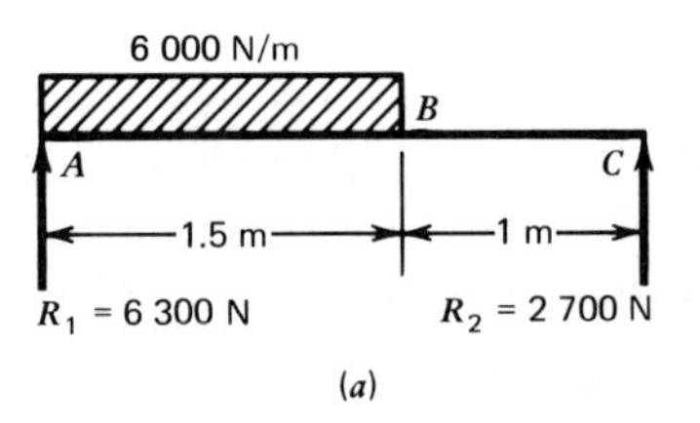

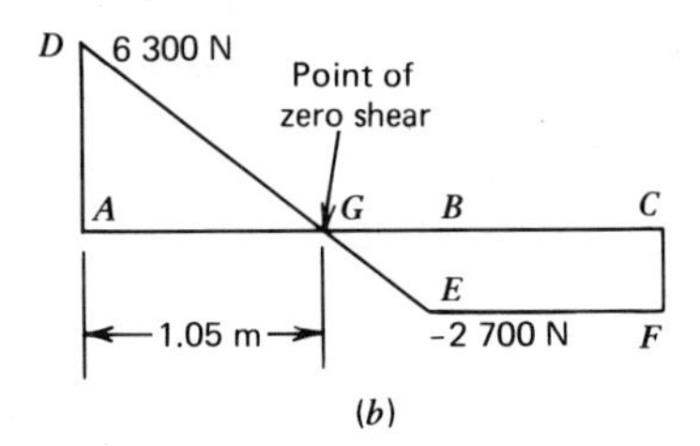

Fig. 6-35. Beam for Illustrative Example 2. (*a*) Free-body diagram. (*b*) Shear diagram.

Solution:

A. The shear diagram is drawn in Fig. 6-35*b*.

1. The base line *ABC* is drawn.
2. The reaction R_1 (6 300 N) is plotted upward from *A* to locate point *D*.
3. From *D* the uniform load is plotted downward as we move across the beam. The sloping line goes from *D* to *E*. The value of the shear at *E* is

$$6\,300 - 6\,000 \times 1.5 = 6\,300 - 9\,000 = -2\,700 \text{ N}$$

4. A horizontal line is drawn from *E* to *F*.
5. The reaction R_2 (2 700 N) is plotted upward from *F* to close the shear diagram at *C*.
6. The shear is zero at point *G*, so we have to locate *G*. Let's call the distance *x* from *A* to *G*. (Remember how you used *x* as an unknown in algebra.) The shear at *G* is the sum of the forces to the left of *G*; the forces to the left are R_1 and *x* ft of the uniform load. So we write

$$V_g = 0$$

$$6\,300 - 6\,000x = 0$$

Transposing,

$$6\,000x = 6\,300$$

$$x = 1.05 \text{ m}$$

B. The maximum bending moment is calculated.

1. The shear is zero at point G, so the maximum bending moment occurs at G.
2. The maximum bending moment is

$$M_g = 6\,300 \times 1.05 - 6\,000 \times 1.05 \times 1.05/2$$
$$= 6\,615 - 3\,307.5 = 3\,307.5 \text{ N·m}$$

TWO POINTS OF ZERO SHEAR

Sometimes there are two points of zero shear. Then, we have to calculate the bending moment at each point to be sure we get the maximum. One point of zero shear will represent the maximum positive bending moment. The other point of zero shear will represent the maximum negative bending moment. What we want to find is the bending moment that has the greatest numerical value. We call that the *maximum bending moment.*

Illustrative Example 3. Figure 6-36*a* shows the free-body diagram of a beam supported at two points that has two concentrated loads. Find the maximum bending moment.

Solution:

A. We draw the shear diagram as in Fig. 6-36*b*.

1. The first step in drawing the shear diagram is to draw the base line $ABCD$ in Fig. 6-36*b*.

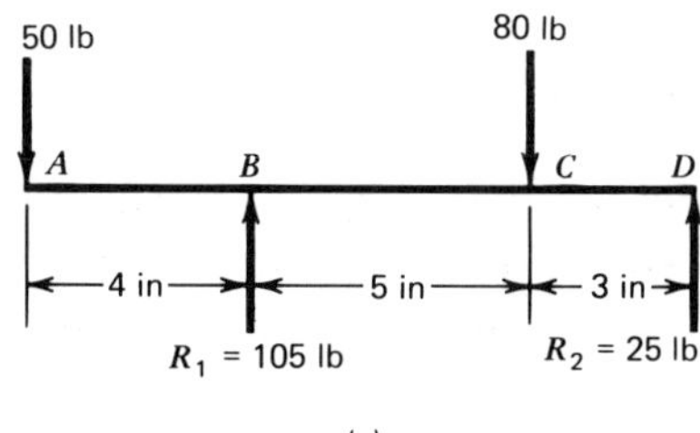

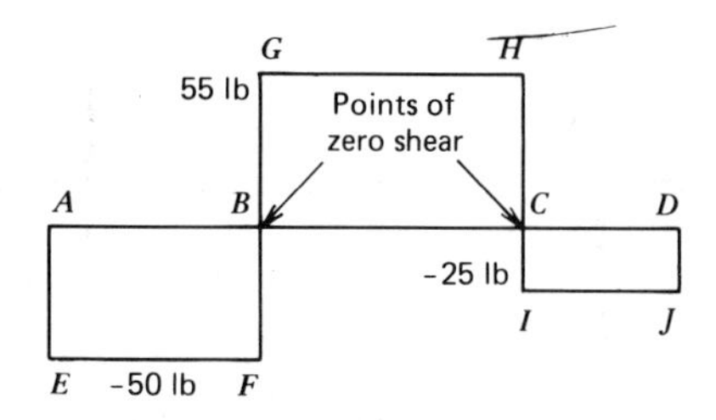

Fig. 6-36. Beam for Illustrative Example 3. (*a*) Free-body diagram. (*b*) Shear diagram.

2. We plot the 50-lb load downward from A to locate E.
3. We draw the horizontal line EF.
4. We plot the reaction R_1 (105 lb) upward from F to G. The value of the shear at G is

$$-50 + 105 = 55 \text{ lb}$$

5. We draw the horizontal line GH.
6. We plot the 80-lb load downward from H to I. The value of the shear at I is

$$55 - 80 = -25 \text{ lb}$$

7. We draw the horizontal line IJ.
8. We finish the shear diagram by plotting the reaction R_2 (25 lb) from J to D.

B. We calculate the maximum bending moment.
1. The first point of zero shear is at B.
2. We calculate the bending moment at B. It is

$$M_b = -50 \times 4 = -200 \text{ lb in.}$$

3. The second point of zero shear is at C.
4. The bending moment at C is

$$M_c = -50 \times 9 + 105 \times 5 = -450 + 525 = 75 \text{ lb in.}$$

5. The bending moment has a greater numerical value at B than at C, so the maximum bending moment is

$$M_b = -200 \text{ lb in.}$$

Practice Problems (Section 6-7). Now see if you can calculate the maximum bending moment for the beams in the following problems. The free-body diagrams are given in the figures.

1. Figure 6-37.
2. Figure 6-38.

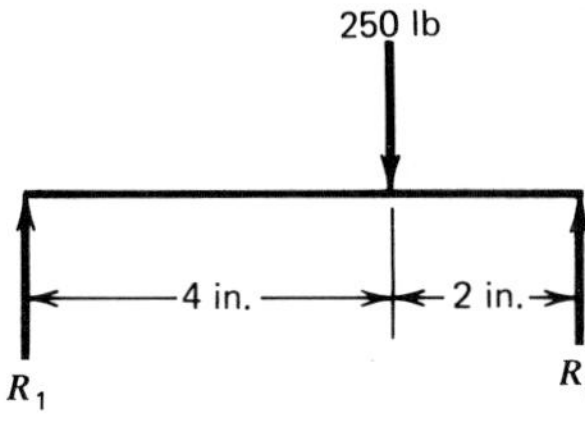

Fig. 6-37. Beam for Problem 1.

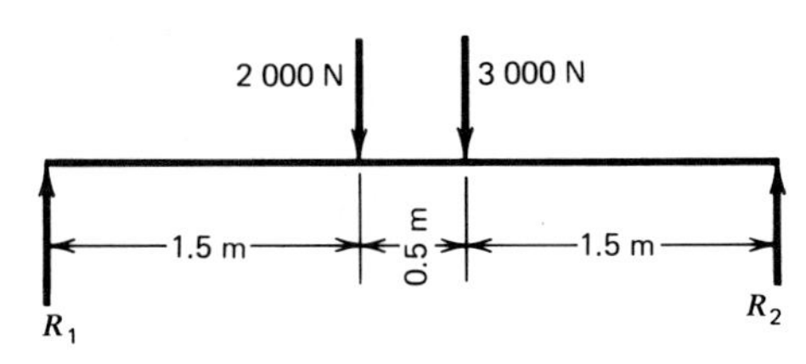

Fig. 6-38. Beam for Problem 2.

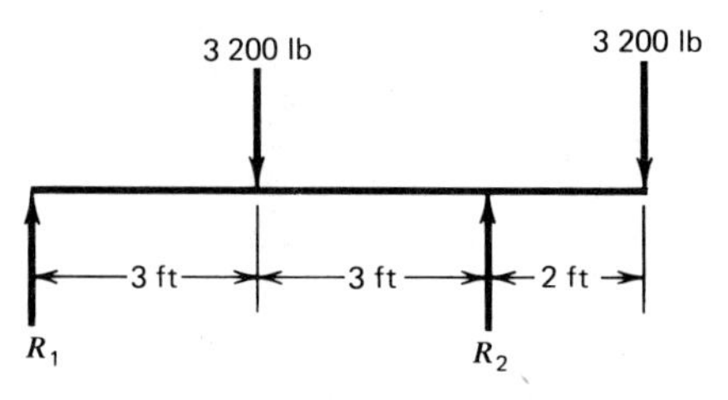

Fig. 6-39. Beam for Problem 3.

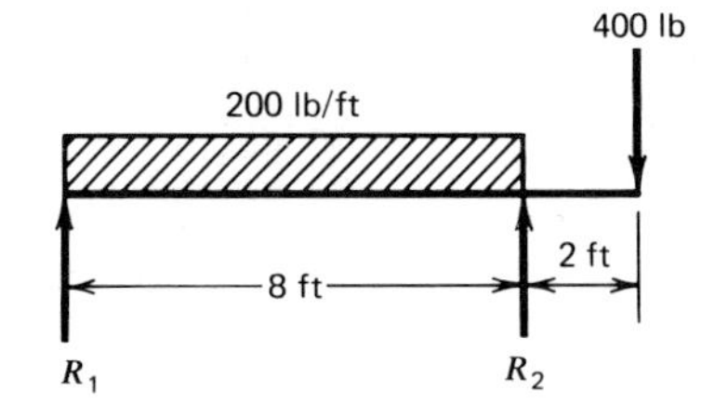

Fig. 6-40. Beam for Problem 4.

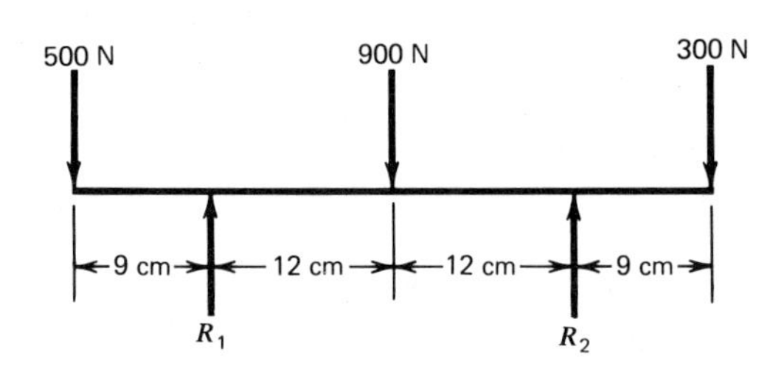

Fig. 6-41. Beam for Problem 5.

3. Figure 6-39.
4. Figure 6-40.
5. Figure 6-41.

6-8. BENDING STRESS, THE FLEXURE FORMULA

Now we are ready to find the stress that is caused by the bending moment. We call this stress *bending stress*. We are going to study the *flexure formula* which gives the relation between the bending stress and the bending moment. The flexure formula is important to designers, because they have so much use for it in designing beams.

Let's look at Fig. 6-42*a* to start. Here we see the free-body diagram of a beam, and we show the shape of the beam because we want to see how the shape changes when the load is applied. Look especially at the short length of beam between the cross sections *B-B* and *C-C*. These cross sections are parallel before the load is applied.

Figure 6-42*b* shows the same beam after the load has been applied and the beam has bent. The cross sections *B-B* and *C-C* are no longer

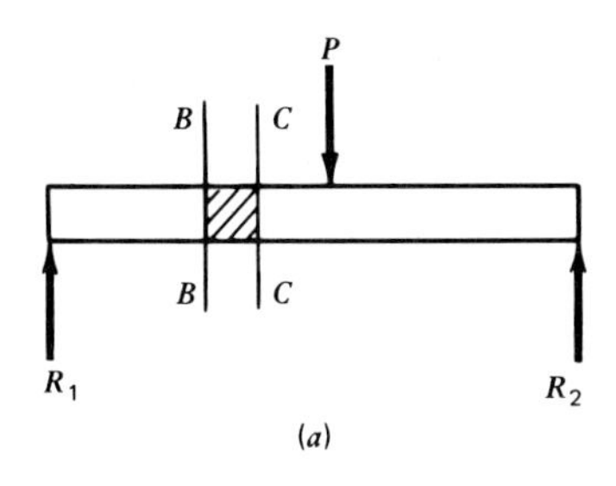

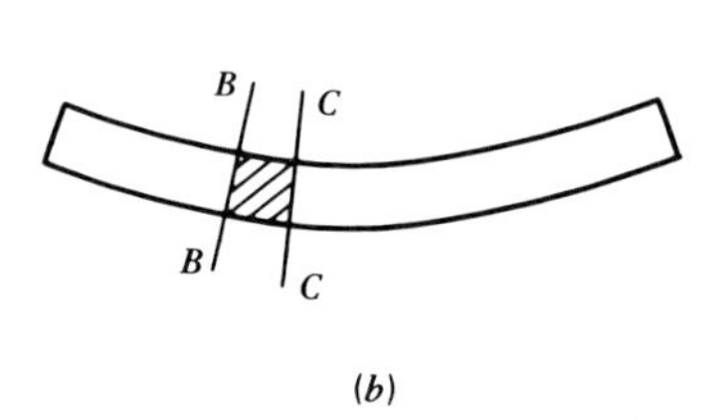

Fig. 6-42. Bending of beam under load. (*a*) Free-body diagram. (*b*) Bent shape of beam.

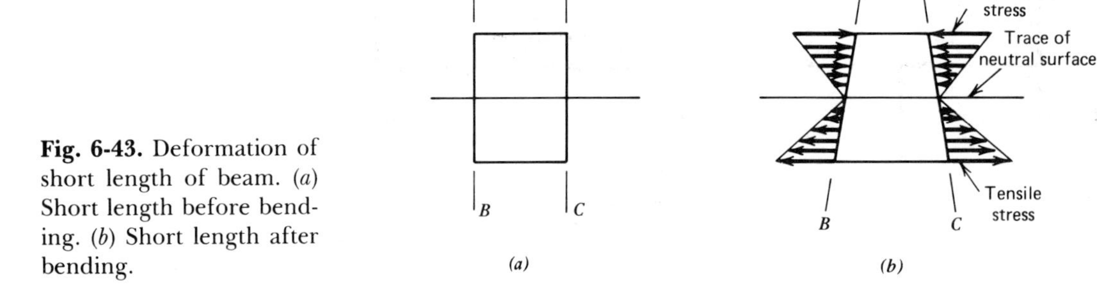

Fig. 6-43. Deformation of short length of beam. (*a*) Short length before bending. (*b*) Short length after bending.

parallel. The short length has been squeezed together at the top and stretched at the bottom.

Figure 6-43*a* shows the same short length of beam before bending, but it is drawn to larger scale. Figure 6-43*b* shows the short length after the beam has bent and also shows the stresses acting on it. There is compressive stress at the top, because the material at the top has been squeezed together, and there is tensile stress at the bottom, because the material at the bottom has been stretched. There is a line *O-O* that is of the same length after bending as before. There isn't any stress along the line *O-O*, because there is no deformation, and we can't have stress without deformation. The line *O-O* is just the trace (projection) of a horizontal surface or plane in the beam where there is no stress. This surface is called the *neutral surface*. Above the neutral surface there is compressive stress, because the material is squeezed together. Below the neutral surface there is tensile stress because the material is stretched.

It has been shown by experiment that the value of the stress is proportional to the distance from the neutral surface. This proportionality is shown by the lengths of the arrows that represent the stress in Fig. 6-43*b*. The maximum stress occurs at the greatest distance from the neutral surface.

The next question is, how is the stress related to the bending moment? Figure 6-44 shows the free-body diagram of the part of the beam to the left of the cross section *C-C*. Here we see the reaction R_1 and the

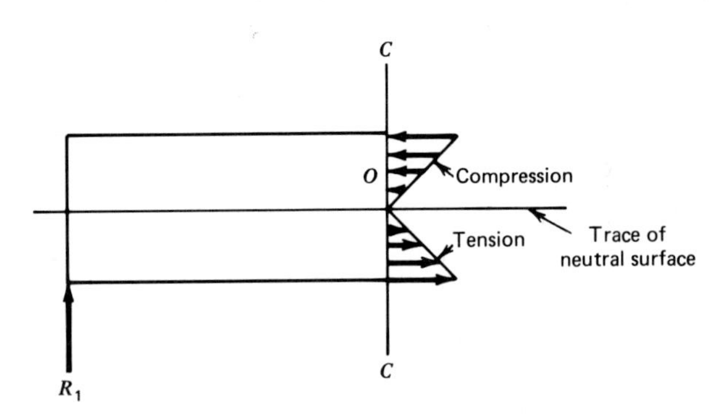

Fig. 6-44. Free-body diagram of part of beam.

stresses on the cross section *C-C*. This part of the beam is in equilibrium, and we can apply an equation of equilibrium. Let's think of a moment equation with center at point *O*. (This would be $\Sigma M_o = 0$.) Two things enter this equation. One is the moment of the forces to the left of the cross section *C-C*, and their moment about point *O* is the bending moment at *O*. The other thing is the moment of the stresses on the cross section. We call the moment of the stresses on the cross section the *resisting moment*. Now if this part of the beam is in equilibrium, the resisting moment must be equal to the bending moment and must be opposite in direction so that $\Sigma M_o = 0$. So we reach the conclusion that the moment of the stresses on the cross section is equal to the bending moment.

Now let's look more closely at the cross section *C-C* and the stresses on it. Figure 6-45*a* shows the front view of the cross section, and here you see the line *O-O*. This line is the intersection of the neutral surface with the plane of the cross section and is called the *neutral axis*. Notice the distance *c* from the neutral axis to the top of the cross section. We will use this distance later.

Figure 6-45*b* shows the side view of the cross section, and here we see the stresses that act on the cross section. The stress is zero at the neutral surface (point *O*) and increases to a maximum at the top and bottom of the cross section. The greatest stress occurs at the greatest distance from the neutral axis, and this greatest distance is represented by *c*. The value of the greatest stress is

$$S = \frac{Mc}{I}$$

where *M* is the bending moment on the cross section and *I* is the moment of inertia (remember moment of inertia from Chapter 5) of the area of the cross section with respect to the neutral axis. This formula is the *flexure formula*.

Another question arises now. It is, where is the neutral axis located? The answer is that the neutral axis passes through the centroid of the area of the cross section.

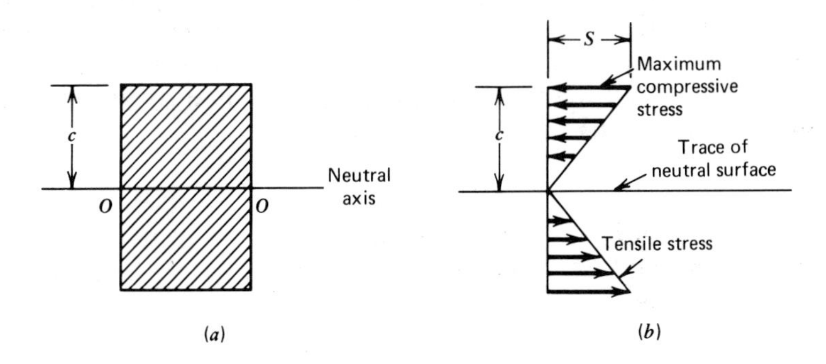

Fig. 6-45. Stresses on cross section of beam. (*a*) Section *C-C* of beam. (*b*) Stress distribution on section.

Now let's look at gravitational units in the flexure formula. We stated it as

$$S = \frac{Mc}{I}$$

but we can change it to a different form in which it is easier to see how the units work out. Thus

$$M = \frac{SI}{c}$$

We want to have S in psi (pounds per square inch). The distance c will be in inches, and the moment of inertia I in in.^{4} (inches to the fourth power). Then the bending moment M must be in lb in. (pound inches) to make the dimensions work out. We can write an equation in dimensions just as easily as an equation in letters or numbers. All we have to do is to replace each letter by the units in which the quantity is expressed. Let's try this with

$$M = \frac{SI}{c}$$

and solve for the dimensions of M. We will replace S by lb/in.^{2} (pounds per square inch), I by in.^{4}, and c by in. Then,

$$M = \frac{\text{lb in.}^4}{\text{in.}^2 \text{ in.}} = \text{lb in.}$$

This answer tells us that M must be expressed in pound inches in the flexure formula when we are working in traditional units. If you calculate the bending moment in pound feet, you must multiply it by 12 to convert it to pound inches. *Don't forget this.*

Now let's go back to the flexure formula as we wrote it first. This is

$$S = \frac{Mc}{I}$$

We can use the formula to solve for the bending stress. We will have to find the bending moment, locate the centroid of the cross section, determine the distance c, and calculate the moment of inertia. It's easiest of all when the cross section is a rectangle or a circle. Then the centroid is at the center, and so the neutral axis goes through the center. The distance c is one half the height of the beam and, best of all, we have a simple formula for the moment of inertia.

Illustrative Example 1. Figure 6-46 shows the cross section of a beam that has a bending moment of 60 000 lb in. Find the maximum bending stress.

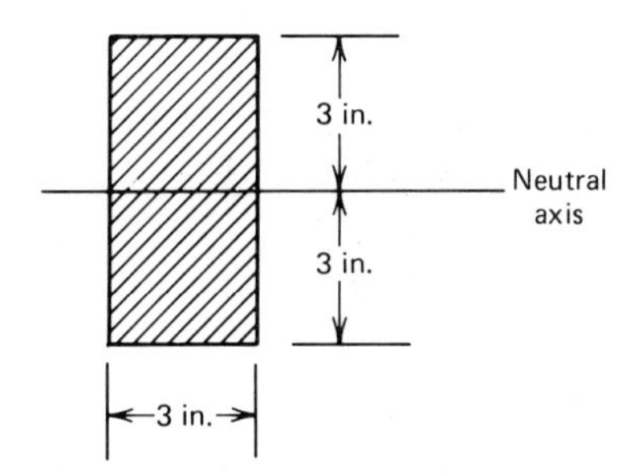

Fig. 6-46. Rectangular cross section.

Solution:

A. The bending moment M is 60 000 lb in.

B. The centroid of a rectangle is at the center of the rectangle, so the neutral axis is at the center.

C. The distance c is 3 in. (one-half the depth).

D. The moment of inertia of a rectangular area, with respect to a centroidal axis parallel to the base, is 1/12 bh^3. (You should remember this from Chapter 5).

   1. The width b is 3 in.
   2. The height h is 6 in.
   3. The moment of inertia is

$$I = \frac{1}{12} bh^3 = \frac{1}{12} \times 3 \times (6)^3 = 54 \text{ in.}^4$$

The bending stress is

$$S = \frac{Mc}{I} = \frac{60\,000 \times 3}{54} = 3\,330 \text{ psi}$$

METRIC UNITS

If we are working in the metric system, we can calculate the bending moment in newton millimeters (N·mm). There may be times when we calculate the bending moment in newton meters first. Then we can convert the bending moment to newton millimeters by multiplying by 10^3. We express the distance c in millimeters and the moment of inertia in millimeters to the fourth power (mm^4). Then we can put the units in the flexure formula. Thus

$$S = \frac{Mc}{I} = \frac{\text{N·mm} \times \text{mm}}{\text{mm}^4} = \frac{\text{N}}{\text{mm}^2} = \text{MPa}$$

and we end with stress in megapascals. Let's try an example.

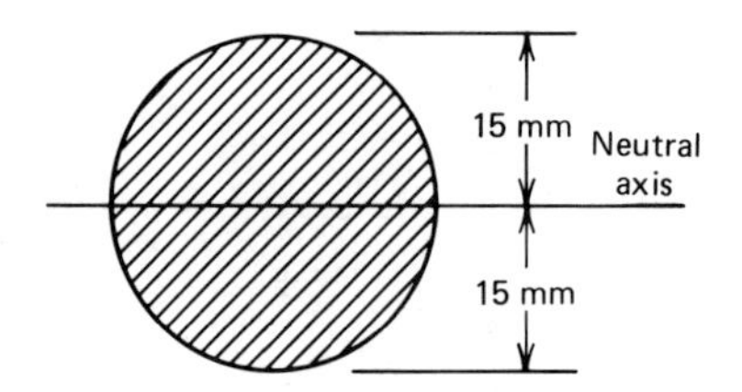

Fig. 6-47. Circular cross section.

Illustrative Example 2. A bending moment of 240 000 N·mm is applied to a steel bar 30 mm in diameter. What is the maximum bending stress?

Solution:

A. The bending moment M is 240 000 N·mm.

B. The centroid is at the center of the circle. Figure 6-47 shows the area.

C. The distance c is 15 mm (one-half the depth).

D. The moment of inertia of a circular area with respect to an axis through the center is $(\pi/64)d^4$.
(Did you remember this from Chapter 5?)

1. The diameter d is 30 mm.
2. The moment of inertia is

$$I = \frac{\pi}{64}d^4 = \frac{\pi}{64}(30)^4 = 39\ 800\ \text{mm}^4$$

E. The maximum bending stress is

$$S = \frac{Mc}{I} = \frac{240\ 000 \times 15}{39\ 800} = 90.4\ \text{MPa}$$

NEGATIVE MOMENT

The flexure formula gives the stress as

$$S = \frac{Mc}{I}$$

When the bending moment M is positive, there is compressive stress above the neutral axis and tensile stress below. When the bending moment is negative, there is tensile stress above the neutral axis and compressive stress below.

On any particular cross section of a beam, the stress varies from zero at the neutral axis to a maximum at the point farthest from the neutral axis. Then, of course, the bending moment varies from one end of the beam to another. We want to find the maximum stress in the beam, and what we must do is to find the maximum bending moment and use it

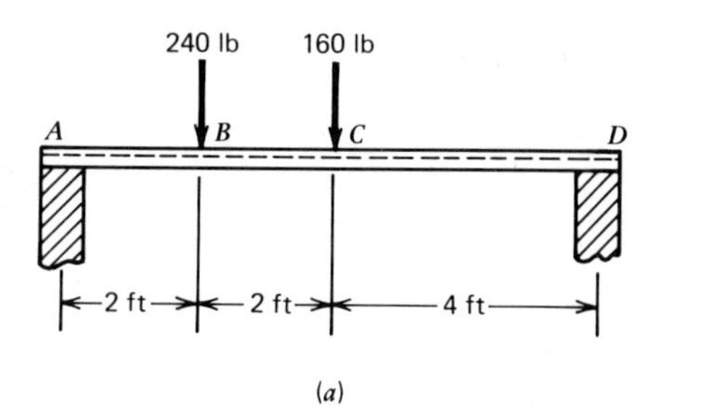

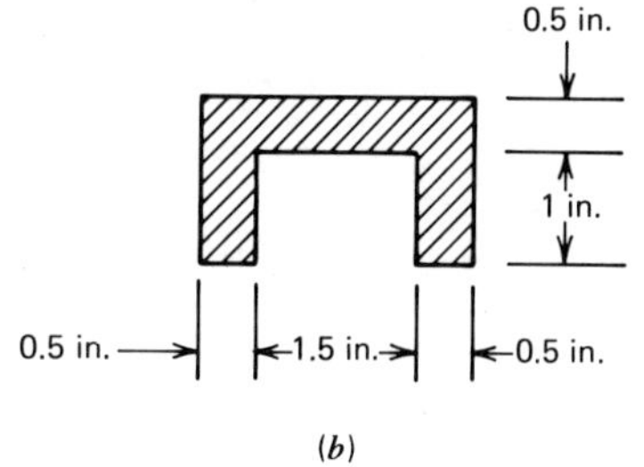

Fig. 6-48. Steel beam with two loads. (*a*) Beam and loads. (*b*) Cross section.

in the flexure formula. Let's take an example now and go all the way through from the beginning. We will start with the beam and the loads. Our procedure will be:

1. Draw the free-body diagram. Find the reactions.
2. Draw the shear diagram and find where the shear is zero.
3. Calculate the maximum bending moment. (The maximum bending moment occurs at the point of zero shear.) Be sure that the bending moment is expressed in pound inches or newton millimeters.
4. Locate the centroid of the cross section. (The neutral axis goes through the centroid.)
5. Calculate the moment of inertia of the cross section with respect to the neutral axis.
6. Apply the flexure formula to calculate the maximum stress.

Illustrative Example 3. Figure 6-48*a* shows the front view of a steel beam supported at the ends that carries two loads. Figure 6-48*b* shows the cross section of the beam. Find the maximum bending stress.

Solution:

A. The reactions.

1. Figure 6-49*a* shows the free-body diagram of the beam. We can find the reaction R_2 by writing a moment equation with

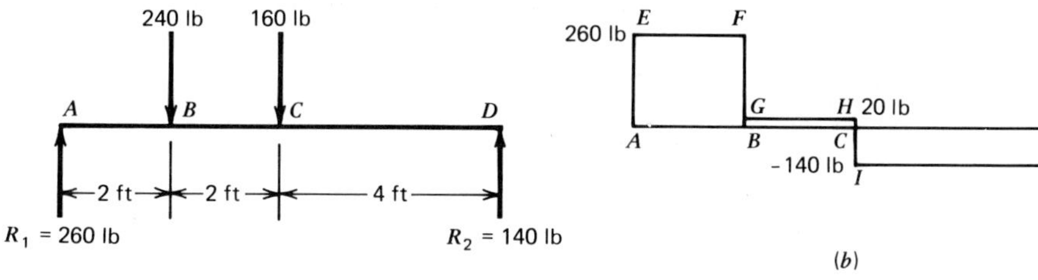

Fig. 6-49. Free-body diagram and shear diagram. (*a*) Free-body diagram. (*b*) Shear diagram.

point A as a center. Thus, taking the forces in order from left to right,

$$\Sigma M_a = 0$$

$$240 \times 2 + 160 \times 4 - 8R_2 = 0$$

$$8R_2 = 480 + 640 = 1\,120$$

$$R_2 = 140 \text{ lb}$$

2. The reaction R_1 is found by writing a moment equation with point D as a center.

$$\Sigma M_d = 0$$

$$8R_1 - 240 \times 6 - 160 \times 4 = 0$$

$$8R_1 = 1\,440 + 640 = 2\,080$$

$$R_1 = 260 \text{ lb}$$

3. Now let's check the reactions by writing an equation for the sum of the vertical forces. Thus

$$\Sigma F_y = 0$$

$$260 - 240 - 160 + 140 = 0$$

$$0 = 0$$

Good! We can see the reactions are correct.

B. The shear diagram, Fig. 6-49*b*.

1. We draw the base line $ABCD$.
2. We start at A and plot the reaction R_1 (260 lb) upward from A to locate point E.
3. We draw the horizontal line EF.
4. We plot the 240-lb load downward from F to locate point G. The shear at G is

$$260 - 240 = 20 \text{ lb}$$

5. We draw the horizontal line GH.
6. We plot the 160-lb load downward from H to locate point I.
7. We draw the horizontal line IJ.
8. We plot the reaction R_2 (140 lb) upward from J to complete the shear diagram at D.
9. The point of zero shear is at C.

C. The maximum bending moment.

1. The maximum bending moment is at C, where the shear is zero.

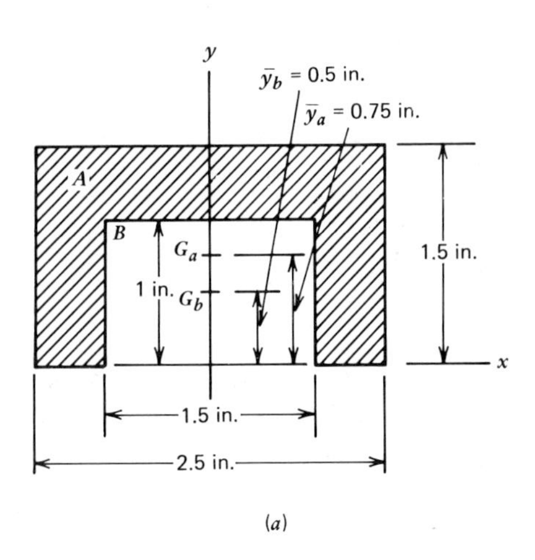

(a)

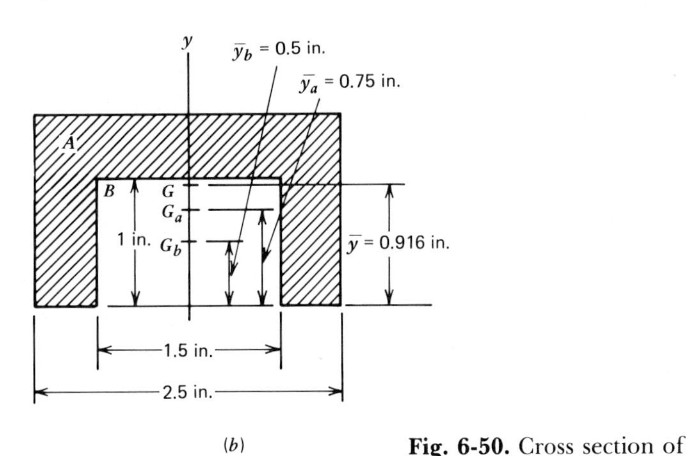

(b)

Fig. 6-50. Cross section of beam. (*a*) Location of centroid. (*b*) Calculation of moment of inertia.

2. The bending moment at C is equal to the moment, with respect to C, of all the forces to the left of C. So

$$M_c = 260 \times 4 - 240 \times 2 = 1\,040 - 480 = 560 \text{ lb ft}$$

3. We must convert the bending moment to pound inches by multiplying by 12. Thus

$$M_c = 560 \times 12 = 6\,720 \text{ lb in.}$$

D. The location of the centroid Fig. 6-50*a*.

1. Let's divide the area into the positive rectangle A, 2.5 in. wide and 1.5 in. high, and the negative rectangle B, 1.5 in. wide and 1 in. high.
2. We will place the x axis at the bottom of the area and the y axis through the center.
3. For the area A
   (a) The area of the rectangle is the product of the width and the height. Then

$$A_a = 2.5 \times 1.5 = 3.75 \text{ in.}^2$$

   (b) The distance from the x axis to the centroid of the rectangle A (at G_a) is one-half the height. Thus

$$\bar{y}_a = \tfrac{1}{2} \times 1.5 = 0.75 \text{ in.}$$

4. For the rectangle B
   (a) The area of the rectangle is

$$A_b = -1.5 \times 1 = -1.5 \text{ in.}^2$$

   The area is negative.
   (b) The distance from the x axis to the centroid of B (at G_b) is

$$\bar{y}_b = \tfrac{1}{2} \times 1 = 0.5 \text{ in.}$$

5. For the entire area,
   (a) The total area is equal to the area of A plus the area of B

$$A = A_a + A_b$$

   We must remember here that B is a negative area. Then

$$A = 3.75 - 1.5 = 2.25 \text{ in.}^2$$

   (b) The moment of the entire area with respect to the x axis is equal to the moment of A plus the moment of B. Thus

$$M_x = A_a\bar{y}_a + A_b\bar{y}_b = 3.75 \times 0.75 + (-1.5) \times 0.5$$

$$= 2.81 - 0.75 = 2.06 \text{ in.}^3$$

   (c) The distance of the centroid of the entire area from the x axis is

$$\bar{y} = \frac{M_x}{A} = \frac{2.06}{2.25} = 0.916 \text{ in.}$$

E. The moment of inertia with respect to an axis through the centroid (Fig. 6-50*b*).
   1. The centroid of the entire area is at G, a distance of 0.916 in. above the x axis. We draw the x_1 axis through G. We need to calculate the moment of inertia with respect to the x_1 axis. We will designate it as I_{x1}.
   2. The area is divided into the same positive rectangle A, 2.5 in. × 1.5 in., and the same negative rectangle B, 1.5 in. × 1 in.
   3. For the rectangle A (we will use the parallel axis theorem),
      (a) The moment of inertia with respect to an axis through the centroid is

$$I_{xg} = \frac{1}{12} bh^3 = \frac{1}{12} \times 2.5 \times (1.5)^3 = 0.703 \text{ in.}^4$$

      (b) The area is

$$A_a = 3.75 \text{ in.}^2$$

      (c) The distance c from the centroid of the rectangle to the x_1 axis is

$$c = 0.916 - 0.75 = 0.166 \text{ in.}$$

      (d) The moment of inertia of A with respect to the x_1 axis is (parallel axis theorem)

$$I_{x1} = I_{xg} + A_a c^2 = 0.703 + 3.75\,(0.166)^2$$

$$= 0.703 + 0.103 = 0.806 \text{ in.}^4$$

4. For the rectangle B,
   (a) The moment of inertia of the rectangle with respect to an axis through the centroid is

$$I_{xg} = \frac{1}{12} bh^3 = \frac{1}{12} \times 1.5 \times (1)^3 = 0.125 \text{ in.}^4$$

   (b) The area is

$$A_b = 1.5 \text{ in.}^2$$

   (c) The distance c from the centroid of B to the x_1 axis is

$$c = 0.916 - 0.500 = 0.416 \text{ in.}$$

   (d) The moment of inertia of B with respect to the x_1 axis is

$$I_{x1} = I_{xg} + Ac^2 = 0.125 + 1.5\,(0.416)^2$$

$$= 0.125 + 0.259 = 0.384 \text{ in.}^4$$

5. The moment of inertia of the entire area is equal to the moment of inertia of A minus the moment of inertia of B. Then

$$I_{x1} = I_a - I_b = 0.806 - 0.384 = 0.422 \text{ in.}^4$$

F. The calculation of the maximum bending stress from the formula is

$$S = \frac{Mc}{I}$$

1. The maximum bending moment is

$$M = 6\,720 \text{ lb in.}$$

2. The distance from the neutral axis to the bottom of the cross section is 0.916 in. The distance from the neutral axis to the top of the cross section is $1.5 - 0.916 = 0.584$ in. The greater distance is the distance to the bottom, so it is there that the maximum bending stress occurs.

$$c = 0.916 \text{ in.}$$

3. The moment of inertia of the cross section with respect to a centroidal axis is

$$I_{x1} = 0.422 \text{ in.}^4$$

4. The maximum bending stress is

$$S = \frac{Mc}{I} = \frac{6\,720 \times 0.916}{0.422} = 14\,600 \text{ psi}$$

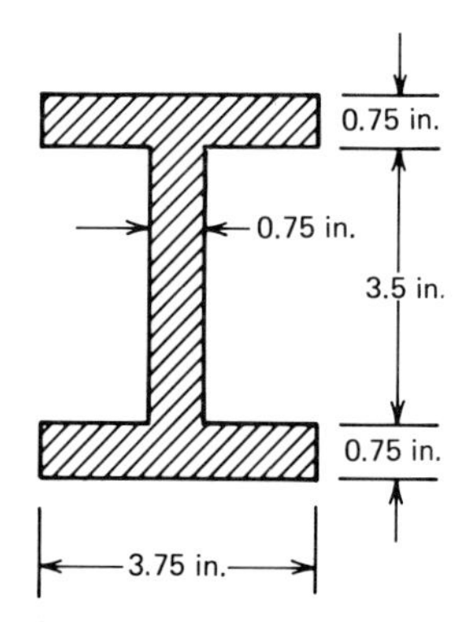

Fig. 6-51. Cross section of I beam.

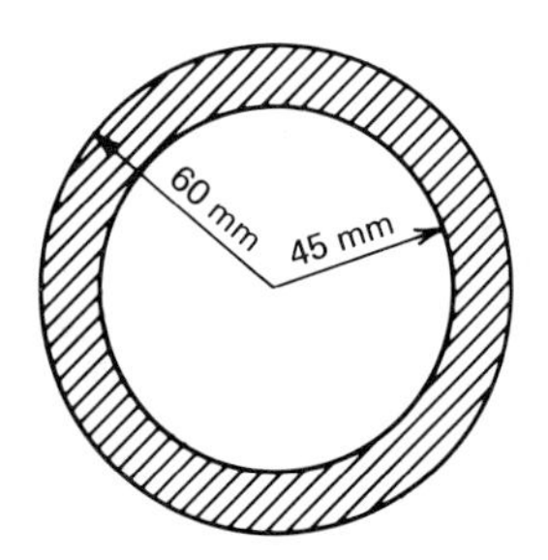

Fig. 6-52. Hollow circular area.

Practice Problems (Section 6-8). Now practice using the flexure formula. The most common mistake is forgetting to change the bending moment to pound inches or to newton millimeters.

1. The maximum bending moment on a square beam, 40 mm by 40 mm in cross section, is 220 000 N·cm. Calculate the maximum bending stress.
2. The maximum bending moment on a circular bar 2.1 in. in diameter is 380 lb ft. What is the maximum bending stress?
3. A bending moment of 10 400 lb ft is applied to a beam with the cross section shown in Fig. 6-51. Find the maximum bending stress.
4. Figure 6-52 shows the cross section of a beam with a maximum bending moment of 480 000 N·cm. What is the maximum bending stress?
5. Figure 6-53 shows the cross section of a beam. The maximum bending moment is 1 960 lb ft. Find the maximum bending stress.
6. A beam having the cross section shown in Fig. 6-54 has a maximum bending moment of 16 000 N·m. Calculate the maximum bending stress.
7. Figure 6-55 shows the free-body diagram of a beam with a circular cross section, 0.84 in. in diameter. Find the maximum bending stress.
8. Figure 6-56 shows the free-body diagram of a beam with a rectangular cross section 200 mm wide and 25 mm high. What is the maximum bending stress?

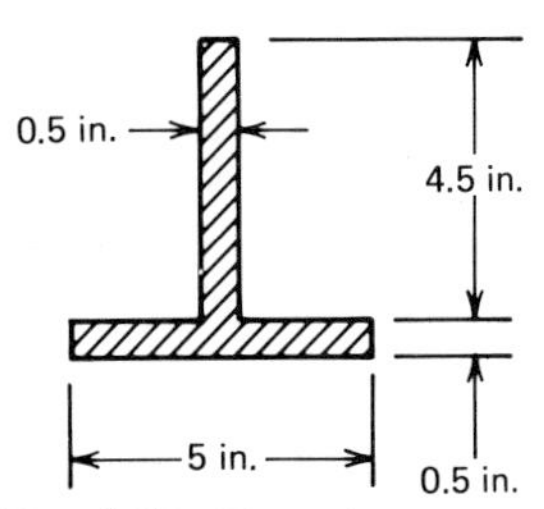

Fig. 6-53. T section.

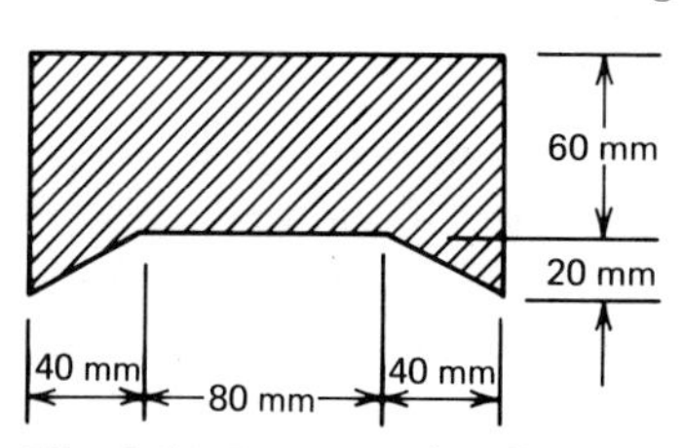

Fig. 6-54. Beam section for Problem 6.

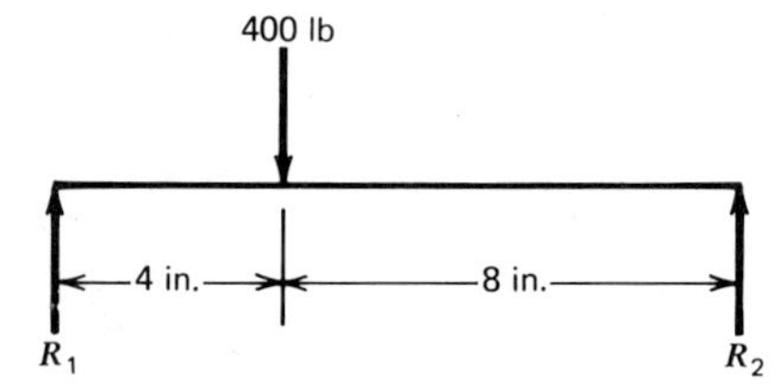

Fig. 6-55. Beam for Problem 7.

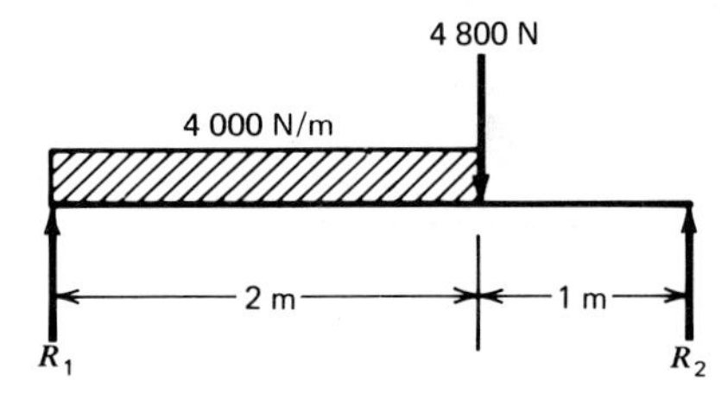

Fig. 6-56. Beam for Problem 8.

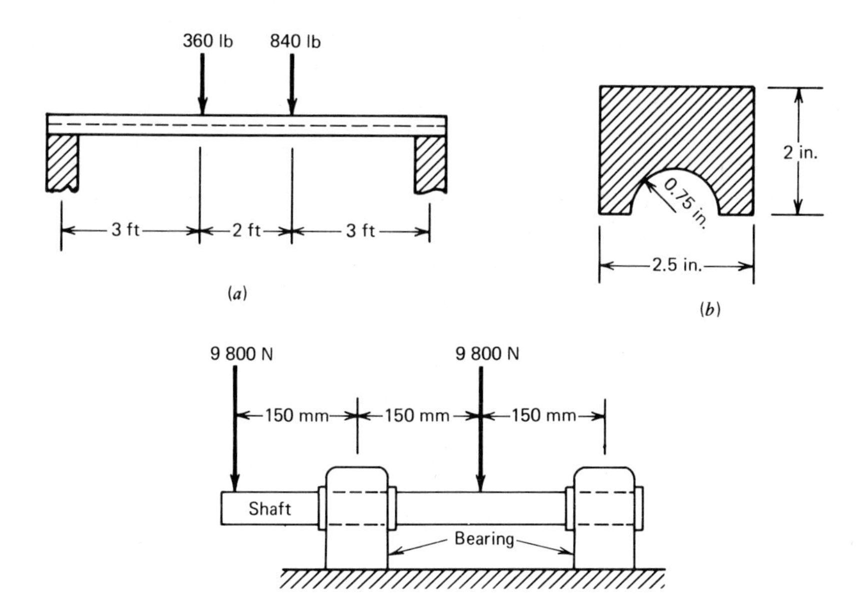

Fig. 6-57. Beam for Problem 9. (*a*) Beam and loads. (*b*) Cross section.

Fig. 6-58. Beam for Problem 10.

9. Figure 6-57*a* shows the loads and supports for a beam, and Fig. 6-57*b* shows the cross section. Find the maximum bending stress.
10. Figure 6-58 shows a shaft supported in bearings that carries two loads. The cross section of the shaft is a solid circle 90 mm in diameter. Find the maximum bending stress.

6-9. THE SECTION MODULUS

Let's look again at the flexure formula. We wrote it in one form as

$$M = \frac{SI}{c}$$

where M is the bending moment, S is the maximum stress on the cross section, I is the moment of inertia of the cross section with respect to the neutral axis, and c is the distance from the neutral axis to the farthest point in the cross section. We can change this to

$$\frac{M}{S} = \frac{I}{c} \qquad \text{or} \qquad \frac{I}{c} = \frac{M}{S}$$

This is the form of the equation used for design. The quantity I/c is called the *section modulus*, and we will represent it by the letter Z.

The unit in which the section modulus is expressed is in.^{3} (inches to the third power) or mm^3 (millimeters to the third power). You see, we are dividing I in in.^{4} by c in in. and the result is Z in in.^{3} (just as $x^4 \div x = x^3$) or mm^4 by m and the result is mm^3.

The first step in designing a beam is to calculate the maximum bending moment M (you know how to do that). Then you divide M by the working stress S. (The working stress, as you recall from Chapter 3, is the stress expected to be developed in service.) The result of dividing M by S is the magnitude of the section modulus the beam must have in order to be strong enough. The last step in designing the beam is to determine the dimensions of the cross section so that the beam will have the required section modulus. We will see how to do this in the next few pages.

6-10. DESIGN OF CIRCULAR BEAMS

We will learn how to design a circular beam first. Figure 6-59 shows the cross section of a circular beam with diameter d. The centroid of the area is at the center of the circle, so the neutral axis goes through the center (the neutral axis is a centroidal axis). The moment of inertia of a circular area, with respect to a diametral axis, is

$$I = \frac{\pi}{64}d^4$$

as you should remember from Chapter 5. Then the distance c from the neutral axis to the farthest point in the cross section is $d/2$. So the section modulus for a circular beam is

$$Z = \frac{I}{c} = \frac{\frac{\pi}{64}d^4}{\frac{d}{2}} = \frac{\pi}{32}d^3$$

which can also be written as

$$d^3 = \frac{32}{\pi}Z$$

We will calculate Z, the section modulus and then solve this equation for d, the diameter.

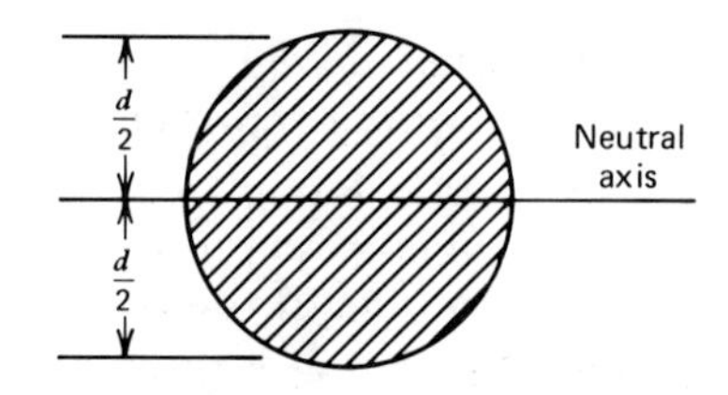

Fig. 6-59. Circular cross section.

Illustrative Example 1. A steel beam of circular cross section is to carry a bending moment of 32 000 lb in. The working stress in bending is 12 000 psi. What diameter is required?

Solution:

1. The bending moment M is 32 000 lb in.
2. The working stress S is 12 000 psi.
3. The required section modulus is

$$Z = \frac{M}{S} = \frac{32\ 000}{12\ 000} = 2.67 \text{ in.}^3$$

4. For the diameter

$$d^3 = \frac{32}{\pi} \times Z = \frac{32}{\pi} \times 2.67 = 27.2 \text{ in.}^3$$

$$d = \sqrt[3]{27.2} = 3.01 \text{ in.}$$

Illustrative Example 2. What diameter is required for a steel beam of circular cross section that is to be subjected to a bending moment of 338 N·m with a working stress of 52 MPa?

Solution:

1. The bending moment must be converted to newton millimeters.

$$M = 338 \times 10^3 = 338\ 000 \text{ N·mm}$$

2. The working stress is $S = 52$ MPa $= 52$ N/mm^2.
3. The required section modulus is

$$Z = \frac{M}{S} = \frac{338\ 000}{52} = 6\ 500 \text{ mm}^3$$

4. The diameter can be calculated from the formula

$$d^3 = \frac{32}{\pi} \times Z = \frac{32}{\pi} \times 6\ 500 = 66\ 200 \text{ mm}^3$$

$$d = \sqrt[3]{66\ 200} = 40.5 \text{ mm}$$

Practice Problems (Sections 6-9 and 6-10). Now design a few circular beams for yourself.

1. A circular beam is to carry a bending moment of 294 lb in. The working stress is 6 000 psi. What diameter is required?
2. The bending moment a certain circular beam must carry is 6 600

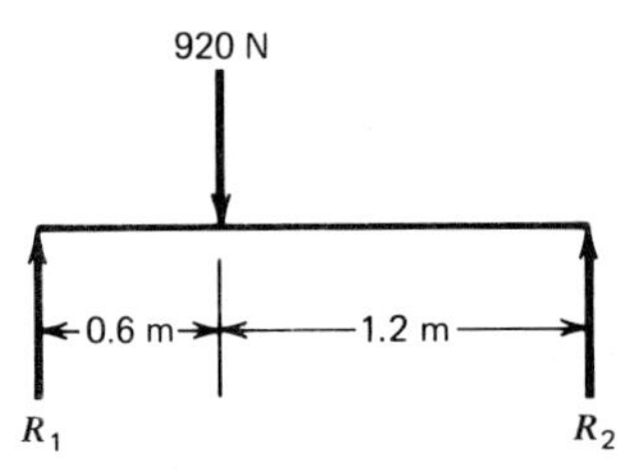

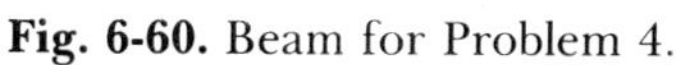

Fig. 6-60. Beam for Problem 4.

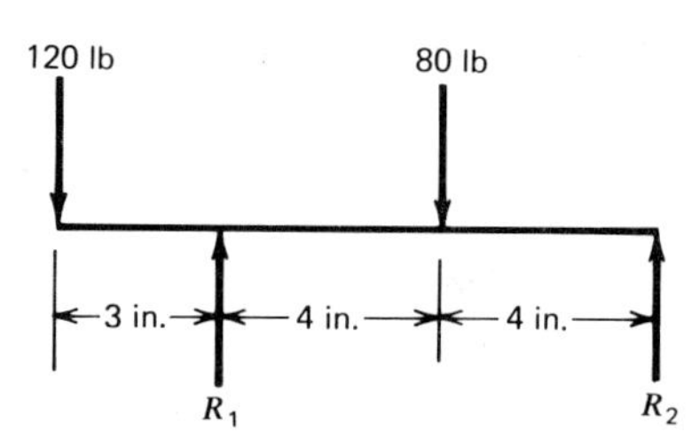

Fig. 6-61. Beam for Problem 5.

lb ft. The working stress is 20 000 psi. What diameter is required?

3. A beam of circular cross section is to be subjected to a bending moment of 570 N·cm. The working stress in bending is 30 MPa. Find the diameter.
4. Figure 6-60 shows the free-body diagram of a beam that is to be circular in cross section. The working stress is 70 MPa. Find the diameter.
5. Figure 6-61 shows the free-body diagram of a beam that is to have a circular cross section. The working stress is to be 18 000 psi. What diameter is required?

6-11. DESIGN OF RECTANGULAR BEAMS

Next, we will see how to design a rectangular beam. Figure 6-62 shows the cross section of a rectangular beam of width b and height h. The centroid of the rectangle is at the center, so the neutral axis passes through the center. The moment of inertia of the cross section, with respect to the neutral axis, is

$$I = \frac{1}{12} bh^3$$

as you should remember, and the distance c from the neutral axis to the farthest point in the cross section is $h/2$. Then the section modulus is

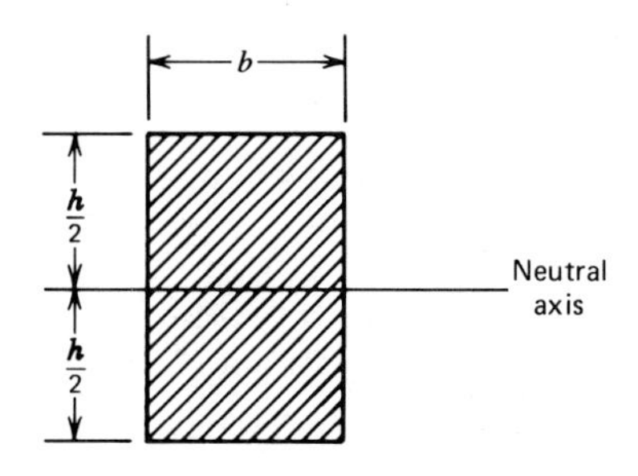

Fig. 6-62. Rectangular cross section.

$$Z = \frac{\frac{1}{12}bh^3}{\frac{h}{2}} = \frac{bh^2}{6}$$

We can multiply the equation by 6 to get

$$bh^2 = 6Z$$

This is the formula from which we must determine the dimensions b and h, after we divide the bending moment M by the working stress S to get the section modulus.

There are several possibilities in designing a rectangular beam. For one, we may have the width known and face the problem of finding the height. (This can happen, for instance, in a case in which the beam must fit in between two other parts, and the width is limited by the space that is available.) Then we simply solve the formula for h.

Rectangular beams are usually made of standard bar stock, which is available in a range of standard dimensions, thus, in sizes from 1/16 in. to 1/2 in. by steps of 1/16 in.; from 1/2 in. to 1 in. by steps of 1/8 in.; from 1 in. to 2 in. by steps of 1/4 in.; from 2 in. to 6 in. by steps of 1/2 in.; and from 6 in. to 10 in. by steps of 1 in. All of this is for the traditional system of units. There are recommended standard dimensions in the metric system for hot-rolled flat steel bars. These are given in terms of the width as the larger dimension of the bar and the thickness as the smaller dimensions. The width ranges from 10 mm to 22 mm in steps of 2 mm; then 25 mm, 28 mm, 30 mm, and 32 mm; from 35 mm to 100 mm in steps of 5 mm; from 100 mm to 200 mm in steps of 10 mm; and from 200 mm to 400 mm in steps of 50 mm. The thickness varies from 3 mm to 8 mm in steps of 1 mm; from 8 mm to 16 mm in steps of 2 mm; it has separate values of 15 mm and 18 mm; it varies from 20 mm to 50 mm in steps of 5 mm, and it has a final value of 60 mm. There is the limitation in this set of recommended dimensions in that the thickness is not usually available in more than about 0.7 of the width. Of course, we can always orient the bar so that the larger dimension is vertical if we choose to do so.

Our procedure in designing a rectangular beam is to calculate the dimensions required and then choose the next larger standard size.

Illustrative Example 1. A rectangular beam 2 in. wide is to be subjected to a bending moment of 45 000 lb in. The working stress is 12 000 psi. What height is required?

Solution:

1. The bending moment M is 45 000 lb in.

2. The working stress S is 12 000 psi.
3. The section modulus that is needed is

$$Z = \frac{M}{S} = \frac{45\ 000}{12\ 000} = 3.75 \text{ in.}^3$$

4. The height of the beam is found from the formula

$$bh^2 = 6Z$$

Substituting for b and for Z

$$2h^2 = 6 \times 3.75 = 22.5$$

$$h^2 = 11.25 \qquad h = \sqrt{11.25} = 3.354 \text{ in.}$$

5. The next larger standard size from 3.354 in. is 3.5 in., so we make that the height of the beam.

$$h = 3.5 \text{ in.}$$

ANOTHER POSSIBILITY

Sometimes the height of the beam is known, and we have the problem of finding the width. (This is also likely to be a case when the beam must fit the available space.) We use the same formula, but solve for b.

A GOOD RULE

It often happens that there is plenty of space for the beam, and the designer can make the width and height of the beam whatever he or she wants, so long as they fit the formula,

$$bh^2 = 6Z$$

There is no definite procedure to be followed in such a case; however, a rule that will always yield good results is to make the width equal to about one half the depth. In the formula,

$$bh^2 = 6Z$$

we can let $b = h/2$. Then,

$$\frac{h}{2}h^2 = 6Z$$

$$h^3 = 12Z$$

We can solve this formula for h. Then we can calculate b, since $b = h/2$. These dimensions probably won't be available in standard sizes, so we will increase each dimension to the next larger standard size.

Illustrative Example 2. A rectangular beam is to be subjected to a bending moment of 760 000 N·mm. The working stress is 108 MPa. Find the size of the beam.

Solution:

1. The bending moment M is 760 000 N·mm.
2. The working stress S is 108 MPa.
3. The required section modulus is

$$Z = \frac{M}{S} = \frac{760\ 000}{108} = 7\ 040 \text{ mm}^3$$

4. Now we will let the width be about one-half the height. We have the formula,

$$h^3 = 12Z = 12 \times 7\ 040 = 84\ 500 \text{ mm}^3$$

$$h = \sqrt[3]{84\ 500} = 43.9 \text{ mm}$$

5. The required width is one-half of this value of h, so

$$b = \tfrac{1}{2} \times 43.9 = 21.95 \text{ mm}$$

6. The next larger standard dimension from 21.95 mm is 25 mm, so

$$b = 25 \text{ mm}$$

7. The next larger standard dimension from 43.9 mm is 45 mm. Thus

$$h = 45 \text{ mm}$$

Practice Problems (Section 6-11). Here are practice problems on designing rectangular beams.

1. A beam of rectangular cross section is to be 2.5 in. wide and carry a bending moment of 30 000 lb in. The working stress is 12 000 psi. Find the height of the beam.
2. A rectangular beam is to be 10 mm wide. The bending moment is 72 000 N·mm, and the working stress is 52 MPa. Find the height of the beam.
3. A rectangular beam, to carry a bending moment of 16 000 N·m, is to be 36 mm in height. The working stress is to be 64 MPa. Find the width of the beam.
4. A rectangular beam 6 mm in height is to be subjected to a bending moment of 12 800 N·mm. The working stress is 130 MPa. Find the width of the beam.

5. A rectangular beam is to carry a bending moment of 4 900 lb in., with a working stress of 8 500 psi. Let the width be about one-half the height, and find the dimensions of the beam.
6. A rectangular beam, for which the width is to be about one-half the height, is to be subjected to a bending moment of 280 lb ft. The working stress is 20 000 psi. Find the dimensions of the beam.

6-12. BEAMS OF STANDARD SHAPES

Standard shapes of beams of steel, aluminum alloys, and magnesium alloys are produced in large quantities by rolling or extruding. Figure 6-63 shows four common types of standard shapes. Rolled shapes are produced by passing the material through rolls that shape it to final form. Extruded shapes are produced by forcing metal through an orifice (hole) having the shape of the desired cross section. Steel-beam sections are rolled, and magnesium-alloy sections are extruded. Some aluminum-alloy sections are rolled, and other aluminum-alloy sections are extruded.

When possible, the designer uses standard shapes for beam sections, because they are economical. Their cost per pound is low because they are produced in large quantities. Also, they are shaped so as to make more efficient use of the material than is done in a rectangular or circular bar. Figure 6-64 shows two beam cross sections that have the same area, but the *I* beam in Fig. 6-64*b* is much stronger in bending than the square

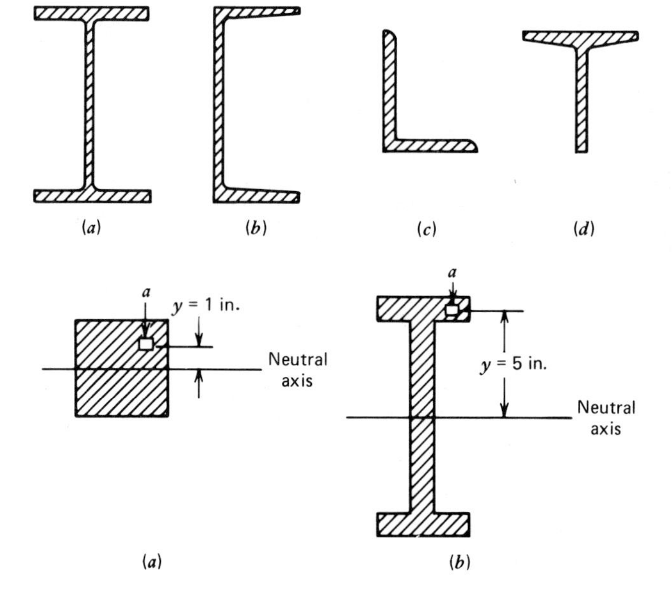

Fig. 6-63. Standard shapes of beams. (*a*) I beam. (*b*) Channel. (*c*) Angle. (*d*) Tee.

Fig. 6-64. Square bar and I beam. (*a*) Square bar. (*b*) I beam.

bar in Fig. 6-64*a*. Let's see why. Remember, we showed in Section 6-8, that the bending moment a beam can carry is equal to

$$M = \frac{SI}{c}$$

where I is the moment of inertia of the cross section with respect to the neutral axis. Now the greater I is, the greater M can be. Let's think of each cross section in Fig. 6-64*a* as divided into small areas a, each at a distance y from the neutral axis. The moment of inertia is (as we defined it in Chapter 5), equal to the sum of the products ay^2; that is,

$$I = \Sigma ay^2$$

We can see that the contribution of each small area a to the moment of inertia depends on the square of the distance of the small area from the neutral axis. For example, when $y = 1$ in., as in Fig. 6-64*a*

$$ay^2 = a(1)^2 = a$$

but when $y = 5$ in. as in Fig. 6-64*b*

$$ay^2 = a(5)^2 = 25a$$

The small area a in Fig. 6-64*b* contributes 25 times as much to the moment of inertia as does the small area a in Fig. 6-64*a*. The total moment of inertia is much greater for the I beam in Fig. 6-64*b* than for the square in Fig. 6-64*a*, even though the two have the same area. In general, a greater moment of inertia is obtained by placing as much of the material as possible as far from the neutral axis as possible.

The companies that manufacture the standard shapes have prepared tables giving such properties of the shapes as the area, moment of inertia, and section modulus. This is a fine thing, because it saves you the trouble of having to calculate a moment of inertia whenever you consider using a standard section for a beam. We are going to give you several samples of such tables so that you can learn how to use them. The tables are not complete, because we haven't space enough for that. However, they are complete enough for our purpose here, which is to study strength of materials as it applies to machine and structural design.

One widely used type of steel beam of standard section is the W shape. Figure 6-65 shows the cross section of a W-shape beam. The depth of the beam is d, the flange width is b_f, the flange thickness is t_f, and the web thickness is t_w. The x and y axes pass through the centroid of the cross section.

Table 6.1 gives the properties of some samples of W-shape beams. The first column of the table gives the designation of the beam. Here, W stands for W shape, the first number following W is the nominal depth of the beam in inches, and the second number is the weight of the beam in pounds per foot of length. Thus, W 24 × 131 is a W-shape

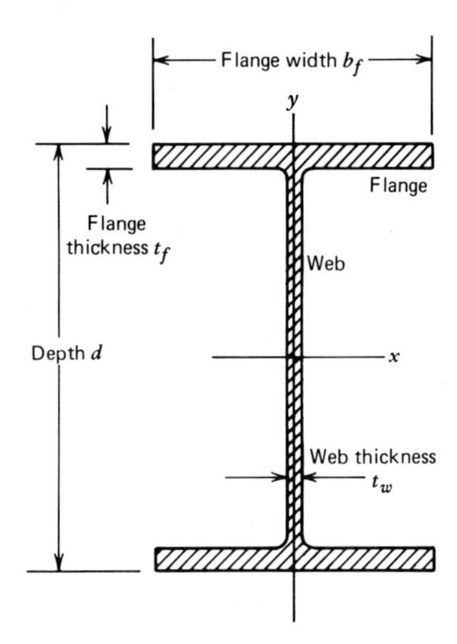

Fig. 6-65. Cross section of W shape.

beam that has a nominal depth of 24 in. and weighs 131 lb/ft of length.

The second column of the table gives the area of the cross section of the beam, and the next four columns give the actual dimensions of the cross section. The next three columns give the moment of inertia I, the section modulus Z, and the radius of gyration r (don't worry about radius of gyration now; it is given here so we can use it later when we study columns) with respect to the x axis. The last three columns give these same properties with respect to the y axis.

Look down the first column now until you see W 16×89. This represents a beam with a nominal depth of 16 in. that weighs 89 lb/ft. The area of the cross section is 26.2 in.^{2}, and the actual depth of the beam is 16.75 in. The flange width is 10.365 in., the flange thickness is 0.875 in., and the web thickness is 0.525 in. The moment of inertia with respect to the x axis is 1 300 in.^{4} and the section modulus is 155 in.^{3} The moment of inertia with respect to the y axis is 163 in.^{4}, and the section modulus is 31.4 in.^{3} It is an advantage to get all of these figures without having to calculate them.

Illustrative Example 1. What is the section modulus with respect to the x axis of a beam designated as W 10×22?

Solution:

Look down the first column of Table 6.1 to find W 10×22. Look across to the column Z (x axis) and read 23.2. The section modulus is

$$Z = 23.2 \text{ in.}^3$$

TABLE 6.1 PROPERTIES OF W-SHAPE STEEL BEAMS

Designation	Area A	Depth d	Flange Width b_f	Flange Thickness t_f	Web Thickness t_w	Elastic Properties x axis I	x axis Z	x axis r	y axis I	y axis Z	y axis r
	$in.^2$	in.	in.	in.	in.	$in.^4$	$in.^3$	in.	$in.^4$	$in.^3$	in.
W 24 × 131	38.5	24.48	12.855	0.960	0.605	4020	329	10.2	340	53.0	2.97
W 24 × 104	30.6	24.06	12.750	0.750	0.500	3100	258	10.1	259	40.7	2.91
W 24 × 84	24.7	24.10	9.020	0.770	0.470	2370	196	9.79	94.4	20.9	1.95
W 18 × 106	31.1	18.73	11.200	0.940	0.590	1910	204	7.84	220	39.4	2.66
W 18 × 55	16.2	18.11	7.530	0.630	0.390	890	98.3	7.41	44.9	11.9	1.67
W 16 × 89	26.2	16.75	10.365	0.875	0.525	1300	155	7.05	163	31.4	2.49
W 16 × 40	11.8	16.01	6.995	0.505	0.305	518	64.7	6.63	28.9	8.25	1.57
W 14 × 159	46.7	14.98	15.565	1.190	0.745	1900	254	6.38	748	96.2	4.00
W 14 × 99	29.1	14.16	14.565	0.780	0.485	1110	157	6.17	402	55.2	3.71
W 14 × 68	20.0	14.04	10.035	0.720	0.415	723	103	6.01	121	24.2	2.46
W 14 × 30	8.85	13.84	6.730	0.385	0.270	291	42.0	5.73	19.6	5.82	1.49
W 12 × 58	17.0	12.19	10.010	0.640	0.360	475	78.0	5.28	107	21.4	2.51
W 12 × 45	13.2	12.06	8.045	0.575	0.335	350	58.1	5.15	50	12.4	1.94
W 12 × 30	8.79	12.34	6.520	0.440	0.260	238	38.6	5.21	20.3	6.24	1.52
W 10 × 100	29.4	11.10	10.340	1.120	0.680	623	112	4.60	207	40.0	2.65
W 10 × 22	6.49	10.17	5.750	0.360	0.240	118	23.2	4.27	11.4	3.97	1.33
W 8 × 28	8.25	8.06	6.535	0.465	0.285	98.0	24.3	3.45	21.7	6.63	1.62
W 8 × 18	5.26	8.14	5.250	0.330	0.230	61.9	15.2	3.43	7.97	3.04	1.23
W 6 × 20	5.87	6.20	6.020	0.365	0.260	41.4	13.4	2.66	13.3	4.41	1.50
W 5 × 16	4.68	5.01	5.000	0.360	0.240	21.3	8.51	2.13	7.51	3.00	1.27

Source: Courtesy American Institute of Steel Construction, Inc.

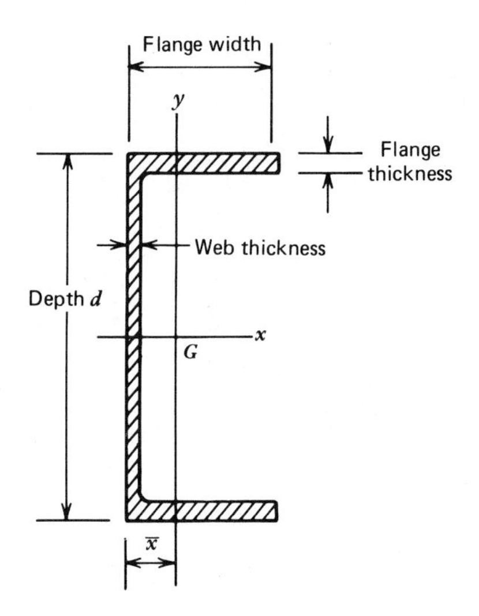

Fig. 6-66. Cross section of channel.

ALUMINUM ALLOY CHANNELS

Figure 6-66 shows the cross section of an aluminum alloy channel. Here the x and y axes intersect at the centroid of the cross section.

Table 6.2 gives the properties of aluminum-alloy standard channels. The information given here is about the same as that in Table 6.1, with one addition. The addition is the location of the centroid. The area of the channel is symmetrical about the x axis, so the centroid is on the x axis. The distance $\bar{x}$, in the last column of the table, is the distance from the back of the channel to the centroid. One method of designating a particular channel is to give the depth in inches, the symbol [to represent channel, and the weight in pounds per foot; for example, 4 [1.738.

Illustrative Example 2. What is the area of cross section of an aluminum-alloy channel designated as 6 [4.030.[2]

Solution:

Look down the first column of Table 6.2 until you see the numbers 6. The first 6 represents the channel, which has a weight of 4.030 lb/ft. Look across to the column headed *Area* and read the area. It is

$$A = 3.427 \text{ in.}^2$$

STRUCTURAL ANGLES

Figure 6-67 shows the cross section of a structural angle. The dimensions L_1 and L_2 are the lengths of the legs, and t is the thickness of the legs. The x and y axes intersect at the centroid of the cross section, that is,

TABLE 6.2. PROPERTIES OF ALUMINUM-ALLOY CHANNELS

			Flange			*Elastic Properties*						
						x axis			*y axis*			
Depth	*Weight per Foot*	*Area*	*Width*	*Thickness*	*Web Thickness*	*I*	*Z*	*r*	*I*	*Z*	*r*	$\bar{x}$
in.	*lb*	*in.²*	*in.*	*in.*	*in.*	*in.⁴*	*in.³*	*in.*	*in.⁴*	*in.³*	*in.*	*in.*
12	11.822	10.053	5.00	0.62	0.35	239.69	39.95	4.88	25.74	7.60	1.60	1.61
12	8.274	7.036	4.00	0.47	0.29	159.76	26.63	4.77	11.03	3.86	1.25	1.14
10	8.360	7.109	4.25	0.50	0.31	116.15	23.23	4.04	13.02	4.47	1.35	1.34
10	6.136	5.218	3.50	0.41	0.25	83.22	16.64	3.99	6.33	2.56	1.10	1.02
9	6.970	5.927	4.00	0.44	0.29	78.31	17.40	3.63	9.61	3.49	1.27	1.25
9	4.983	4.237	3.25	0.35	0.23	54.41	12.09	3.58	4.40	1.89	1.02	0.93
8	5.789	4.923	3.75	0.41	0.25	52.69	13.17	3.27	7.13	2.82	1.20	1.22
8	4.147	3.526	3.00	0.35	0.19	37.40	9.35	3.26	3.25	1.57	0.96	0.93
7	4.715	4.009	3.50	0.38	0.21	33.79	9.65	2.90	5.13	2.23	1.13	1.20
7	3.205	2.725	2.75	0.29	0.17	22.09	6.31	2.85	2.10	1.10	0.88	0.84
6	4.030	3.427	3.25	0.35	0.21	21.04	7.01	2.48	3.76	1.76	1.05	1.12
6	2.834	2.410	2.50	0.29	0.17	14.35	4.78	2.44	1.53	0.90	0.80	0.79
5	3.089	2.627	2.75	0.32	0.19	11.14	4.45	2.06	2.05	1.14	0.88	0.95
5	2.212	1.881	2.25	0.26	0.15	7.88	3.15	2.05	0.98	0.64	0.72	0.73
4	2.331	1.982	2.25	0.29	0.19	5.21	2.60	1.62	1.02	0.69	0.72	0.78
4	1.738	1.478	2.00	0.23	0.15	3.91	1.95	1.63	0.60	0.45	0.64	0.65
3	1.597	1.358	1.75	0.26	0.17	1.97	1.31	1.20	0.42	0.37	0.55	0.62
3	1.135	0.965	1.50	0.20	0.13	1.41	0.94	1.21	0.22	0.22	0.47	0.49
2	1.071	0.911	1.25	0.26	0.17	0.546	0.546	0.774	0.139	0.178	0.391	0.471
2	0.577	0.491	1.00	0.13	0.13	0.288	0.288	0.766	0.045	0.064	0.303	0.298

Source: Courtesy of the Aluminum Association.

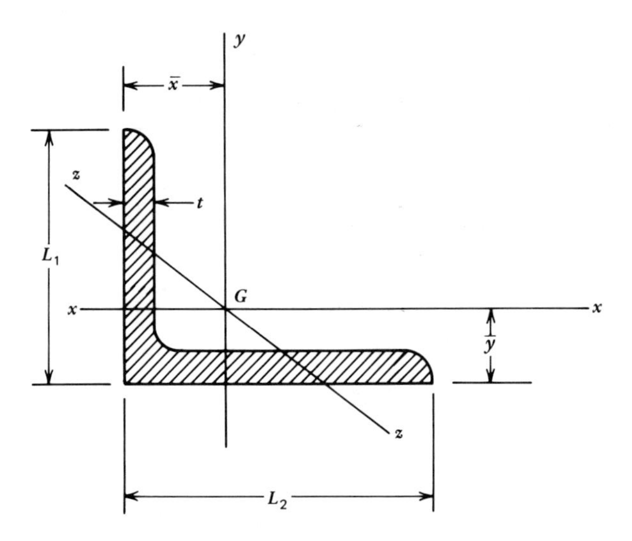

Fig. 6-67. Cross section of angle.

point G. The location of the centroid is given by the coordinates $\bar{x}$ and $\bar{y}$. There is a third axis, z in Fig. 6-67. This is because the minimum moment of inertia is with respect to this oblique axis.

The common way of designating an angle is by giving the dimensions in this manner, $L_1 \times L_2 \times t$, for example, $40 \times 40 \times 5$, where the dimensions are in millimeters, which designates an angle for which each leg is 40 mm long and 5 mm thick.

Table 6.3 gives the properties of some equal-leg angles. Notice that these properties are in metric units. The elastic properties are the same with respect to the y axis as with respect to the x axis, because the legs of each angle are equal.

The dimensions of each angle are given in millimeters, but the other properties are given in centimeters to some power.

Illustrative Example 3. What is the moment of inertia, with respect to the x-x axis of an angle designated as $80 \times 80 \times 12$?

Solution:

Look down the first column of Table 6.3 until you see the numbers $80 \times 80 \times 12$. Look across and read its moment of inertia as

$$I = 102 \text{ cm}^4$$

Practice Problems (Section 6-12). You ought to be acquainted with Tables 6.1, 6.2, and 6.3 before you go on, so here are a few practice problems which require you to use the tables.

TABLE 6.3. PROPERTIES OF EQUAL-LEG ANGLES

		Elastic Properties				
		x and y axes				*Z axis*
Designation	*Area*	*I*	*Z*	*r*	$\bar{x}$	*r*
mm × mm × mm	*cm*^{2}	*cm*^{4}	*cm*^{3}	*cm*	*cm*	*cm*
30 × 30 × 3	1.74	1.40	0.65	0.90	0.84	0.58
30 × 30 × 5	2.78	2.16	1.04	0.88	0.92	0.57
40 × 40 × 4	3.08	4.47	1.55	1.21	1.12	0.78
40 × 40 × 6	4.48	6.31	2.26	1.19	1.20	0.77
50 × 50 × 4	3.89	8.97	2.46	1.52	1.36	0.98
50 × 50 × 7	6.56	14.6	4.16	1.49	1.49	0.96
60 × 60 × 5	5.82	19.4	4.45	1.82	1.64	1.17
60 × 60 × 10	11.1	34.9	8.41	1.78	1.85	1.16
70 × 70 × 6	8.13	36.9	7.27	2.13	1.93	1.37
70 × 70 × 8	10.6	47.5	9.52	2.11	2.01	1.36
80 × 80 × 8	12.3	72.2	12.6	2.43	2.26	1.56
80 × 80 × 12	17.9	102	18.2	2.39	2.41	1.55
90 × 90 × 6	10.6	80.3	12.2	2.76	2.41	1.78
90 × 90 × 10	17.1	127	19.8	2.72	2.58	1.76
100 × 100 × 10	19.2	177	24.6	3.04	2.82	1.95
100 × 100 × 15	27.9	249	35.6	2.98	3.02	1.93
120 × 120 × 8	18.7	255	29.1	3.69	3.23	2.37
120 × 120 × 15	33.9	445	52.4	3.62	3.51	2.33
150 × 150 × 10	29.3	624	56.9	4.62	4.03	2.97
150 × 150 × 18	51.0	1050	98.7	4.54	4.57	2.92
180 × 180 × 15	52.1	1590	122	5.52	4.98	3.54
180 × 180 × 20	68.3	2040	159	5.47	5.18	3.51
200 × 200 × 18	69.1	2600	181	6.13	5.60	3.93
200 × 200 × 24	90.6	3330	235	6.06	5.84	3.90

Source: Reprinted with permission from the American National Standards Institute, the copyright holder.

1. What is the area of cross section of a steel beam W 6 × 20?
2. What is the moment of inertia, with respect to the y axis, of a steel beam W 10 × 22?
3. What is the section modulus, with respect to the x axis, of a steel beam W 14 × 68?
4. What is the moment of inertia, with respect to the x axis, of an aluminum-alloy channel 8 [5.789?
5. What is the area of cross section of an aluminum-alloy channel 5 [2.212?

6. What is the section modulus, with respect to the x axis, of an aluminum-alloy channel 3 [1.597?
7. What is the area of cross section of an angle $40 \times 40 \times 4$?
8. What is the moment of inertia, with respect to the y axis, of an angle $70 \times 70 \times 6$?
9. What is the section modulus, with respect to the x axis, of an angle $120 \times 120 \times 15$?

6-13. DESIGN OF BEAMS OF STANDARD SHAPES

It's easy to design beams of standard shape if you know how to calculate bending moment (you are supposed to know how by this time). You can find the section modulus required by dividing the maximum bending moment M by the working stress S. Thus

$$Z = \frac{M}{S}$$

Then you look down the column headed Z until you see a number equal to or a little greater than the section modulus you need. You pick this section modulus and look across to the columns on the left to get the size of the beam. Usually, the section modulus you require is in between two standard sizes, and you take the next larger standard size from the figure you need.

You might wonder whether to use the section modulus for the x axis or the y axis. The answer to this is that the x axis is the neutral axis when it is perpendicular to the loads. Thus if the loads are vertical and the x axis is horizontal, the x axis is the neutral axis and is the one in which you are interested. The y axis is the neutral axis if it is perpendicular to the loads. However, these beams are almost always used with the x axis as the neutral axis, because they are stronger that way. When nothing is said to the contrary, you are to assume that the x axis is the neutral axis.

Illustrative Example 1. A steel I beam is to be subjected to a bending moment of 820 000 lb in. The working stress is 12 000 psi. Find the proper size of beam.

Solution:

1. The bending moment M is 820 000 lb in.
2. The working stress S is 12 000 psi.
3. The required section modulus is

$$Z = \frac{M}{S} = \frac{820\ 000}{12\ 000} = 68.3 \text{ in.}^3$$

4. Use Table 6.1. In the column headed Z (x axis), look down. You find the number 58.1 and the number 78.0. The number 78.0 is the next larger standard size from 68.3, so you choose 78.0.
5. Look across to the left from 78.0 to the first and second columns. The beam is designated as

$$\text{W } 12 \times 58$$

Illustrative Example 2. An aluminum-alloy channel is to carry a bending moment of 2 400 lb ft. The working stress is 10 000 psi. Select the proper beam.

Solution:

1. The bending moment M is 2 400 lb ft., but this must be multiplied by 12 to convert it to lb in.

$$M = 2\ 400 \times 12 = 28\ 800 \text{ lb in.}$$

2. The working stress S is 10 000 psi.
3. The required section modulus is

$$\frac{I}{c} = \frac{M}{S} = \frac{28\ 800}{10\ 000} = 2.88 \text{ in.}^3$$

4. Use Table 6.2. Look down the column headed Z (x axis) to the figure 3.15. This is the next larger number from 2.88.
5. Look across to the left from 3.15 to the first two columns in the table. The depth of the channel is 5 in., and the weight is 2.212 lb/ft. The channel is designated as

5 [2.212

Illustrative Example 3. An equal-leg angle is to be subjected to a bending moment of 254 000 N·mm. The working stress is 86 MPa. Select the proper angle.

Solution:

1. The bending moment M is 254 000 N·mm.
2. The working stress S is 86 MPa.
3. The required section modulus is

$$Z = \frac{M}{S} = \frac{254\ 000}{86} = 2\ 953 \text{ mm}^3$$

4. The section modulus must be converted to cubic centimeters to use Table 6.3. So,

$$Z = 2\,953 \times 10^{-3} = 2.953 \text{ cm}^3$$

5. Use Table 6.3. Look down the column headed Z to the number 4.16, which is the next larger than 2.953. The angle is 50 × 50 × 7.

6. You could go a step farther. Notice the next number in this column. It is 4.45, which is for an angle 60 × 60 × 5. This angle is preferable in that it has a smaller area of cross section than 50 × 50 × 7, so it would weigh less and be cheaper.

A FINAL WORD

The designer is free to choose the type of beam, whether I beam, channel, or angle, and also to choose the material. The materials (steel, aluminum alloys, and magnesium alloys) are all available in I beams, channels, and angles, as well as other shapes. Each case is a problem in itself, and we can take only enough space here to state a few general points. If it is important to have the beam light in weight, it is desirable to use a magnesium alloy or aluminum alloy, because these alloys are much lighter than steel. On the other hand, they are more expensive, so if economy is the chief factor, it is probably best to use steel. The choice between an I beam, channel, or angle is likely to depend on how the beam is to be fastened to other members.

We have been giving you the maximum bending moment in a lot of examples, but this isn't the way the designer gets the problem. He or she has to start from the beginning, so let's do just that in an example.

Illustrative Example 4. Figure 6-68*a* shows a beam which is welded to two vertical members as part of a machine frame. The beam carries a uniformly distributed load of 1 200 lb/ft over 6 ft of the length and a concentrated load of 6 000 lb. Select the proper W-shape steel beam for this use, with a working stress of 16 000 psi.

Solution:

A. The reactions:

1. Figure 6-68*b* shows the free-body diagram of the beam, with the heavy line representing the beam. The loads are drawn and then the reactions.

2. The reaction R_2 can be calculated by writing a moment equation with center at A. This way,

$$\Sigma M_A = 0$$

$$1\,200 \times 6 \times 3 + 6\,000 \times 7 - 10R_2 = 0$$

$$10R_2 = 1\,200 \times 6 \times 3 + 6\,000 \times 7$$
$$= 21\,600 + 42\,000$$
$$= 63\,600$$
$$R_2 = 6\,360 \text{ lb}$$

3. The reaction R_1 is calculated by writing a moment equation with center at D.

$$\Sigma M_D = 0$$
$$10R_1 - 1\,200 \times 6 \times 7 - 6\,000 \times 3 = 0$$
$$10R_1 = 1\,200 \times 6 \times 7 + 6\,000 \times 3$$
$$= 50\,400 + 18\,000$$
$$= 68\,400$$
$$R_1 = 6\,840 \text{ lb}$$

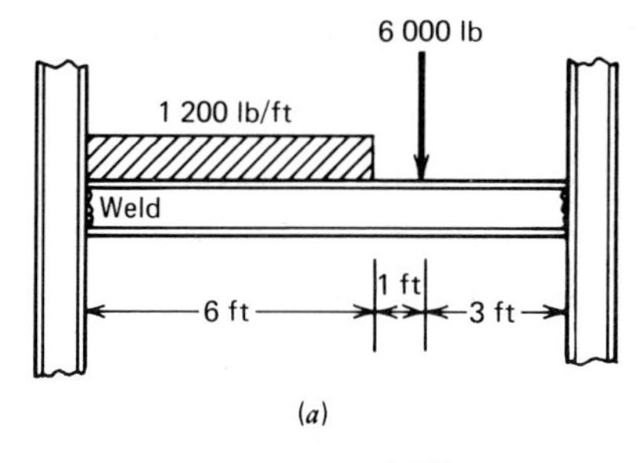

(a)

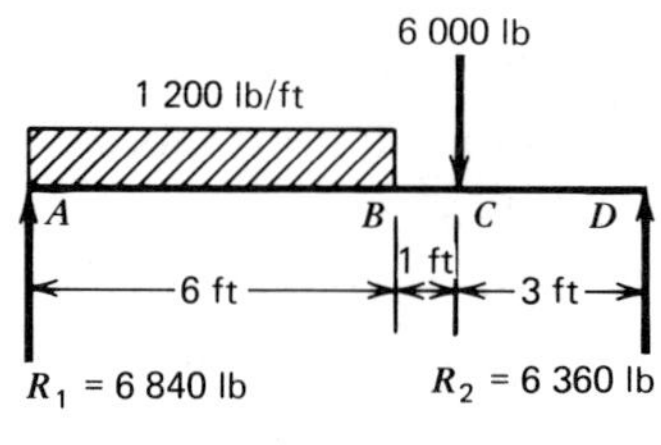

(b)

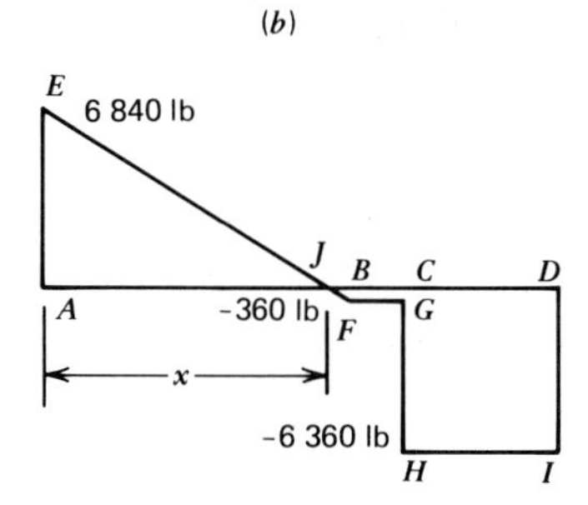

(c)

Fig. 6-68. Beam for Illustrative Example 4. (*a*) Beam. (*b*) Free-body diagram. (*c*) Shear diagram.

4. Then, the reactions are checked by summing the vertical forces. This way,

$$\Sigma F_y = 0$$

$$R_1 - 1\,200 \times 6 - 6\,000 + R_2 = 0$$

The values of R_1 and R_2 are substituted to get

$$6\,840 - 7\,200 - 6\,000 + 6\,360 = 0$$
$$13\,200 - 13\,200 = 0$$

The reactions check, so we can go ahead.

B. The shear diagram. Look at Fig. 6-68*c*.

1. *ABCD* is drawn for the baseline.
2. The reaction R_1 (6 840 lb) is plotted upward from A to locate point E in the shear diagram. On the diagram 6 840 lb is written.
3. The uniform load is plotted downward from E to locate point F. The shear at F is

$$V = 6\,840 - 1\,200 \times 6 = 6\,840 - 7\,200 = -360 \text{ lb}$$

and this value is written on the diagram.

4. The horizontal line FG is drawn.
5. The 6 000-lb load is plotted downward from G to locate point H, where the shear is

$$V = -360 - 6\,000 = -6\,360 \text{ lb}$$

This value is written on the diagram.

6. The horizontal line HI is drawn.
7. The reaction R_2 (6 360 lb) is plotted upward from I to close the shear diagram at D.
8. The point of zero shear is at J:

$$V_j = 6\,840 - 1\,200x = 0 \qquad x = 5.7 \text{ ft}$$

C. The maximum bending moment.

1. The maximum bending moment occurs at J, where the shear is zero.
2. The bending moment at J is equal to the moment, with respect to J of all forces to the left of J. So,

$$M = 6\,840 \times 5.7 - 1\,200 \times 5.7 \times 5.7/2$$
$$= 39\,900 - 19\,500$$
$$= 19\,400 \text{ lb ft}$$

3. The bending moment is multiplied by 12 to convert it to lb in.

$$M = 19\,400 \times 12 = 233\,000 \text{ lb in.}$$

D. Selecting the beam.
   1. The maximum bending moment M is 233 000 lb in.
   2. The working stress S is 16 000 psi.
   3. The section modulus required is

$$Z = \frac{M}{S} = \frac{233\,000}{16\,000} = 14.56 \text{ in.}^3$$

   4. The properties of W-shape steel beams are given in Table 6.1. Looking down the column headed Z (x axis), the number 15.2 is the next larger number from 14.56.
   5. Looking across to the left from the number 15.2 to the first column, the depth of the beam is 8 and the weight is 18 lb/ft. The beam is designated as

$$\text{W } 8 \times 18$$

Practice Problems (Section 6-13). You can practice now in designing beams of standard shape.

1. A steel beam is to be subjected to a bending moment of 47 000 lb ft. The working stress is 18 000 psi. Select the proper beam.
2. An aluminum-alloy channel is to be subjected to a bending moment of 68 000 lb in. The working stress is 12 000 psi. Select the proper beam.
3. An equal-leg angle is to be subjected to a bending moment of 120 000 N·mm. The working stress is 66 MPa. Select the proper angle.
4. Select the proper equal-leg angle for the loading in Fig. 6-69. The working stress is 82 MPa.

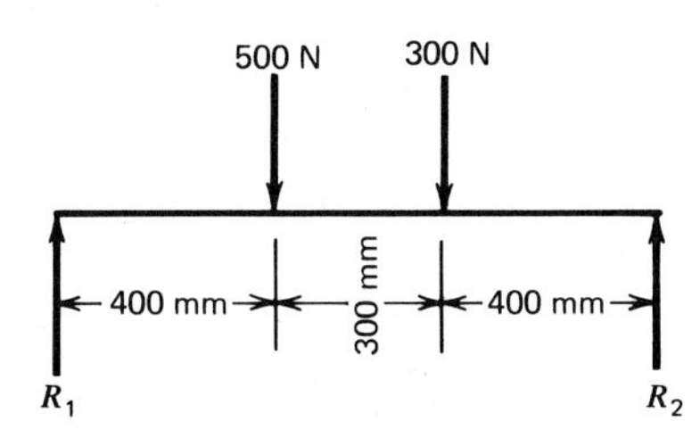

Fig. 6-69. Beam for Problem 4.

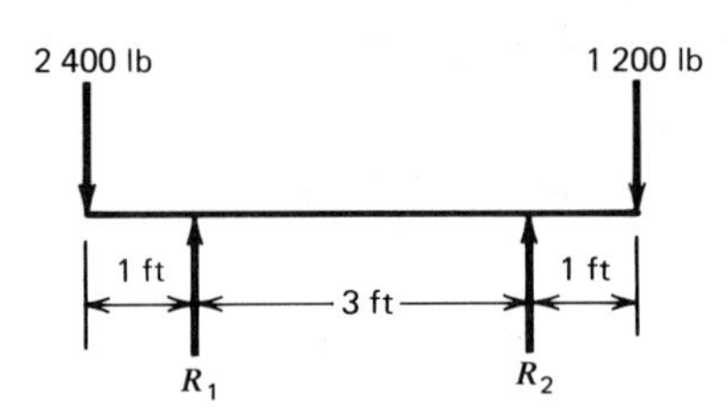

Fig. 6-70. Beam for Problem 5.

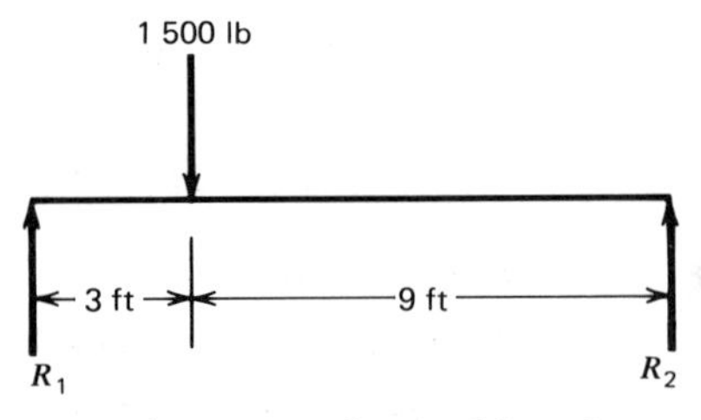

Fig. 6-71. Beam for Problem 6.

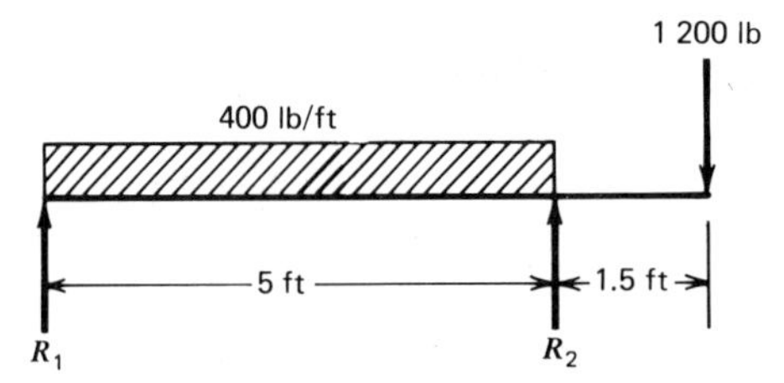

Fig. 6-72. Beam for Problem 7.

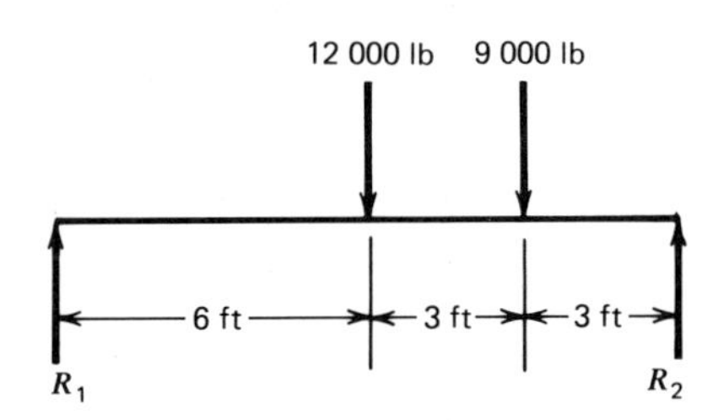

Fig. 6-73. Beam for Problem 8.

5. Select the proper aluminum-alloy channel for the loading in Fig. 6-70. The working stress is 12 000 psi.
6. Select the proper steel beam for the loading in Fig. 6-71. The working stress is 10 000 psi.
7. Select the proper aluminum-alloy channel for the loading in Fig. 6-72. The working stress is 9 000 psi.
8. Select the proper steel beam for the loading in Fig. 6-73. The working stress is 18 000 psi.

6-14. SHEARING STRESS IN BEAMS

The designer has to know something about shearing stress in beams, but it isn't as important in a metal beam as bending stress, because the shearing stress is usually so small. The usual procedure is to select the beam on the basis of bending moment, as we have been doing, and then to check the shearing stress. It usually works out that the actual value of the shearing stress is well below the working stress in shear. This is all right, of course, because the working stress is a value that is not to be exceeded; it's all right if the actual stress is below the working stress. It's only when the beam is short (say, the length is about equal to the depth) that shearing stress is likely to be serious. Most of the time for a metal beam the actual shearing stress is below the working stress in shear, and the calculation of shearing stress is just a formality.

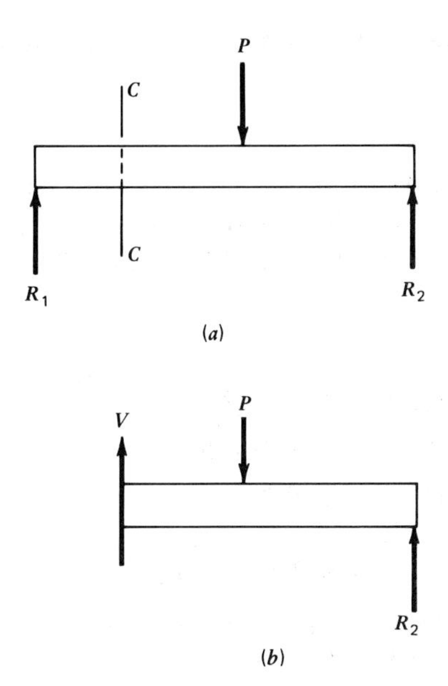

Fig. 6-74. Beam and shear force on beam. (*a*) Free-body diagram of beam. (*b*) Free-body diagram of part of beam.

Let's look into shearing stress in beams and see why it occurs. Figure 6-74*a* shows the free-body diagram of a beam that carries one load and is supported at the ends. We know that there is a shearing force at each point along the beam, and we know how to calculate the shearing force at any point. (The shearing force at any point is equal to the sum of the forces to the left of the point.) Figure 6-74*b* shows the free-body diagram of the part of the beam to the right of the cross section *C-C*, and we see the shearing force *V* exerted on this part of the beam by the part to the left of the cross section *C-C*. The shearing force *V* is distributed over the area of the cross section of the beam. However, the shearing stress is not distributed uniformly over the cross section.

6-15. SHEARING STRESS IN RECTANGULAR BEAMS

The maximum shearing stress on the cross section of a rectangular beam is given by the formula,

$$S_s = \frac{3}{2}\frac{V}{bh}$$

where V is the shearing force on the cross section, b is the width of the cross section, and h is the height of the cross section. (This formula is given without proof, because the derivation is long and complicated and you can be a good designer without it.) Now, the shearing force V is not the same for all points along the beam. We must use the maximum shearing force to calculate the maximum stress. It doesn't matter whether

V is positive or negative. We just take the greatest numerical value and use it in the formula.

Illustrative Example 1. Figure 6-75*a* shows the free-body diagram of a beam of rectangular cross section 1.5 in. wide and 2 in. high. (The reactions are given to save time.) Calculate the maximum shearing stress.

Solution:

A. The shear diagram (Fig. 6-75*b*).

1. The base line ABC is drawn.
2. The reaction R_1 (360 lb) is plotted upward from A to locate point D.
3. The horizontal line DE is drawn.
4. The 600-lb load is plotted downward from E to locate point F. The shearing force at F is

$$V = 360 - 600 = -240 \text{ lb}$$

5. The horizontal line FG is drawn.
6. The reaction R_2 (240 lb) is plotted upward to C. This finishes the shear diagram.
7. The maximum value of the shear is seen to be

$$V = 360 \text{ lb}$$

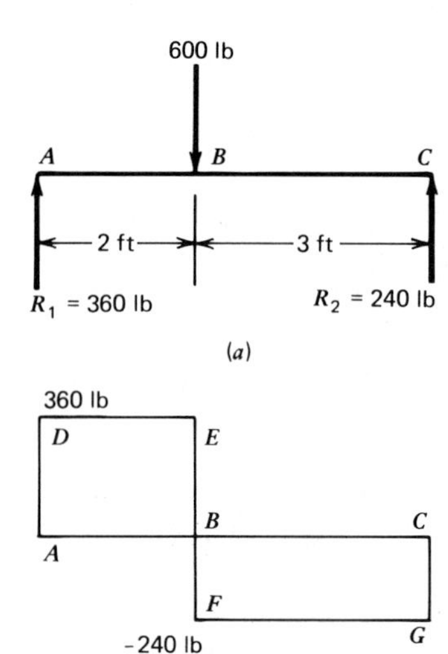

Fig. 6-75. Beam for Illustrative Example 1. (*a*) Free-body diagram. (*b*) Shear diagram.

B. The maximum shearing stress.

1. The maximum shear V is 360 lb.
2. The width b of the beam is 1.5 in.
3. The height h of the beam is 2 in.
4. The maximum shearing stress is

$$S_s = \frac{3\,V}{2\,bh} = \frac{3}{2} \times \frac{360}{1.5 \times 2} = 180 \text{ psi}$$

6-16. SHEARING STRESS IN CIRCULAR BEAMS

The maximum shearing stress on a cross section of a circular beam is given by the formula

$$S_s = \frac{4\,V}{3\,\pi r^2}$$

where V is the shearing force on the cross section and r is the radius. (We can use this formula without going to the trouble to derive it.) All we have to do is to find the maximum value of the shearing force and substitute in the formula.

Illustrative Example 1. Figure 6-76*a* shows the free-body diagram of a circular beam 30 mm in diameter. Calculate the maximum shearing stress.

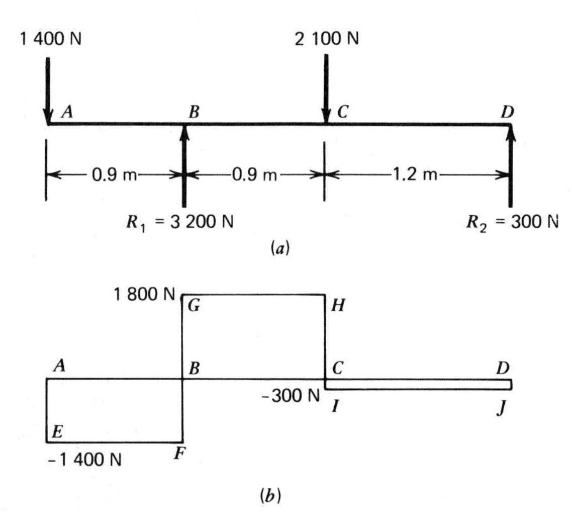

Fig. 6-76. Beam for shear stress on circular cross section. (*a*) Free-body diagram. (*b*) Shear diagram.

Solution:

A. The shear diagram (Fig. 6-76*b*).

1. The base line *ABCD* is drawn.
2. The 1 400-N load is plotted downward from *A* to locate point *E*. The shear at *E* is

$$V_e = -1\,400 \text{ N}$$

3. The horizontal line *EF* is drawn.
4. The reaction R_1 (3 200 N) is plotted upward from *F* to locate point *G*. The shear at *G* is

$$V_g = -1\,400 + 3\,200 = 1\,800 \text{ N}$$

5. The horizontal line *GH* is drawn.
6. The 2 100-N load is plotted downward from *H* to locate point *I*. The shear at *I* is

$$V_i = 1\,800 - 2\,100 = -300 \text{ N}$$

7. The horizontal line *IJ* is drawn.
8. The reaction R_2 (300 N) is plotted upward from *J* to *D*. This completes the shear diagram.
9. The maximum numerical value of the shearing force is 1 800 N.

B. The maximum shearing stress.

1. The maximum shearing force *V* is 1 800 N.
2. The radius of the beam is one-half the diameter

$$r = \frac{d}{2} = \frac{30}{2} = 15 \text{ mm}$$

3. The maximum shearing stress is

$$S_s = \frac{4\,V}{3\,\pi r^2} = \frac{4}{3} \times \frac{1\,800}{\pi(15)^2} = 3.395 \text{ MPa}$$

6-17. SHEARING STRESS IN BEAMS OF STANDARD SHAPES

An exact formula for the shearing stress in a beam of standard shape, such as an I beam or channel, would be too complicated for practical use, so an approximate formula is used instead. The approximate formula is simple, but we need to look at the standard shapes again before we state it. Figure 6-77 shows four of the most common standard shapes. The vertical part of each is called the *web*. The thickness of the web is

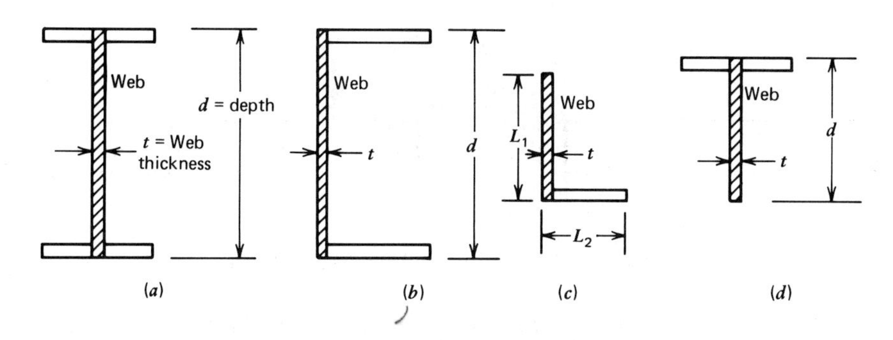

Fig. 6-77. Web areas of standard shapes. (*a*) I beam. (*b*) Channel. (*c*) Angle. (*d*) T section.

t, and the height of the web is the same as the depth of the beam. Now, the approximate formula gives the shearing stress at any cross section of the beam as

$$S_s = \frac{V}{td}$$

where V is the shearing force on the cross section, t is the thickness of the web, and d is the depth of the beam. What we really do here is to assume that the shearing force V is uniformly distributed over the area of the web. The shaded area in each section of Fig. 6-77 is the area of the web.

We remember that the shearing force is not the same at all cross sections of the beam. We must use the maximum shearing force V to calculate the maximum shearing stress.

Illustrative Example 1. Calculate the maximum shearing stress in the beam for the problem in Illustrative Example 4, Section 6-13. Figure 6-68*b* shows the free-body diagram and Fig. 6-68*c* shows the shear diagram. The beam is a steel W shape, W 8 × 18.

Solution:

1. The maximum shearing force V is seen in Fig. 6-68*c* to be 6 840 lb.
2. Table 6.1 gives the thickness t of the web as 0.230 in.
3. The depth d of the I beam is 8.14 in.
4. The maximum shearing stress is

$$S_s = \frac{V}{td} = \frac{6.840}{0.230 \times 8.14} = 3\,650 \text{ psi}$$

Practice Problems (Sections 6-14 to 6-17). Now you can practice calculating shearing stress. Every designer must know how to do it.

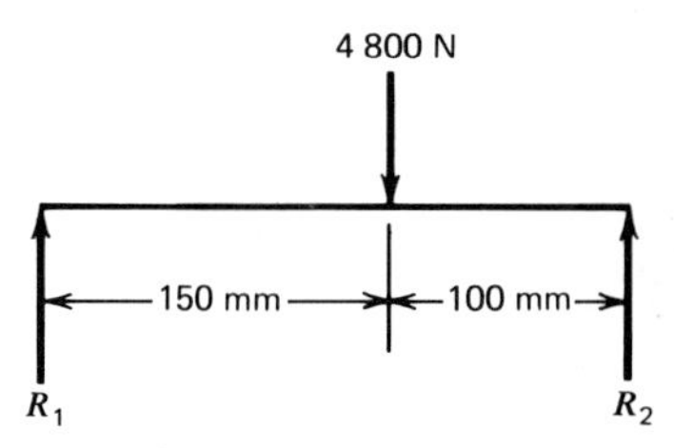

Fig. 6-78. Beam for Problem 3.

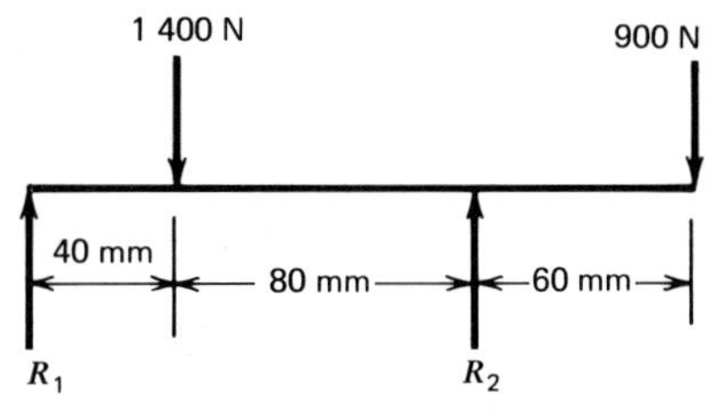

Fig. 6-79. Beam for Problem 6.

1. The maximum shearing force on a square beam ⅞ in. on a side is 750 lb. Calculate the maximum shearing stress.
2. The maximum shearing force on a rectangular beam 1.30 in. wide and 2 in. high is 3 760 lb. Calculate the maximum shearing stress.
3. Figure 6-78 shows the free-body diagram of a rectangular beam 30 mm wide and 50 mm deep. Calculate the maximum shearing stress.
4. The maximum shearing force on a circular beam 1.5 in. in diameter is 16 000 lb. Calculate the maximum shearing stress.
5. The maximum shearing force on a circular beam 14 mm in diameter is 2 600 N. Calculate the maximum shearing stress.
6. Figure 6-79 shows the free-body diagram of a circular beam 22 mm in diameter. Calculate the maximum shearing stress.
7. The maximum shearing force on a steel beam W 10 × 22 is 23 600 lb. Calculate the maximum shearing stress.
8. The maximum shearing force on an aluminum-alloy channel 7 [4.715 lb is 11 700 lb. Calculate the maximum shearing stress. (Hint: Table 6.2 gives the web thickness for aluminum-alloy channels.)
9. Figure 6-80 shows the free-body diagram of a beam which is a metric angle, 50 × 50 × 4. Calculate the maximum shearing stress.
10. Figure 6-81 shows the free-body diagram of a steel beam, W 6 × 20. Calculate the maximum shearing stress.

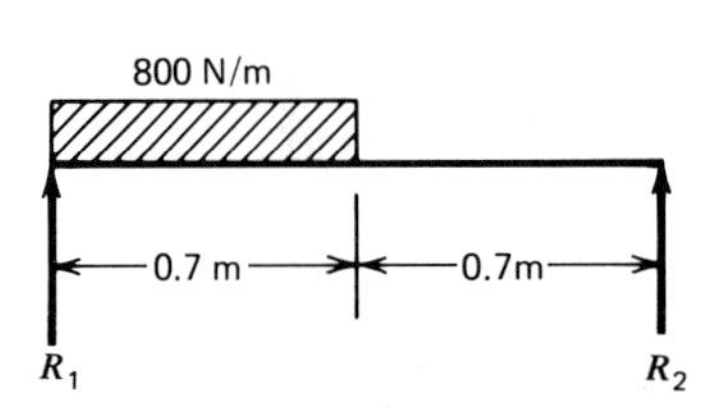

Fig. 6-80. Beam for Problem 9.

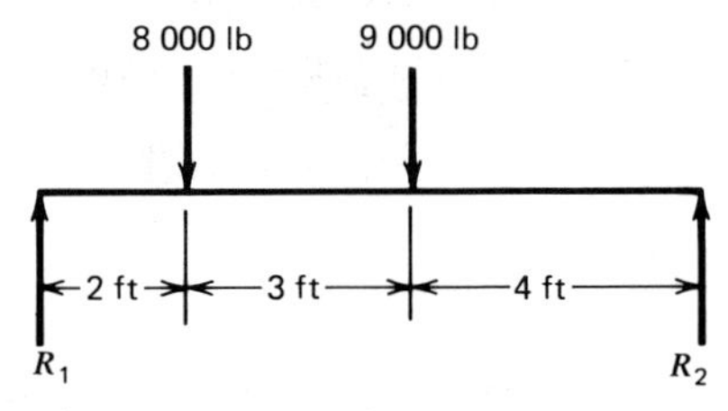

Fig. 6-81. Beam for Problem 10.

SUMMARY Here is a summary of the material in this chapter.

1. A beam is a member that is bent by forces perpendicular to it.
2. The reactions on a beam are the forces exerted on the beam by the supports.
3. The shear at any point in a beam is the sum of the forces to the left of the point. The shear diagram is a diagram that shows the value of the shear at any point along the beam.
4. The bending moment at any point in a beam is the moment, with respect to the point, of all forces to the left of the point.
   (a) The moment diagram is a diagram that shows the value of the bending moment at each point along the beam.
   (b) The bending moment is maximum where the shear is zero.
5. The flexure formula gives the relation between bending stress and bending moment. It is

$$S = \frac{Mc}{I}$$

   (a) M is the bending moment in lb in. or N·mm.
   (b) c is the distance from the neutral axis to the farthest point in the cross section.
   (c) I is the moment of inertia of the area of the cross section with respect to the neutral axis.
   (d) The neutral axis is a centroidal axis.
6. For the purpose of design, the flexure formula is rewritten as

$$Z = \frac{M}{S}$$

   (a) Z is called the *section modulus.*
   (b) For a rectangular beam, $Z = bh^2/6$.
   (c) For a circular beam, $Z = \pi d^3/32$.
   (d) The section modulus of a standard shape is found from tables.
7. The maximum shearing stress in a beam is given by

   (a) $S_s = \dfrac{3\,V}{2\,bh}$, for a rectangular beam.

   (b) $S_s = \dfrac{4\,V}{3\pi r^2}$, for a circular beam.

   $S_s = \dfrac{V}{dt}$, for a standard shape.

REVIEW QUESTIONS See if you can answer these review questions without looking at the earlier part of the chapter.

1. What is a beam?
2. What is a reaction on a beam?
3. How are the reactions calculated?
4. How are the reactions checked?
5. What is the *shear* at any point in a beam?
6. What is a *shear diagram*?
7. What is the *bending moment* at any point in a beam?
8. What is a *moment diagram*?
9. Where is the *bending moment* a maximum?
10. What is the *flexure formula*?
11. What is the *neutral axis*?
12. How would you write the flexure formula if you wanted to use it to calculate *bending stress*?
13. How would you write the *flexure formula* if you wanted to use it to design a beam?
14. What is the *section modulus*?
15. What is the formula for the section modulus of a rectangular beam?
16. What is the formula for the section modulus of a circular beam?
17. What is the formula for the maximum shearing stress in a rectangular beam?
18. What is the formula for the maximum shearing stress in a circular beam?
19. What is the *web* of an I beam?
20. How do you calculate the shearing stress in a beam of standard shape?

REVIEW PROBLEMS Here is a list of review problems to test your knowledge of this chapter. There are only a few problems, but each is comprehensive. If you can work them, you know the material of this chapter. If you can't work them, you need to study the chapter again.

1. Figures 6-82*a* and 6-82*b* are two views of a push rod that slides in the bearings *A* and *B*. Figure 6-82*c* shows the cross section of the rod.
   (a) Calculate the reactions.
   (b) Draw the shear diagram.
   (c) Draw the moment diagram.
   (d) Calculate the maximum bending stress.

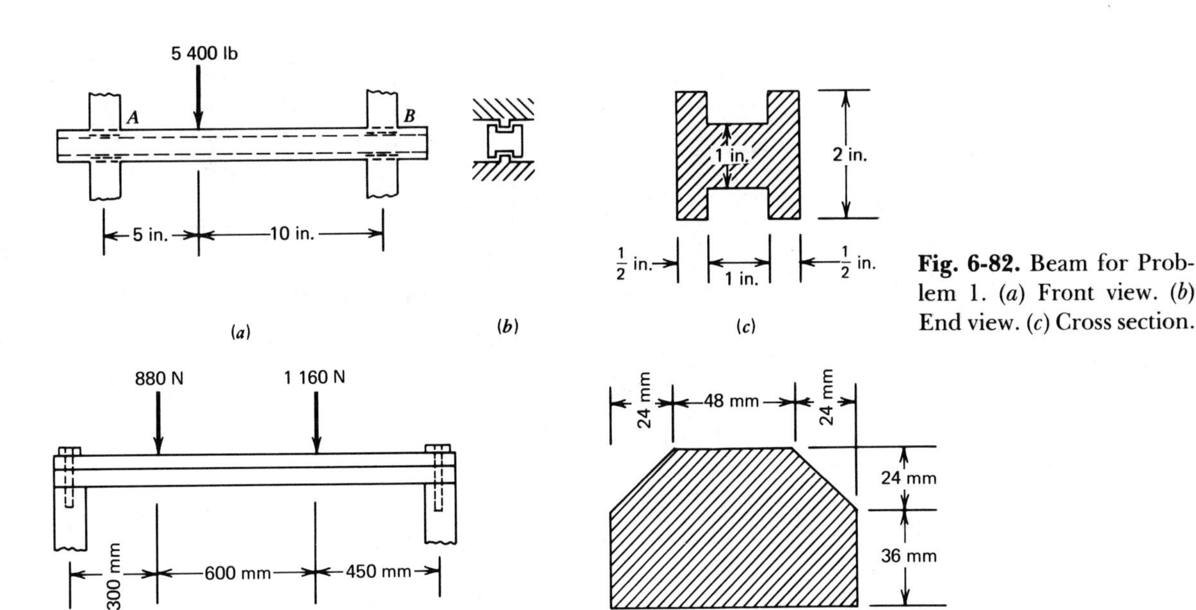

Fig. 6-82. Beam for Problem 1. (*a*) Front view. (*b*) End view. (*c*) Cross section.

Fig. 6-83. Beam for Problem 2. (*a*) Loading. (*b*) Cross section.

2. Figure 6-83 shows the loading and cross section of a beam that is fastened by cap screws at the ends.
   (a) Calculate the reactions.
   (b) Draw the shear diagram.
   (c) Draw the moment diagram.
   (d) Calculate the maximum bending stress.
3. Figure 6-84 shows the loading and cross section of a beam that carries a uniformly distributed load and a concentrated load. The supports are 12 mm from the ends.

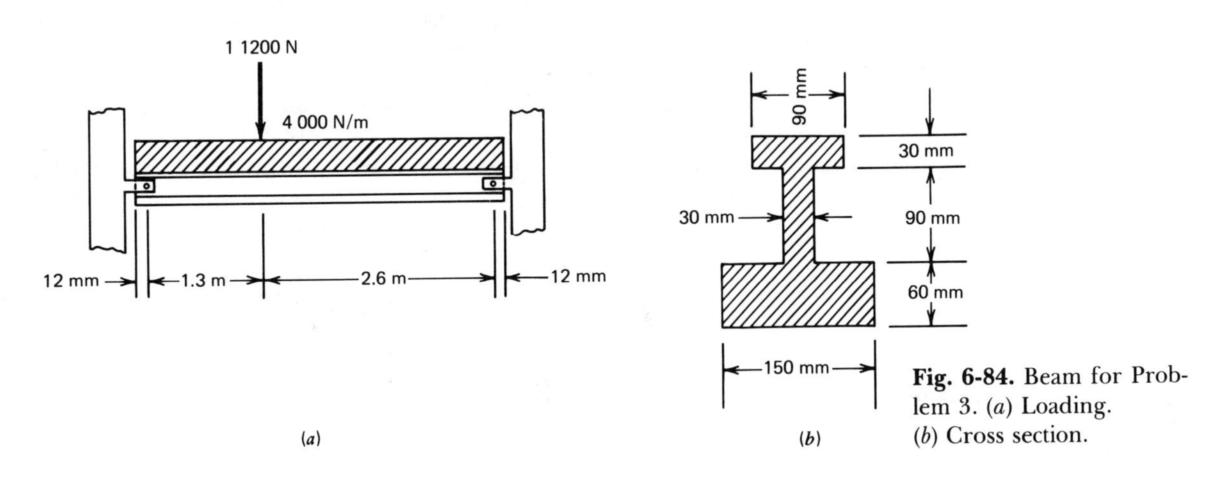

Fig. 6-84. Beam for Problem 3. (*a*) Loading. (*b*) Cross section.

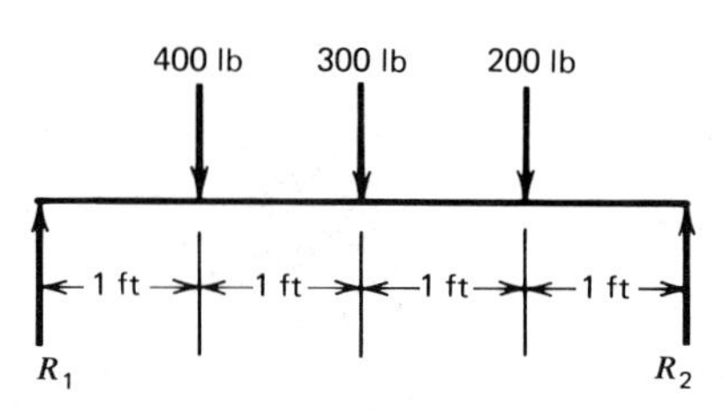

Fig. 6-85. Beam for Problem 4.

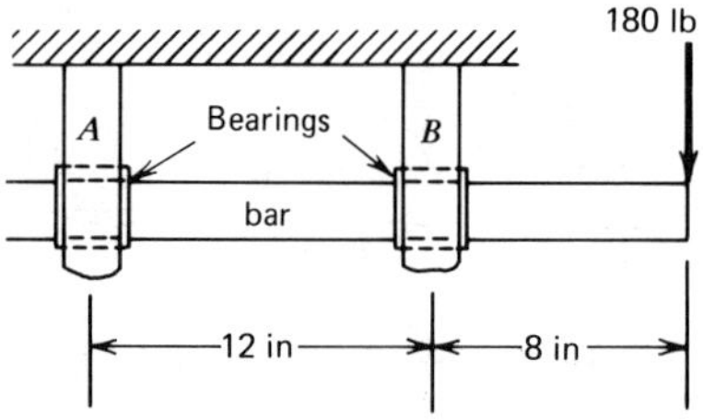

Fig. 6-86. Beam for Problem 5.

(a) Calculate the reactions.
(b) Draw the shear diagram.
(c) Draw the moment diagram.
(d) Calculate the maximum bending stress.

4. Figure 6-85 shows the free-body diagram of a beam that is supported at the ends and carries three concentrated loads.
   (a) Calculate the reactions.
   (b) Draw the shear diagram.
   (c) Calculate the maximum bending moment.
   (d) Select a rectangular beam of width equal to about one-half the height. Use a working stress of 4 000 psi in bending.
   (e) Calculate the maximum shearing stress in the beam you select.
5. Figure 6-86 shows a circular bar supported by the bearings *A* and *B*.
   (a) Calculate the reactions.
   (b) Draw the shear diagram.
   (c) Calculate the maximum bending moment.
   (d) Determine the proper diameter of circular bar. Use a working stress of 9 000 psi.
   (e) Calculate the maximum shearing stress in the bar you select.
6. Figure 6-87 shows the free-body diagram of a beam.
   (a) Calculate the reactions.
   (b) Draw the shear diagram.
   (c) Draw the moment diagram.
   (d) Select the proper steel beam. Use a working stress in bending of 16 000 psi.
   (e) Calculate the maximum shearing stress in the beam you select.

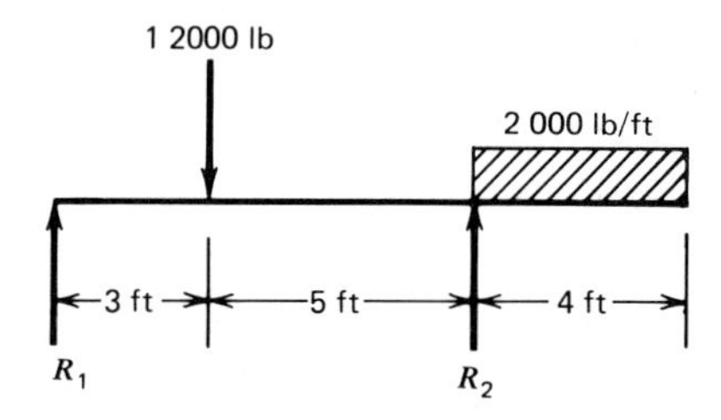

Fig. 6-87. Beam for Problem 6.

7 BENDING MOMENT COMBINED WITH TENSION OR COMPRESSION

PURPOSE OF THIS CHAPTER. You remember that we studied problems of simple tension and compression in Chapter 3. We found that the stress in a bar that is subjected to a tensile force or a compressive force is

$$S = \frac{P}{A}$$

where S is the stress, P is the total force on the cross section, and A is the area of cross section of the bar. You remember also that the force that causes this stress P/A is perpendicular to the cross section of the bar. Hereafter we will call this stress *axial stress*, and we will call the force that is perpendicular to the cross section an *axial force*.

You remember that we studied beams in Chapter 6. We learned to calculate bending moment and to calculate the bending stress from the formula

$$S = \frac{Mc}{I}$$

where S is the stress, M is the bending moment, c is the distance from the neutral axis to the farthest point in the cross section, and I is the moment of inertia of the area of the cross section with respect to the neutral axis.

In this chapter we are going to study problems of machine and structural members that are subjected to a tensile or compressive force on the cross section and to a bending moment as well, and we are going to learn how to calculate the maximum stress in such members. We will find that there is an axial stress P/A and a bending stress Mc/I on each cross section, and we will see how to combine these stresses to find the maximum stress. This is something that designers must know so they can be sure that the maximum stress in the member does not exceed the working stress.

7-1. REVIEW OF PRECEDING CHAPTERS

This will be a new kind of chapter, because we are going to apply nearly all the principles you have learned in the previous chapters of this book. You will have to know the material of the preceding chapters, or you will find yourself in difficulty here. If you have studied the earlier chapters carefully, you will have no trouble, but if you have skipped through them, you need a review. Now just to make certain you are ready for this chapter, suppose you answer the following review questions. If you can answer them, fine! If you can't, you had better go back and study those portions of the book where you can find the answers. The chapter where the answer can be found is given in parenthesis after each question.

Review Questions for Chapters 1–6

1. What is a *component of a force*? (Chapter 1)
2. How do you find the *horizontal component* of a force when you know the magnitude and direction of the force? (Chapter 1)
3. What is the moment of a force with respect to a point? (Chapter 1)
4. What is the *principle of moments*? (Chapter 1)
5. State how you would use the principle of moments to calculate the moment of a force. (Chapter 1)
6. What is *equilibrium*? (Chapter 2)
7. What are the *equations of equilibrium*? (Chapter 2)
8. How do you choose the *moment center* when you write a moment equation? (Chapter 2)
9. What is a free-body diagram? (Chapter 2)
10. What is the advantage of drawing the free-body diagram? (Chapter 2)
11. What is *stress*? (Chapter 3)
12. How can you tell whether a stress is tension or compression? (Chapter 3)
13. What is the *moment of an area*? (Chapter 4)
14. What is the *centroid* of an area? (Chapter 4)

15. Where is the centroid of the area of a right triangle? (Chapter 4)
16. Where is the centroid of the quadrant of a circle? (Chapter 4)
17. How would you calculate the moment of a composite area? (Chapter 4)
18. What is *moment of inertia* of area? (Chapter 5)
19. Give the formula for the moment of inertia of a rectangular area with respect to a centroidal axis that is parallel to the base. (Chapter 5)
20. What is the *moment of inertia* of the area of a right triangle with respect to a centroidal axis parallel to the base? (Chapter 5)
21. What is the moment of inertia of a circular area with respect to a diameter? (Chapter 5)
22. State the parallel axis theorem. (Chapter 5)
23. How would you use the parallel axis theorem to find the moment of inertia of a composite area? (Chapter 5)
24. What is a *beam*? (Chapter 6)
25. What is *bending moment*? (Chapter 6)
26. State the *flexure formula*. (Chapter 6)
27. What is the *neutral axis* of a beam? (Chapter 6)
28. What is the distance c in the flexure formula? (Chapter 6)
29. How can you tell whether the stress in a beam is tension or compression? (Chapter 6)
30. How can you tell whether the bending moment is positive or negative? (Chapter 6)

7-2. CALCULATION OF THE AXIAL STRESS

Let's look at an axial force or two. Figure 7-1*a* shows a straight bar, and we are interested in the cross section *C-C*. Figure 7-1*b* shows the part of the bar to the right of the cross section *C-C*, and we see the axial force

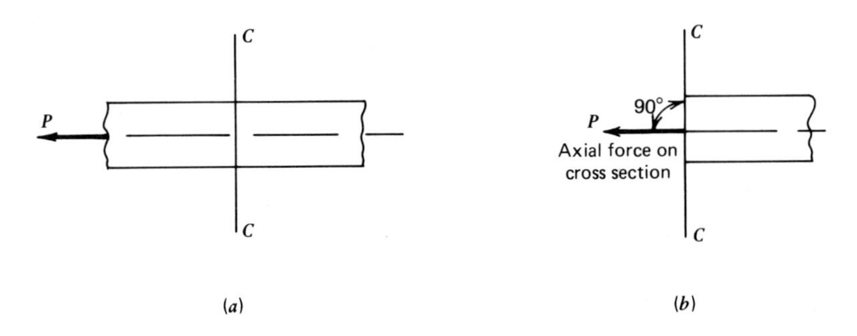

Fig. 7-1. Axial force on straight bar. (*a*) Part of straight bar. (*b*) Axial force on straight bar.

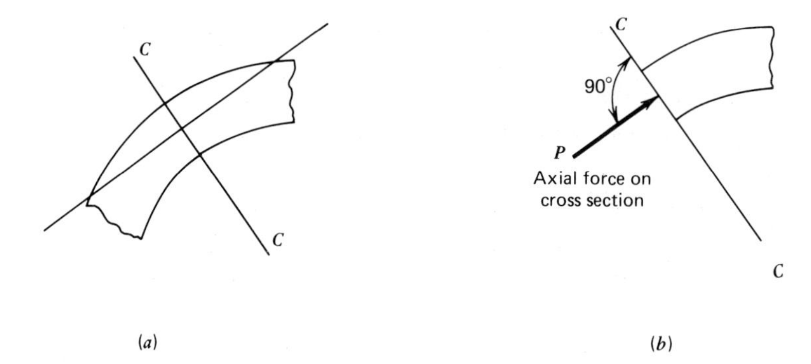

Fig. 7-2. Axial force on curved bar. (*a*) Part of curved bar. (*b*) Axial force on curved bar.

P that acts on the cross section C-C. Notice that the force is perpendicular to the cross section.

Figure 7-2*a* shows a curved bar, and it is the cross section C-C that interests us. Figure 7-2*b* shows the part of the bar to the right of the cross section C-C. Here the force P is perpendicular to the cross section C-C and is an axial force.

Now let's see how to calculate the axial force on the cross section of a bar. Figure 7-3*a* shows the free-body diagram of a bar that is subjected to three loads, P_1, P_2, and P_3, and a reaction R. We want to establish a basis for calculating the axial force on such a cross section as C-C. Figure 7-3*b* shows the free-body diagram of the part of the bar to the left of the cross section C-C, and it is from this picture that we find the value of the axial force. We see the loads P_1 and P_2 and the axial force P. Notice that the axial force is a tensile force; that is, it tends to stretch the material. This part of the bar is in equilibrium (if it weren't in equilibrium it would move), and we can find the force P by writing an equation of equilibrium. The logical equation to write is the one that says that the

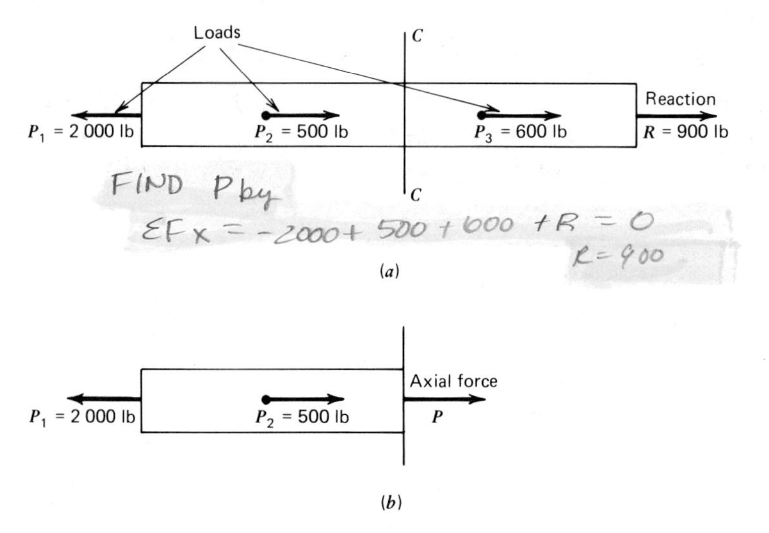

Fig. 7-3. Free-body diagram for calculating axial force. (*a*) Free-body diagram. (*b*) Free-body diagram of left part of bar.

sum of the horizontal forces is equal to zero, because the force we want to find is horizontal. Thus

$$\Sigma F_x = 0$$

$$-P_1 + P_2 + P = 0$$

We can transpose the terms P_1 and P_2 in this equation to get

$$P = P_1 - P_2$$

We can use this equation to set up a rule for calculating the axial force. We notice that the force P_1 causes tensile stress on the cross section *C-C* (it tends to pull the bar in two), and the force P_2 causes compressive stress on the cross section *C-C* (it tends to squeeze the material together at the cross section *C-C*). We say that P_1 is a tensile force and that P_2 is a compressive force. We notice also that P_1 and P_2 are the only forces to the left of the cross section *C-C*. Now, if we take the forces to one side of the cross section and add them (letting a tensile force be positive and a compressive force be negative), we will get the value of the axial force. That's what we have in the result

$$P = P_1 - P_2$$

A positive result shows that the axial force is a tensile force, and a negative result shows that the axial force is a compressive force. The axial force is perpendicular to the cross section *C-C* and, in this case, so are the forces P_1 and P_2. In other problems we will have forces that are not perpendicular to the cross section, but we will only take forces or components that are perpendicular to the cross section in calculating the axial force.

Rule. *The axial force on a cross section of a bar is equal to the sum of the forces or components (to one side of the cross section) that are perpendicular to the cross section; tension is positive, and compression is negative.*

We don't have to draw the free-body diagram of a part of the bar to find the axial force but can work from the free-body diagram of the entire bar. When we have found the axial force, we can just divide by the area of the cross section to get the axial stress:

$$S = \frac{P}{A}$$

Now we are ready for an example.

Illustrative Example 1. Figure 7-4*a* shows a clamp used for lifting boxes. The curved bar *BD* is fastened to the curved bar *ADE* by a pin at *D*. When the upward force R_1 is exerted on the clamp, the ends *A*

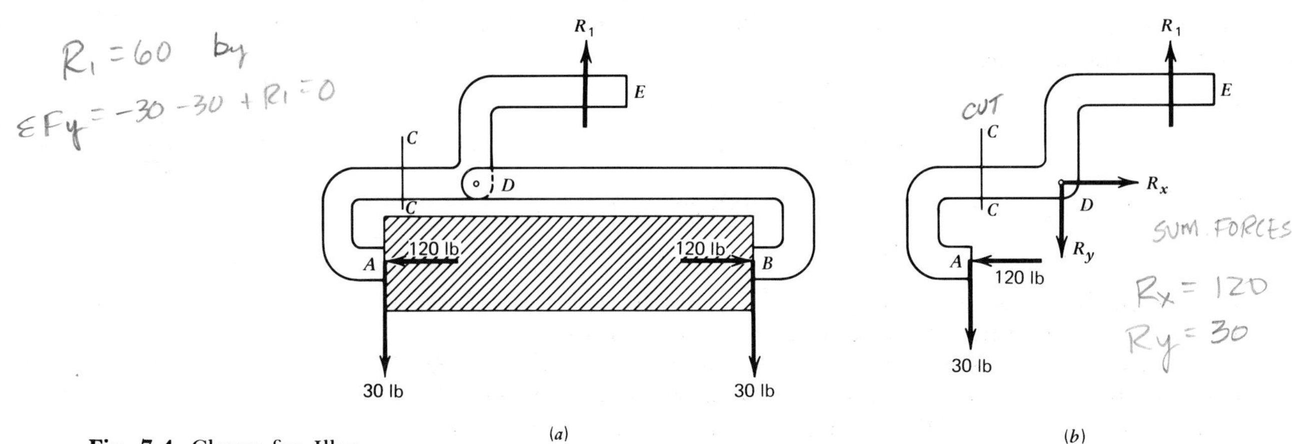

Fig. 7-4. Clamp for Illustrative Example 1. (*a*) Lifting device. (*b*) Free-body diagram of left part of device.

and B tend to move together and clamp the box between them; friction keeps the box from sliding out. The box exerts a horizontal force of 120 lb and a vertical force of 30 lb on the clamp at A. Find the axial stress on the cross section C-C. The cross section is a circle 0.5 in. in diameter.

Solution:

A. The free-body diagram of the curved bar ADE is shown in Fig. 7-4*b*. The forces at A are loads, and R_1 is a reaction. Also, there is the reaction of the pin at D; we designate this reaction as R and represent it by its two components R_x and R_y.

B. The axial force. Remember that we take only the forces that are to one side of the cross section and are perpendicular to the cross section. Let's take the forces to the left side of the cross section, because we know the values of these forces. Then we have for the axial force

$$P = 120 \text{ lb}$$

The 120-lb force is positive because it tends to pull the material away from the cross section; that is, it is a tensile force.

C. The axial stress.

1. The force P is 120 lb.
2. The area of the cross section is circular, so

$$A = \frac{\pi}{4}d^2 = \frac{\pi}{4}(0.5)^2 = 0.1963 \text{ in.}^2$$

Here d is the diameter.

3. The axial stress is

$$S = \frac{P}{A} = \frac{120}{0.1953} = 611 \text{ psi tension}$$ + BENDING STRESS ALREADY CALCULATED

OBLIQUE FORCE

It often happens that a force is neither parallel nor perpendicular to the cross section. Then we just replace the force by components (you learned to do this in Chapter 1), one parallel and one perpendicular to the cross section. Only the component that is perpendicular to the cross section contributes anything to the axial force.

Illustrative Example 2. Figure 7-5*a* shows a *pump rod* that is driven by a connecting rod and pushes the piston of the pump. The pump rod is supported in bearings at A and B. The connecting rod exerts a force of 16 200 N on the pump rod as shown, and the piston exerts a reaction R. The cross section of the pump rod is a rectangle 24 mm wide and 18 mm in height. Calculate the axial stress on the cross section C-C.

Solution:

A. The free-body diagram, Fig. 7-5*b*. The 16 200-N load is replaced by horizontal and vertical components.

1. The horizontal component is

$$16\ 200 \cos 60° = 16\ 200 \times 0.5 = 8\ 100 \text{ N}$$

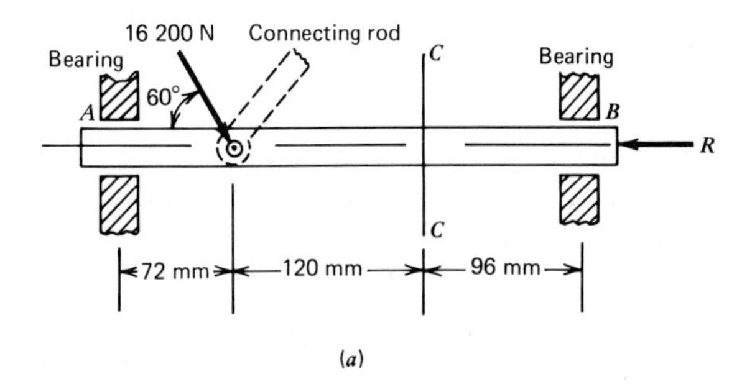

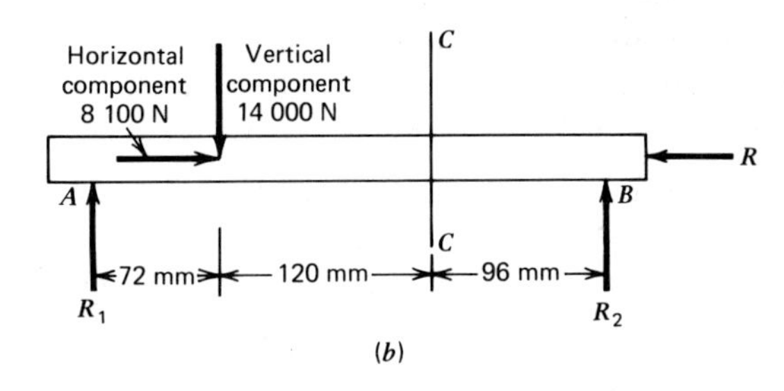

Fig. 7-5. Pump rod for Illustrative Example 2. (*a*) Pump rod. (*b*) Free-body diagram.

2. The vertical component is

$$16\,200 \sin 60° = 16\,200 \times 0.866 = 14\,000 \text{ N}$$

The reaction R is exerted by the piston of the pump, and the bearing reactions R_1 and R_2 are vertical.

B. The axial force. The axial force is equal to the sum of the forces to the left that are perpendicular to the cross section *C-C*. The only such force is the compressive force of 8 100 N, so

$$P = -8\,100 \text{ N}$$

The negative sign indicates compression.

C. The axial stress.

1. The axial force P is 8 100 N.
2. The area of the cross section is rectangular.

$$A = bh = 24 \times 18 = 432 \text{ mm}^2$$

3. The axial stress is

$$S = \frac{P}{A} = \frac{8\,100}{432} = 18.75 \text{ MPa compression}$$

VERTICAL BAR

Would it cause you any trouble if the bar were vertical instead of horizontal? It doesn't need to disturb you. You can just look at the bar from the side, and then you see one end of the bar as the right end and the other end as the left end.

Illustrative Example 3. Figure 7-6*a* shows the loading on the arm of a small crane. Figure 7-6*b* shows the cross section *C-C*. Calculate the axial stress on the cross section *C-C*.

Solution:

A. The free-body diagram. Figure 7-6*a* can serve as the free-body diagram, because we are going to look at the figure from the left side as indicated by the arrow in Fig. 7-6*a*. Then we see the upper part of the arm on the left. The 600-lb load is the only force on the left part and is a known force.

B. The axial force. The only force to the left of the cross section *C-C* is the 600-lb load. It pushes the bar down against the cross section *C-C* so it is a compressive force. Then

$$P = -600 \text{ lb}$$

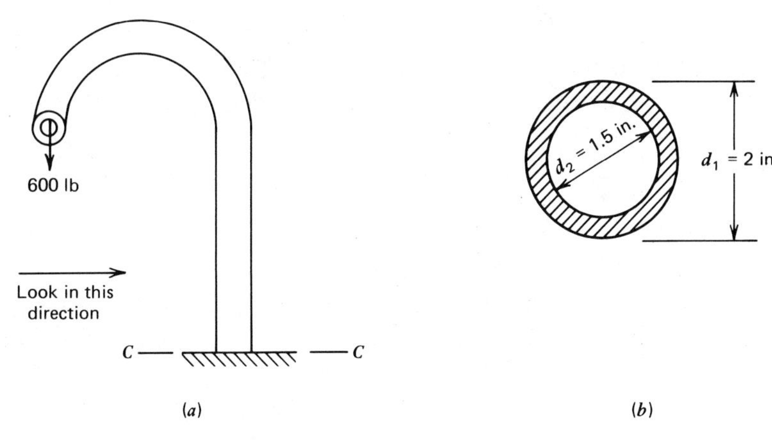

Fig. 7-6. Bent bar for Illustrative Example 3. (*a*) Loading. (*b*) Section *C-C*.

C. The axial stress.
   1. The axial force P is -600 lb.
   2. The area of the cross section is equal to the area of the large circle minus the area of the small circle. Thus

$$A = \frac{\pi}{4}d_1^2 - \frac{\pi}{4}d_2^2 = \frac{\pi}{4}(2)^2 - \frac{\pi}{4}(1.5)^2$$

$$= 3.14 - 1.77 = 1.37 \text{ in.}^2$$

   3. The axial stress is

$$S = \frac{P}{A} = \frac{600}{1.37} = 438 \text{ psi compression}$$

Illustrative Example 4. Figure 7-7*a* shows a bracket subjected to two loads that has the cross section shown in Fig. 7-7*b*. Calculate the axial stress on section *C-C*.

Solution:

A. The free-body diagram. Figure 7-7*a* shows all of the forces above

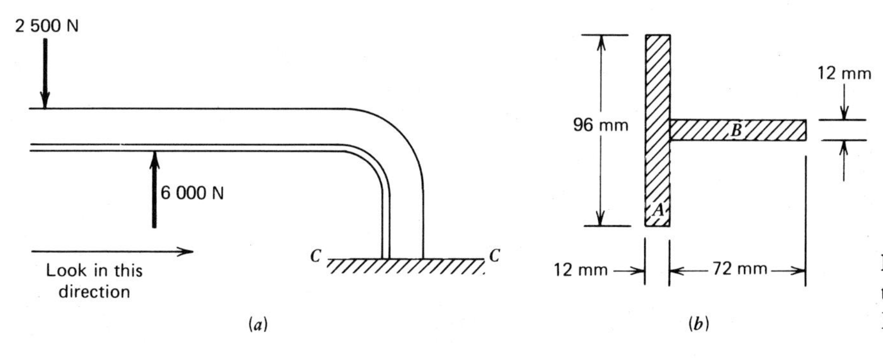

Fig. 7-7. Bracket for Illustrative Example 4. (*a*) Loading. (*b*) Section *C-C*.

section C-C, so Fig. 7-7a can serve as the free-body diagram.

B. The axial force. We look at Fig. 7-7a from left to right as the arrow indicates. Then we see the two known forces to the left of section C-C. The axial force is

$$P = -2\,500 + 6\,000 = 3\,500 \text{ N tension}$$

C. The axial stress.

1. The axial force is 3 500 N.
2. The area of the cross section is calculated by dividing the area into the two rectangles A and B in Fig. 7-7b. The area of A is $96 \times 12 = 1\,152$ mm², and the area of B is $72 \times 12 = 864$ mm². The total area is $A = 1\,152 + 864 = 2\,016$ mm².
3. The axial stress is

$$S = \frac{P}{A} = \frac{3\,500}{2\,016} = 1.736 \text{ MPa tension}$$

Practice Problems (Section 7-2). Now you can try a few problems to practice calculating axial stress.

1. Figure 7-8 shows a bent lever subjected to two known forces of 940 lb and supported by a pin at A. The cross section is rectangular 2 in. by ⅜ in. Calculate the axial stress on the cross section C-C.
2. Figure 7-9 shows a brake arm supported by a pin at A. The bar is circular and is ⅞ in. in diameter. Calculate the axial stress on section C-C.
3. The yoke in Fig. 7-10a is subjected to the three forces shown. The cross section is in the shape of an I as shown in Fig. 7-10b. Find the axial stress on section C-C.

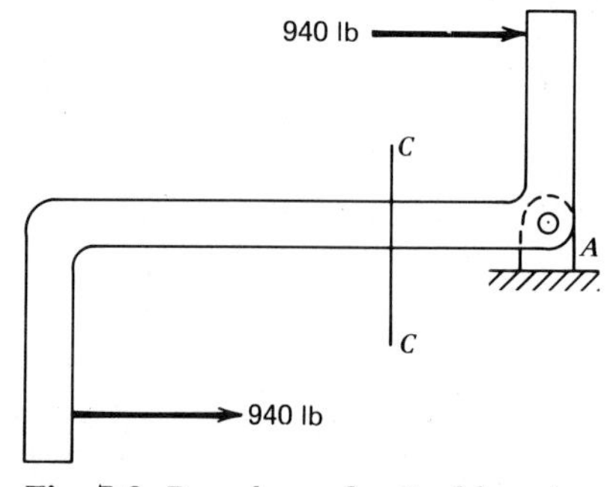

Fig. 7-8. Bent lever for Problem 1.

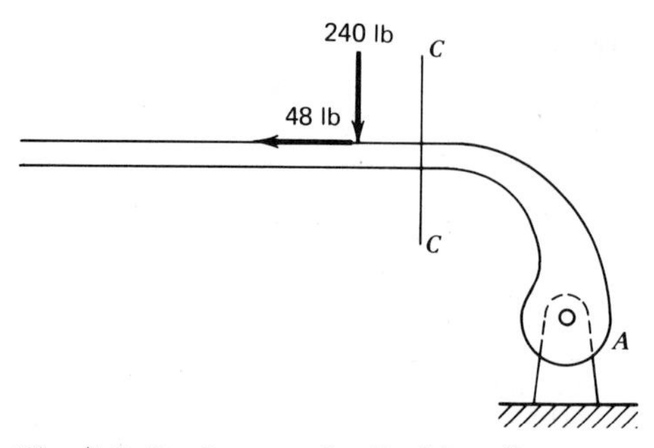

Fig. 7-9. Brake arm for Problem 2.

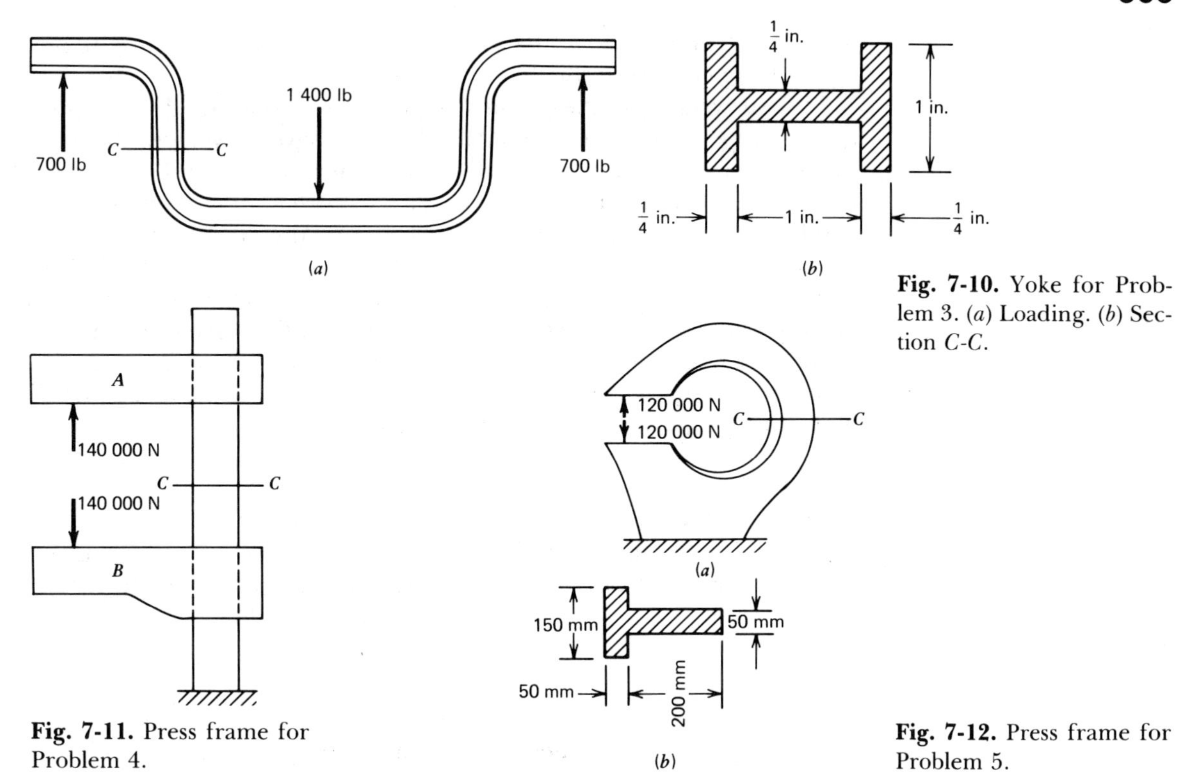

Fig. 7-10. Yoke for Problem 3. (*a*) Loading. (*b*) Section *C-C*.

Fig. 7-11. Press frame for Problem 4.

Fig. 7-12. Press frame for Problem 5.

4. Figure 7-11 shows a press frame. The upper bracket *A* and the lower bracket *B* are clamped to the vertical post which is circular and 75 mm in diameter. What is the axial stress on section *C-C*?
5. Figure 7-12*a* shows the loading on a press frame, and Fig. 7-12*b* shows the cross section *C-C*. Calculate the axial stress on section *C-C*.

7-3. CALCULATION OF THE BENDING STRESS

Now let's see how to calculate the bending stress. If we can find the bending moment, we can use the flexure formula

$$S = \frac{Mc}{I}$$

where S is the stress in psi or MPa, M is the bending moment in lb in. or N·mm, c is the distance from the neutral axis to the farthest point in the cross section, and I is the moment of inertia of the area of the cross section with respect to the neutral axis. You remember this from Chapter

6, and you remember also that the neutral axis passes through the centroid of the cross section.

When we studied beams in Chapter 6, we considered only forces that were perpendicular to the bar. Now we have to work problems in which some of the forces may be parallel to the bar, and we have to calculate the bending moment. We will still say that the bending moment is the moment of all the forces to the left of the cross section we are interested in, and we will take the moment center at the centroid of the cross section. It won't matter whether a force is parallel or perpendicular to the plane of the cross section. The moment of the force is the product of the force and the perpendicular distance of the force from the moment center. A clockwise moment is positive, and a counterclockwise moment is negative. Let's try an example.

Illustrative Example 1. Figure 7-13 shows the free-body diagram of part of the clamp we studied in Example 1, Section 7-2. The curved bar is circular and is 0.5 in. in diameter. Calculate the bending stress at section *C-C*.

Solution:

1. The free-body diagram is already drawn in Fig. 7-13.
2. The bending moment. The centroid of the circular section is at the center, and so we choose point O as a moment center. The moment arm of the 120-lb force is 4 in. and the moment arm of the 30-lb force is 2 in., so

$$M = 120 \times 4 - 30 \times 2 = 480 - 60 = 420 \text{ lb in.}$$

3. The moment of inertia and the distance c. The moment of inertia of the circular area is

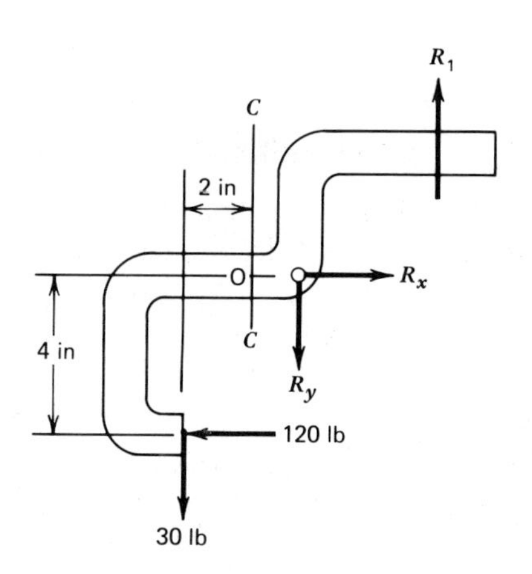

Fig. 7-13. Free-body diagram for part of clamp.

$$I = \frac{\pi}{64}d^4 = \frac{\pi}{64}(0.5)^4 = 0.00307 \text{ in.}^4$$

The distance c is half the diameter, so

$$c = \tfrac{1}{2}(0.5) = 0.25 \text{ in.}$$

4. The bending stress is

$$S = \frac{Mc}{I} = \frac{420 \times 0.25}{0.00307} = 34\,200 \text{ psi}$$

TENSION OR COMPRESSION

We must know whether the bending stress is tension or compression, because later we are going to add it to the axial stress. You should remember from Chapter 6 that when the bending moment is positive, the bending stress at the top of the beam is compression and the bending stress at the bottom of the beam is tension. When the bending moment is negative, the stress at the top of the beam is tension, and the stress at the bottom of the beam is compression.

Illustrative Example 2. Figure 7-14*a* shows a brake arm that is subjected to three known forces and is supported by a pin at *A*. Figure 7-14*b* shows the cross section *C-C*. Calculate the bending stress at section *C-C*.

Solution:

1. The free-body diagram. Figure 7-14*a* shows enough data to serve as the free-body diagram.
2. The bending moment is the moment, with respect to point *O*, of

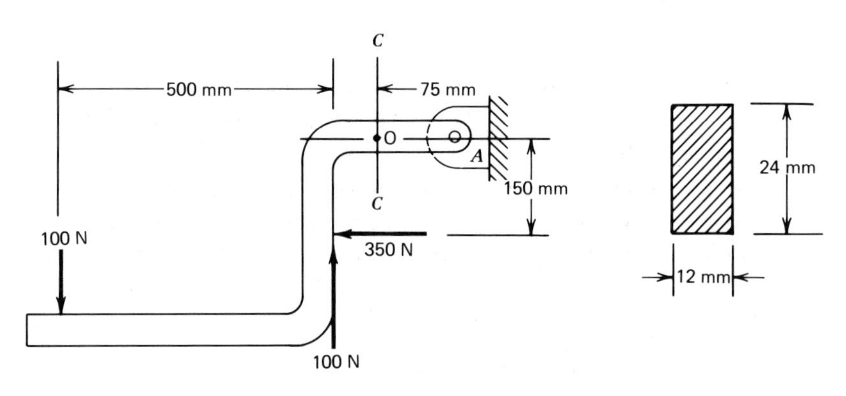

Fig. 7-14. Brake arm. (*a*) Loading. (*b*) Section *C-C*.

all forces to the left of section *C-C*. Point *O* is at the centroid of the cross section. Then,

$$M = -100 \times 575 + 100 \times 75 + 350 \times 150$$
$$= -57\,500 + 7\,500 + 52\,500 = 2\,500 \text{ N·mm}$$

3. The moment of inertia and the distance c. The moment of inertia of the rectangle is

$$I = \frac{1}{12} bh^3 = \frac{1}{12}(12)(24)^3 = 13\,800 \text{ mm}^4$$

The distance c is one half the depth, so

$$c = \tfrac{1}{2}(24) = 12 \text{ mm}$$

4. The bending stress is

$$S = \frac{Mc}{I} = \frac{2\,500 \times 12}{13\,800} = 2.17 \text{ MPa}$$

The bending moment is positive, so the stress at the top of the section is compression, and the stress at the bottom is tension.

VERTICAL BAR

It needn't cause you any trouble if a bar is vertical instead of horizontal. Then we can just look from the side. If we look from the left side of the member, we will see the upper end to our left and the lower end to our right.

Illustrative Example 3. Figure 7-15*a* shows the loading on a post and bracket. Figure 7-15*b* shows the cross section of the post. Calculate the bending stress on section *AB*.

Solution:

1. The free-body diagram. Figure 7-15*a* can be used for the free-body diagram, because there are no unknown forces above the cross section *AB*.
2. The bending moment. We look at the post from the left side, as indicated by the arrow, and then we see the top of the post to our left. Point *O* is the centroid of the cross section *AB* and is the moment center for calculating the bending moment. The moment arm of the 600-lb force is 20 + 2.5 = 22.5 in. Then,

$$M = 600 \times 22.5 = 13\,500 \text{ lb in.}$$

This is a positive bending moment, because it is clockwise.

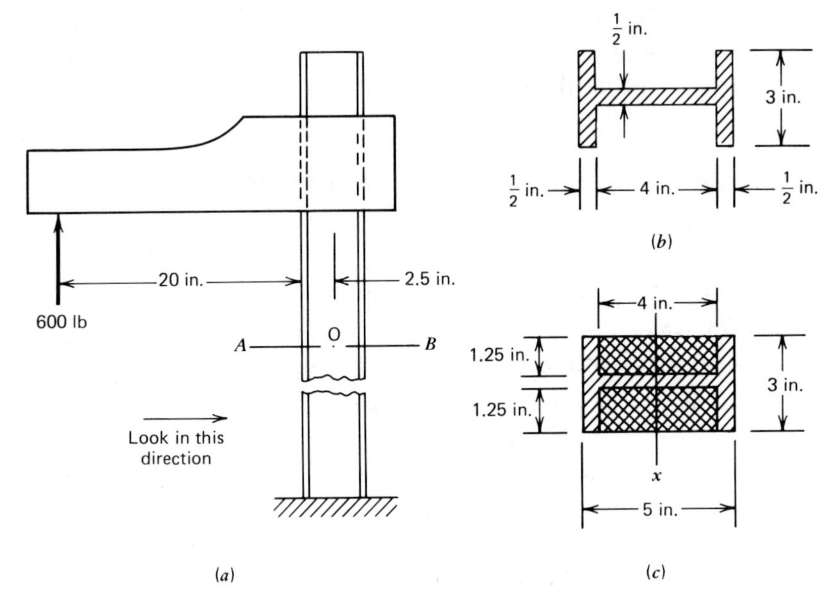

Fig. 7-15. Post and bracket. (*a*) Loading on post. (*b*) Section *A-B*. (*c*) Division of area for calculating moment of inertia.

3. The moment of inertia. The neutral axis is the x axis, shown in Fig. 7-15*c*. The simplest way to calculate the moment of inertia of the area is to regard it as a positive rectangle 3 in. wide and 5 in. deep, and two negative rectangles, each 1.25 in. wide and 4 in. deep, as shown in Fig. 7-15*c*. (Remember that the width is the dimension parallel to the axis and the depth is the dimension perpendicular to the axis.) The moment of inertia of the positive rectangle is

$$I = \frac{1}{12}bh^3 = \frac{1}{12} \times 3 \times (5)^3 = 31.25 \text{ in.}^4$$

The moment of inertia of one of the negative rectangles is

$$I = \frac{1}{12}bh^3 = \frac{1}{12} \times 1.25 \times (4)^3 = 6.67 \text{ in.}^4$$

Then the moment of inertia of the composite area is

$$I = 31.25 - 2 \times 6.67 = 17.91 \text{ in.}^4$$

4. The distance c. The distance c is one-half the depth of the cross section, and this is

$$c = \tfrac{1}{2} \times 5 = 2.5 \text{ in.}$$

5. The bending stress, given by the flexure formula, is

$$S = \frac{Mc}{I} = \frac{13\,500 \times 2.5}{17.91} = 1\,884 \text{ psi}$$

The bending moment is positive, so, as we look from the left side, we see a compressive stress at the top of the bar (on side B) and a tensile stress at the bottom of the bar (on side A).

Practice Problems (Section 7-3). Now try it for yourself.

1. Figure 7-16 shows a coupling subjected to the two 180-lb forces shown. The cross section is circular and is ¾ in. in diameter. Calculate the bending stress on section C-C.
2. Figure 7-17 shows a link that has a square cross section 25 mm on a side. Find the bending stress on section C-C.
3. The hollow cylinder in Fig. 7-18 is subjected to the eccentric load of 52 000 lb. What is the bending stress at the base?
4. Figure 7-19a shows the load applied to a bracket, and Fig. 7-19b shows the cross section. Calculate the bending stress on section C-C.
5. Figure 7-20a shows the load applied to part of a machine frame, and Fig. 7-20b shows the cross section. Find the bending stress at the top and at the bottom of section A-B.

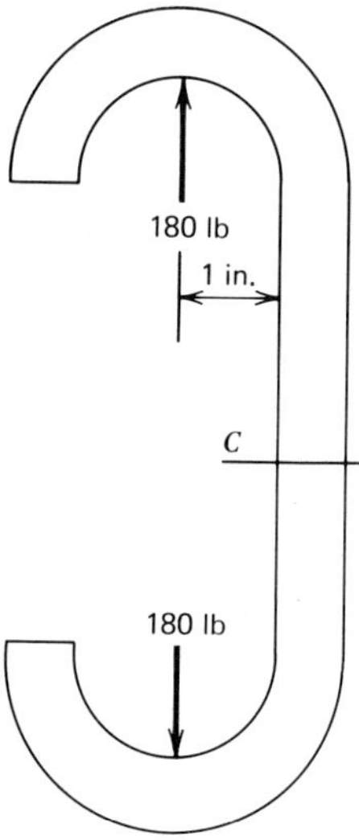

180 lb
1 in.
C C
180 lb

Fig. 7-16. Coupling for Problem 1.

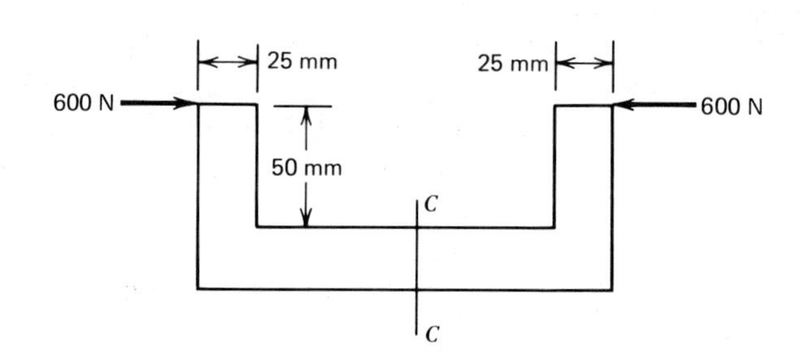

Fig. 7-17. Link for Problem 2.

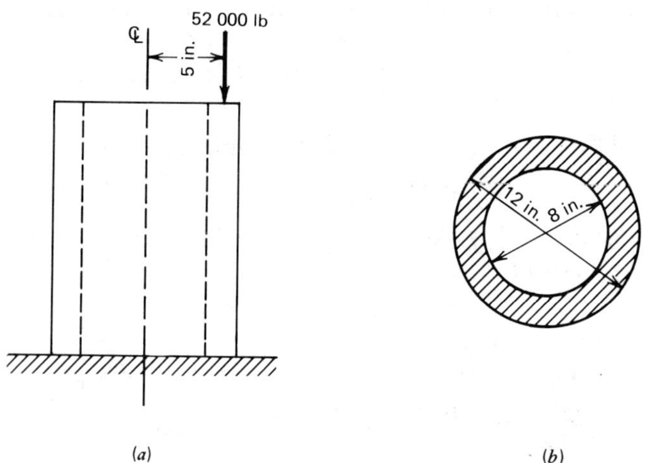

Fig. 7-18. Hollow cylinder for Problem 3. (a) Front view. (b) Section.

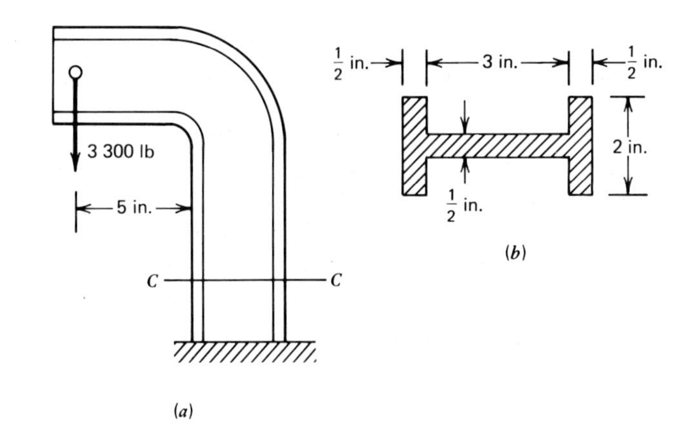

Fig. 7-19. Bracket for Problem 4. (*a*) Loading. (*b*) Section *C-C*.

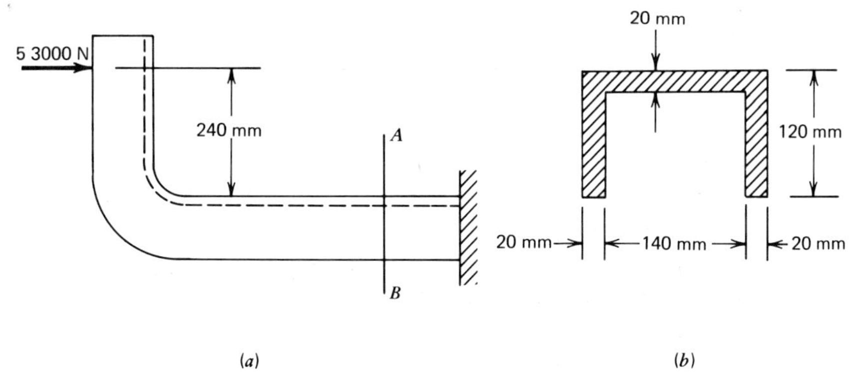

Fig. 7-20. Part of machine frame for Problem 5. (*a*) Loading. (*b*) Section *A-B*.

7-4. BENDING STRESS COMBINED WITH AXIAL STRESS

There are many cases in which a member has an axial force and a bending moment on the same cross section, and we want to be able to calculate the maximum stress, to be sure that the maximum stress is not too high.

Let's suppose that the part of a machine member shown in Fig. 7-21 has an axial force P and a bending moment M on the cross section A-B. We know how to find the axial stress by using the formula,

$$S = \frac{P}{A}$$

This axial stress is the same at all points in the cross section, and we can determine whether it is tension or compression. Also, we know how to find the bending stress by using the formula,

$$S = \frac{Mc}{I}$$

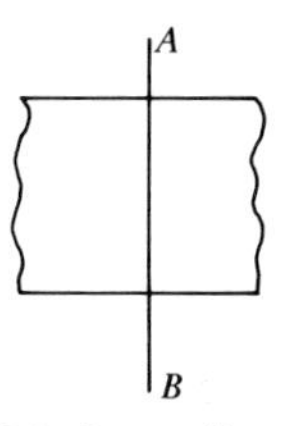

Fig. 7-21. Part of machine member.

The bending stress is not the same for all points in the cross section, but we can find its maximum value, and we can find whether it is tension or compression.

The axial stress and the bending stress are exerted at the same point, say point A in Fig. 7-21, and besides, each is tension or compression. Then we can add the axial stress at the point to the bending stress at the point to get the total stress. For example, if the axial stress at A is 2 000 psi tension and the bending stress at A is 12 000 psi tension, the total stress is

$$2\,000 + 12\,000 = 14\,000 \text{ psi tension}$$

Illustrative Example 1. Figure 7-22*a* shows a link that is subjected to two eccentric forces as shown; the forces are 1 in. above the center of the link. Figure 7-22*b* shows the cross section of the link. Calculate the stress at point A.

Solution:

The free-body diagram is already drawn in Fig. 7-22*a*.

A. The axial stress.

1. The axial force is the sum of the forces, to the left of the section line A-B, that are perpendicular to the section. The sum is

$$P = 17\,000 \text{ lb tension}$$

2. The cross section is rectangular and its area is

$$A = 1 \times 4 = 4 \text{ in.}^2$$

3. The axial stress is

$$S = \frac{P}{A} = \frac{17\,000}{4} = 4\,250 \text{ psi tension}$$

B. The bending stress.

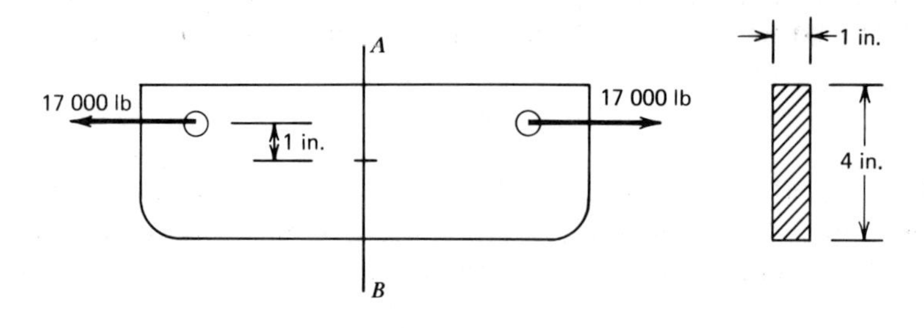

Fig. 7-22. Link for Illustrative Example 1. (*a*) Loading. (*b*) Section.

1. The *bending moment* is the moment, with respect to the centroid of the cross section, of all forces to the left of the cross section. So

$$M = -17\,000 \times 1 = -17\,000 \text{ lb in.}$$

2. The moment of inertia and the distance c. The moment of inertia of the rectangular area is

$$I = \frac{1}{12} bh^3 = \frac{1}{12}(1)(4)^3 = 5.33 \text{ in.}^4$$

The distance c is one half the depth and is

$$c = \tfrac{1}{2}(4) = 2 \text{ in.}$$

3. The bending stress is

$$S = \frac{Mc}{I} = \frac{17\,000 \times 2}{5.33} = 6\,380 \text{ psi tension}$$

The bending stress is tension at A, because a negative bending moment produces tension at the top of the bar.

C. The total stress at A is the sum of the axial stress and the bending stress. Both are tension and the sum is

$$S = 4\,250 + 6\,380 = 10\,630 \text{ psi tension}$$

TENSION AND COMPRESSION

There is more to it than this. The axial stress is the same for all points in the cross section, but the bending stress is tension on one side of the bar and compression on the other side. We are likely to have a case in which the axial stress at a point is tension and the bending stress is compression. Now *tension* stretches the material, and *compression* squeezes it. Tension is the opposite of compression and tends to cancel compression, so if one stress is tension and the other compression, we will take the difference between the two. For example, if the axial stress at a point is 4 000-psi tension and the bending stress is 1 800 psi compression, the total stress is

$$4\,000 - 1\,800 = 2\,200 \text{ psi tension}$$

Illustrative Example 2. Figure 7-23 shows the free-body diagram of a circular bar 1 in. in diameter that is supported at points A and B. The bar is subjected to a vertical force of 210 lb and to two horizontal forces of 3 300 lb. Find the total stress at the top and at the bottom of the cross section C-C. (The reactions at A and B are given.)

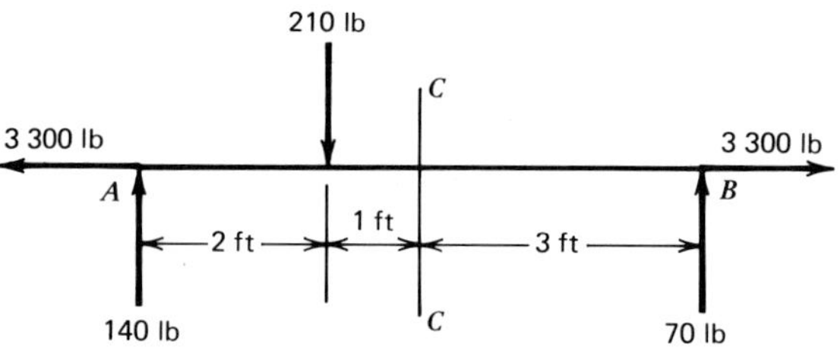

Fig. 7-23. Circular bar for Illustrative Example 2.

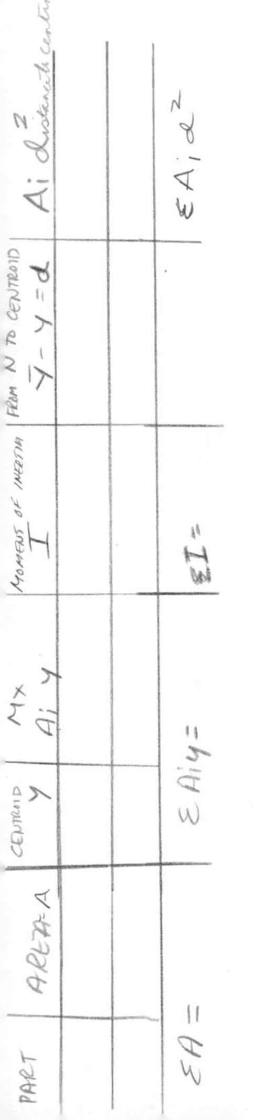

Solution:

A. The axial stress.

1. The axial force on the section is

$$P = 3\,300 \text{ lb tension}$$

2. The area of the circular bar is

$$A = \frac{\pi}{4}d^2 = \frac{\pi}{4}(1)^2 = 0.785 \text{ in.}^2$$

3. The axial stress is

$$S = \frac{P}{A} = \frac{3\,300}{0.785} = 4\,200 \text{ psi tension}$$

The axial stress has the same value for all points in the cross section.

B. The bending stress.

1. *The bending moment* is the moment of the forces to the left of the section. The moment arm of the reaction at A is 3 ft and the moment arm of the 210-lb force is 1 ft. Then,

$$M = 140 \times 3 - 210 \times 1 = 420 - 210 = 210 \text{ lb ft}$$

Notice that the bending moment is positive.

2. The moment of inertia and the distance c. The moment of inertia of the circular area is

$$I = \frac{\pi}{64}d^4 = \frac{\pi}{64}(1)^4 = 0.0491 \text{ in.}^4$$

The distance c is one half the diameter, so

$$c = \tfrac{1}{2}(1) = 0.5 \text{ in.}$$

3. The bending stress

$$S = \frac{Mc}{I} = \frac{210 \times 12 \times 0.5}{0.0491} = 25\,700 \text{ psi}$$

Notice that we multiplied the bending moment by 12 to convert it to pound inches (Chapter 6). The positive bending moment produces compression in the top of the bar and tension in the bottom.

C. The total stress.

1. At the top of the bar the axial stress is 4 200 psi tension and the bending stress is 25 700 psi compression, so the total stress is

$$S = 25\,700 - 4\,200 = 21\,500 \text{ psi compression}$$

2. At the bottom of the bar the axial stress is 4 200 psi tension, and the bending stress is 25 700 psi tension, so the total stress is

$$S = 25\,700 + 4\,200 = 29\,900 \text{ psi tension}$$

THE WHOLE WORKS

Let's take one real comprehensive example before we leave this subject. This one will have everything. We will have to draw a free-body diagram, calculate reactions, locate the centroid of the cross section, and calculate the moment of inertia. You will need to know all that we have studied so far.

Illustrative Example 3. Figure 7-24*a* shows a push rod that is supported in smooth bearings at *A* and *B*. The push rod is driven by the 3 600-N force at the left end and pushes a piston at the right end. The piston exerts the force *R* in reaction. The cross section of the rod is shown in Fig. 7-24*b*; the groove in the bottom serves as a guide against the bearings. Calculate the total stress at the top and at the bottom of the bar at the cross section *C-C*.

Solution:

A. The free-body diagram is shown in Fig. 7-24*c*. The 3 600-N force at the left end is resolved into *x* and *y* components to make calculation easier. The *x* component is

$$3\,600 \cos 30° = 3\,600 \times 0.866 = 3\,120 \text{ N}$$

and the *y* component is

$$3\,600 \sin 30° = 3\,600 \times 0.5 = 1\,800 \text{ N}$$

The bearing reactions are R_1 and R_2.

B. The reactions.

1. The reaction R_2 is calculated by writing a moment equation with point *A* as a center; this is logical, because the unknown forces R_1 and *R* pass through *A* and have no moment. Notice

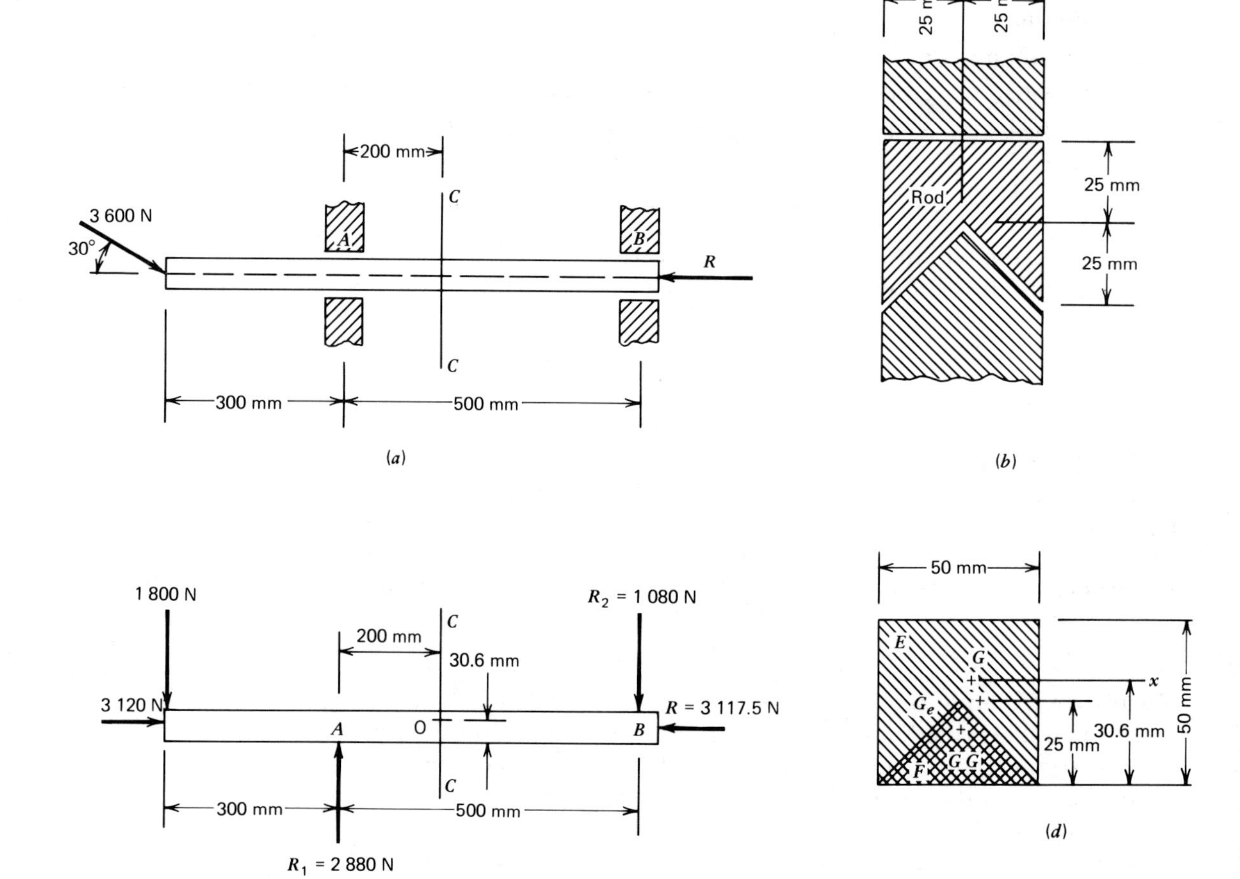

Fig. 7-24. Push rod for Illustrative Example 3. (*a*) Loading. (*b*) Section. (*c*) Free-body diagram. (*d*) Division of area into positive rectangle and negative triangle.

that the force of 3 120 N also passes through A. Then, taking the forces in order from left to right,

$$-1\,800 \times 300 + 500R_2 = 0$$

Transposing,

$$500R_2 = 1\,800 \times 300 = 540\,000$$

$$R_2 = 1\,080 \text{ N}$$

2. With R_2 known, R_1 can be found from the equation $\Sigma F_y = 0$. Thus,

$$\Sigma F_y = 0$$

$$-1\,800 + R_1 - 1\,080 = 0$$

Transposing,

$$R_1 = 1\,800 + 1\,080 = 2\,880 \text{ N}$$

3. The reaction R can be found by using $\Sigma F_x = 0$

$$\Sigma F_x = 0$$

$$3\,120 - R = 0 \qquad R = 3\,120 \text{ N}$$

C. The centroid of the area. Figure 7-24*d* shows the area divided into a positive rectangle E 50 mm wide and 50 mm deep and a negative triangle F 50 mm wide and 25 mm deep. This is to be used for locating the centroid and calculating the moment of inertia. We draw the x_1 axis at the base of the figure.

1. For the positive rectangle E
   (a) The area is

$$A_e = 50 \times 50 = 2\,500 \text{ mm}^2$$

   (b) The distance from the x_1 axis to the centroid of E is one-half the depth, so

$$\bar{y}_e = \tfrac{1}{2}(50) = 25 \text{ mm}$$

2. For the negative triangle F
   (a) The area is one-half the product of the width and the depth, so

$$A_f = \tfrac{1}{2} \times 50 \times 25 = 625 \text{ mm}^2$$

   (b) The distance from the x axis to the centroid of F is one-third the depth, so

$$\bar{y}_f = \tfrac{1}{3}(25) = 8.33 \text{ mm}$$

3. For the entire area
   (a) The area is

$$A = A_e + A_f = 2\,500 - 625 = 1\,875 \text{ mm}^2$$

   (b) The moment of the area with respect to the x_1 axis is

$$M_{x1} = A_e\bar{y}_e + A_f\bar{y}_f$$

$$= 2\,500 \times 25 - 625 \times 8.33 = 62\,500 - 5\,200$$

$$= 57\,300 \text{ mm}^3$$

   (c) The distance from the x_1 axis to the centroid of the entire area is

$$\bar{y} = \frac{M_{x1}}{A} = \frac{57\,300}{1\,875} = 30.6 \text{ mm}$$

D. The axial stress.

1. The axial force on section *C-C* is

$$P = 3\,120 \text{ N compression}$$

2. The area, as calculated previously, is

$$A = 1\,875 \text{ mm}^2$$

3. The axial stréss is

$$S = \frac{P}{A} = \frac{3\,120}{1\,875} = 1.66 \text{ MPa compression}$$

The axial stress is the same for all points in the cross section.

E. The bending moment. The moment center for the bending moment is taken at the centroid of the cross section; this is at point *O* in Fig. 7-25*c* and is 30.6 mm ($\bar{y}$ = 30.6 mm) above the bottom of the section. Hence the 3 120-N force at the left end of the bar has a moment arm of 5.6 mm. The moment arm of the 1 800-N force is 500 mm, and that of the reaction R_1 is 200 mm. Then,

$$M = 3\,120 \times 5.6 - 1\,800 \times 500 + 2\,880 \times 200$$

$$= 17\,400 - 900\,000 + 576\,000 = -306\,600 \text{ N·mm}$$

F. The moment of inertia. The moment of inertia must be calculated with respect to the centroidal axis, which in this case is the *x* axis. It is necessary to use the parallel-axis theorem for each part of the area, but we have done this several times before, and we can do it again.

1. For the positive rectangle *E*

(a) $I_{xg} = \frac{1}{12} bh^3 = \frac{1}{12}(50)(50)^3 = 521\,000 \text{ mm}^4$

(b) $A_e = 50 \times 50 = 2\,500 \text{ mm}^2$

(c) $c = \bar{y} - \bar{y}_e = 30.6 - 25 = 5.6 \text{ mm}$

(d) $I_e = I_{xg} + A_c^2 = 521\,000 + 2\,500(5.6)^2$
$= 521\,000 + 78\,400 = 599\,400 \text{ mm}^4$

2. For the negative triangle *F*

(a) $I_{xg} = \frac{1}{36} bh^3 = \frac{1}{36}(50)(25)^3 = 21\,700 \text{ mm}^4$

(b) $A_f = ½(50)(25) = 625 \text{ mm}^2$

(c) $c = \bar{y} - \bar{y}_f = 30.6 - 8.33 = 22.3 \text{ mm}$

(d) $I_f = I_{xg} + A_c^2 = 21\,700 + 625 \times (22.3)^2$
$= 21\,700 + 311\,000 = 332\,700 \text{ mm}^4$

3. For the entire area,

$$I = I_e - I_f = 599\,400 - 332\,700 = 266\,700 \text{ mm}^4$$

G. The bending stress.

1. At the top of the rod the distance c from the neutral axis is

$$50 - 30.6 = 19.4 \text{ mm}$$

Then,

$$S = \frac{Mc}{I} = \frac{306\,600 \times 19.4}{266\,700} = 22.3 \text{ MPa tension}$$

The bending stress at the top is tension, because the bending moment is negative.

2. At the bottom of the rod the distance c from the neutral axis is 30.6 mm, so

$$S = \frac{Mc}{I} = \frac{306\,600 \times 30.6}{266\,700} = 35.2 \text{ MPa compression}$$

The *bending stress* is compression at the bottom of the bar, because the bending moment is negative.

H. The total stress.

1. At the top of the bar the axial stress is 1.66 MPa compression and the bending stress is 22.3 tension; so the total stress is

$$22.3 - 1.66 = 20.64 \text{ MPa tension}$$

2. At the bottom of the bar the axial stress is 1.66-MPa compression and the bending stress is 35.2-MPa compression, so the total stress is

$$35.2 + 1.66 = 36.86 \text{ MPa compression}$$

Practice Problems (Section 7-4). Here are a few practice problems to help you remember what you have just studied.

1. Figure 7-25 shows a curved bar that is subjected to the force of 28 000 lb. At the support, the cross section of the bar is a rectangle 2 in. wide and 6 in. deep. Calculate the stress at the top and bottom of the section at the support.
2. The ring in Fig. 7-26 is a circular wire 0.1 in. in diameter. Find the stress at the top and at the bottom of section C-C.

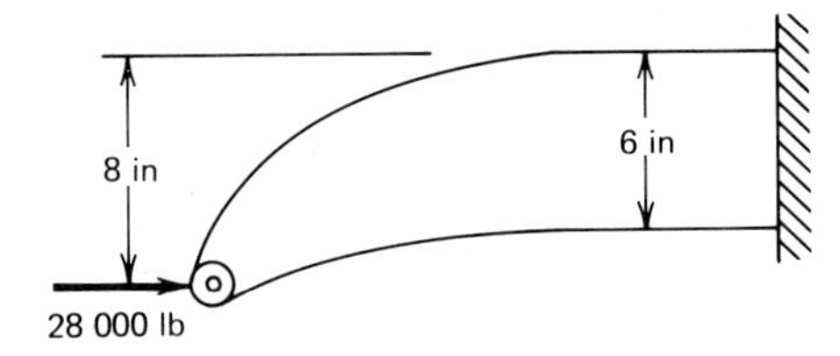

Fig. 7-25. Curved bar for Problem 1.

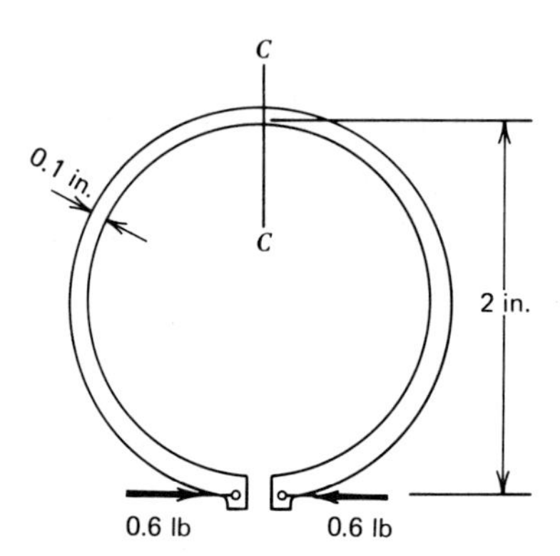

Fig. 7-26. Wire ring for Problem 2.

3. Figure 7-27*a* shows the loading on a large press frame, and Fig. 7-27*b* shows the cross section. Calculate the stress at each side of section *C-C*.
4. Figure 7-28 shows a connecting rod that is supported in a smooth bearing at *A* and by a pin at *B*. The cross section is square 30 mm on a side. Find the stress on each side of section *C-C*.
5. Figure 7-29*a* shows the loading on a frame that is supported by knife-edges at *A* and *B*, and Fig. 7-29*b* shows the cross section. Calculate the stress at each side of the section *C-C*.

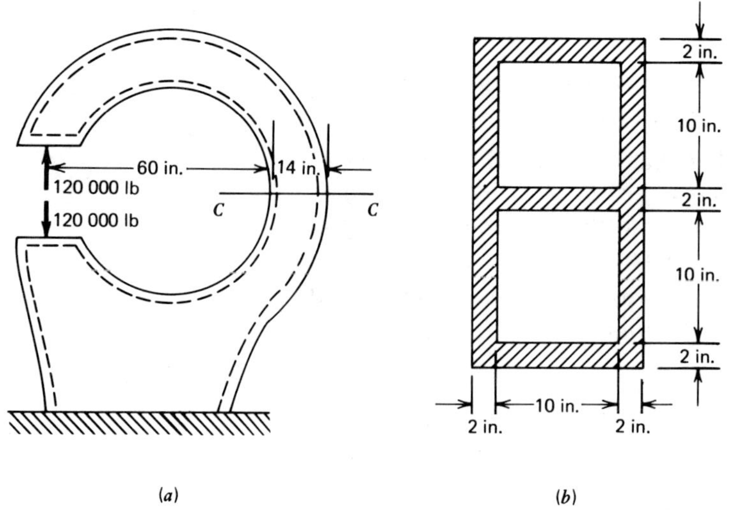

Fig. 7-27. Press frame for Problem 3. (*a*) Loading. (*b*) Section *C-C*.

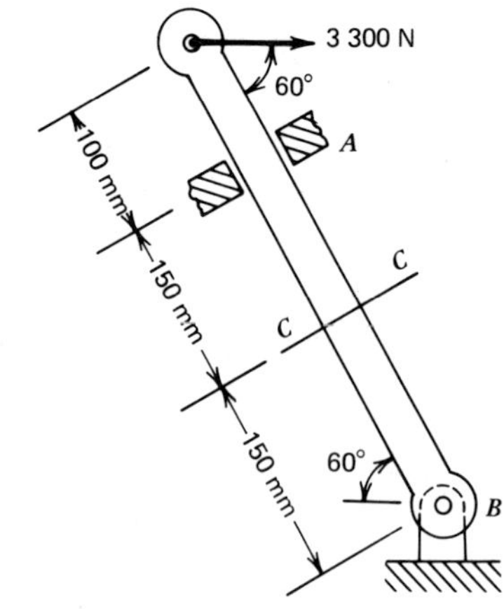

Fig. 7-28. Connecting rod for Problem 4.

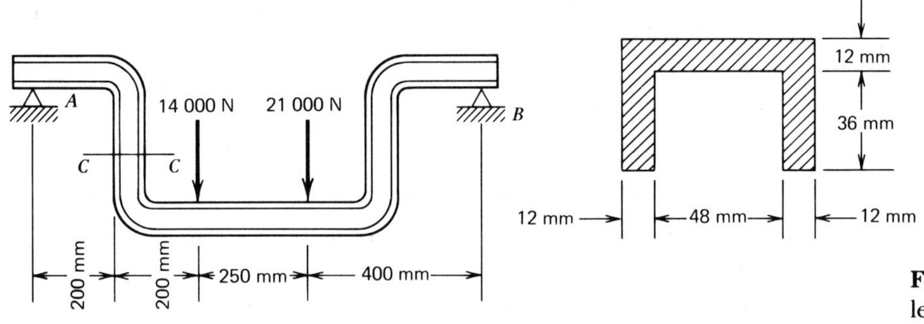

Fig. 7-29. Frame for Problem 5. (*a*) Loading. (*b*) Section *C-C*.

SUMMARY Here is a brief summary of Chapter 7.

1. An axial force is a force of tension or compression that is perpendicular to the cross section of a member or bar. The axial force is equal to the sum of all forces to the left of the section which are perpendicular to the section.
   (a) The magnitude of the axial stress is

$$S = \frac{P}{A}$$

   (b) The axial stress is the same at all points in the cross section.
2. The bending stress is the stress due to the bending moment. The bending moment is equal to the moment, with respect to the centroid of the section, of all forces to the left of the section. Clockwise moments are positive, and counterclockwise moments are negative.
   (a) The magnitude of the bending stress at the edge of the cross section is

$$S = \frac{Mc}{I}$$

   (b) The bending stress is compression at the top of the section and tension at the bottom when the bending moment is positive.
   (c) The bending stress is tension at the top of the section and compression at the bottom when the bending moment is negative.
3. The total stress at any point in the section is equal to the sum of the axial stress and the bending stress.

REVIEW QUESTIONS Try to answer these questions without looking at the preceding pages.

1. What is the direction of an axial force with respect to the cross section of a bar?

2. How do you calculate the axial force at a cross section when there are several forces acting on a machine part?
3. What do you do with a force that is neither parallel nor perpendicular to the cross section?
4. Give the formula for axial stress.
5. How do you calculate the bending moment on a cross section?
6. Where do you choose the moment center when you calculate the bending moment at a cross section?
7. Give the formula for bending stress.
8. What is the *centroid* of an area?
9. How do you locate the centroid of a composite area?
10. Where is the centroid of:
    (a) The area of a right triangle?
    (b) The area of a quadrant of a circle?
11. With respect to what axis do you calculate the moment of inertia of the area of the cross section?
12. How do you use the parallel-axis theorem to calculate the moment of inertia of a composite area?
13. What is the *parallel-axis theorem*?
14. What is the *moment of inertia* of
    (a) A rectangular area with respect to an axis through the centroid and parallel to the base?
    (b) The area of a right triangle with respect to an axis through the centroid and parallel to the base.
    (c) A circular area with respect to a diameter?
15. How do you calculate the moment of inertia of a hollow circular area?
16. How can you tell whether the bending stress at the top of the cross section is tension or compression?
17. How can you tell whether the bending moment is positive or negative?
18. What is the *total stress*?
19. When do you take the sum of the axial stress and the bending stress, and when do you take the difference?
20. What is the distance c in the flexure formula?

REVIEW PROBLEMS Test your knowledge with these review problems before you go on.

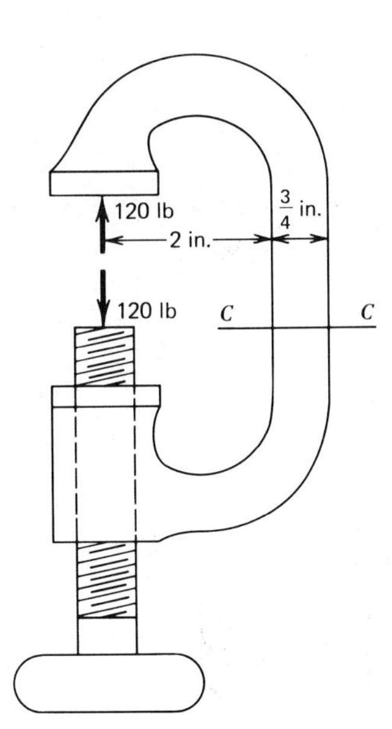

Fig. 7-30. Clamp for Problem 1.

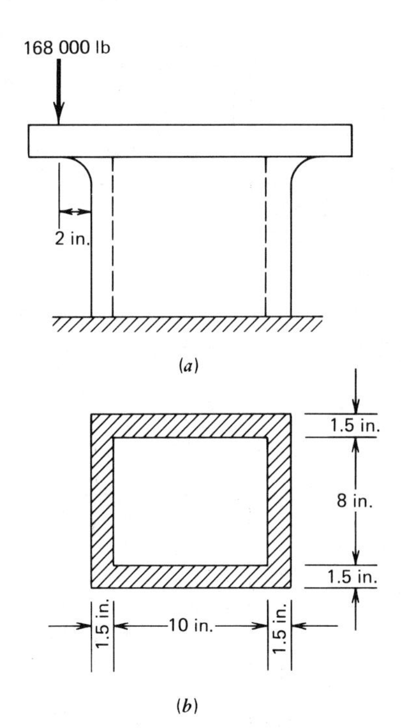

Fig. 7-31. Machine base for Problem 2. (*a*) Loading. (*b*) Section.

1. Figure 7-30 shows a *C* clamp. The section *C-C* is rectangular, ¾ in. wide and ¼ in. thick. Find the stress at each side of the section *C-C*.
2. Figure 7-31*a* shows the loading on a machine base, and Fig. 7-31*b* shows the cross section of the base. Find the stress at each side of the base.
3. Figure 7-32 shows a machine part that is driven by the 2 400-lb force and pushes a heavy carriage. The carriage exerts the force *R*, and the machine part is supported in smooth bearings at *A* and *B*. The cross section is a square 2 in. on a side. Calculate the stress at the top and at the bottom of the section *C-C*.

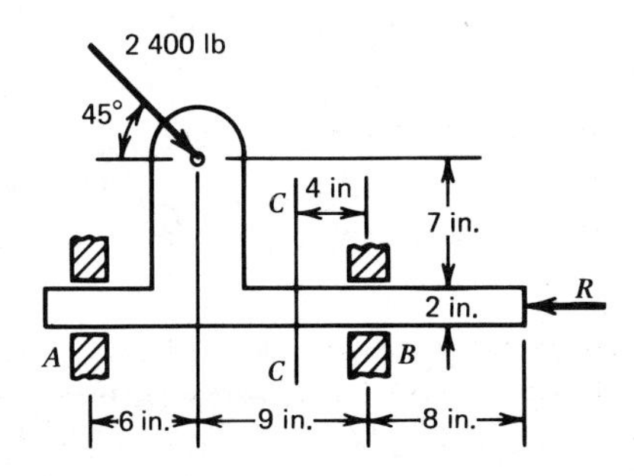

Fig. 7-32. Machine part for Problem 3.

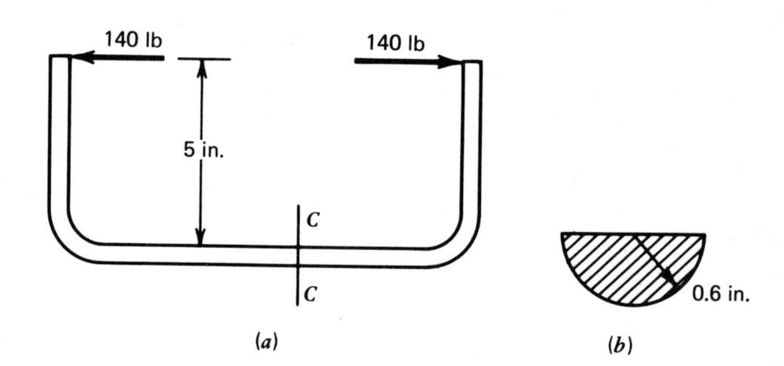

Fig. 7-33. Bent bar for Problem 4. (*a*) Loading. (*b*) Section *C-C*.

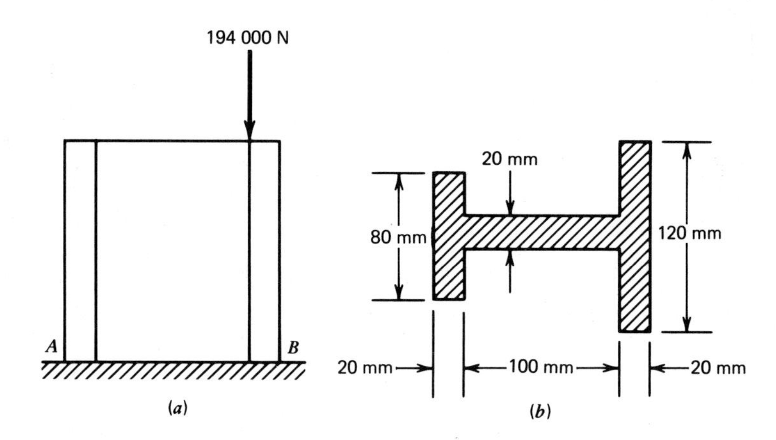

Fig. 7-34. Compression member for Problem 5. (*a*) Loading. (*b*) Cross section.

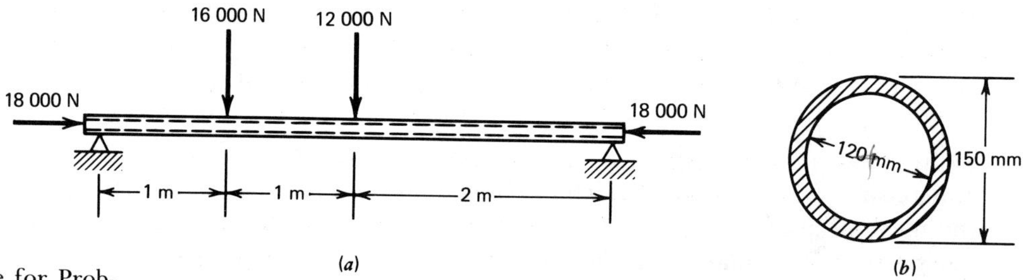

Fig. 7-35. Tube for Problem 6. (*a*) Loading. (*b*) Cross section.

4. Figure 7-33*a* shows the loading on a bent bar and Fig. 7-33*b* shows the cross section, which is a semicircle 1.2 in. in diameter. Find the stress at the top and at the bottom of the section *C-C*.
5. Figure 7-34*a* shows the loading of a short compression member, and Fig. 7-34*b* shows the cross section. Calculate the stress at each side of the cross section *A-B*.
6. Figure 7-35*a* shows the loading on a circular tube, and Fig. 7-35*b* shows the cross section. Find the stress at the top and at the bottom of the section where the 12 000-N load is applied.

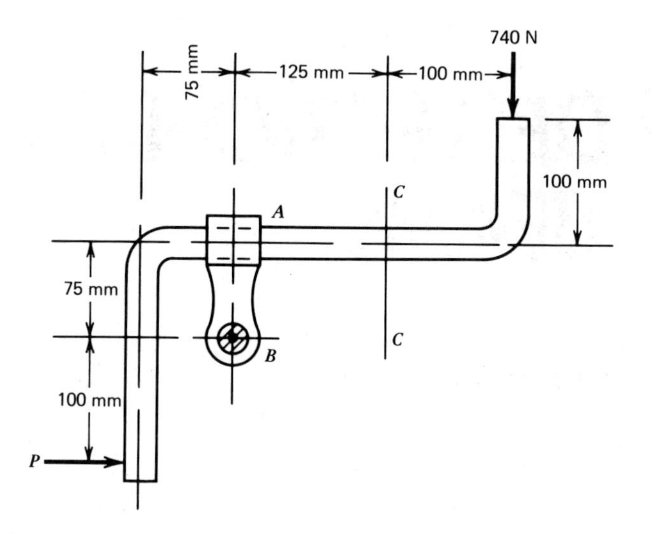

Fig. 7-36. Link for Problem 7.

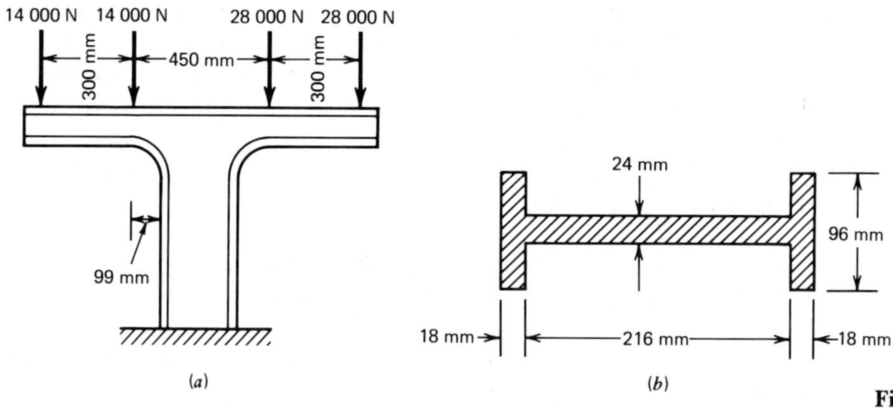

Fig. 7-37. Casting for Problem 8. (*a*) Loading. (*b*) Cross section.

7. Figure 7-36 shows a link in a mechanism. A fitting is fastened to the link at A, and the fitting is supported by a pin at B. The link is circular in cross section and is 27 mm in diameter. Find the stress at the top and at the bottom of the section C-C.
8. A casting carries the loads shown in Fig. 7-37*a*, and the lower part has the cross section shown in Fig. 7-37*b*. Find the stress at each side of the base.

8 DEFLECTION OF BEAMS

PURPOSE OF THIS CHAPTER. The purpose of this chapter is to teach you how to calculate the deflection of a beam. *Deflection* is the bending or sag of a beam under load.

It is important for a designer to be able to calculate the deflection of a beam, to be sure to design a beam that will not deflect too much. A beam may be strong enough to carry the loads but may still deflect too much to be suitable. For example, it would be possible to design a floor system that would not fall down under the loads but would deflect several inches whenever a person walked across the floor. You should learn enough about deflection so you can design beams properly.

8-1. WHAT DEFLECTION IS

Figure 8-1 illustrates *deflection.* Figure 8-1*a* shows a steel beam before the loads are applied, and Fig. 8-1*b* shows the cross section. The beam is welded to steel angles at the ends, and the angles are welded to steel columns. Figure 8-1*c* shows the beam as it looks when it is supporting the loads. The line AB in Fig. 8-1*c* goes through the centroid of each cross section of the beam. The upper part of the beam shortens and is in compression; the lower part of the beam stretches and is in tension. There is no change of length and no stress along the line AB; this line is called the *elastic curve of the beam.* Figure 8-1*d* shows the free-body diagram of the beam. The heavy line AB is the center line of the beam before the loads are applied, and the curved line is the elastic curve; the loads cause the beam to bend; the center line is called the *elastic curve* after the beam is bent. Any point C moves downward as the beam bends; the distance that the point moves is called its *deflection.*

Some points in a beam deflect more than other points. Usually, we are interested in the point that deflects the most; that is, we want to calculate the *maximum deflection* of the beam. The formulas you can use for this purpose will be given to you. Some of these formulas will be exact, and some will be only good approximations. However, even the approximate formulas will be close enough for practical purposes. The

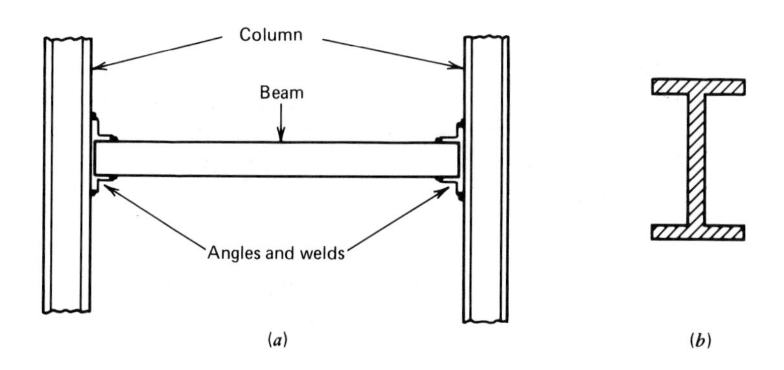

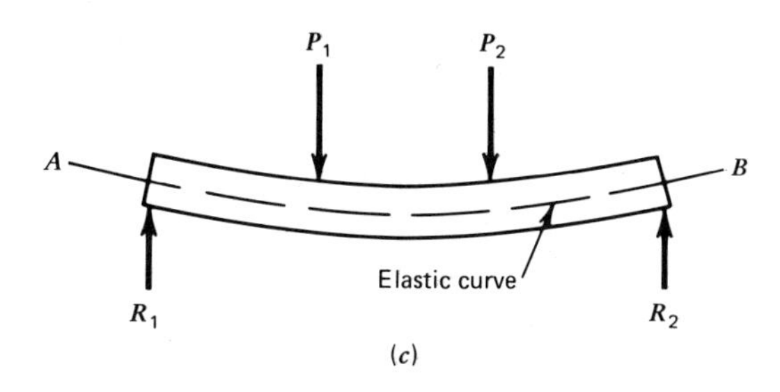

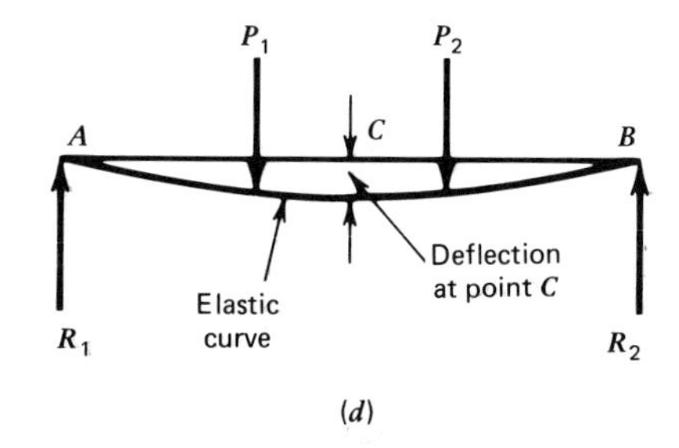

Fig. 8-1. Deflection of a beam. (*a*) Beam before loading. (*b*) Cross section. (*c*) Beam after loading. (*d*) Free-body diagram.

derivation of the formulas will not be given, because that would take too much space, but you will be shown how to use the formulas.

8-2. MAXIMUM DEFLECTION OF A BEAM WITH A CONCENTRATED LOAD AT THE CENTER

One of the simplest formulas is for a beam that is supported at the ends and carries a concentrated load at the center. Figure 8-2 shows a beam with this loading; here we use the symbol P for the load and the symbol

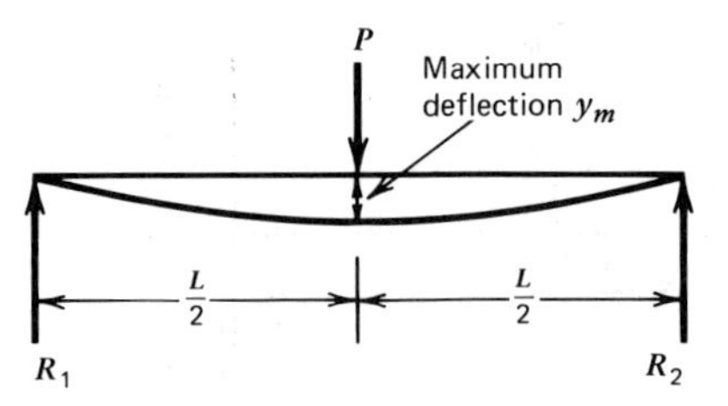

Fig. 8-2. Beam with concentrated load.

L for the length of the beam. The maximum deflection is represented by y_m (where the subscript $_m$ signifies maximum); it occurs at the center of the beam and is

$$y_m = \frac{1}{48} \frac{PL^3}{EI}$$

where E is the *modulus of elasticity* of the material in tension and compression, and I is the *moment of inertia* of the cross section of the beam. Watch the units in this equation: take P in *pounds*, L in *inches*, E in *pounds per square inch*, and I in *inches to the fourth power*, if you want to get y_m in inches. The equation can be written in terms of units as

$$y_m = \frac{\text{lb. in.}^3}{\dfrac{\text{lb.}}{\text{in.}^2}\ \text{in.}^4} = \text{in.}$$

all this is for the gravitational system of units.

In the metric system, you would take P in *newtons*, L in *millimeters*, E in *gigapascals*, and I in *millimeters to the fourth power*. Thus, in terms of units,

$$y_m = \frac{\text{N}\cdot\text{mm}^3}{\text{GPa mm}^4}$$

but the simplest way is probably to multiply E in *gigapascals* by 10^3 to convert it to newtons per square millimeter. Thus we can write

$$y_m = \frac{\text{N}\cdot\text{mm}^3}{\dfrac{\text{N}}{\text{mm}^2}\ \text{mm}^4} = \text{mm}$$

Let's try some examples,

Illustrative Example 1. A steel beam, W 12 × 45, is 16 ft long and is supported at the ends. The beam is subjected to a concentrated load of 18 000 lb at the center. Calculate the maximum deflection.

Solution:

1. The load P is 18 000 lb.

2. The length L is 16 ft, and this must be expressed in inches

$$L = 16 \times 12 = 192 \text{ in.}$$

3. The *modulus of the elasticity* of steel is (see Table 3.1)

$$E = 30 \times 10^6 \text{ psi}$$

4. Look up I in column 7 of Table 6.1

$$I = 350 \text{ in.}^4$$

5. The maximum deflection is

$$y_m = \frac{1}{48}\frac{PL^3}{EI} = \frac{1}{48}\frac{18\,000 \times (192)^3}{30\,000\,000 \times 350} = 0.253 \text{ in.}$$

Illustrative Example 2. A rectangular aluminum alloy beam 20 mm wide and 40 mm deep is 0.8 m long and is supported at the ends. The beam is subjected to a concentrated load of 1 200 N at the center. Calculate the maximum deflection.

Solution:

1. The load P is 1 200 N.
2. The length L is 0.8 m, and this must be expressed in millimeters

$$L = 0.8 \times 10^3 = 800 \text{ mm}$$

3. The *modulus of elasticity* is (See Table 3.1)

$$E = 69 \text{ GPa} = 69 \times 10^3 \text{ N/mm}^2$$

4. $I = \frac{1}{12} bh^2 = \frac{1}{12} \times 20(40)^3 = 107\,000 \text{ mm}^4$.
5. The maximum deflection is

$$y_m = \frac{1}{48}\frac{1\,200 \times (800)^3}{69\,000 \times 107\,000} = 1.734 \text{ mm}$$

8-3. MAXIMUM DEFLECTION OF A BEAM WITH A UNIFORMLY DISTRIBUTED LOAD

Figure 8-3 shows a beam with a uniformly distributed load. Represent the load in pounds per foot by w (small letter) and the total load by W (capital letter). The maximum deflection occurs at the center and is

$$y_m = \frac{5}{384}\frac{WL^3}{EI}$$

Notice here that W is the total load in pounds. You know how to obtain L, E, and I.

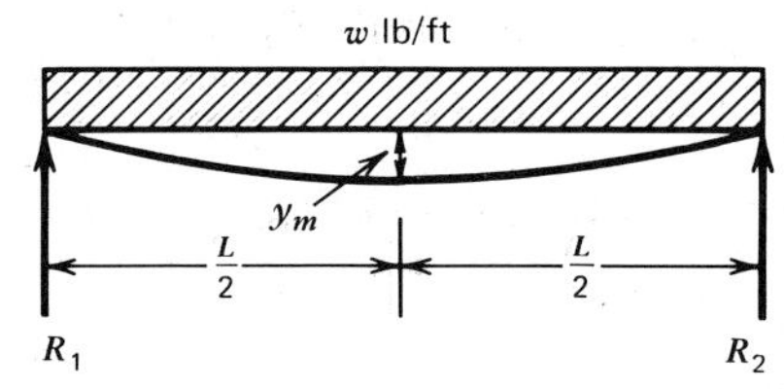

Fig. 8-3. Beam with uniform load.

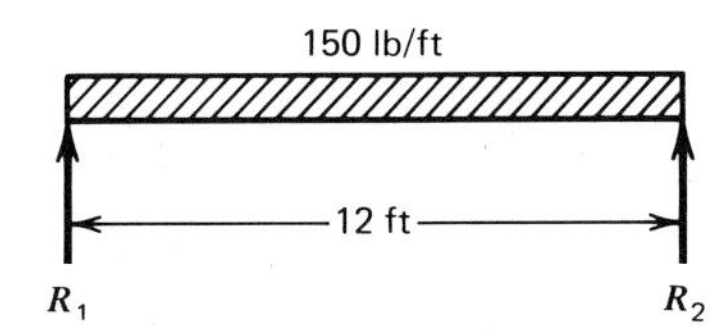

Fig. 8-4. Loading for Illustrative Example 1.

Illustrative Example 1. A 2 by 10 (actually 1½ in. by 9¼ in.) wood joist is subjected to the load shown in Fig. 8-4. The modulus of elasticity is 1.6×10^6 psi. Calculate the maximum deflection.

Solution:

1. The total load W is equal to the load w in pounds per foot times the length of the beam

$$W = 150 \times 12 = 1\,800 \text{ lb}$$

2. The length L is 12 ft. In inches, this is

$$L = 12 \times 12 = 144 \text{ in.}$$

3. $E = 1.6 \times 10^6$ psi
4. Calculate I

$$I = \frac{1}{12} bh^3 = \frac{1}{12} \times 1.5 \times (9.25)^3$$

$$= 98.9 \text{ in.}^4$$

5. The maximum deflection is

$$y_m = \frac{5}{384} \frac{WL^3}{EI} = \frac{5}{384} \frac{1\,800(144)^3}{1.6 \times 10^6 \times 98.9}$$

$$= 0.442 \text{ in.}$$

METRIC UNITS

In metric units, the distributed load w might be in newtons per meter. Then, the total load W would be equal to the distributed load w times the length. In any case, the total load should be calculated in newtons. Let's try this in an example.

Illustrative Example 2. A steel pipe, 40 mm outer diameter and 30 mm inner diameter, is 3.7 m long. The pipe is horizontal and is supported at the ends. The weight of steel is 76.97×10^{-6} N/mm³. What is the maximum deflection of the pipe?

Solution:

1. The area of the cross section of the pipe is

$$A = \frac{\pi}{4}(40)^2 - \frac{\pi}{4}(30)^2 = 1\,257 - 707$$

$$= 550 \text{ mm}^2$$

The volume of a 1-mm length of the pipe is

$$V = 550 \times 1 = 550 \text{ mm}^3$$

The weight of a 1-mm length of pipe is

$$w = 550 \times 76.97 \times 10^{-6} = 42.3 \times 10^{-3} \text{ N}$$

The length of the pipe is 3.7 m or 3.7×10^3 mm. The total weight of the pipe is

$$W = 42.3 \times 10^{-3} \times 3.7 \times 10^3 = 156.5 \text{ N}$$

The pipe is a beam. This is the total distributed load on the beam.

2. The length L is 3.7 m or 3 700 mm.
3. $E = 207$ GPa or 207×10^3 N/mm².
4. $I = \frac{\pi}{64}(40)^4 - \frac{\pi}{64}(30)^4 = 125\,700 - 39\,800$

$$= 85\,900 \text{ mm}^4$$

5. The maximum deflection is

$$y_m = \frac{5}{384}\frac{WL^3}{EI} = \frac{5}{384}\frac{156.5 \times (3\,700)^3}{207\,000 \times 85\,900} = 5.805 \text{ mm}$$

Practice Problems (Sections 8-2 and 8-3). Calculate the following deflections so you will become familiar with the procedure.

1. A steel beam, W 8 × 28, 16 ft long is supported at the ends and subjected to a uniformly distributed load of 1 200 lb/ft. Calculate the maximum deflection.
2. An aluminum alloy channel 5[3.089, 8 ft long, is supported at the ends and subjected to a concentrated load of 1 300 lb at the center. Find the maximum deflection.
3. A steel beam W 10 × 22 is 20 ft long and is supported at the ends. The beam carries a concentrated load of 12 000 lb at the center. What is the maximum deflection?

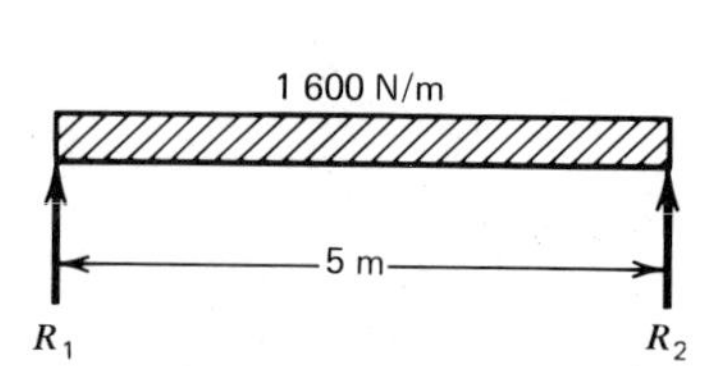

Fig. 8-5. Free-body diagram for Problem 4.

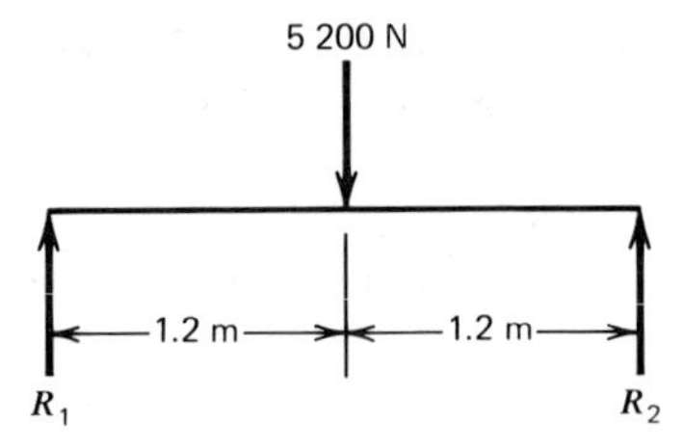

Fig. 8-6. Free-body diagram for Problem 5.

4. A rectangular wood beam 40 mm wide and 290 mm deep carries the load shown in Fig. 8-5; $E = 11$ GPa. Calculate the maximum deflection.
5. A hollow steel beam is subjected to the loading shown in Fig. 8-6. The cross section is a square; the outer dimension is 80 mm, and the inner dimension is 60 mm. Find the maximum deflection.

8-4. A COLLECTION OF FORMULAS FOR BEAM DEFLECTIONS

Now you will be given a collection of formulas for beam deflections. These formulas give you the deflections due to concentrated and uniformly distributed loads placed in different positions on beams. Table 8.1 presents seven special cases of beam loadings and gives formulas for deflections. The problem now is to learn to use the table. Later you will be shown how to apply these formulas to almost any beam-loading problem you need to solve.

The first beam in Table 8.1 is a beam supported at the ends with a concentrated load at the center, which was discussed at the beginning of this chapter. Case 2 of Table 8.1 is a beam supported at the ends that carries a concentrated load at any point. Take b as the distance from the load to the closest support. Then, as you see in the table, the deflection at the center is

$$y_m = \frac{1}{16}\frac{Pb}{EI}\left(L^2 - \frac{4}{3}b^2\right)$$

(If you take L and b in inches in this formula; then the deflection y_m is in inches. If you take L and b in millimeters, the deflection is in millimeters) The maximum deflection does not occur at the center for this loading. However, this formula is simple and gives a result that is approximately equal to the maximum deflection; the error is usually less than 2%, and this is close enough for practical purposes.

TABLE 8.1. BEAM DEFLECTIONS

Case 1. *Beam Supported at Ends with Concentrated Load at Center*

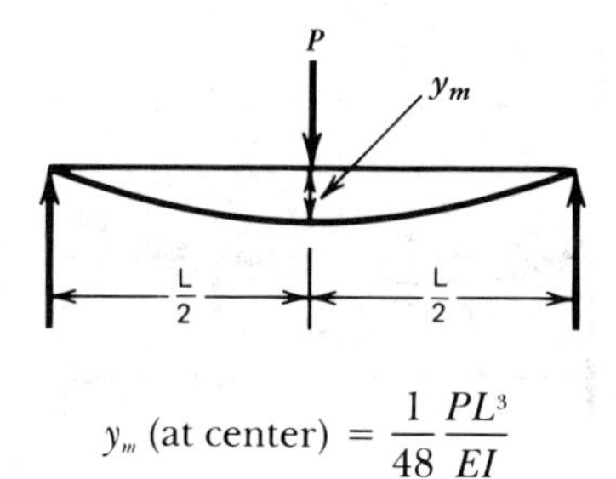

$$y_m \text{ (at center)} = \frac{1}{48}\frac{PL^3}{EI}$$

Case 2. (Take b as distance from load to closest support)

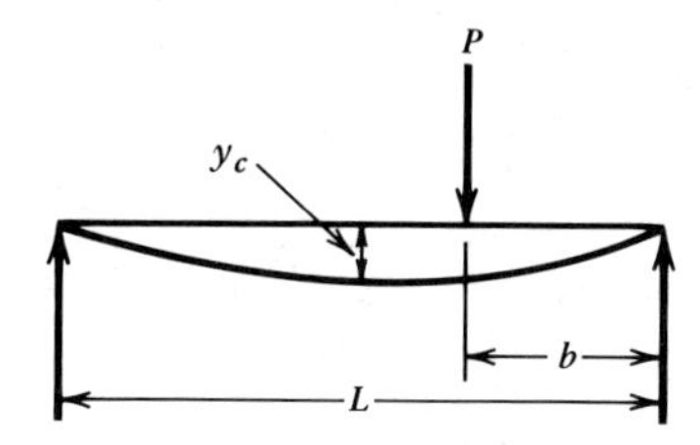

$$y_c \text{ (at center)} = \frac{1}{16}\frac{Pb}{EI}\left(L^2 - \frac{4}{3}b^2\right)$$

Case 3. *Beam Overhanging One Support with Concentrated Load at End of Overhang*

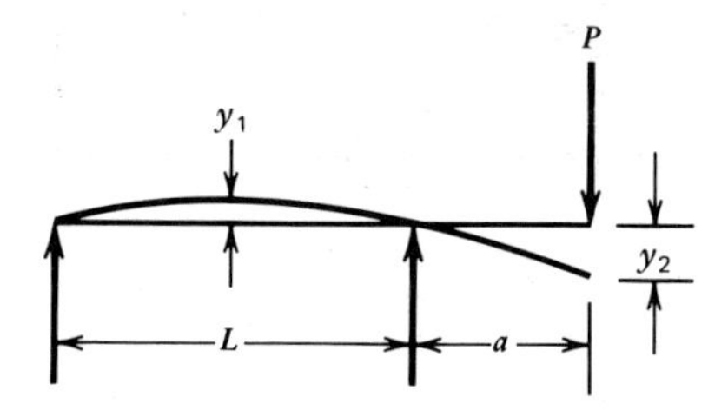

y_1 (halfway between supports)

$$= \frac{1}{16}\frac{PaL^3}{EI}$$

This is approximately equal to the maximum deflection between supports.

$$y_2 = \frac{1}{3}\frac{Pa^2}{EI}(L + a)$$

TABLE 8.1 ***(continued)***

Case 4. *Beam Overhanging One Support with Concentrated Load at Any Point Between Supports*

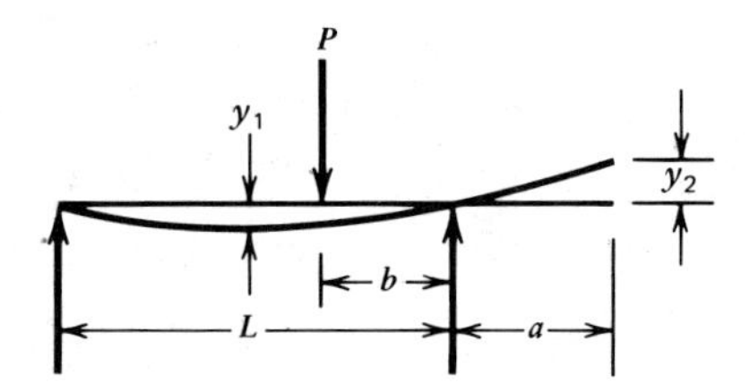

(Take b as distance from load to closest support for y_1)

y_1 (halfway between supports)

$$= \frac{1}{16}\frac{Pb}{EI}\left(L^2 - \frac{4}{3}b^2\right)$$

This is approximately equal to the maximum deflection between supports.

$$y_2 = \frac{1}{6}\frac{Pab}{EIL}(L - b)(2L - b)$$

Case 5. *Beam Supported at Ends with Uniformly Distributed Load*

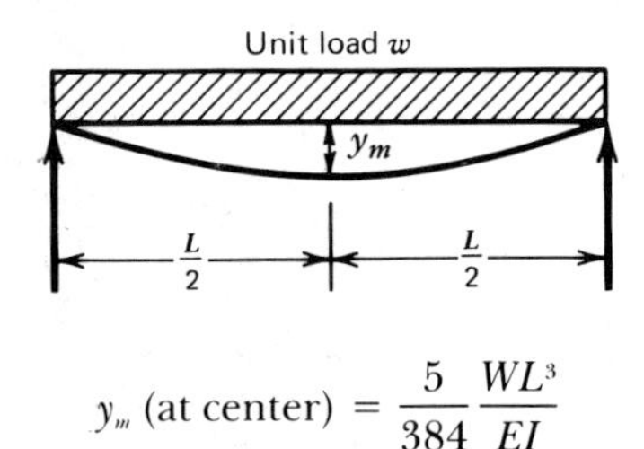

$$y_m \text{ (at center)} = \frac{5}{384}\frac{WL^3}{EI}$$

This is the maximum deflection.

Case 6. *Beam Overhanging One Support with Uniformly Distributed Load between Supports*

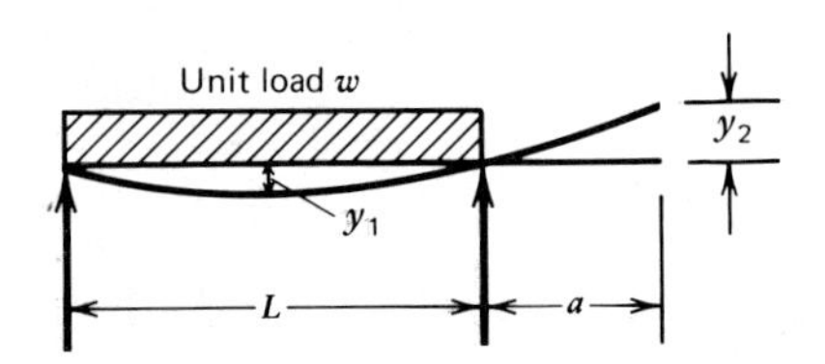

TABLE 8.1 ***(continued)***

y_1 (halfway between supports)

$$= \frac{5}{384} \frac{WL^3}{EI}$$

This is the maximum deflection between supports.

$$y_2 = \frac{1}{24} \frac{WL^2a}{EI}$$

Case 7. *Beam Overhanging One Support with Uniformly Distributed Load on Overhang*

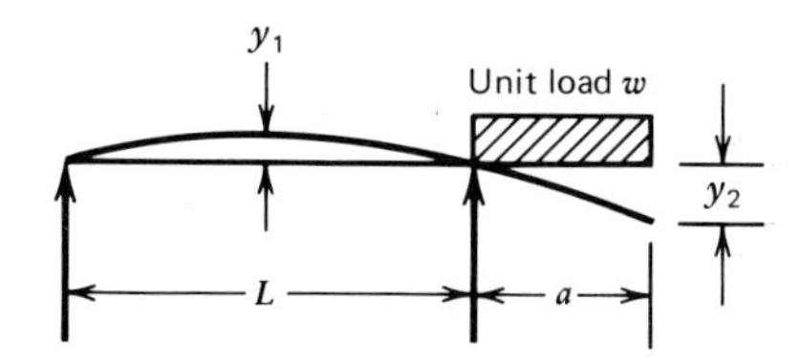

y_1 (halfway between supports)

$$= \frac{1}{32} \frac{Wa^2L}{EI}$$

This is approximately equal to the maximum deflection between supports.

$$y_2 = \frac{1}{6} \frac{Wa}{EIL} (a^2L + \frac{3}{4}a^3)$$

Illustrative Example 1. A steel beam, W 8 × 28, is 18 ft long and is supported at the ends. The beam carries a concentrated load of 9 800 lb at a distance of 6 ft from the left end. Calculate the deflection at the center.

Solution:

Figure 8-7 shows the free-body diagram of the beam.

1. The load P is 9 800 lb.
2. The length b is the distance from the load to the closest support and this is 6 ft, but we must change b to inches

$$b = 6 \times 12 = 72 \text{ in.}$$

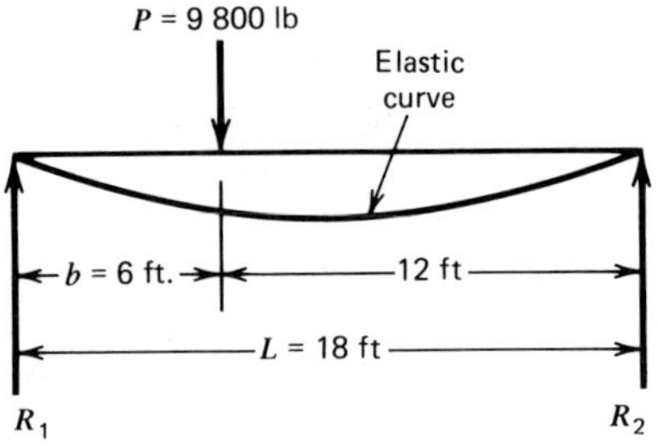

Fig. 8-7. Free-body diagram for Illustrative Example 1.

3. The modulus of elasticity for a steel beam is

$$E = 30 \times 10^6 \text{ psi}$$

4. Look up I in column 7 of Table 6.1

$$I = 98 \text{ in.}^4$$

5. The length L is 18 ft. Change L to inches:

$$L = 18 \times 12 = 216 \text{ in.}$$

6. The deflection at the center is

$$y = \frac{1}{16}\frac{Pb}{EI}\left(L^2 - \frac{4}{3}b^3\right)$$

$$= \frac{1}{16}\frac{9\,800 \times 72}{30 \times 10^6 \times 98}\left[(216)^2 - \frac{4}{3}(72)^2\right]$$

$$= 0.596 \text{ in.}$$

OTHER CASES

Case 3 in Table 8.1 is a beam that overhangs one support and carries a concentrated load at the end of the overhang.

Here you are given two formulas. One is for the upward deflection y_1, halfway between the supports, and the other formula is for the downward deflection y_2 at the end of the overhang. The formulas contain the length L between the supports and the distance a from the end of the overhang to the support.

Case 4 of Table 8.1 is a beam that overhangs one support and carries a concentrated load at any point between the supports. Here L is the distance between supports and a is the length of the overhang. Again two formulas are given, one for the downward deflection y_1, halfway between the supports, and the other for the upward deflection y_2, at the end of the overhang. Notice that you are to take b as the distance from the load to the support that is next to the overhang. The formula for y_2 looks a little complicated, so study carefully the following example.

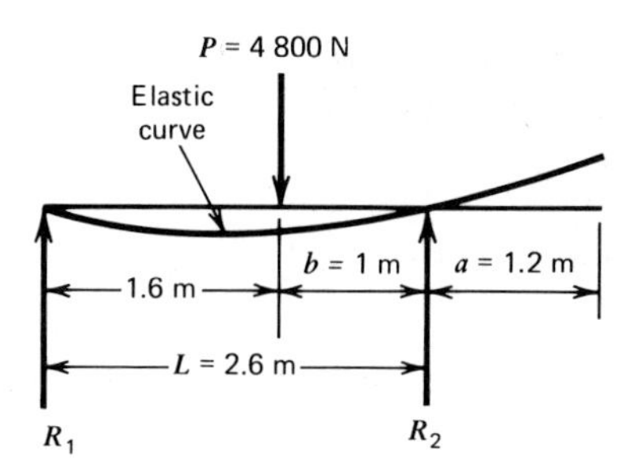

Fig. 8-8. Free-body diagram for Illustrative Example 2.

Illustrative Example 2. A 50 mm × 200 mm wood beam, 3.8 m long, is supported at the left end and at a distance of 1.2 m from the right end. The beam carries a concentrated load of 4 800 N at a distance of 1.6 m from the left end, and $E = 11$ GPa. Calculate the maximum deflection.

Solution:

Figure 8-8 shows the free-body diagram of the beam.

1. The load P is 4 800 N.
2. Notice that $b = 1$ m, but we must express b in millimeters:

$$b = 1 \times 10^3 \text{ mm}$$

3. The distance a is 1.2 m

$$a = 1.2 \times 10^3 \text{ mm}$$

4. $E = 11 \text{ GPa} = 11 \times 10^3$ MPa.
5. $I = \frac{1}{12} \times 50 \times (200)^3 = 33.3 \times 10^6 \text{ mm}^4$.
6. The length L between supports is 2.6 m:

$$L = 2.6 \times 10^3 \text{ mm}$$

7. The upward deflection at the end of the overhang is

$$y_2 = \frac{1}{6}\frac{Pab}{EIL}(L - b)(2L - b)$$

$$= \frac{1}{6}\frac{4\,800 \times 1.2 \times 10^3 \times 1 \times 10^3}{11 \times 10^3 \times 33.3 \times 10^6 \times 2.6 \times 10^3}$$

$$(2.6 \times 10^3 - 1 \times 10^3)(2 \times 2.6 \times 10^3 - 1 \times 10^3)$$

$$= 6.77 \text{ mm}$$

8. The downward deflection at a point halfway between the supports is

$$y_1 = \frac{1}{16}\frac{Pb}{EI}\left(L^2 - \frac{4}{3}b^2\right)$$

$$= \frac{1}{16} \frac{4\,800 \times 1 \times 10^3}{11 \times 10^3 \times 33.3 \times 10^6} \left[(2.6 \times 10^3)^2 - \frac{4}{3} (1 \times 10^3)^2 \right]$$

$$= 4.45 \text{ mm}$$

9. The deflection y_2 is greater than y_1, so y_2 is the maximum deflection. It is

$$\textit{max deflection} = 6.77 \text{ mm}$$

MORE CASES

You can see that Case 5 of Table 8.1, is a beam supported at the ends and subjected to a uniformly distributed load we have studied previously.

The last two cases in Table 8.1 are beams with uniformly distributed loads. You have had enough experience in calculating deflection to understand them. The main things to notice are that *W* (capital letter) is the total load and that all the dimensions are to be taken in inches or millimeters.

You should notice that in Cases 2, 3, 4, and 7 of Table 8.1, we gave you formulas for the deflection at a point halfway between the supports of the beam, and in each of these cases we told you this deflection was approximately equal to the maximum deflection, for any point between the supports. The difference between the formulas we gave you and the maximum deflection is never more than about 2%; this small difference is not important in practical problems. It means that you may calculate the deflection to be 0.51 in. when the maximum deflection is 0.52 in., but do not worry about the variance. The important part of the work is not to get such an answer as 0.112 in. or 2.38 in. when it ought to be 0.52 in. An error of no more than 2% is satisfactory for structural work, so you can use these formulas with confidence when you want to calculate the maximum deflection of a beam. The real gain from using the formulas in Table 8.1 will show up in the next section.

Practice Problems (Section 8-4). Practice calculating beam deflections by working the following problems.

1. A beam, W 12 × 45, is 24 ft long, the beam is supported at the left end and at a point 8 ft from the right end. The beam carries a uniformly distributed load of 1 500 lb/ft over the portion between the supports. Calculate the deflection of a point halfway between the supports and at the end of the overhang.
2. A 6 by 12 wood beam (actually 5½ in. × 11½ in.), 18 ft long, is supported at the right end and at a point 6 ft from the left end; the beam carries a concentrated load of 3 000 lb at the end of the overhang; $E = 1.2 \times 10^6$ psi. Calculate the deflection at the

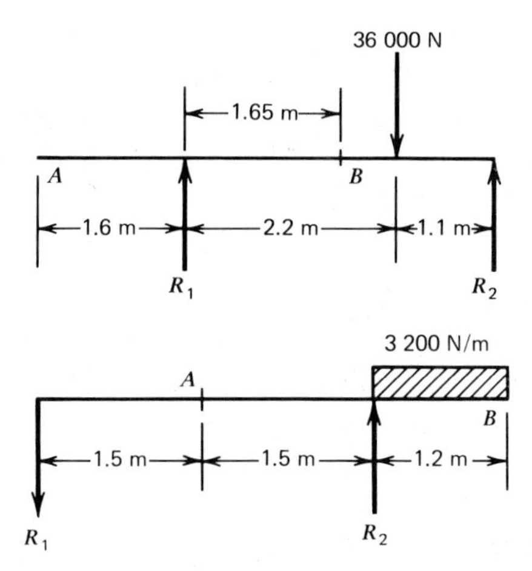

Fig. 8-9. Free-body diagram for Problem 3.

Fig. 8-10. Free-body diagram for Problem 4.

end of the overhang and at a point halfway between the supports.

3. A steel pipe 150 mm outer diameter and 120 mm inner diameter is subjected to the loading shown in Fig. 8-9. Calculate the deflections at A and B.

4. Figure 8-10 shows the loading for a 40 mm × 300 mm wood beam; $E = 10.5$ GPa. Find the deflections at A and B.

8-5. BEAM DEFLECTION BY SUPERPOSITION

In this section you will be shown a method of calculating an approximate value of the maximum deflection of a beam with several loads. This method is called *superposition*, and as we use it here, the word *superposition* means *putting together*.

Figure 8-11 shows a beam with three loads, P_1, P_2, and P_3. We can use this beam as an example in explaining the method of superposition. Suppose we want to calculate the maximum deflection of the beam. Now we have, in Case 2 of Table 8.1, a formula that gives the deflection at the center of a beam that carries a concentrated load at any point; this deflection at the center is approximately equal to the maximum deflection of the beam. We could use this formula to calculate the deflection that would occur at the center if P_1 were the only load on the beam.

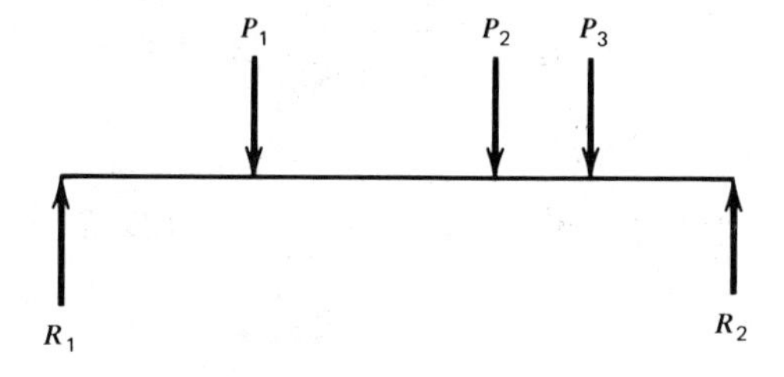

Fig. 8-11. Beam with three loads.

Next, we could calculate the deflection that would occur at the center if P_2 were the only load on the beam, and then we could calculate the deflection that would occur at the center if P_3 were the only load on the beam. This would give us the deflection at the center due to each of the loads acting separately and, if we added the deflections, the result would be the deflection at the center with all three of the loads applied to the beam.

Each of the separate deflections at the center that we calculate in this method of superposition is approximately equal to the maximum deflection due to the particular load. When we add the deflections, the result is so close to the maximum deflection of the beam, under the three loads, that the difference is not worth worrying about.

Illustrative Example 1. Figure 8-12*a* shows the loading for an aluminum alloy channel 12 [11.882. Calculate the maximum deflection.

Solution:

1. First, get the information that will be needed in every part of the problem.
   (a) $E = 10 \times 10^6$ psi
   (b) Look up I in column 7 of Table 6.2.

$$I = 239.69 \text{ in.}^4$$

   (c) The length L is 25 ft; change the length to inches:

$$L = 25 \times 12 = 300 \text{ in.}$$

2. For the 6 000-lb load draw the picture in Fig. 8-12*b*. Take the formula from Case 2 in Table 8.1.
   (a) The load P is 6 000 lb
   (b) The distance b is the distance from the load to the nearest support

$$b = 6 \times 12 = 72 \text{ in.}$$

   (c) The deflection at the center, due to the 6 000-lb load is

$$y = \frac{1}{16}\frac{Pb}{EI}\left(L^2 - \frac{4}{3}b^2\right)$$

$$= \frac{1}{16}\frac{6\,000 \times 72}{10 \times 10^6 \times 239.69}\left[(300)^2 - \frac{4}{3}(72)^2\right]$$

$$= 0.936 \text{ in.}$$

3. For the 4 000-lb load we can use the picture in Fig. 8-12*c*.
   (a) $P = 4\,000$ lb
   (b) $b = 12$ ft; $b = 12 \times 12 = 144$ in.

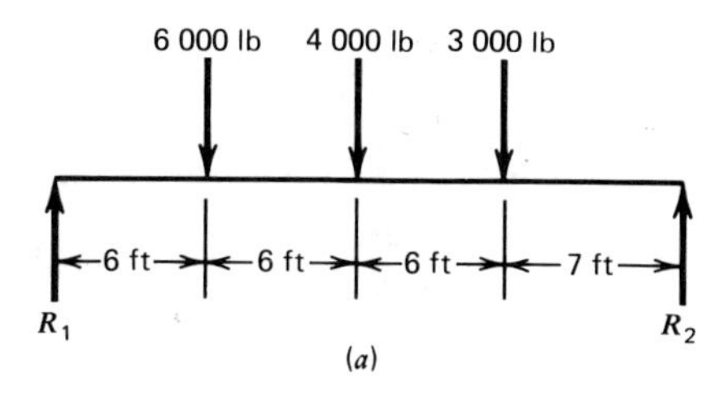

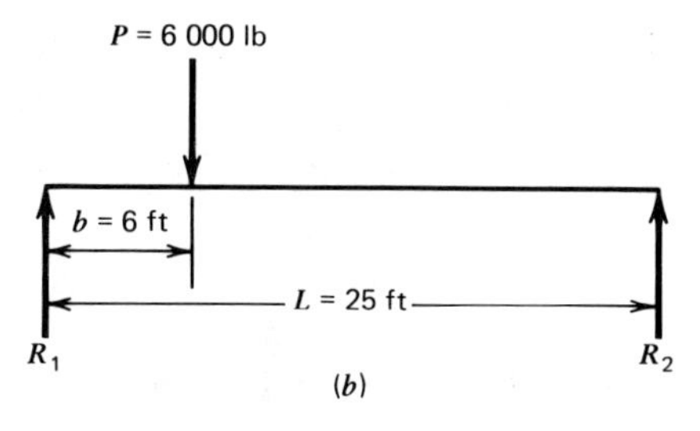

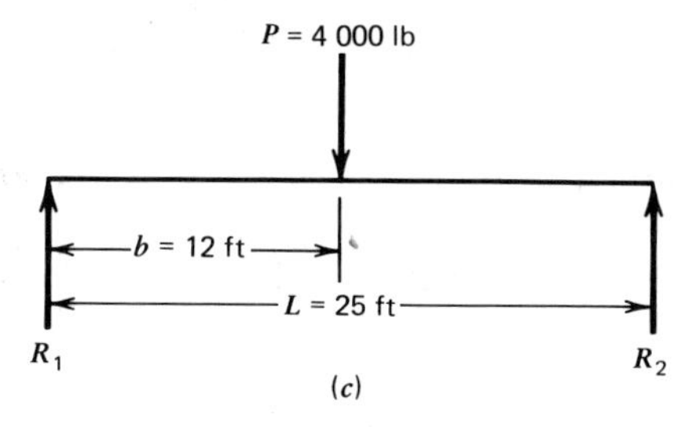

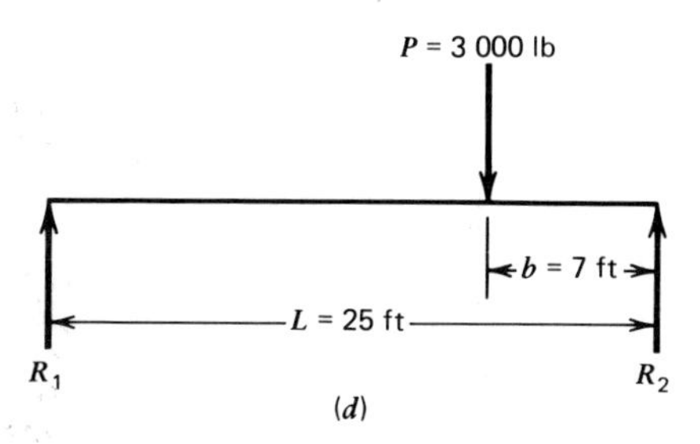

Fig. 8-12. Beam for Illustrative Example 1. (*a*) Free-body diagram of beam. (*b*) 6 000-lb load alone. (*c*) 4 000-lb load alone. (*d*) 3 000-lb load alone.

(c)
$$y = \frac{1}{16}\frac{Pb}{EI}\left(L^2 - \frac{4}{3}b^2\right)$$
$$= \frac{1}{16}\frac{4\ 000 \times 144}{10 \times 10^6 \times 239.69}\left[(300)^2 - \frac{4}{3}(144)^2\right]$$
$$= 0.936 \text{ in.}$$

4. For the 3 000-lb load, we work with the picture in Fig. 8-12*d*.
   (a) $P = 3\ 000$ lb

(b) $b = 7$ ft; $b = 7 \times 12 = 84$ in.

(c) $$y = \frac{1}{16}\frac{Pb}{EI}\left(L^2 - \frac{4}{3}b^2\right)$$

$$= \frac{1}{16}\frac{3\,000 \times 84}{10 \times 10^6 \times 239.69}\left[(300)^2 - \frac{4}{3}(84)^2\right]$$

$$= 0.530 \text{ in.}$$

5. The total deflection is the sum of the deflections due to the separate loads:

$$y = 0.936 + 0.936 + 0.530 = 2.402 \text{ in.}$$

MORE

Table 8.1 gives enough formulas for deflection so that you can calculate the deflection due to each part of the load separately. Then you can add the separate deflections to get the total deflection.

Illustrative Example 2. Figure 8-13*a* shows the free-body diagram, and Fig. 8-13*b* shows the cross section of a wood beam; $E = 11$ GPa. Calculate the maximum deflection for a point between the supports.

Fig. 8-13. Beam for Illustrative Example 2. (*a*) Free-body diagram of beam. (*b*) Cross section. (*c*) Uniform load alone. (*d*) Concentrated load alone.

Solution:

1. Obtain the general information first.

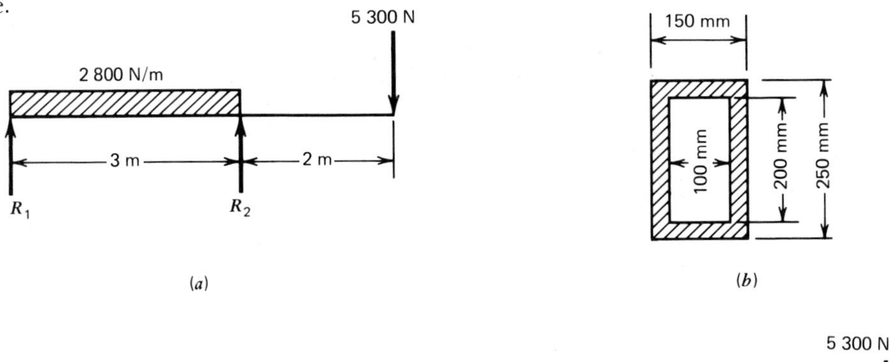

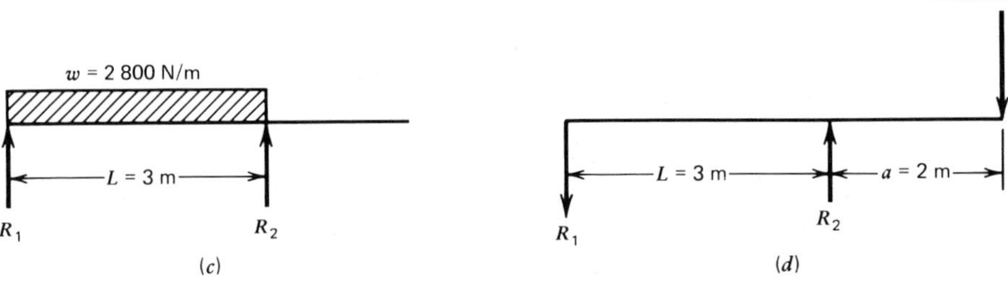

(a) $E = 11 \text{ GPa} = 11 \times 10^3 \text{ MPa}$

(b) $I = \dfrac{1}{12}(150)(250)^3 - \dfrac{1}{12}(100)(200)^3$

$= 195.3 \times 10^6 - 66.7 \times 10^6 = 128.6 \times 10^6 \text{ mm}^4$

(c) The length L between supports is 3 m

$$L = 3 \times 10^3 \text{ mm}$$

(d) The length of the overhang is 2 m

$$a = 2 \times 10^3 \text{ mm}$$

2. For the uniform load alone, we use the picture in Fig. 8-13*c*. This is Case 6 in Table 8.1.
   (a) The total load is

$$W = 2\,800 \times 3 = 8\,400 \text{ N}$$

   (b) The deflection at a point halfway between the supports is

$$y = \frac{5}{384}\frac{WL^3}{EI} = \frac{5}{384}\frac{8\,400(3 \times 10^3)^3}{11 \times 10^3 \times 128.6 \times 10^6}$$

$$= 2.088 \text{ mm}$$

   Notice that this deflection is downward.

3. For the concentrated load alone, look at Fig. 8-13*d*. This is Case 3 of Table 8.1.
   (a) $P = 5\,300$ N
   (b) The deflection at a point halfway between the supports is

$$y = \frac{1}{16}\frac{PaL^2}{EI} = \frac{1}{16}\frac{5\,300 \times 2 \times 10^3(3 \times 10^3)^2}{11 \times 10^3 \times 128.6 \times 10^6}$$

$$= 4.215 \text{ mm}$$

   Notice that this deflection is upward.

4. For the total deflection, we combine the two deflections we calculated separately. We have an upward deflection of 4.215 mm and a downward deflection of 2.088 mm, so we must subtract one from the other. The total deflection is

$$y = 4.215 - 2.088 = 2.127 \text{ mm}$$

   and it is upward. This is the deflection of a point halfway between the supports and, for practical purposes, it is the maximum deflection.

Practice Problems (Section 8-5). Work the following problems to learn how to figure deflections.

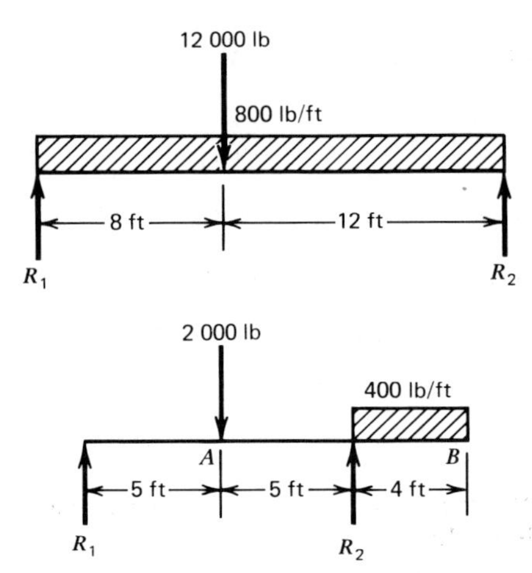

Fig. 8-14. Free-body diagram for Problem 1.

Fig. 8-15. Free-body diagram for Problems 2 and 3.

1. Figure 8-14 shows the loading for a steel beam, W 14 × 68. Calculate the maximum deflection.
2. A 4 by 10 (actually 3½ in. by 9¼ in.) wood beam carries the loads shown in Fig. 8-15; $E = 1.2 \times 10^6$ psi. Calculate the deflection at point A.
3. Figure 8-15 shows the loading on an aluminum-alloy channel 8 [5.789. Find the deflection at point B.
4. A hollow rectangular steel beam 140 mm wide and 240 mm deep on the outside has a wall thickness of 10 mm and carries the loads shown in Fig. 8-16. Find the maximum deflection.
5. Figure 8-17 shows the loading for a rectangular wood beam 68 mm wide and 300 mm deep; $E = 12$ GPa. Calculate the maximum deflection.

Fig. 8-16. Free-body diagram for Problem 4.

Fig. 8-17. Free-body diagram for Problem 5.

SUMMARY The main points of this chapter are summarized here.

1. The line that passes through the centroid of each cross section of a beam is called the *elastic curve of the beam.* The elastic curve is straight before the loads are applied to the beam and is bent after the loads are applied.
2. The deflection of any point in a beam is the distance that this point moves as the beam bends.
3. Table 8.1 gives formulas for deflections for several types of beam loading.
   (a) The lengths L, a, and b in these formulas are to be taken in inches or millimeters.
   (b) The load W is the total of the uniformly distributed load.
   (c) The deflection y is in inches or millimeters.
   (d) A deflection at the center of a beam that is supported at two points is approximately equal to the maximum deflection between the supports.
4. When several loads are applied to a beam, the deflection can be calculated separately for each load. The separate deflections can be added to give the total deflection. This method is called *superposition.*

REVIEW QUESTIONS These questions may trip you if you have not studied this chapter thoroughly.

1. What is the *elastic curve of a beam*?
2. What is meant by the *deflection of a beam*?
3. What property of the material of which a beam is made affects the deflection? How?
4. What property of the beam cross section affects the deflection?
5. How is the method of superposition used to calculate the maximum deflection of a beam?

REVIEW PROBLEMS Complete the chapter by solving the following problems.

1. A 48 mm by 96 mm wood beam, 2.2 m long, is supported at the ends and carries a uniformly distributed load of 1 400 N/m; $E = 9.6$ GPa. What is the maximum deflection?
2. A steel beam, W 10 × 22, 18 ft long, is supported at the ends and carries a concentrated load of 4 700 lb at the center. Find the maximum deflection.

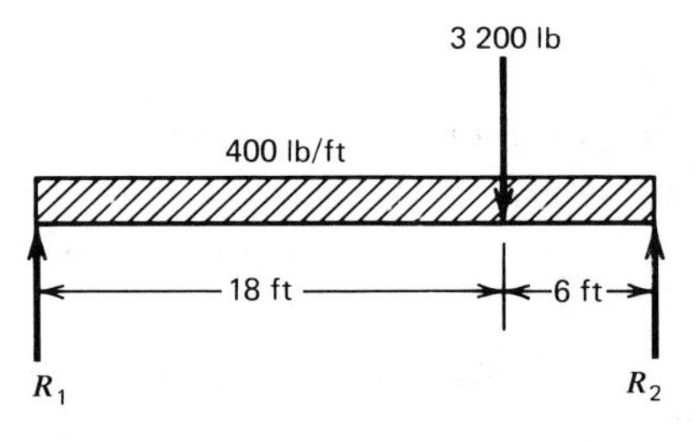

Fig. 8-18. Free-body diagram for Review Problem 3.

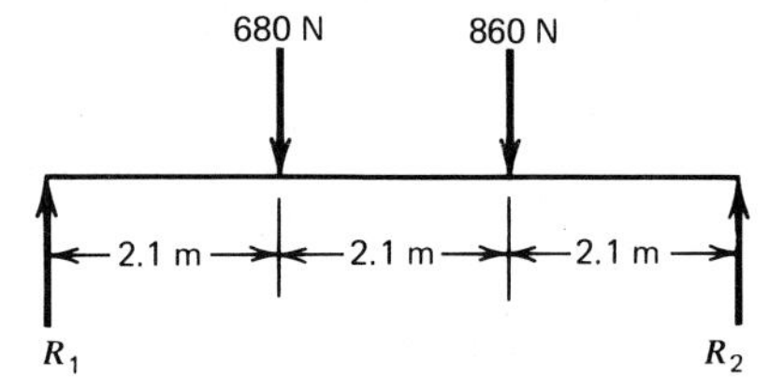

Fig. 8-19. Free-body diagram for Review Problem 4.

3. Figure 8-18 shows the loading for an aluminum-alloy channel 12 [11.822. Calculate the maximum deflection.
4. What is the maximum deflection for a circular steel rod 44 mm in diameter that carries the loads shown in Fig. 8-19?

9 STATICALLY INDETERMINATE BEAMS

PURPOSE OF THIS CHAPTER. The purpose of this chapter is to study a type of beam that is a little different from the beams we have studied so far. This type of beam is called a *statically indeterminate beam.*

In all of the beams studied in previous chapters we have been able to calculate the reactions by writing moment equations, because there were only two reactions. However, when there are more than two reactions on the beam, you cannot find the reactions by writing moment equations. Such a beam is called a *statically indeterminate beam.*

Moment equations are equations of equilibrium (remember from Chapter 2), and the equations of equilibrium are also called the *equations of statics* (because a body at rest is static). If the reactions on a beam can be calculated by using the equations of statics (as we have been doing), the beam is said to be *statically determinate*; that is, the reactions can be determined by means of the equations of statics. However, if the reactions on a beam cannot be calculated by means of the equations of statics, the beam is said to be statically indeterminate; that is, the reactions cannot be determined by means of the equations of statics.

In this chapter you will be shown how to recognize a statically indeterminate beam. Then you will be given diagrams and formulas from which you can calculate the reactions, shearing forces, and bending moments on the types of beams you are likely to encounter in engineering work. You will be able to calculate the stresses if you know the cross section of the beam, and you will be able to design a beam for any ordinary loading.

There are several theorems and basic methods for analyzing a statically indeterminate beam. However, it would take too much time and space to explain and teach you how to use these methods. Instead, you will be given the formulas and diagrams without deriving or proving them. You should accept the formulas as correct and learn to use them.

After the reactions have been found, a statically indeterminate beam is just like any other beam. You can calculate shearing forces and bending moments and use the flexure formula to calculate a bending stress or to design a beam.

9-1. TYPES OF SUPPORTS AND REACTIONS

The first step in the study of indeterminate beams is to learn something about the different types of supports and the sort of reactions these supports exert on beams. Figure 9-1*a* shows a beam supported by a pivot at the left end and a wall at the right end. Supports of these types only exert forces on the beam; Fig. 9-1*b* shows the forces in the free-body diagram of the beam.

Figure 9-2*a* shows a steel beam supported by riveted connections at the ends. The beam is riveted to short lengths of steel angles; in turn, the angles are riveted to steel columns. There may be some room for argument about the reactions exerted by such riveted connections, but it is usually assumed that the only reaction is a force such as you see in the free-body diagram in Fig. 9-2*b*.

Beams that have supports of the types shown in Figs. 9-1 and 9-2 are called *simple beams*.

Figure 9-3*a* shows a beam with built-in ends. A beam of this type is called a *fixed beam*. The distinguishing feature of a fixed beam is that a

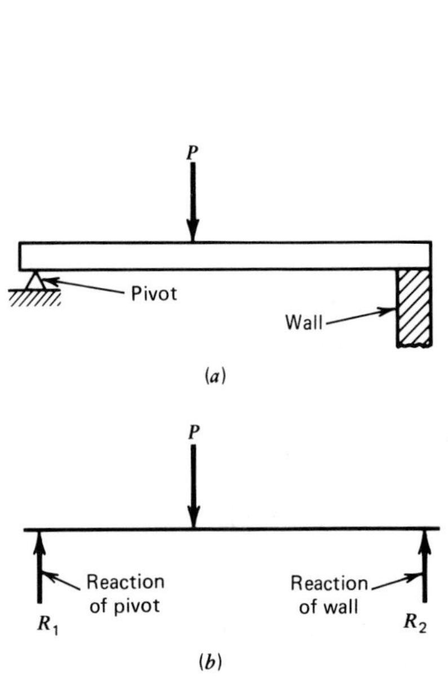

Fig. 9-1. Reactions of pivot and wall. (*a*) Beam. (*b*) Free-body diagram.

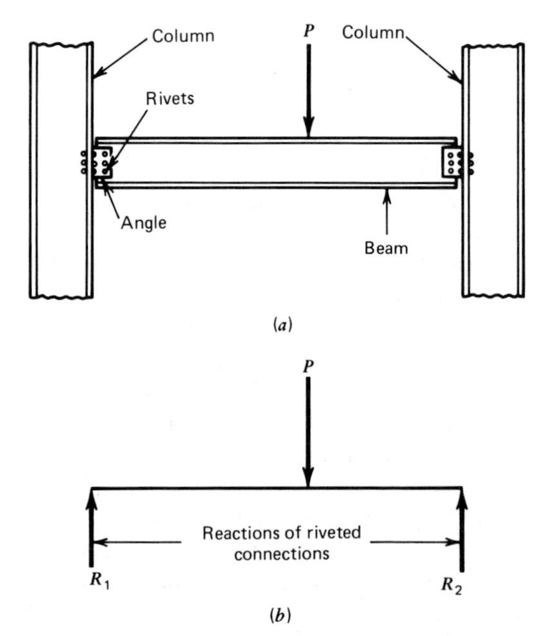

Fig. 9-2. Reactions of riveted connections on beam. (*a*) Beam. (*b*) Free-body diagram.

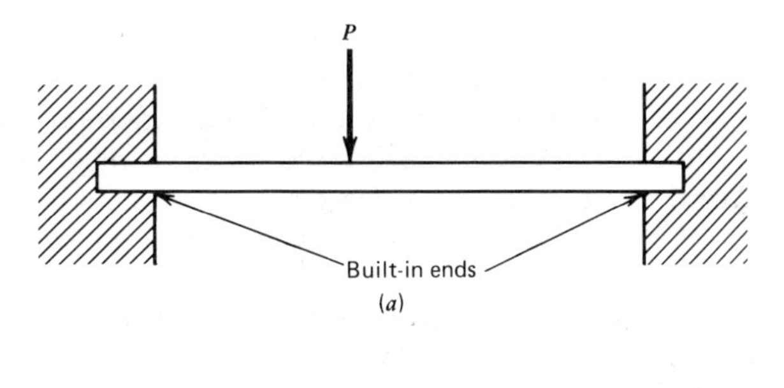

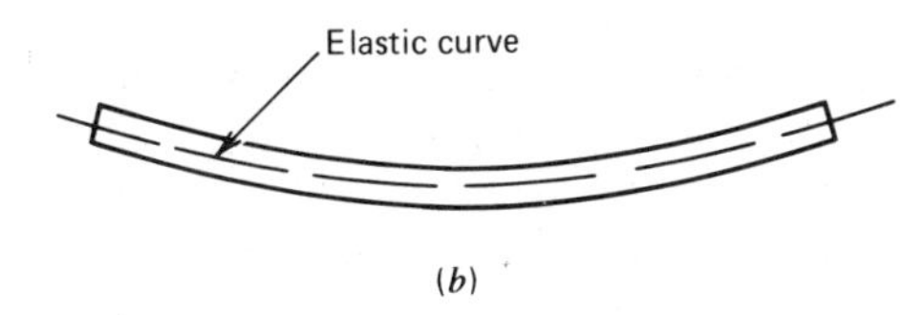

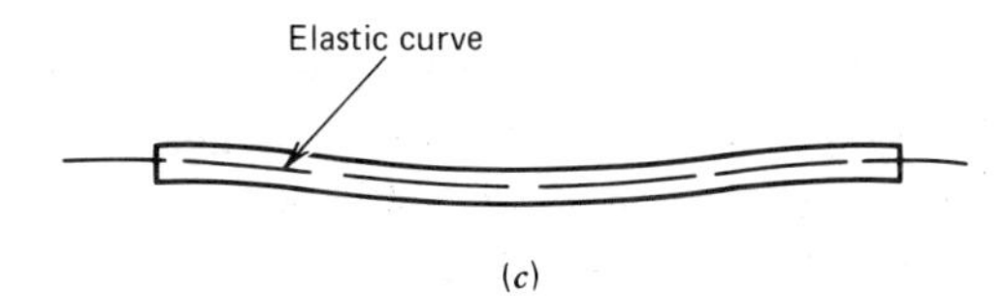

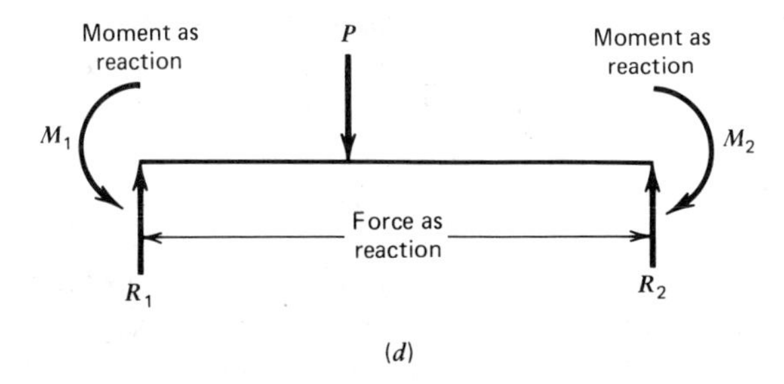

Fig. 9-3. Reactions on fixed beam. (*a*) Beam. (*b*) Bent shape of simple beam. (*c*) Bent shape of fixed beam. (*d*) Free-body diagram.

support not only holds up the beam, but, also, prevents the beam from rotating or turning at the support. To illustrate, Fig. 9-3*b* shows the bent shape of a simple beam; you have seen this before in Chapter 8, on deflection. Notice that the ends of the beam actually do turn as the beam bends. Figure 9-3*c* shows the bent shape of a fixed beam; the support of a fixed beam prevents the end of the beam from turning.

A support at a built-in end has a lot to do. Not only must the support hold up the beam, but also the support must keep the beam from turning. The support must exert a moment on the beam to keep the end of the beam from turning. Figure 9-3*d* shows the free-body diagram of the fixed beam of Fig. 9-3*a*. Here you see the forces R_1 and R_2 exerted by the supports and the moments M_1 and M_2 exerted by the supports. Notice how these moments are represented. This method is not in use everywhere, but it is convenient. A force tends to move a body in a straight line, and so we represent a force by a straight arrow; a moment tends to turn a body, and we represent a moment by a curved (turning) arrow.

We count the moments M_1 and M_2 in Fig. 9-3*d* as reactions along with the forces R_1 and R_2. This make four reactions on the beam, so the beam is statically indeterminate.

The reactions on a beam consist of a force at each point at which the beam is supported and a moment at each built-in end. If the total number of reactions is more than two, the beam is statically indeterminate. You could write as many moment equations as your patience would allow, but you could not find the reactions that way.

9-2. FIXED BEAMS

In this section you will study several examples of fixed beams; values of the reactions, shearing forces, and bending moments will be given. Figure 9-4*a* shows a fixed beam (a beam with built-in ends) that carries a uniformly distributed load w; the length of the beam is L. The total load on the beam is W:

$$W = wL$$

Figure 9-4*b* shows the free-body diagram of the fixed beam with the uniform load. Here you see the forces R_1 and R_2 at the ends of the beam and the moments M_1 and M_2 at the ends. The values are

$$R_1 = \frac{W}{2} \qquad R_2 = \frac{W}{2} \qquad M_1 = \frac{WL}{12} \qquad M_2 = \frac{WL}{12}$$

If you are using gravitational units in these formulas, W is the total load on the beam and is expressed in pounds and R_1 and R_2 are also in pounds. If you express the length L in feet, the moments M_1 and M_2 are in pound feet. If you are using metric units in these formulas, W is in newtons, as are R_1 and R_2. If you express the length L in meters, then M_1 and M_2 are in newton meters.

Figure 9-4*c* shows the shear diagram of the beam. (The shearing force at any point in a beam is a force that tends to shear the beam in two at that point; the shearing force is equal to the sum of the forces to the left of the point. The shear diagram shows the value of the shearing force at each point along the beam.)

Notice in Fig. 9-4*c* that the maximum value of the shearing force is $W/2$ and the shearing force is *zero* at point B. Point B is at the center of the beam.

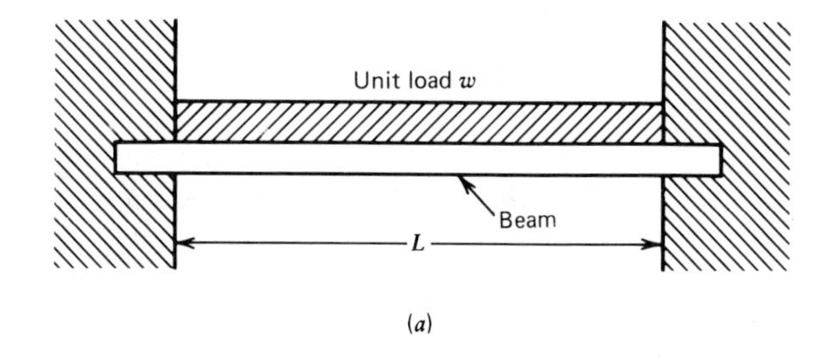

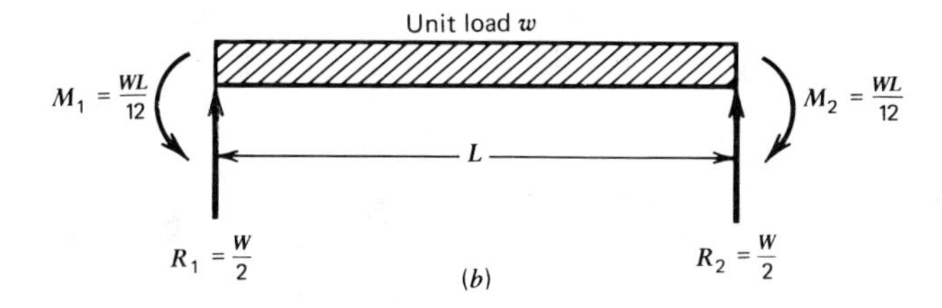

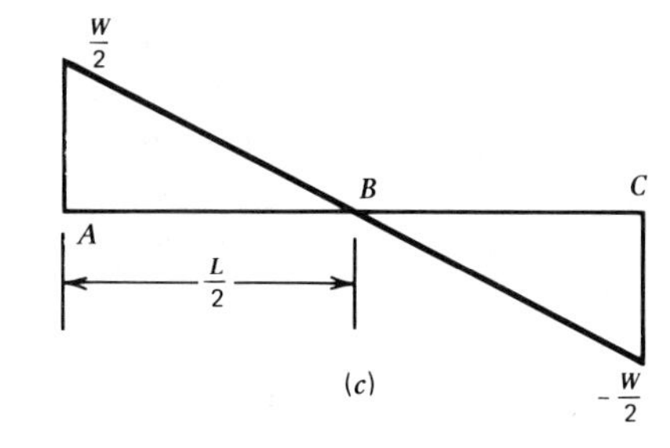

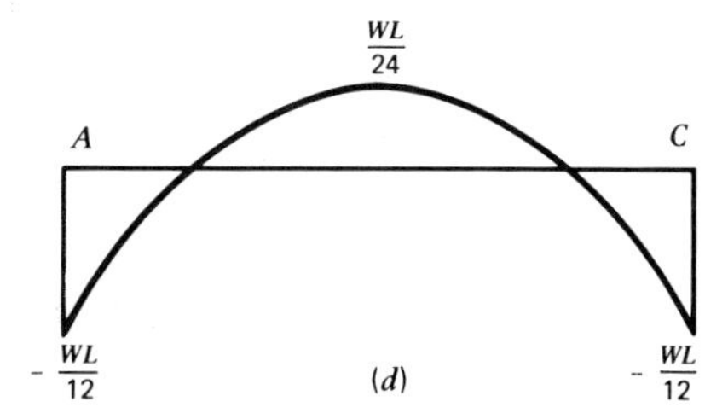

Fig. 9-4. Fixed beam with uniform load. (*a*) Beam. (*b*) Free-body diagram of beam. (*c*) Shear diagram. (*d*) Moment diagram.

Figure 9-4*d* shows the moment diagram of the beam. (The bending moment tends to bend the beam; the bending moment at any point is equal to the moment, with respect to that point, of all of the forces to the left of the point. The moment diagram shows the value of the bending moment at each point along the beam.) You can see in Fig. 9-4*d* that the bending moment at each end of the beam is

$$-\frac{WL}{12}$$

and the bending moment at the center is

$$\frac{WL}{24}$$

The maximum numerical value of the bending moment occurs at the end and is

$$\frac{WL}{12}$$

This is the value of most interest to you, as it causes the maximum bending stress; pay no attention to the negative sign of the bending moment at the ends.

The bending stress in the beam is given by the flexure formula. You probably remember this formula,

$$S = \frac{Mc}{I}$$

where M is the bending moment in pound inches, c is the distance from the neutral axis to the farthest point in the cross section, and I is the moment of inertia of the cross section about the neutral axis. You can find the section modulus $Z = I/c$ of a standard section in the tables in Chapter 6.

Illustrative Example 1. A steel beam W 12 × 30 is 20 ft long and is built in at the ends. The beam carries a uniformly distributed load of 1 500 lb/ft. Calculate the maximum bending stress.

Solution:

1. The total load is

$$W = 1\,500 \times 20 = 30\,000 \text{ lb}$$

2. The length L is 20 ft.
3. The maximum bending moment is (see Fig. 9-4*d*)

$$M = \frac{WL}{12} = \frac{30\,000 \times 20}{12} = 50\,000 \text{ lb ft}$$

This must be changed to lb in.

$$M = 50\,000 \times 12 = 600\,000 \text{ lb in.}$$

4. The section modulus is obtained from column 8 of Table 6.1.

$$\frac{I}{c} = 38.6 \text{ in.}^3$$

5. The maximum bending stress is

$$S = \frac{Mc}{I} = \frac{600\ 000}{38.6} = 15\ 500 \text{ psi}$$

CONCENTRATED LOAD AT CENTER

Figure 9-5*a* shows a fixed beam that carries a concentrated load at the center. The load is represented by P and the length by L. Figure 9-5*b* shows the free-body diagram of the beam. Here you see that the reactions consist of the forces R_1 and R_2 and the moments M_1 and M_2.

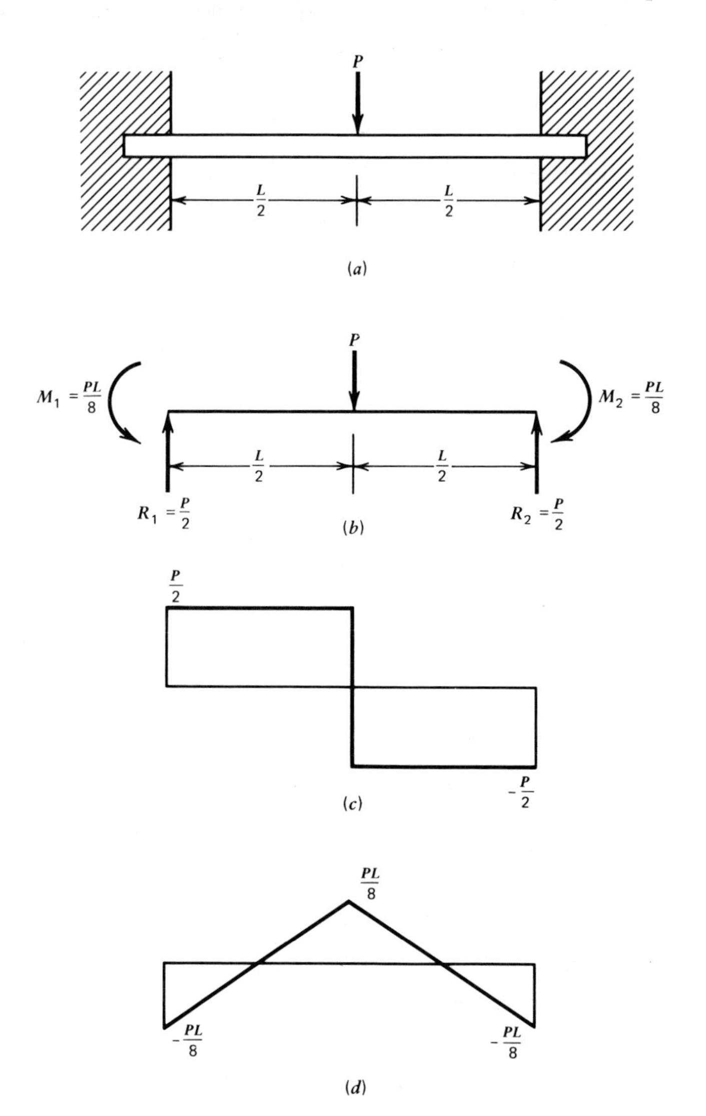

Fig. 9-5. Fixed beam with concentrated load at center. (*a*) Beam. (*b*) Free-body diagram of beam. (*c*) Shear diagram. (*d*) Moment diagram.

Figures 9-5*c* and 9-5*d* show the shear and moment diagrams for the fixed beam with a concentrated load at the center. The bending moment is

$$\frac{PL}{8}$$

at the center and

$$-\frac{PL}{8}$$

at each end. The numerical values of these bending moments are the same, so there is no doubt that the maximum bending moment is

$$\frac{PL}{8}$$

We can write the flexure formula as

$$S = \frac{Mc}{I}$$

if we want to calculate a bending stress, or we can write the formula as

$$Z = \frac{I}{c} = \frac{M}{S}$$

if we want to design a beam. Here S is the *working stress.*

Illustrative Example 2. A fixed beam 16 ft long is to carry a concentrated load of 8 700 lb at the center. Select a suitable steel beam, using a working stress of 18 000 psi.

Solution:

1. The load P is 8 700 lb.
2. The length L is 16 ft.
3. The maximum bending moment is (see Fig. 9-5*d*)

$$M = \frac{PL}{8} = \frac{8\,700 \times 16}{8} = 17\,400 \text{ lb ft}$$

$$M = 17\,400 \times 12 = 208\,800 \text{ lb in.}$$

4. The working stress S is 18 000 psi.
5. The section modulus required is

$$Z = \frac{M}{S} = \frac{208\,800}{18\,000} = 11.6 \text{ in.}^3$$

6. Look in column 8 of Table 6-1 to pick the beam. Read up column 8 until you see 13.4. This is as near to 11.6 as we can find in the table. Then look to the left to see that the depth of the beam is 6 in. and the weight is 20 lb/ft. Then a suitable beam is W 6 × 20. We could also use W 8 × 18 as a beam for this loading; this beam has a section modulus of 15.2 in.^{3}, so it is strong enough. Also, it is a bit lighter in weight than W 6 × 20.

CONCENTRATED LOAD AT ANY POINT

Figure 9-6*a* shows a fixed beam of length L with a concentrated load P at a distance a from the left end. Here a can be anything less than L, so we can take this as a fixed beam with a concentrated load at any point. Figure 9-6*b* shows the free-body diagram of this beam, and you can see what the reactions are. Figures 9-6*c* and 9-6*d* show the shear and moment diagrams. The maximum bending moment occurs at the support that is closest to the load. If the load is closer to the left support than the right, the maximum bending moment is

$$\frac{Pab^2}{L^2}$$

but if the load is closer to the right support than the left, the maximum bending moment is

$$\frac{Pa^2b}{L^2}$$

If you are working in gravitational units, it is best to take the lengths a, b, and L in feet to calculate one of these bending moments. This gives the bending moment in pound feet, and you can multiply by 12 to convert it to pound inches.

If you are using metric units, you can take the lengths in meters and calculate the bending moment in newton meters. Then you can convert it to newton millimeters by multiplying by 10^3.

Illustrative Example 3. A 100 mm by 250 mm rectangular wood beam is 4.8 m long and fixed at the ends. The beam carries a concentrated load of 9 600 N at a distance of 1.2 m from the left end. Calculate the maximum bending stress.

Solution:

Figure 9-7 shows the beam and load.

1. The load P is 9 600 N.
2. The length a is 1.2 m.
3. The length b is 3.6 m.

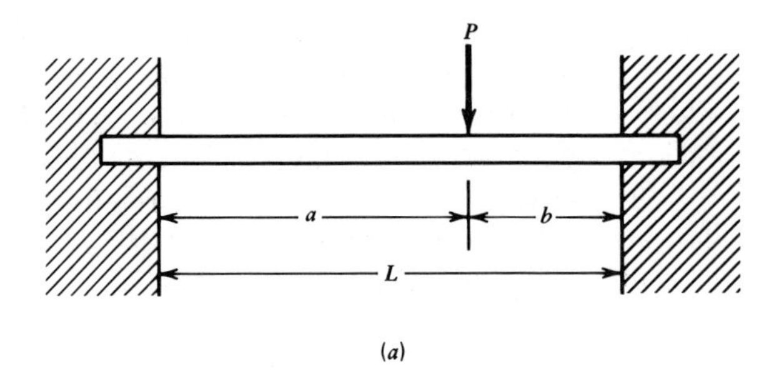

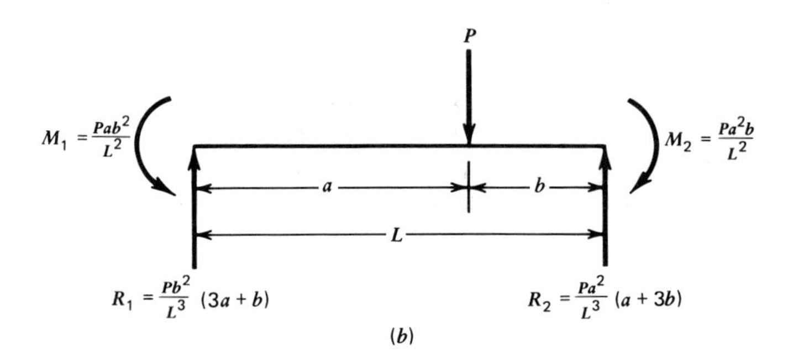

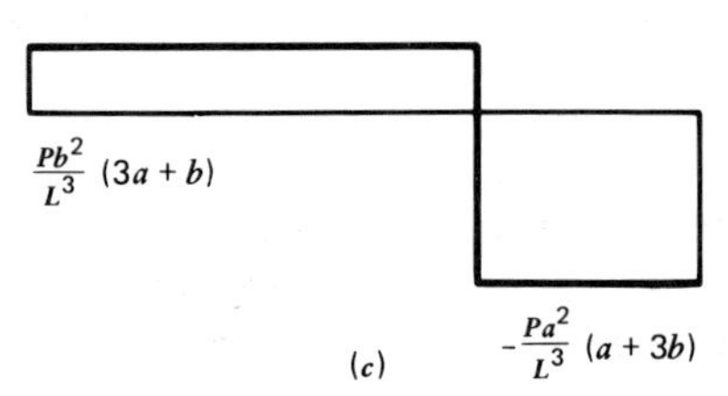

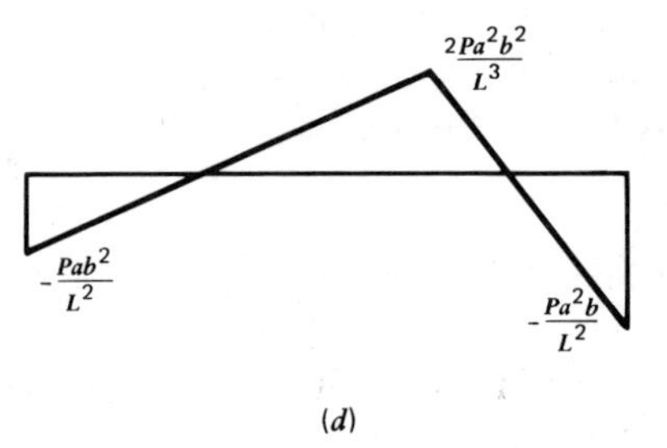

Fig. 9-6. Fixed beam with concentrated load at any point. (*a*) Beam. (*b*) Free-body diagram of beam. (*c*) Shear diagram. (*d*) Moment diagram.

4. The length L is 4.8 m.
5. The load is closer to the left support than the right, so the maximum bending moment occurs at the left end

$$M = \frac{Pab^2}{L^2} = \frac{9\,600 \times 1.2(3.6)^2}{(4.8)^2} = 6\,480 \text{ N·m}$$

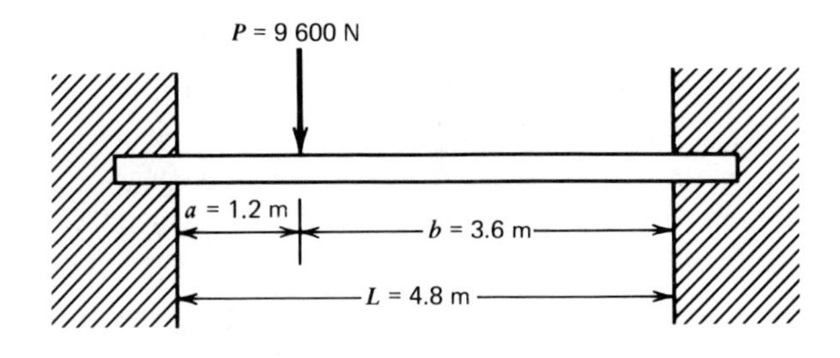

Fig. 9-7. Beam for Illustrative Example 3.

$$M = 6\,480 \times 10^3 \text{ N·mm}$$

6. The moment of inertia is

$$I = \frac{1}{12} \times 100(250)^3 = 130.2 \times 10^6 \text{ mm}^4$$

7. The distance c is 125 mm.
8. The maximum bending stress is

$$S = \frac{Mc}{I} = \frac{6\,480 \times 10^3 \times 125}{130.2 \times 10^6} = 6.22 \text{ MPa}$$

CONSTANT CROSS SECTION

Here is a word of caution in using the formulas given in this section. The formulas apply only to beams in which the cross section is the same from one end to the other. However, most beams in structures do have the same cross section from one end to the other, so the formulas should suit your needs.

Practice Problems (Section 9-2). These problems will help you remember fixed beams.

1. A steel beam, W 12 × 45 is 16 ft long and is fixed at the ends. The beam carries a uniformly distributed load of 1 600 lb/ft. Calculate the maximum bending stress.
2. A 40 mm by 240 mm rectangular wood beam 4.75 m long and fixed at the ends is subjected to a concentrated load of 5 100 N at the center. What is the maximum bending stress?
3. A fixed beam 20 ft long carries a concentrated load of 10 300 lb at a distance of 14 ft from the left end. Select a suitable steel beam, using a working stress of 20 000 psi.
4. A fixed beam 3.3 m long carries a uniformly distributed load of 840 N/m. Find the required depth of a rectangular wood beam 48 mm wide using a working stress of 8 MPa.

5. A fixed beam 18 ft long is to carry a uniformly distributed load of 900 lb/ft. Select a suitable aluminum-alloy channel, using a working stress of 18 000 psi.
6. A fixed beam 14 ft long is to carry a concentrated load of 4 900 lb at the center. Select a suitable steel beam, using a working stress of 16 000 psi.
7. A beam, W 14 × 30, is 21 ft long and is a fixed beam. It is subjected to a concentrated force of 15 800 lb at a distance of 9 ft from the left end. Calculate the maximum bending stress.

9-3. CONTINUOUS BEAMS

Continuous beams make up an important class of statically indeterminate beams; they are important to the structural engineer because continuous beams are used so commonly in building and bridge structures.

A *continuous beam* is a beam that continues over more than two supports. For example, Fig. 9-8*a* shows a beam that is continuous over three supports and carries a uniformly distributed load. The part of the beam between two adjacent supports is called a *span*; thus the beam in Fig. 9-8*a* extends over two spans.

A beam has a reaction at each support, and supports of the type shown in Fig. 9-8*a* only exert forces. Nevertheless, there are three reactions on this beam, and so the beam is statically indeterminate. (Any beam that has more than two reactions is statically indeterminate; the reactions cannot be calculated by means of moment equations.)

Formulas for the values of the reactions, for the beam in Fig. 9-8, cannot be worked out without taking the time and space to develop special methods, so the values will be given without proof. The reactions are shown in Fig. 9-8*b*.

You could draw the shear and moment diagrams for yourself, after you know the reactions; that is, you could if you applied what you learned in Chapter 6. However, the diagrams will be given to save your time. Figure 9-8*c* shows the shear diagram and Fig. 9-8*d* shows the moment diagram.

In Fig. 9-8*c* W is the total load on one span, and L is the length of a span; w is the unit load, so

$$W = wL$$

The bending moment has its maximum value at the center support of this continuous beam.

$$\text{Max. } M = -\frac{WL}{8}$$

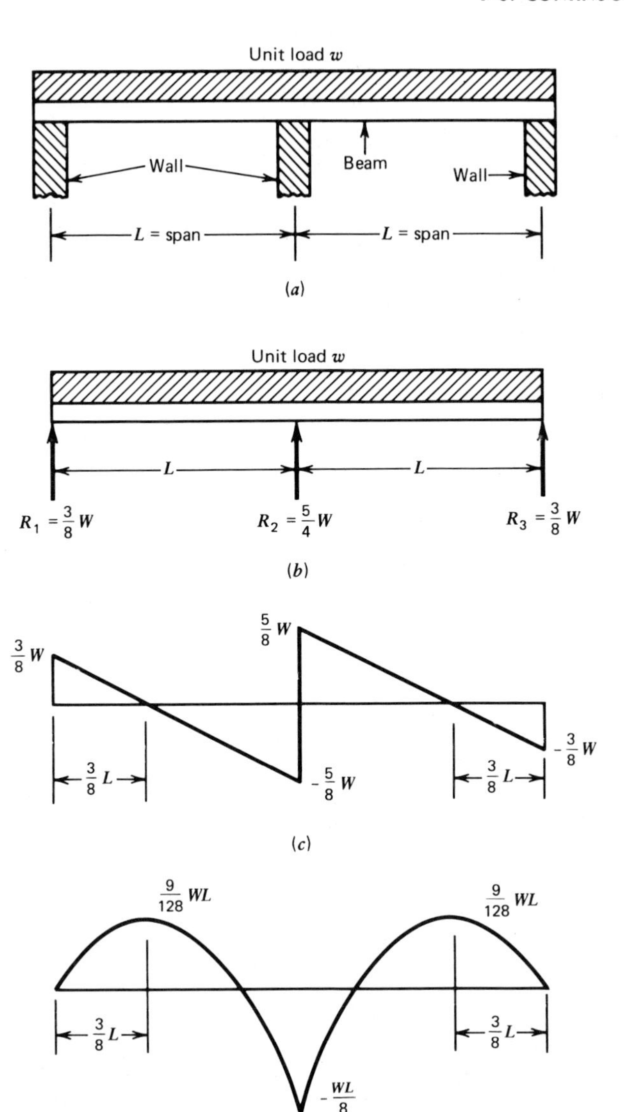

Fig. 9-8. Continuous beam of two spans. (*a*) Beam. (*b*) Free-body diagram of beam. (*c*) Shear diagram. (*d*) Moment diagram.

This figure is probably the most important one for the beam and is what you would use in the flexure formula if you wrote it as

$$S = \frac{Mc}{I}$$

to calculate a stress or as

$$Z = \frac{I}{c} = \frac{M}{S}$$

to design a beam.

Illustrative Example 1. A steel beam, W 8 × 18, is used as a two-span continuous beam. Each span is 15 ft long, and the beam carries a uniformly distributed load of 600 lb/ft. Calculate the maximum bending stress.

Solution:

1. The length L is 15 ft.
2. The total load on one span is

$$W = wL = 600 \times 15 = 9\,000 \text{ lb}$$

3. The maximum bending moment is (see Fig. 9-8*d*)

$$M = \frac{WL}{8} = \frac{9\,000 \times 15}{8} = 16\,900 \text{ lb ft}$$

$$M = 16\,900 \times 12 = 203\,000 \text{ lb in.}$$

4. The section modulus is (from column 8 of Table 6.1)

$$Z = \frac{I}{c} = 15.2 \text{ in.}^3$$

5. The maximum bending stress is

$$S = \frac{Mc}{I} = \frac{203\,000}{15.2} = 13\,400 \text{ psi}$$

MORE SPANS

You will now be given a considerable amount of information about continuous beams, and you will be expected to study this information until you understand it thoroughly. You should be able to understand the problems, because you have already worked with reactions and moment diagrams. Figures 9-9, 9-10, and 9-11 show you the free-body diagrams and moment diagrams for three different continuous beams. In each case the length of each span is L, the beam is subjected to a uniformly distributed load of w, and the total load in each span is W. The maximum bending moment for each beam is also shown. You should now be able to obtain the necessary information from these figures and use it.

Illustrative Example 2. A steel beam 85 ft long is continuous over five

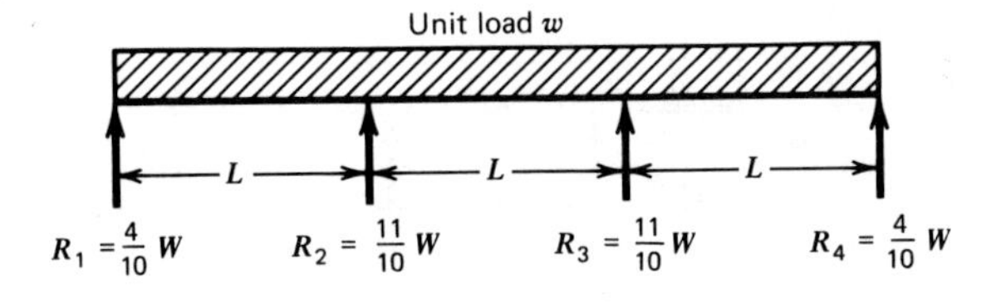

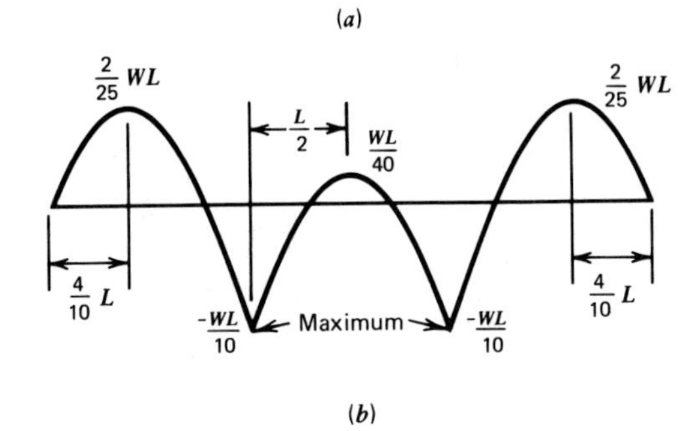

Fig. 9-9. Continuous beam of three spans. (*a*) Free-body diagram of beam. (*b*) Moment diagram.

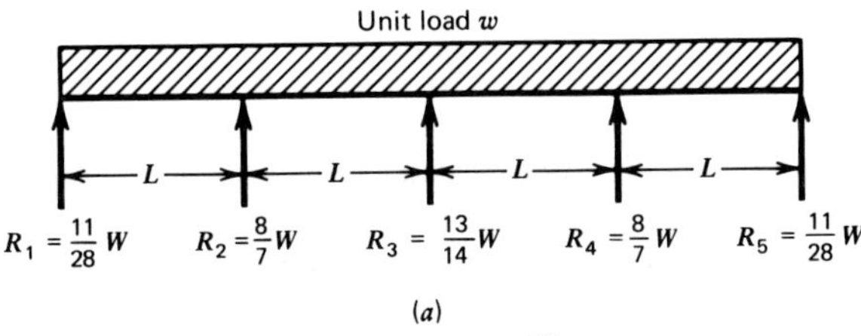

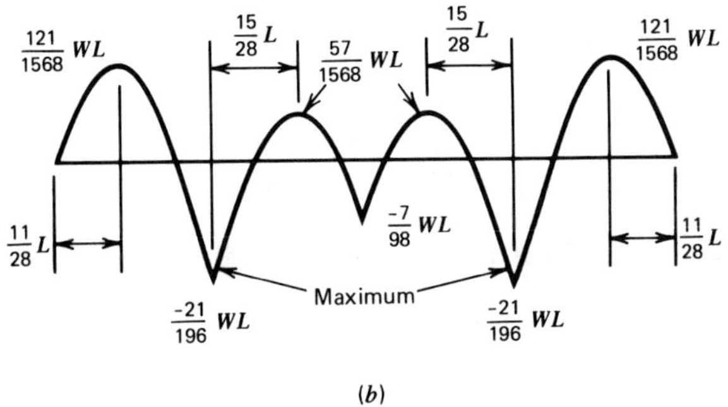

Fig. 9-10. Continuous beam of four spans. (*a*) Free-body diagram of beam. (*b*) Moment diagram.

spans of 17 ft each. The beam carries a uniformly distributed load of 450 lb/ft. Calculate the maximum bending moment.

Solution:

See Fig. 9-11*b*

1. The length of each span is $L = 17$ ft.
2. The total load in each span is

$$W = wL = 450 \times 17 = 7\,650 \text{ lb}$$

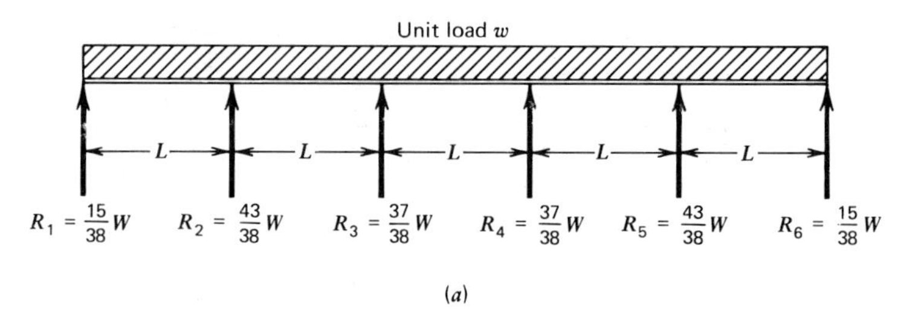

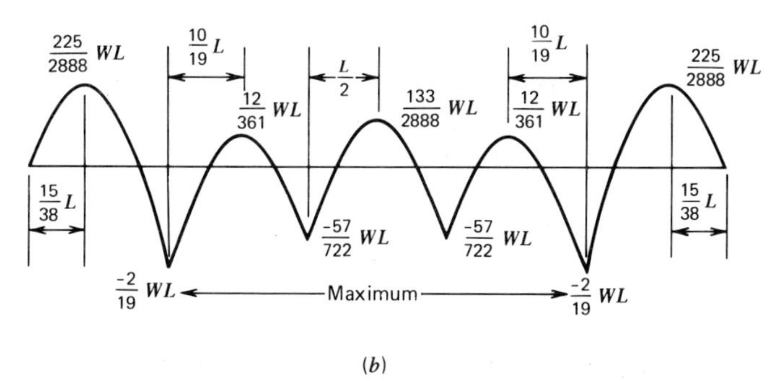

Fig. 9-11. Continuous beam of five spans. (*a*) Free-body diagram of beam. (*b*) Moment diagram.

3. The maximum bending moment is

$$M = \frac{2}{19}WL = \frac{2}{19} \times 7\,650 \times 17 = 13\,700 \text{ lb ft}$$

WARNING

These formulas and moment diagrams are only correct for beams that have the same cross section from one end to the other. They are not right for beams that do not have the same cross section at all points. However, there is not much reason for you to design beams of varying cross section, so these formulas and moment diagrams should serve your purpose.

Practice Problems (Section 9-3). Solving the following problems will help you to learn more about continuous beams.

1. A steel beam W 10 × 22 is 80 ft long and is used as a continuous beam of four equal spans; the beam carries a uniformly distributed load of 1 200 lb/ft. Calculate the maximum bending stress.
2. Select a suitable rectangular wood beam to be used as a continuous beam over three spans of 3.8 m each; the beam is to carry a uniformly distributed load of 4 600 N/m, and the working stress is 11 MPa.
3. Pick a suitable steel beam to carry a uniformly distributed load

of 1 600 lb/ft over five equal spans of 19 ft each. The working stress is 18 000 psi.

4. Select a suitable steel beam to carry a uniformly distributed load of 1 760 lb/ft over two equal spans of 20 ft each. The working stress is 20 000 psi.
5. A 48 mm by 288 mm rectangular wood beam 16.8 m long is used as a continuous beam over four equal spans of 4.2 m each. The beam carries a uniformly distributed load of 2 800 N/m. Calculate the maximum bending stress.

SUMMARY Here are the main points for you to remember about statically indeterminate beams.

1. A reaction is a force or moment exerted at a point at which a beam is supported.
   (a) Any support that holds up a beam exerts a force on the beam.
   (b) A built-in or fixed support exerts a moment on a beam in addition to a force.
   (c) The total number of reactions on a beam is equal to the sum of the number of forces and the number of moments exerted on the beam by the supports.
2. If the total number of reactions (including both forces and moments) is more than two, the beam is statically indeterminate; the reactions cannot be found by writing moment equations.
   (a) If the reactions are known, the bending moments can be calculated and the flexure formula can be applied.
   (b) You have been given the reactions and moment diagrams for a number of statically indeterminate beams. You should be able now to use the reactions and moment diagrams.
3. The maximum bending moment in a statically indeterminate beam is usually negative, but you can disregard the negative sign for steel and wood beams. The important fact you should remember about the bending moment is its *numerical value*.

REVIEW QUESTIONS See if you can answer the following questions.

1. What is a *reaction*?
2. What is a *fixed end of a beam*?
3. What kinds of reactions are exerted on beams?
4. What kind of reaction does a riveted connection exert on a beam?
5. What kind of reaction is exerted at a built-in support?

6. What is a *statically indeterminate beam*?
7. How can you use the flexure formula for an indeterminate beam?
8. Does the maximum bending moment on a statically indeterminate beam usually occur at a support or at a point between supports?
9. State the meaning of W and L in the formula

$$M = \frac{WL}{10}$$

for the bending moment in a statically indeterminate beam.

REVIEW PROBLEMS As a test of what you have learned in this chapter, solve the following problems.

1. Select a suitable steel beam for the loading shown in Fig. 9-12. The working stress is 20 000 psi.
2. A 150 mm by 300 mm rectangular wood beam 5.2 m long is fixed at the ends and carries a uniformly distributed load of 5 200 N/m. Find the maximum bending stress.
3. Figure 9-13 shows the loading for a steel beam W 12 × 45. Calculate the maximum bending stress.
4. A fixed beam 20 ft long carries a concentrated load of 6 200 lb at the center. Select a suitable steel beam, using a working stress of 18 000 psi.
5. Select a suitable rectangular wood beam for the loading shown in Fig. 9-14. Use a working stress of 12 MPa.
6. An aluminum alloy channel, 10 [8.360 76 ft long is used as a continuous beam over four equal spans. The beam carries a uniformly distributed load of 840 lb/ft. What is the maximum bending stress?

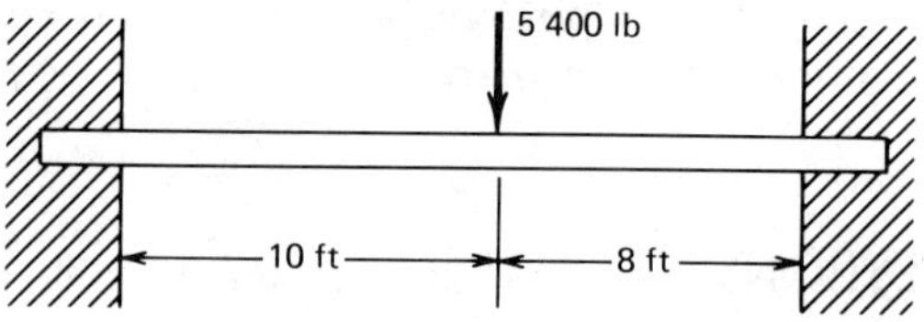

Fig. 9-12. Beam for Problem 1.

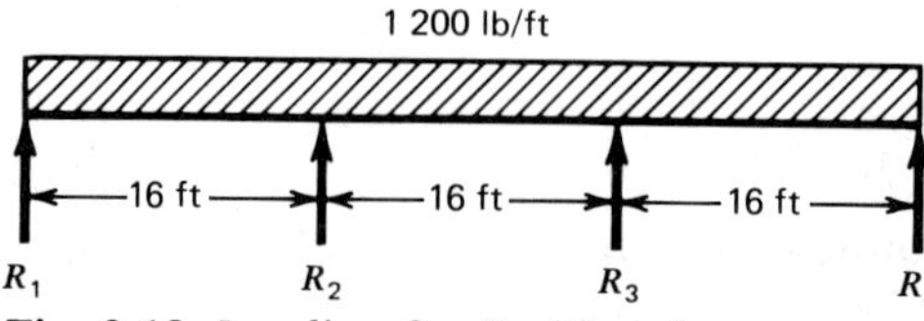

Fig. 9-13. Loading for Problem 3.

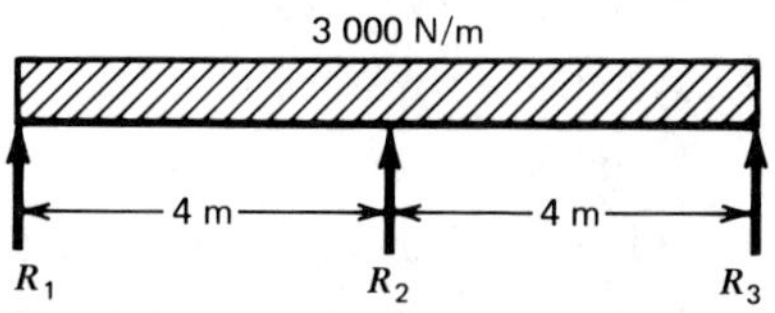

Fig. 9-14. Loading for Problem 5.

10 MEMBERS SUBJECTED TO TORSION

PURPOSE OF THIS CHAPTER. The purpose of this chapter is to learn to find the stress in a bar that is subjected to torsion. Torsion is twist; that is, one end of the bar is twisted with respect to the other end. The twisting is caused by torque, which is a moment in a plane perpendicular to the axis of the bar. You will learn how to calculate torque and then how to calculate the stresses that result from it.

A common type of torsion member is the shaft. A shaft is a circular bar that transmits power and motion from one part of a machine to another. A shaft rotates; that is, it turns about its axis and is supported in bearings. In a few cases the torques involved are small, and the function of the shaft is primarily to transmit motion from one end to the other, for example, to turn the hands of an electric clock.

Sometimes there are torsion members in the structures of buildings and bridges. These members are likely to be structural shapes, for example, I-beams, channels, or angles. You will learn how to calculate the stresses in them, too.

10-1. TORQUE

Torque is *twisting moment*; that is, a moment that tends to twist a bar. Let's see how it works. Figure 10-1 shows the free-body diagram of a

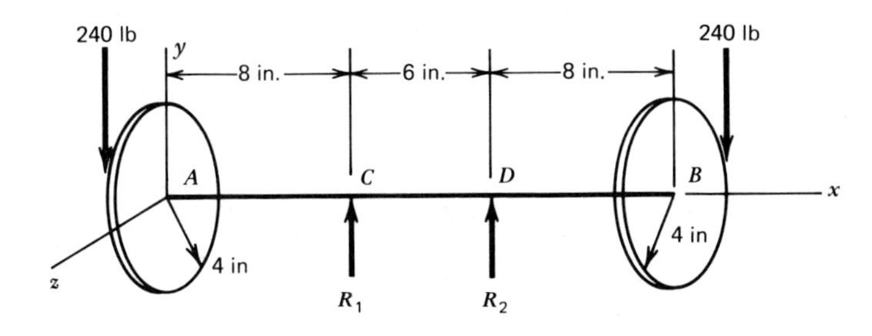

Fig. 10-1. Free-body diagram of shaft and gears.

shaft that carries the gears A and B; other gears exert the 240-lb forces on A and B. The shaft is supported in bearings at C and D, and the bearings exert the reactions R_1 and R_2. Here we are going to use a three-dimensional coordinate system as we did in studying equilibrium of shafts in Chapter 2. Notice the x axis coincides with the axis of the shaft; the y axis is vertical, and the z axis is horizontal. Now you can see in Fig. 10-1 that each of the 240-lb forces has moment about the center of the shaft. The force that acts on gear A has a counterclockwise moment as you look along the x-axis from right to left, and the force that acts on gear B has a clockwise moment. These moments tend to twist the shaft, and so they are called *twisting moments*. (*Torque* is just another name for twisting moment.) The moment of the force that acts on gear A is counterclockwise and tends to turn or twist the left end of the shaft in a counterclockwise direction.

The amount of the twisting moment at any point in the shaft is the moment with respect to the x axis of all forces to the left of the point. Bearing reactions, such as R_1 and R_2 in Fig. 10-1 go through the center of the shaft and have no moment (you see, their moment arms are zero). When you calculate torque, you multiply each force in pounds by a distance in inches, and the result for torque is in pound inches (lb in.), if you are working in gravitational units. If you are working in the metric system, you would multiply each force in newtons by a distance in millimeters, and then the torque would be in newton millimeters (N·mm).

Illustrative Example 1. Calculate the torque at point C for the shaft in Fig. 10-1.

Solution:

The only force to the left of point C is the 240-lb force that is applied to the gear A. You can see in Fig. 10-1 that the moment arm of this force is 4 in. Then the torque at point C is the moment of the 240-lb force with respect to the axis of the shaft:

$$T = -240 \times 4 = -960 \text{ lb in.}$$

where T stands for *torque*. The negative sign is used for the answer to signify that the torque is counterclockwise.

Illustrative Example 2. Figure 10-2 shows a shaft that carries the pulleys A and B and is supported in bearings at C and D. The forces on the pulleys are exerted by V belts that pass over the pulleys. Calculate the torque on the shaft at point E.

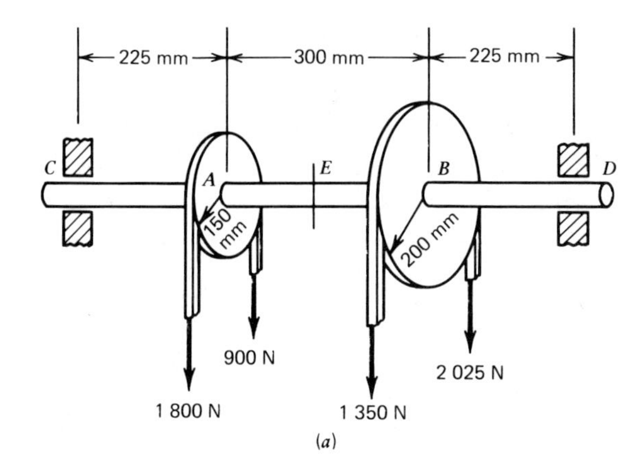

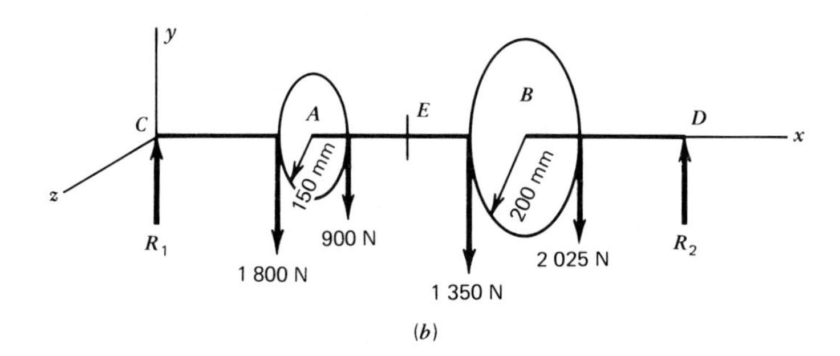

Fig. 10-2. Shaft and free-body diagram. (*a*) Picture of shaft. (*b*) Free-body diagram.

Solution:

1. Figure 10-2 shows the free-body diagram of the shaft. The forces that act on the pulleys are shown and the bearing reactions R_1 and R_2.
2. The torque at E is the moment with respect to the center of the shaft of all of the forces to the left of E. There are three forces to the left of E; the bearing reaction R_1 and the belt pulls on pulley A. The bearing reaction R_1 goes through the center of the shaft, so it has no moment, and you can see the moment arms of the forces on pulley A in Fig. 10-2. The moment arm of the 1 800-N force is 150 mm, and the moment is counterclockwise; the moment arm of the 900-N force is also 150 mm, but the moment is clockwise. So,

$$\begin{aligned} T &= -1\,800 \times 150 + 900 \times 150 \\ &= -270\,000 + 135\,000 \\ &= -135\,000 \text{ N·mm} \end{aligned}$$

Practice Problems (Section 10-1). Now practice on calculating torque.

1. Figure 10-3 shows a shaft with a lever on each end supported in bearings at C and D. Calculate the torque at C.
2. Figure 10-4 shows a shaft that carries the pulleys A and B and is supported in bearings at C and D. The forces on the pulleys are exerted by belts that pass over the pulleys. What is the torque at point D?
3. Figure 10-5 shows a pipe turned by a wrench. Calculate the torque on the pipe at A.
4. The shaft in Fig. 10-6 carries the pulleys A and B and the gear C. The shaft is supported in bearings at D, E, and F. Find the torque at point D.
5. Calculate the torque at point E in Fig. 10-6.

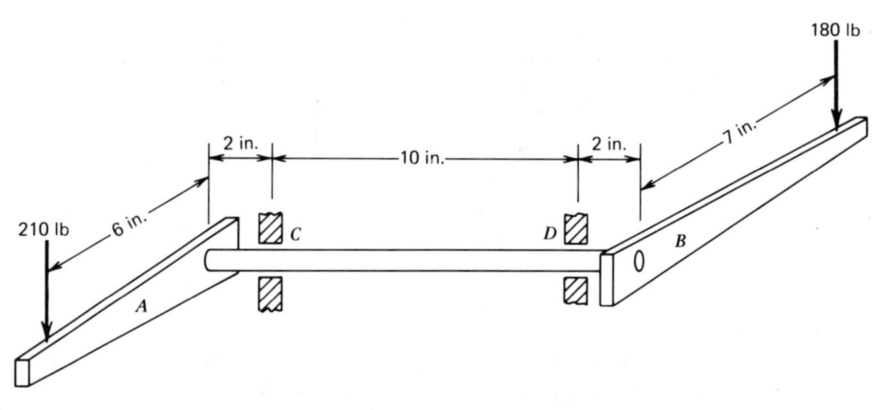

Fig. 10-3. Shaft and levers for Problem 1.

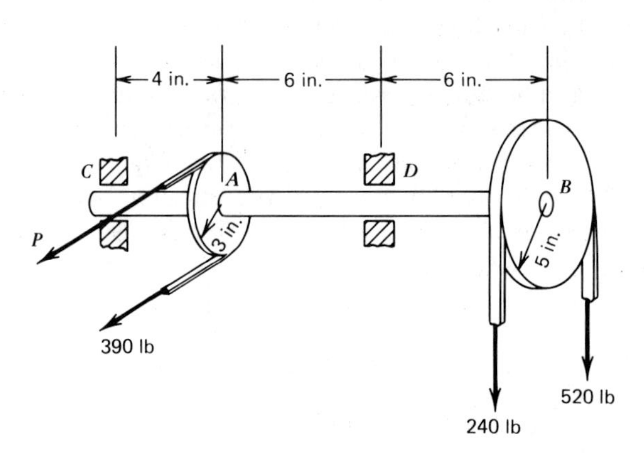

Fig. 10-4. Shaft and pulleys for Problem 2.

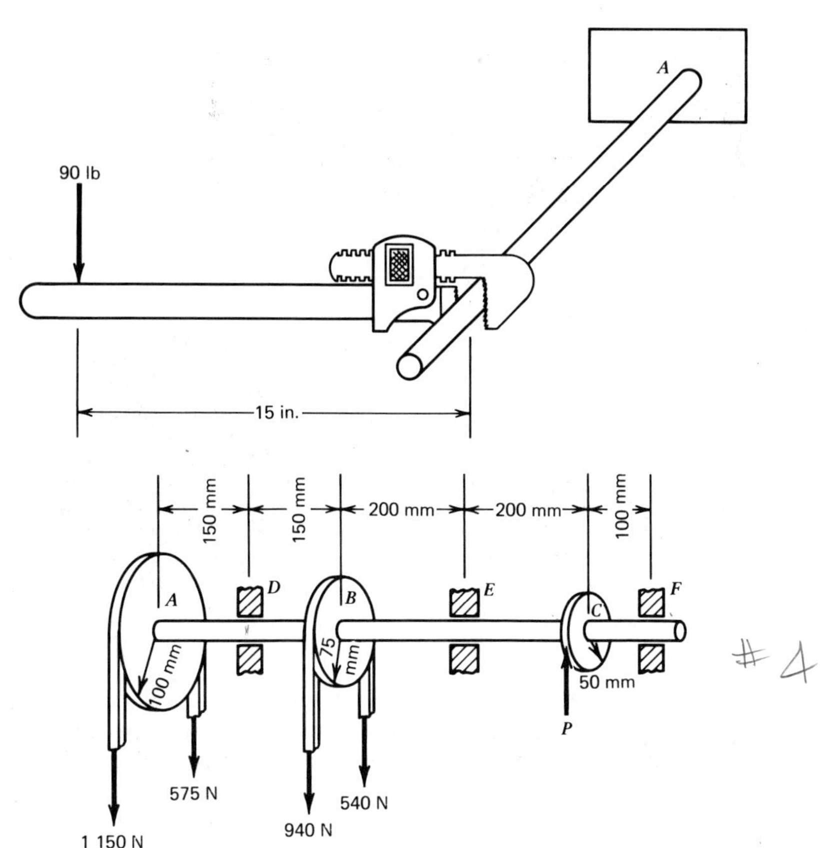

Fig. 10-5. Pipe twisted by wrench for Problem 3.

Fig. 10-6. Shaft and pulleys for Problems 4 and 5.

10-2. TORSIONAL SHEARING STRESS IN A SOLID CIRCULAR SHAFT

Now let's learn what kind of stress there is in a circular shaft that is twisted and what the value of the maximum stress is. This stress is called *torsional shearing stress*. You will see the formula that gives the relation between stress and torque, and you will see how the stress appears. The formula will be stated without proof, but you will be able to use it just as well as if we had proved it. Looking at Fig. 10-7*a*, you can see that the forces tend to twist the shaft, and we want to find the stress on such a cross section as *E-E*. Let's imagine for a minute that we are not dealing with a single shaft but with a row of washers such as you see in Fig. 10-7*b*. If a torque were applied to the row of washers, one washer would turn or slide with respect to the next washer, as, for instance, on such a plane as *E-E*. But in a single shaft this tendency to turn or slide is resisted by shearing stress developed on the cross section. Figure 10-7*c* shows how the stresses occur in a shaft of circular cross section. The

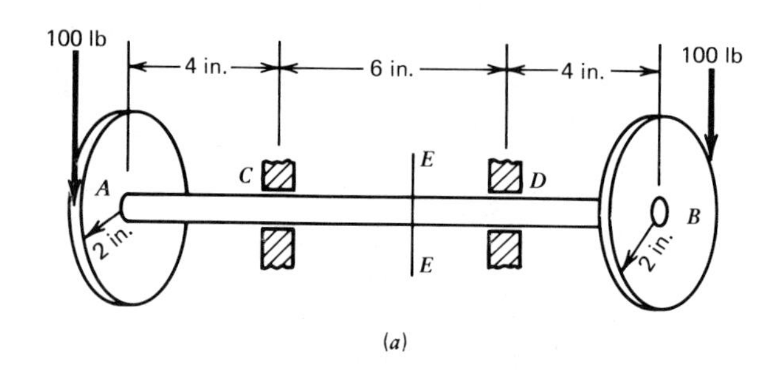

(a)

(b)

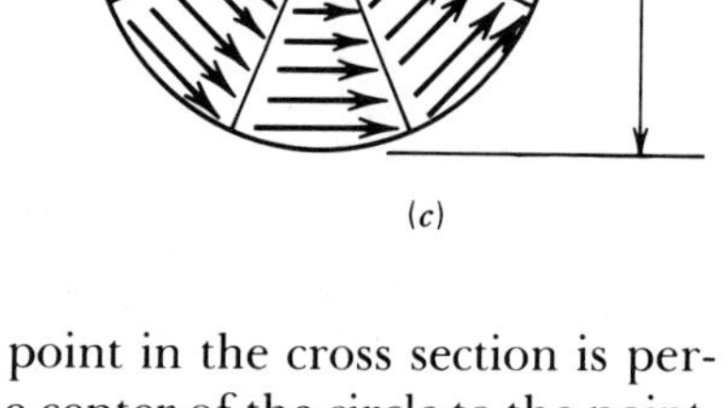

(c)

Fig. 10-7. Torsional shearing stress. (*a*) Shaft and gears. (*b*) Row of washers. (*c*) Shearing stress on cross section.

direction of the shearing stress at any point in the cross section is perpendicular to the radius drawn from the center of the circle to the point, and the maximum stress occurs at the outside or rim of the cross section. The value of the maximum shearing stress in a solid circular shaft is

$$S = \frac{16T}{\pi d^3}$$

where S is the shearing stress in psi, T is the torque in pound inches, and d is the diameter of the shaft in inches, if we are working with gravitational units. Let's look at the units of the equation. If we substitute psi (lb/in.^{2}) for S, lb in. for T, and in. for d, we get,

$$\frac{\text{lb}}{\text{in.}^2} = \frac{\text{lb in.}}{\text{in.}^3}$$

We can cancel in. on the right side of the equation to get (just as $x/x^3 = 1/x^2$)

$$\frac{\text{lb}}{\text{in.}^2} = \frac{\text{lb}}{\text{in.}^2}$$

This final equality of dimensions shows that the dimensions given for the quantities in the equation are correct. You *must* remember that the torque T is to be expressed in pound inches. You may calculate it first in pound feet, but if you do you must multiply by 12 to convert it to pound inches.

Illustrative Example 1. A torque of 2 640 lb in. is applied to a solid circular shaft 1 in. in diameter. Calculate the maximum torsional shearing stress.

Solution:

1. The torque T is 2 640 lb in.
2. The diameter d is 1 in.
3. The maximum torsional shearing stress is

$$S = \frac{16}{\pi}\frac{T}{d^3} = \frac{16}{\pi}\frac{2\,640}{1^3} = 13\,400 \text{ psi}$$

METRIC UNITS

If you are working in the metric system, you would express the torque in newton millimeters and the shaft diameter in millimeters. Then you would get the stress in megapascals. Thus

$$S = \frac{16}{\pi}\frac{T}{d^3} = \frac{\text{N}\cdot\text{mm}}{\text{mm}^3} = \frac{\text{N}}{\text{mm}^2} = \text{MPa}$$

Let's try this.

Illustrative Example 2. A torque of 48 000 000 N·mm is applied to a solid circular shaft 150 mm in diameter. What is the maximum torsional shearing stress?

Solution:

1. The torque is 48 000 000 N·mm.
2. The diameter d is 150 mm.
3. The maximum torsional shearing stress is

$$S = \frac{16}{\pi}\frac{T}{d^3} = \frac{16}{\pi}\frac{48\,000\,000}{(150)^3} = 72.4 \text{ MPa}$$

CALCULATING TORQUE

Now let's turn the formula around and try calculating the torque that will develop a given stress. We wrote the formula as,

$$S = \frac{16}{\pi}\frac{T}{d^3}$$

and we can rewrite it as

$$\frac{\pi d^3 S}{16} = T \qquad \text{or}$$

$$T = \frac{\pi}{16} d^3S$$

If we know the diameter of the shaft and the maximum stress, we can find the torque.

Illustrative Example 3. What torque must be applied to a solid circular shaft 2.5 in. in diameter to cause a maximum torsional shearing stress of 12 000 psi?

Solution:

1. The diameter d is 2.5 in.
2. The maximum torsional shearing stress is 12 000 psi.
3. The torque is

$$T = \frac{\pi}{16} d^3S = \frac{\pi}{16}(2.5)^3\ 12\ 000 = 36\ 800 \text{ lb in.}$$

Illustrative Example 4. The maximum torsional shearing stress in a solid circular shaft 14 mm in diameter is 66.3 MPa. Calculate the torque.

Solution:

1. The diameter d is 14 mm.
2. The maximum torsional shearing stress is 66.3 MPa.
3. The torque is

$$T = \frac{\pi}{16} d^3S = \frac{\pi}{16}(14)^3\ 66.3 = 35\ 700 \text{ N·mm}$$

Practice Problems (Section 10-2). You should have learned to use the formula by this time. Now practice so you will remember it.

1. A solid circular shaft 3 in. in diameter is subjected to a torque of 25 000 lb in. What is the maximum torsional shearing stress?
2. The torque applied to a solid circular shaft 12 in. in diameter is 220 000 lb ft. Calculate the maximum torsional shearing stress.
3. A torque of 630 N·mm is applied to a solid circular shaft 6 mm in diameter. Find the maximum torsional shearing stress.
4. The bent circular shaft in Fig. 10-8 is supported in bearings at A and B. The diameter of the shaft is 36 mm. Calculate the maximum torsional shearing stress.

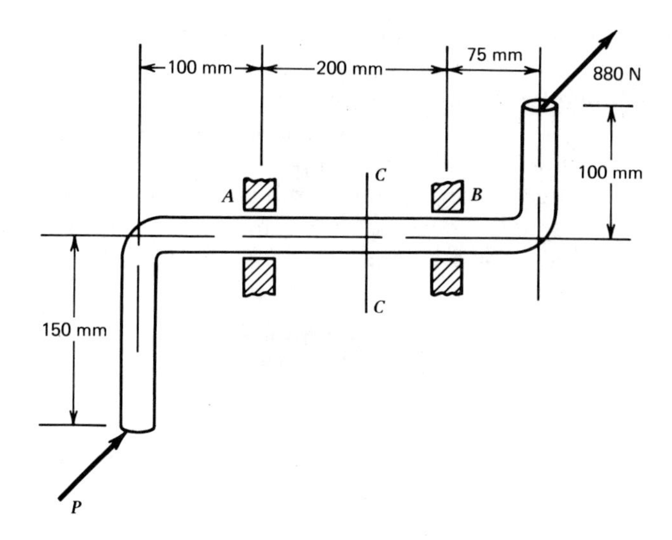

Fig. 10-8. Bent bar for Problem 4.

5. What torque must be applied to a solid circular shaft 2 in. in diameter to cause a maximum torsional shearing stress of 8 000 psi?
6. The maximum torsional shearing stress in a solid circular shaft 8 in. in diameter is 10 800 psi. Calculate the torque.
7. Calculate the torque that must be applied to a solid circular shaft 20 mm in diameter to cause a maximum torsional shearing stress of 49 MPa.
8. The maximum torsional shearing stress in a solid circular shaft 80 mm in diameter is 112 MPa. What is the torque?

10-3. TORSIONAL SHEARING STRESS IN A HOLLOW CIRCULAR BAR

Next we will study a hollow circular shaft and learn to find the maximum torsional shearing stress. Figure 10-9 shows the cross section of a hollow circular shaft; let's call the outer diameter d_2 and the inner diameter d_1.

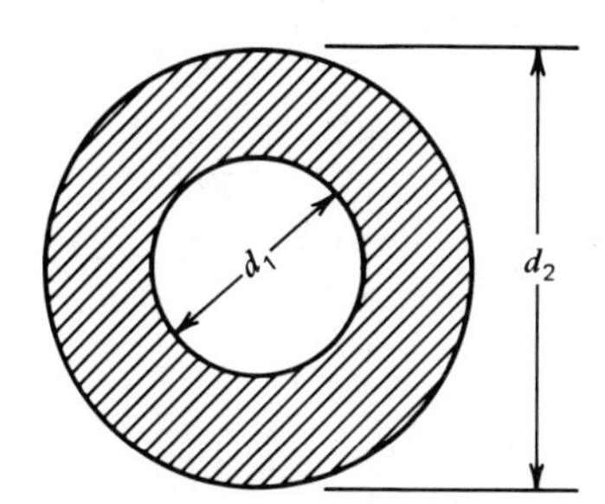

Fig. 10-9. Cross section of hollow shaft.

There are several reasons for using hollow shafts in machines. For one thing, it may be necessary to have more than one shaft in the same place; this can be done by placing one shaft inside another; naturally, then, the outer shaft must be hollow. In other cases the shaft is made hollow to provide a channel for oil to be pumped through. Also, a hollow shaft is somewhat lighter in weight than a solid shaft that carries the same torque and is desirable when it is important to save weight.

The formula for the maximum torsional shearing stress in a hollow circular shaft is

$$S = \frac{16}{\pi} \frac{Td_2}{(d_2^4 - d_1^4)}$$

where S is the stress in psi, T is the torque in pound inches, d_2 is the outer diameter in inches, and d_1 the inner diameter in inches. (Proof of this formula requires the use of calculus, and we won't bother to prove it here.) You have the formula. Now let's use it.

Illustrative Example 1. A torque of 24 200 lb in. is applied to a hollow circular shaft of 3 in. outer diameter and 2 in. inner diameter. What is the maximum torsional shearing stress?

Solution:

1. The torque T is 24 200 lb in.
2. The outer diameter d_2 is 3 in.
3. The inner diameter d_1 is 2 in.
4. The maximum torsional shearing stress is

$$S = \frac{16}{\pi} \frac{Td_2}{(d_2^4 - d_1^4)} = \frac{16}{\pi} \frac{24\ 200 \times 3}{[(3)^4 - (2)^4]} = 5\ 690 \text{ psi}$$

METRIC UNITS

When you are working in metric units, you would express the torque in newton millimeters and the shaft diameters in millimeters. Then, the stress would be in megapascals.

Illustrative Example 2. A torque of 30×10^6 N·mm is applied to a hollow circular shaft of 160 mm outer diameter and 120 mm inner diameter. Calculate the maximum torsional shearing stress.

Solution:

1. The torque T is 30×10^6 N·mm.

2. The outer diameter d_2 is 160 mm.
3. The inner diameter d_1 is 120 mm.
4. The maximum torsional shearing stress is

$$S = \frac{16}{\pi}\frac{Td_2}{(d_2^4 - d_1^4)} = \frac{16}{\pi}\frac{30 \times 10^6 \times 160}{[(160)^4 - (120)^4]} = 54.6 \text{ MPa}$$

TORQUE

Now let's turn the formula around so that we can calculate the torque when we know the diameters and the stress. We had the formula as

$$S = \frac{16}{\pi}\frac{Td_2}{(d_2^4 - d_1^4)}$$

but we can write it as,

$$T = \frac{\pi}{16}\frac{(d_2^4 - d_1^4)}{d_2}S$$

Illustrative Example 3. The maximum torsional shearing stress in a hollow circular shaft of 5 in. outer diameter and 4 in. inner diameter is 10 000 psi. What is the torque?

Solution:

1. The outer diameter d_2 is 5 in.
2. The inner diameter d_1 is 4 in.
3. The maximum torsional shearing stress S is 10 000 psi.
4. The torque is

$$T = \frac{\pi}{16}\frac{(d_2^4 - d_1^4)}{d_2}S = \frac{\pi}{16}\frac{[(5)^4 - (4)^4]}{5}10\,000$$

$$= 145\,000 \text{ lb in.}$$

Illustrative Example 4. Calculate the torque that will cause a maximum torsional shearing stress of 112 MPa in a hollow circular shaft of 15 mm outer diameter and 12.5 mm inner diameter.

Solution:

1. The outer diameter d_2 is 15 mm.
2. The inner diameter d_1 is 12.5 mm.
3. The maximum torsional shearing stress S is 112 MPa.

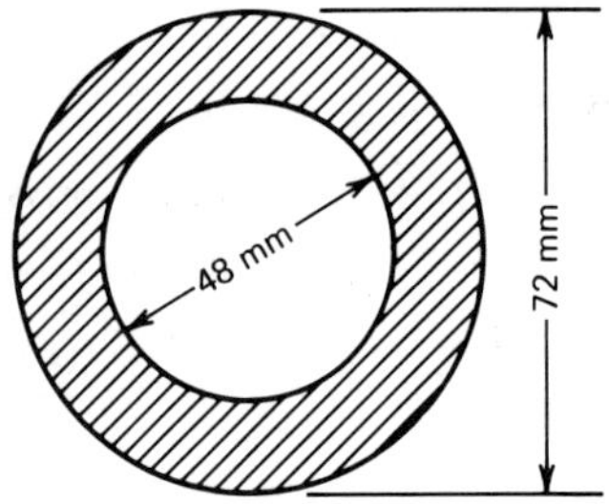

Fig. 10-10. Cross section of hollow shaft for Problem 6.

4. The torque is

$$T = \frac{\pi}{16}\frac{[d_2^4 - d_1^4]}{d_2}S = \frac{\pi}{16}\frac{[(15)^4 - (12.5)^4]}{15}112$$

$$= 38\ 000 \text{ N·mm}$$

Practice Problems (Section 10-3). Now you are ready to practice on hollow shafts.

1. Calculate the maximum torsional shearing stress in a hollow circular shaft 2.4 in. outer diameter and 1.6 in. inner diameter that is subjected to a torque of 14 000 lb in.
2. A hollow circular shaft 10 in. outer diameter and 8 in. inner diameter is subjected to a torque of 97 000 lb ft. What is the maximum torsional shearing stress?
3. What is the maximum torsional shearing stress resulting from a torque of 6.75×10^6 N·mm applied to a hollow circular shaft of 90 mm outer diameter and 75 mm inner diameter?
4. The maximum torsional shearing stress in a hollow circular shaft 0.8 in. outer diameter and 0.5 in. inner diameter is 6 600 psi. Calculate the torque.
5. What torque will cause a maximum torsional shearing stress of 84 MPa in a hollow circular shaft of 400 mm outer diameter and 250 mm inner diameter?
6. The maximum torsional shearing stress in a shaft that has the cross section shown in Fig. 10-10 is 98 MPa. What is the torque?

10-4. HOW TO CHOOSE THE DIAMETER OF A SOLID CIRCULAR BAR

One of the real problems of design is the problem of finding the diameter of a shaft when the torque and the working stress in shear are known. (The working stress is the maximum stress that is expected to be developed in service.) We will have to change the torsion formula so that we can solve for the diameter. You remember that we wrote the formula as

$$S = \frac{16}{\pi}\frac{T}{d^3}$$

and we can rewrite it as

$$d^3 = \frac{16}{\pi}\frac{T}{S}$$

and this is what we need. If we know T and S we can solve for d. Thus taking the cube root of each side of the equation

$$d = \sqrt[3]{\frac{16}{\pi}\frac{T}{S}}$$

The torque T must be in pound inches in this formula and the working stress S must be in psi; the torque may be in newton millimeters, and the working stress is in megapascals.

Illustrative Example 1. A solid circular shaft is to be subjected to a torque of 27 000 lb in., and the working stress in shear is 9 000 psi. What diameter of shaft is required?

Solution:

1. The torque T is 27 000 lb in.
2. The working stress S is 9 000 psi.
3. The diameter required is

$$d = \sqrt[3]{\frac{16}{\pi}\frac{T}{S}} = \sqrt[3]{\frac{16}{\pi}\frac{27\,000}{9\,000}} = 2.48 \text{ in.}$$

Illustrative Example 2. Find the diameter of a solid circular shaft that is to carry a torque of 225×10^6 N·mm with a working stress in shear of 102 MPa.

Solution:

1. The torque T is 225×10^6 N·mm.
2. The working stress S is 102 MPa.
3. The diameter required is

$$d = \sqrt[3]{\frac{16}{\pi}\frac{T}{S}} = \sqrt[3]{\frac{16}{\pi}\frac{225 \times 10^6}{102}} = 224 \text{ mm}$$

Practice Problems (Section 10-4). There's nothing difficult about finding the diameter of a shaft when you know the torque and the working stress, but you had better practice it.

1. A solid circular shaft is to carry a torque of 442 lb in. with a working stress in shear of 6 200 psi. Find the required diameter.
2. What diameter of solid circular shaft is required to carry a torque of 89 000 lb in. with a working stress in shear of 10 000 psi?
3. The working stress for a solid circular shaft is 70 MPa, and the torque is 560 000 N·mm. Calculate the required diameter.
4. A torque of 4.98×10^6 N·mm is to be applied to a solid circular shaft for which the working stress in shear is 56 MPa. What diameter of shaft is required?

10-5. ANGLE OF TWIST OF A CIRCULAR BAR

We saw in Chapter 3 that all objects deform when subjected to force. A shaft is no exception, and the deformation of a shaft is *twist*, or *turning*, of one end with respect to the other. The amount of the twist is important in many machines, and our problem here is to learn to calculate it.

Figure 10-11*a* shows a shaft that carries the gears *A* and *B* and is supported in bearings at *C* and *D*. You can see that the forces tend to twist the shaft. Then Fig. 10-11*b* shows a larger view of part of the shaft. Now let's imagine for a minute that the left end of the part in Fig. 10-11*b* is held so that it cannot turn. Under this condition, the right end of the part of Fig. 10-11*b* will turn, and the line *E-E* will turn to the position *F-F*; the right end of this part of the shaft actually turns or twists through the angle θ (Greek letter *theta*). The question is, *how much is* θ? We can answer the question by giving you the formula. (We won't bother to derive it, because that would take too much space and would require using calculus.) The formula is

$$\theta = \frac{32}{\pi}\frac{TL}{E_s d^4}$$

Here T is the torque in pound inches, L is the length of the shaft in inches, E_s is the shearing modulus of elasticity in psi of the material of which the shaft is made, and d is the diameter of the shaft in inches, if you are working with gravitational units. In the metric system T is in newton millimeters, L is in millimeters, E_s is in megapascals, and d is in

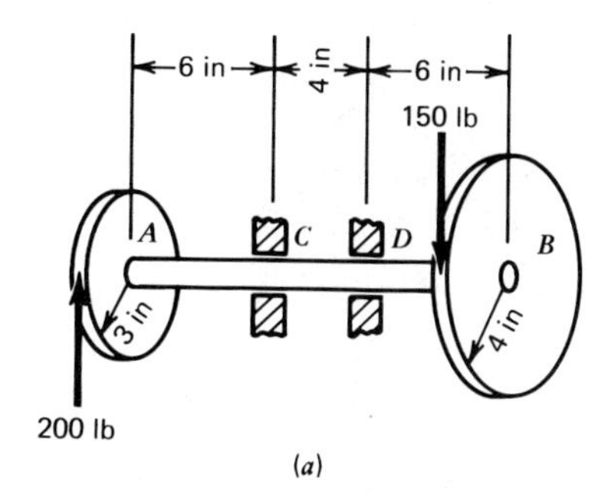

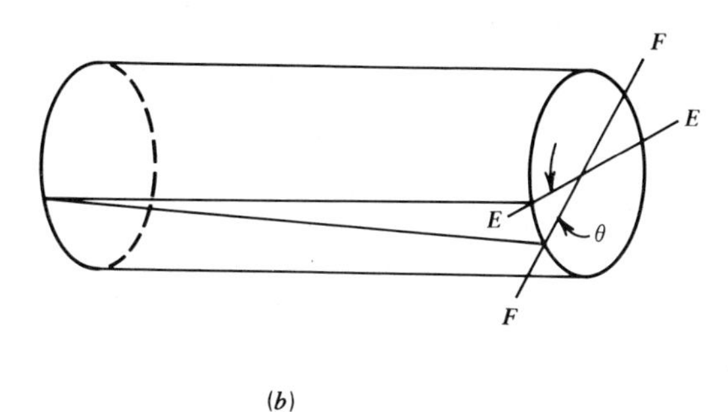

Fig. 10-11. Twist of shaft. (*a*) Shaft and gears. (*b*) Twist.

millimeters. Then the angle θ is given in radians. (A *radian* is a unit of angular measure and is equal to 57.3°.)

You should remember that Table 3.1 gives values of E_s for many of the materials of engineering. You can refer to this table for the values you need. Now let's try the formula.

Illustrative Example 1. A solid circular steel shaft 2 in. in diameter and 40 in. long is subjected to a torque of 15 000 lb in. Calculate the angle of twist.

Solution:

1. The torque T is 15 000 lb in.
2. The length L is 40 in.
3. From Table 3.1, the shearing modulus of elasticity for steel is 12 000 000 psi.
4. The diameter d is 2 in.
5. The angle of twist is

$$\theta = \frac{32}{\pi}\frac{TL}{E_s d^4} = \frac{32}{\pi}\frac{15\,000 \times 40}{12\,000\,000(2)^4} = 0.0318 \text{ radians}$$

6. A *radian* is equal to 57.3°, and we can change the answer to degrees by multiplying by 57.3. Thus

$$\theta = 0.0318 \times 57.3 = 1.82°$$

Illustrative Example 2. A solid circular brass shaft 18 mm in diameter and 400 mm long is subjected to a torque of 85 000 N·mm. What is the angle of twist?

Solution:

1. The torque T is 85 000 N·mm.
2. The length L is 400 mm.
3. From Table 3.1 the shearing modulus of elasticity for brass is 34.5 GPa or 34.5×10^3 MPa.
4. The diameter d is 18 mm.
5. The angle of twist is

$$\theta = \frac{32}{\pi}\frac{TL}{E_s d^4} = \frac{32}{\pi}\frac{85\,000 \times 400}{34.5 \times 10^3(18)^4} = 0.0956 \text{ radians}$$

6. We can convert the answer to degrees by multiplying by 57.3, so

$$\theta = 0.0956 \times 57.3 = 5.48°$$

ANGLE OF TWIST

HOLLOW CIRCULAR SHAFTS

The formula for the angle of twist of a hollow circular shaft is almost the same as the formula for a solid shaft. It is

L = LENGTH
T= TORQUE IN IN LBS
E = STRESS MODULUS
d_2 = O.D.
d_1 = I.D.

$$\theta = \frac{32}{\pi}\frac{TL}{E_s(d_2^4 - d_1^4)}$$

Here T, L, and E_s mean the same as before; d_2 is the outer diameter of the hollow shaft, and d_1 is the inner diameter. Figure 10-12 shows the cross section of the hollow shaft.

Illustrative Example 3. A hollow, circular, gray cast-iron shaft, 6 in. outer diameter, 4 in. inner diameter, and 30 in. long, is subjected to a torque of 140 000 lb in. Calculate the angle of twist.

Solution:

1. The torque T is 140 000 lb in.

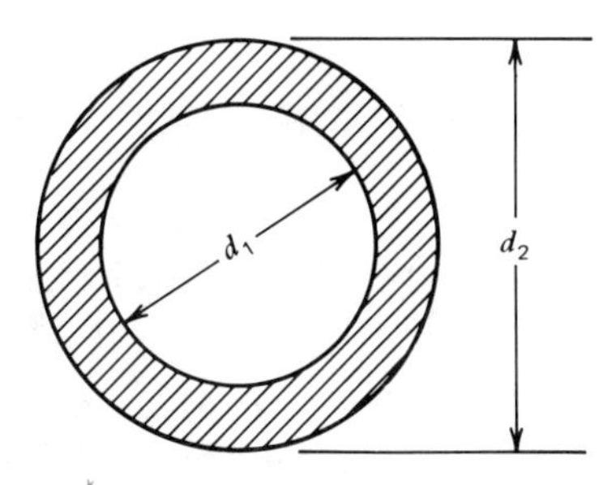

Fig. 10-12. Cross section of hollow shaft.

2. The length L is 30 in.
3. From Table 3.1, the shearing modulus of elasticity for gray cast-iron is 6 000 000 psi.
4. The outer diameter d_2 is 6 in.
5. The inner diameter d_1 is 4 in.
6. The angle of twist is

$$\theta = \frac{32}{\pi}\frac{TL}{E_s(d_2^4 - d_1^4)} = \frac{32}{\pi}\frac{140\,000 \times 30}{6\,000\,000[(6)^4 - (4)^4]}$$

$$= 0.00686 \text{ radians}$$

7. If we want the answer in degrees

$$\theta = 0.00686 \times 57.3 = 0.393°$$

Illustrative Example 4. A hollow circular aluminum tube is subjected to a torque of 54 000 N·mm. The outer diameter of the tube is 25 mm, and the inner diameter is 20 mm. The length is 600 mm. What is the angle of twist?

Solution:

1. The torque T is 54 000 N·mm.
2. The length L is 600 mm.
3. From Table 3.1, the shearing modulus of elasticity for aluminum is 27.6 GPa or 27.6×10^3 MPa.
4. The outer diameter d_2 is 25 mm.
5. The inner diameter d_1 is 20 mm.
6. The angle of twist is

$$\theta = \frac{32}{\pi}\frac{TL}{E_s(d_2^4 - d_1^4)} = \frac{32}{\pi}\frac{54\,000 \times 600}{27.6 \times 10^3[(25)^4 - (20)^4]}$$

$$= 0.0518 \text{ radians}$$

7. In degrees,

$$\theta = 0.0518 \times 57.3 = 2.97°$$

Practice Problems (Section 10-5). Now try calculating the angle of twist for yourself. Look in Table 3.1 for the shearing modulus of elasticity.

1. A solid circular shaft 6 in. in diameter and 8 ft long is made of malleable cast iron. What is the angle of twist due to a torque of 190 000 lb in.?
2. A solid circular steel shaft 22 mm in diameter and 360 mm long is subjected to a torque of 175 000 N·mm. Calculate the angle of twist.
3. A solid circular brass shaft 1.6 in. in diameter and 12 in. long carries a torque of 16 000 lb in. Find the angle of twist.
4. A hollow circular steel shaft 2 in. outer diameter, 1.5 in. inner diameter, and 20 in. long is subjected to a torque of 11 200 lb in. What is the angle of twist?
5. A hollow circular gray cast-iron shaft has outer diameter of 300 mm, inner diameter of 250 mm, and is 2.3 m long. Find the angle of twist due to a torque of 110×10^6 N·mm.
6. A magnesium alloy tube with 3 in. outer diameter and 2 in. inner diameter is 3 ft long. The tube carries a torque of 26 000 lb in. Calculate the angle of twist.

10-6. TORSIONAL SHEARING STRESS IN A RECTANGULAR BAR

A designer of machines and structures sometimes has occasion to calculate the torsional shearing stress in a rectangular bar. It isn't that the rectangular bar is used as a shaft and turns in bearings; instead, the bar is usually a part of a structure and is subjected to loads that produce a torque. For example, Fig. 10-13 shows the loading on a bent rectangular bar fastened to a machine frame by fillet welds; the 200-lb force applied at the end of the bar causes a torque on any cross section *C-C* between points *A* and *B*. It is helpful to use the *xyz* set of axes here in calculating the torque. The torque is equal to the moment of the 200-lb force with respect to the *x* axis, and this is

$$T = 200 \times 10 = 2\,000 \text{ lb in.}$$

since the moment arm of the force is 10 in.

We can't begin to derive a formula for the torsional shearing stress in a rectangular bar, because we would have to use mathematics far

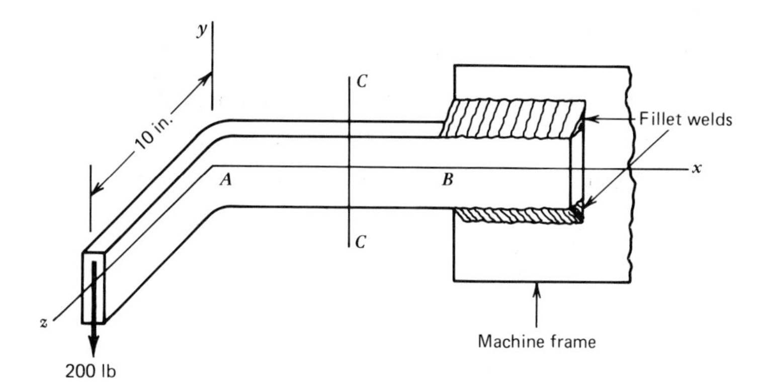

Fig. 10-13. Rectangular bar in torsion.

beyond calculus, but we can just give you an approximate formula and show you how to use it. Even though the formula is only approximate, it still gives results that are almost exactly right. (The maximum error is only about 3% and this is usually close enough for design purposes.)

Figure 10-14 shows the cross section of a rectangular bar; we will represent the short side by b and the long side by h. Then the approximate formula for the maximum torsional shearing stress is

$$S = \frac{T(3h + 2b)}{b^2h^2}$$

where S is the maximum torsional shearing stress in psi, T is the torque in pound inches, h is the long side of the rectangle, and b is the short side; h and b are in inches, if you are working in gravitational units. In the metric system, S is in megapascals, T is in newton millimeters, and h and b are in millimeters. Well, there's the formula, and you can see that it won't be too hard to use. Remember to take h as the long side of the rectangle and b as the short side. The maximum stress occurs at the middle of the long side of the rectangle, at points A in Fig. 10-14.

Illustrative Example 1. A torque of 7 200 lb in. is applied to a rectangular bar 2.5 in. by 1 in. Calculate the maximum torsional shearing stress.

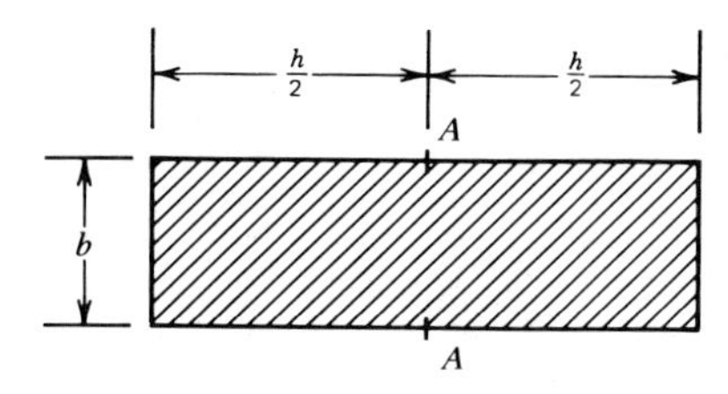

Fig. 10-14. Cross section of rectangular bar.

Solution:

1. The torque T is 7 200 lb in.
2. The long side of the rectangle is 2.5 in. This is h.
3. The short side b of the rectangle is 1 in.
4. The maximum torsional shearing stress is

$$S = \frac{T(3h + 2b)}{b^2h^2} = \frac{7\,200(3 \times 2.5 + 2 \times 1)}{(1)^2(2.5)^2}$$

$$= 10\,900 \text{ psi}$$

Illustrative Example 2. A rectangular bar 12 mm by 3 mm in cross section is subjected to a torque of 1 750 N·mm. What is the maximum torsional shearing stress?

Solution:

1. The torque T is 1 750 N·mm.
2. The long side h of the rectangle is 12 mm.
3. The short side b of the rectangle is 3 mm.
4. The maximum torsional shearing stress is

$$S = \frac{T(3h + 2b)}{b^2h^2} = \frac{1\,750(3 \times 12 + 2 \times 3)}{(3)^2(12)^2} = 56.7 \text{ MPa}$$

SQUARE BAR

In the special case of a square bar, the dimensions h and b are equal; each is equal to the length of one side of the square. This makes it easy.

Illustrative Example 3. A square bar 1.5 in. on a side is subjected to a torque of 12 000 lb in. Find the maximum torsional shearing stress.

Solution:

1. The torque T is 12 000 lb in.
2. The dimensions h and b are each 1.5 in.
3. The maximum torsional shearing stress is

$$S = \frac{T(3h + 2b)}{b^2h^2} = \frac{12\,000(3 \times 1.5 + 2 \times 1.5)}{(1.5)^2(1.5)^2} = 17\,800 \text{ psi}$$

Practice Problems (Section 10-6). This is just one more step on the way to learning strength of materials. Now practice it.

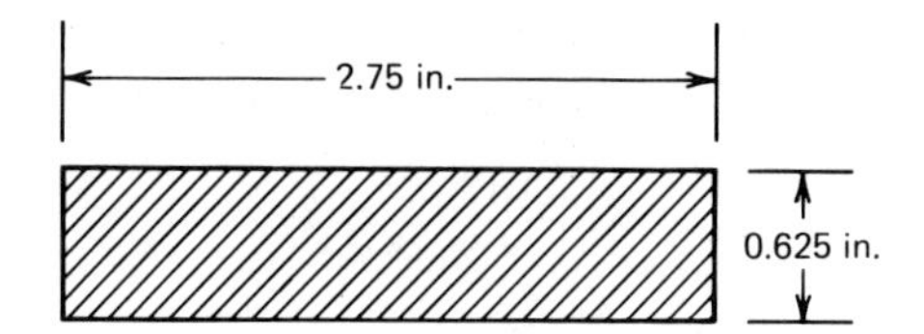

Fig. 10-15. Cross section of rectangular bar for Problem 1.

1. A torque of 4 700 lb in. is applied to a bar that has the cross section shown in Fig. 10-15. Calculate the maximum torsional shearing stress.
2. A square bar 1.25 in. on a side is subjected to a torque of 3 300 lb in. What is the maximum torsional shearing stress?
3. A rectangular bar 6 in. by 1 in. is subjected to a torque of 4 800 lb ft. Find the maximum torsional shearing stress.
4. A torque of 80 000 N·mm is applied to a rectangular bar 75 mm by 15 mm. Calculate the maximum torsional shearing stress.
5. Calculate the maximum torsional shearing stress due to a torque of 90×10^6 N·mm applied to a square bar 200 mm on a side.
6. A rectangular bar 150 mm by 40 mm is subjected to a torque of 7.5×10^6 N·mm. Find the maximum torsional shearing stress.

10-7. TORSIONAL SHEARING STRESS IN THIN-WALLED OPEN SECTIONS

Figure 10-16 shows three types of thin-walled open cross sections; here the word *open* means that the cross section does not completely enclose any area. On many occasions structural members that have such cross sections are subjected to torque. We need a method of calculating the stress in such a case.

Figure 10-16*b* shows a channel section divided into narrow rectan-

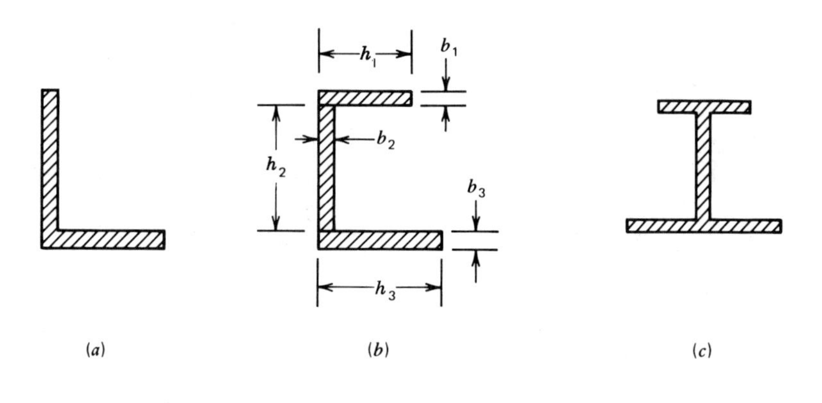

Fig. 10-16. Thin-walled open sections. (*a*) Angle. (*b*) Channel. (*c*) I section.

gles. The maximum shear stress will occur at the center of the long side of the thickest rectangle. This stress is given by the formula

$$S_s = \frac{3T\,b_{max}}{\Sigma hb^3}$$

Here T is the torque, b_{max} is the thickness of the thickest rectangle, and Σhb^3 is just the sum of hb^3 for all of the narrow rectangles. For the channel section in Fig. 10-16*b*:

$$\Sigma hb^3 = h_1b_1^3 + h_2b_2^3 + h_3b_3^3$$

In the gravitational system of units, take T in pound inches and h and b in inches to get the stress in psi. In the metric system, take T in newton millimeters and h and b in millimeters to get S in MPa.

Illustrative Example 1. Figure 10-17*a* shows the cross section of a steel bar subjected to a torque of 10 000 lb in. Calculate the maximum shear stress.

Solution:

1. Figure 10-17*b* shows the division of the cross section into narrow rectangles.
2. The torque T is 10 000 lb in.
3. The greatest thickness b_{max} is ½ in.
4. The quantity Σhb^3 is

$$3.5(\tfrac{1}{2})^3 + 6(\tfrac{1}{2})^3 = 0.438 + 0.750$$
$$= 1.188 \text{ in.}^4$$

5. The maximum stress is

$$S = \frac{3 \times 10\,000 \times \tfrac{1}{2}}{1.1875} = 12\,600 \text{ psi}$$

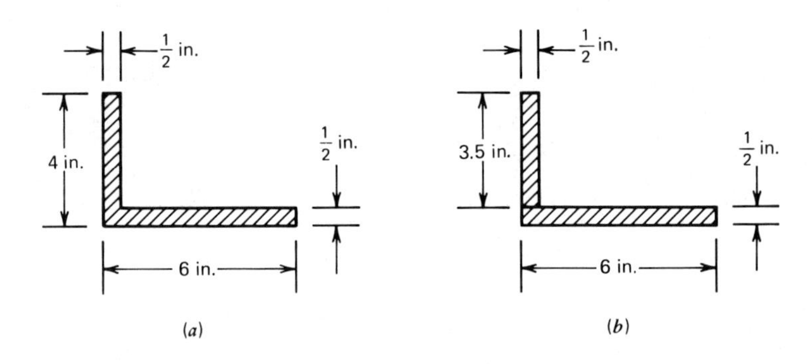

Fig. 10-17. Angle section. (*a*) Angle section. (*b*) Division into rectangles.

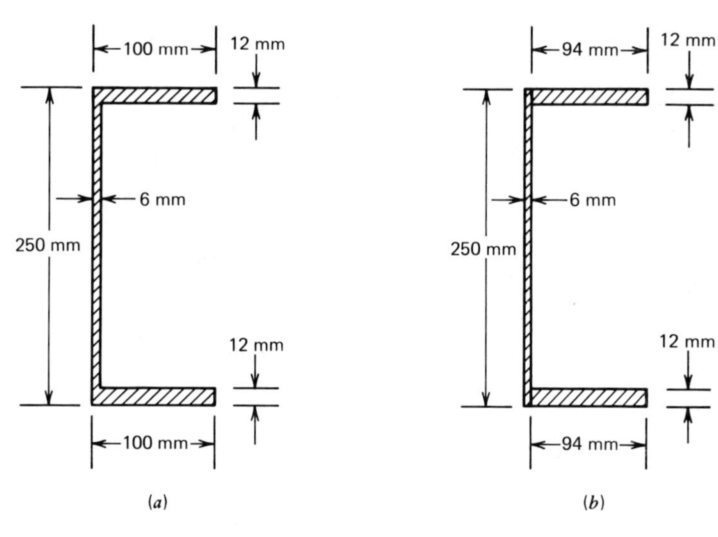

Fig. 10-18. Channel section. (*a*) Channel section. (*b*) Division into rectangles.

Illustrative Example 2. Figure 10-18*a* shows the cross section of a channel subjected to a torque of 1 000 000 N·mm. What is the maximum shear stress?

Solution:

1. Figure 10-18*b* shows the cross section divided into narrow rectangles.
2. The torque T is 1 000 000 N·mm.
3. The greatest thickness is $b_{max} = 12$ mm.
4. The quantity Σhb^3 is

$$94(12)^3 + 250(6)^3 + 94(12)^3 = 162\ 000 + 54\ 000 + 162\ 000$$
$$= 378\ 000 \text{ mm}^4$$

5. The maximum shear stress is

$$S_s = \frac{3 \times 1\ 000\ 000 \times 12}{378\ 000} = 95.2 \text{ MPa}$$

Practice Problems (Section 10-7). Here are problems to practice what you have just learned.

1. A torque of 380 lb in. is applied to the section of Fig. 10-19. What is the maximum shear stress?
2. Calculate the maximum shear stress due to a torque of 57 000 lb in. applied to the section of Fig. 10-20.
3. What maximum shear stress results from a torque of 150 000 N·mm on the section of Fig. 10-21?

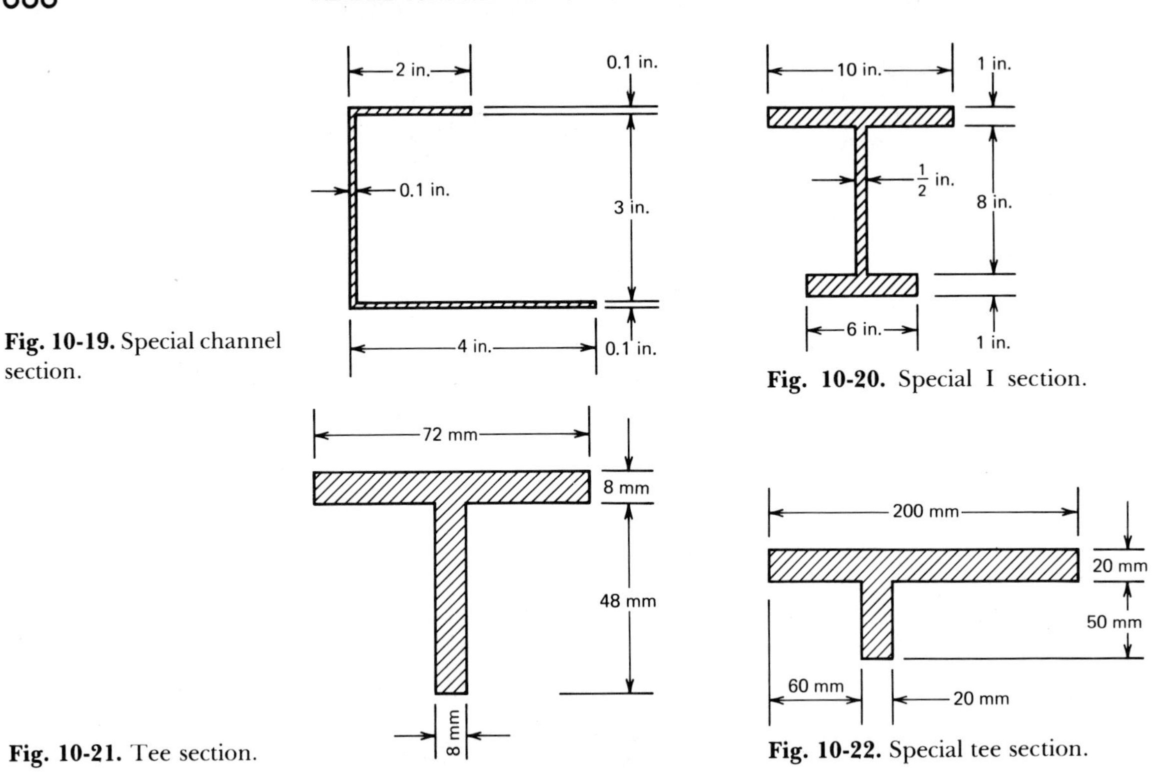

Fig. 10-19. Special channel section.

Fig. 10-20. Special I section.

Fig. 10-21. Tee section.

Fig. 10-22. Special tee section.

4. Calculate the maximum shear stress when a torque of 2 400 000 N·mm is applied to the section of Fig. 10-22.

10-8. TORSIONAL SHEAR STRESS IN THIN-WALLED CLOSED SECTIONS

Figure 10-23*a* shows a thin-walled closed section. The word closed means that the wall of the section completely encloses an area. Bars of such cross sections are often used in torsion. Now we will give you a method of calculating the maximum shearing stress.

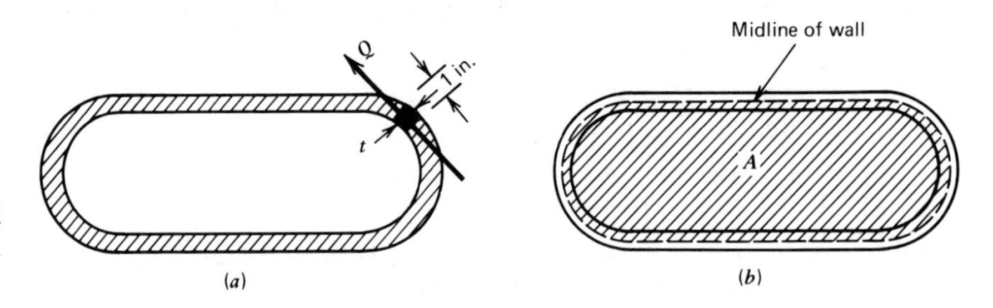

Fig. 10-23. Thin-walled closed section. (*a*) Section. (*b*) Section with midline.

The force Q in Fig. 10-23*a* is the *shear flow* which is the shear force (in pounds or newtons) per unit length of the wall of the section. The area to which Q is applied is $1 \times t$. Then the shearing stress is equal to the force Q divided by the area, and this is

$$S_s = \frac{Q}{t}$$

which shows that the shearing stress is maximum where the wall is thinnest if the wall thickness is not constant.

Figure 10-23*b* shows the cross section again. Here, the dotted line is the midline of the wall. The area enclosed by the midline is designated as A. Then the maximum shearing stress is

$$S_s = \frac{T}{2At}$$

where T is the torque, A is the area enclosed by the midline of the wall, and t is the thickness of the thinest part of the wall.

If you are working in gravitational units, you should express T in pound inches, A in square inches, and t in inches to get S_s in psi. If you are working in the metric system, you should express T in N·mm, A in mm^2, and t in millimeters to get S_s in MPa.

How about some examples?

Illustrative Example 1. A torque of 11 600 lb in. is applied to a bar with the cross section shown in Fig 10-24*a*. Find the maximum shearing stress.

Solution:

1. The torque T is 11 600 lb in.
2. Figure 10-24*b* shows the midline of the wall. The area enclosed by the midline is a rectangle 2.9 in. × 1.9 in. So,

$$A = 2.9 \times 1.9 = 5.51 \text{ in.}^2$$

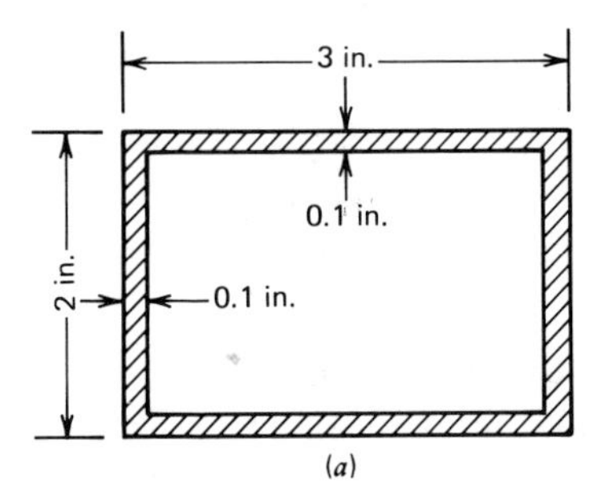

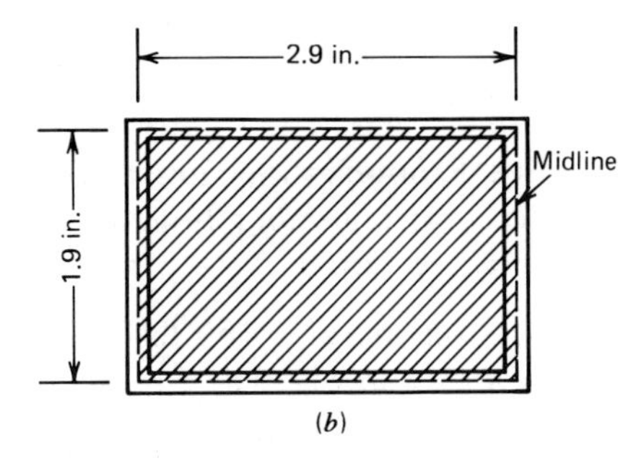

Fig. 10-24. Thin-walled closed section for Illustrative Example 1. (*a*) Section. (*b*) Section showing midline.

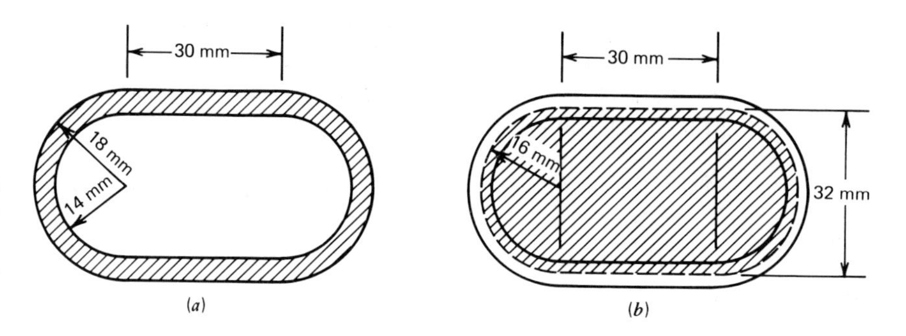

Fig. 10-25. Thin-walled closed section for Illustrative Example 2. (*a*) Section. (*b*) Section showing midline.

3. The wall thickness t is 0.1 in.
4. The maximum shearing stress is

$$S = \frac{11\ 600}{2 \times 5.51 \times 0.1} = 10\ 500 \text{ psi}$$

Illustrative Example 2. A torque of 956 000 N·mm is applied to a bar that has the cross section shown in Fig. 10-25*a*. What is the maximum shearing stress?

Solution:

1. The torque T is 956 000 N·mm.
2. The midline of the wall is shown in Fig. 10-25*b*. The area within the midline is a rectangle 30 mm × 32 mm plus two semicircles 32 mm in diameter. So,

$$A = 30 \times 32 + \frac{\pi}{4}(32)^2 = 960 + 804 = 1\ 764 \text{ mm}^2$$

3. The wall thickness t is 4 mm.
4. Then the maximum shearing stress is

$$S = \frac{956\ 000}{2 \times 1\ 764 \times 4} = 67.7 \text{ MPa}$$

Practice Problems (Section 10-8). Now practice on these problems so you will remember how to calculate shearing stress in thin-walled closed sections.

1. A torque of 9 160 lb in. is applied to a bar of the cross section shown in Fig. 10-26. What is the maximum shearing stress?
2. Calculate the maximum shearing stress when a torque of 31 700

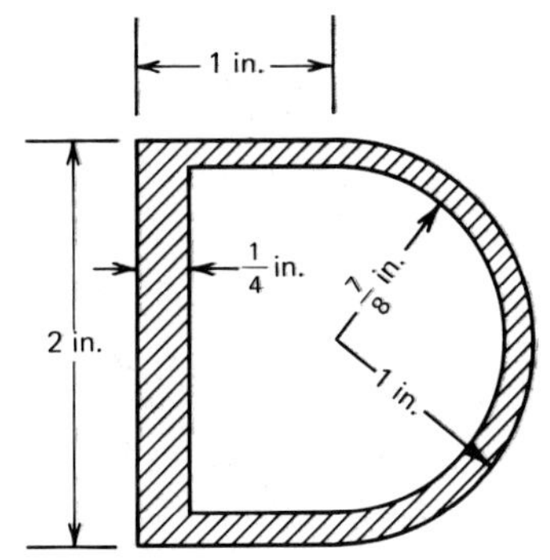

Fig. 10-26. Section for Problem 1.

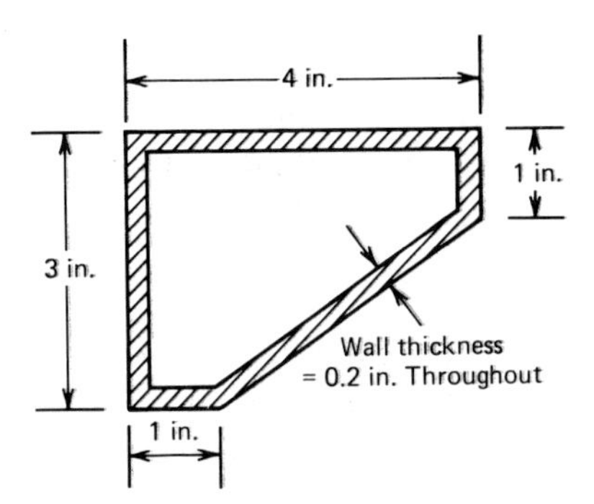

Fig. 10-27. Section for Problem 2.

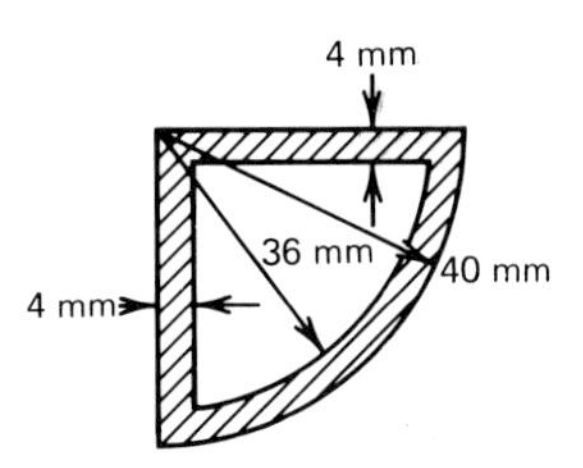

Fig. 10-28. Section for Problem 3.

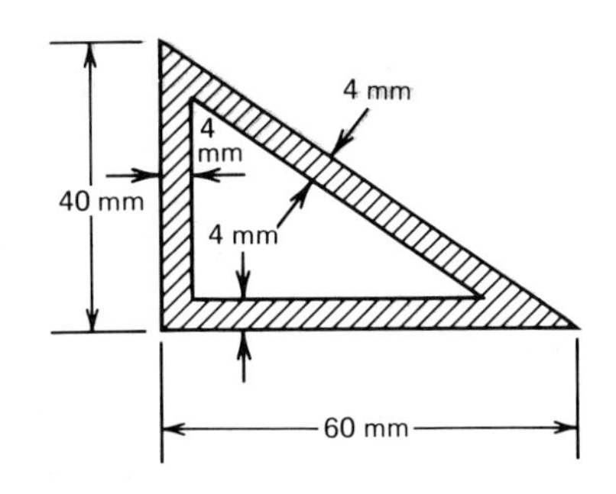

Fig. 10-29. Section for Problem 4.

lb in. is applied to a bar that has the cross section shown in Fig. 10-27.

3. What maximum shearing stress will result from a torque of 200 000 N·mm applied to a bar of the cross section shown in Fig. 10-28?

4. Figure 10-29 shows the cross section of a bar subjected to a torque of 492 000 N·mm. What is the maximum shearing stress?

SUMMARY

Study this summary of the main points in this chapter.

1. A *torque* is a twisting moment that tends to turn or rotate one end of a bar or shaft with respect to the other end. The amount of the torque at any cross section of a shaft is the moment with respect to the center line of the shaft of all forces to one side of the cross section.

2. The maximum torsional shearing stress in a solid circular shaft is

$$S_s = \frac{16}{\pi}\frac{T}{d^3}$$

where S_s is the stress in psi or megapascals, T is the torque in pound inches or newton millimeters, and d is the diameter of the shaft in inches or millimeters.

(a) The formula can be rewritten as,

$$T = \frac{\pi}{16} d^3 S_s$$

for the purpose of calculating the torque when the diameter of the shaft and the maximum torsional shearing stress are known.

(b) The formula can be expressed as,

$$d = \sqrt[3]{\frac{16}{\pi} \frac{T}{S}}$$

for the purpose of calculating the required diameter when the torque and stress are known.

3. The maximum torsional shearing stress in a hollow circular shaft is

$$S_s = \frac{16}{\pi} \frac{T d_2}{(d_2^4 - d_1^4)}$$

where d_2 is the outer diameter of the shaft and d_1 is the inner diameter.

4. The angle of twist of a solid circular shaft is

$$\theta = \frac{32}{\pi} \frac{TL}{E_s d^4}$$

where θ is the angle in radians (1 radian = 57.3°), T is the torque in pound inches or newton millimeters, L is the length of the shaft in inches or millimeters, E_s is the shearing modulus of elasticity of the material of the shaft in psi or megapascals (consult Table 3.1 for values of E_s) and d is the diameter of the shaft in inches or millimeters.

(a) The angle of twist of a hollow circular shaft of outer diameter d_2 and inner diameter d_1 is

$$\theta = \frac{32}{\pi} \frac{TL}{E_s (d_2^4 - d_1^4)}$$

5. The maximum torsional shearing stress in a rectangular bar is approximately

$$S_s = \frac{T (3h + 2b)}{b^2 h^2}$$

where S is the stress, T is the torque, h is the long side of the rectangle, and b is the short side of the rectangle.

6. The maximum torsional shearing stress in a thin-walled open section is

$$S_s = \frac{3Tb_{max}}{\Sigma hb^3}$$

where b_{max} is the maximum thickness of the section.

7. The maximum torsional shearing stress in a thin-walled closed section is

$$S_s = \frac{T}{2At}$$

where A is the area enclosed by the midline of the wall of the section and t is the smallest thickness of the wall.

REVIEW QUESTIONS You should be able to answer these questions without referring to the preceding pages.

1. What is a *torque*?
2. Distinguish between *bending moment* and *twisting moment.*
3. Why don't bearing reactions affect the torque on a straight shaft?
4. What formula would you use to calculate the maximum torsional shearing stress in a solid circular shaft?
5. How would you calculate the torque on a solid circular shaft if you knew the maximum stress and the diameter?
6. Give the formula you would use to find the required diameter for a solid circular shaft if you knew the torque and the working stress.
7. What is the formula for calculating the maximum torsional shearing stress in a hollow circular shaft?
8. State the formula for the angle of twist in a solid circular shaft.
9. What property of the material does the angle of twist depend on?
10. Give the approximate formula for the maximum torsional shearing stress in a rectangular bar.
11. What is the formula for the angle of twist in a hollow circular shaft?
12. How would you calculate the maximum torsional shearing stress in a thin-walled open section?
13. How would you calculate the maximum torsional shearing stress in a thin-walled closed section?

REVIEW PROBLEMS If you have studied this chapter thoroughly and really know it, you should be able to work these problems. If you can't work them, you need to study the chapter again.

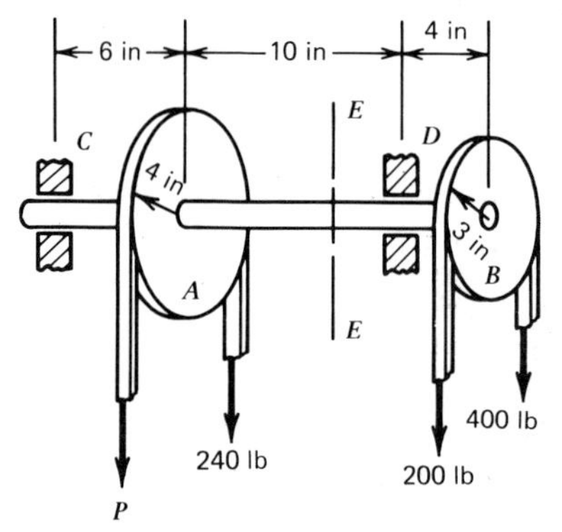

Fig. 10-30. Shaft and pulleys for Problem 1.

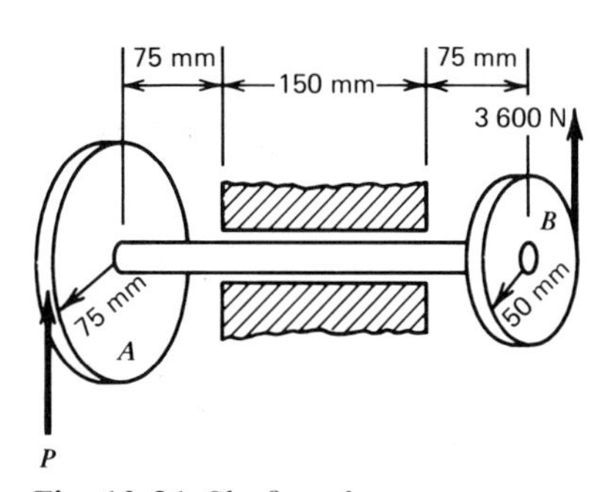

Fig. 10-31. Shaft and gears for Problem 2.

1. Figure 10-30 shows a shaft that carries the two pulleys A and B and is supported in bearings at C and D. The shaft is solid and is 1.2 in. in diameter. Calculate the maximum torsional shearing stress.
2. Figure 10-31 shows a solid circular shaft 24 mm in diameter that carries the gears A and B and is supported in a long bearing at the center. What is the maximum torsional shearing stress?
3. A solid circular shaft is to be subjected to a torque of 9 400 lb in., and the maximum torsional shearing stress is 11 000 psi. Find the required diameter.
4. What diameter is required for a solid circular shaft that is to carry a torque of 8 000 N·mm with a maximum torsional shearing stress of 42 MPa?
5. What torque must be applied to a solid circular shaft 2 in. in diameter to cause a maximum torsional shearing stress of 12 000 psi?
6. The maximum torsional shearing stress in a solid circular shaft 150 mm in diameter is 70 MPa. Find the torque.
7. A hollow circular shaft 3.8 in. outer diameter and 2.2 in. inner diameter is subjected to a torque of 2 200 lb ft. What is the maximum torsional shearing stress?
8. What is the maximum torsional shearing stress in a hollow circular shaft 12 mm outer diameter and 9 mm inner diameter from a torque of 16 000 N·mm?
9. A solid circular steel shaft 2.25 in. in diameter and 8 in. long is subjected to a torque of 18 000 lb in. What is the angle of twist?
10. A solid circular brass shaft 6 mm in diameter and 500 mm long carries a torque of 12 000 N·mm. Calculate the angle of twist.
11. A hollow circular gray cast-iron shaft 7.5 in. outer diameter, 5 in.

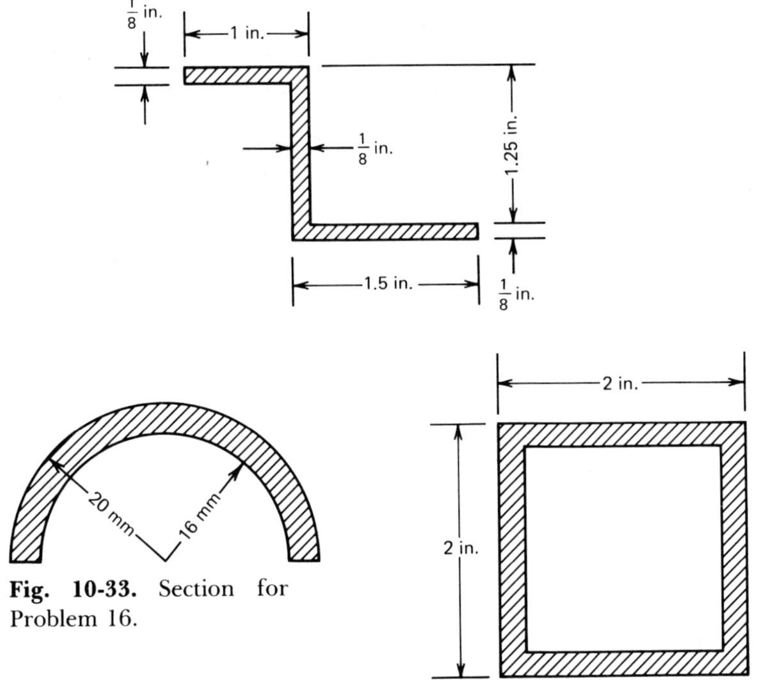

Fig. 10-32. Section for Problem 15.

Fig. 10-33. Section for Problem 16.

Fig. 10-34. Section for Problem 17.

inner diameter, and 8 ft long is subjected to a torque of 92 000 lb ft. Find the angle of twist.

12. What is the angle of twist for an aluminum-alloy tube 50 mm outer diameter, 36 mm inner diameter, and 1 100 mm long from a torque of 1.25×10^6 N·mm?
13. A rectangular bar 4 in. by 1.6 in. is subjected to a torque of 14 000 lb in. Calculate the maximum torsional shearing stress.
14. What is the maximum torsional shearing stress in a square bar 12 mm on a side, from a torque of 76 000 N·mm?
15. What is the maximum torsional shearing stress when a torque of 224 lb in. is applied to a bar which has the cross section shown in Fig. 10-32?
16. A torque of 9 700 N·mm is applied to a bar that has the cross section shown in Fig. 10-33. What is the maximum torsional shearing stress?
17. Figure 10-34 shows the cross section of a square bar. The wall thickness is 0.2 in. A torque of 8 500 lb in. is applied to the bar. What is the maximum torsional shearing stress?
18. Figure 10-35 shows the cross section of a bar. The wall thickness is 6 mm. A torque of 5 640 000 N·mm is applied to the bar. Calculate the maximum torsional shearing stress.

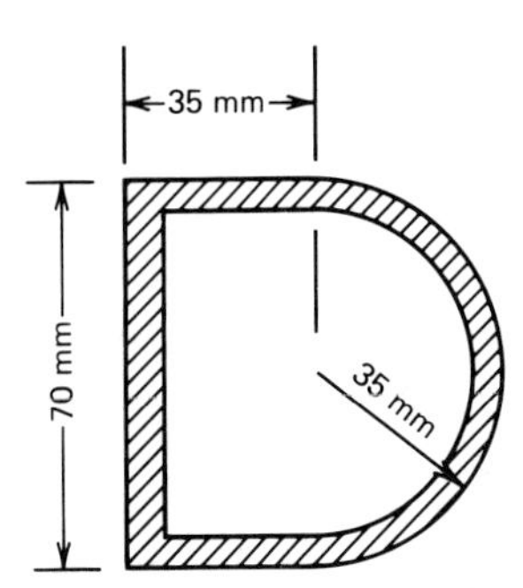

Fig. 10-35. Section for Problem 18.

11 BENDING MOMENT COMBINED WITH TORQUE

PURPOSE OF THIS CHAPTER. You remember that we studied beams and bending moment in Chapter 6. You learned what bending moment is, how to calculate it, how to find the maximum bending stress, and even how to design a beam. Then in Chapter 10 we studied torsion and torsional shearing stress. You learned how to calculate the torque and the maximum torsional shearing stress and how to design a shaft. Now in this chapter we will study problems in which a bar or shaft is subjected to both bending moment and torque at the same time. This is an important topic, because there are so many examples in design in which a bar carries both bending moment and torque. You will learn how to calculate the most important stress in the bar, and it will turn out to be maximum shearing stress. There is the bending stress due to the bending moment and the torsional shearing stress due to the torque; but neither of these is the most important stress, the one that really counts. Instead, the most important stress is usually considered to be the maximum shearing stress; we will study it in detail a little later.

Also in this chapter you are going to learn how to design a shaft that is subjected to both moment and torque. However, we will study only shafts of circular cross section.

11-1. REVIEW OF BENDING MOMENT AND MOMENT DIAGRAMS

Right here is a good place to review bending moment and moment diagrams. You need to know these things to understand this chapter, and there is no point in going on unless you do. Let's look at the free-body diagram of a beam in Fig. 11-1. The heavy line represents the beam; a load of 600 lb is applied at C, and a load of 400 lb at D; the beam is supported at points A and B, and you see the reactions R_1 (at A) and R_2 (at B).

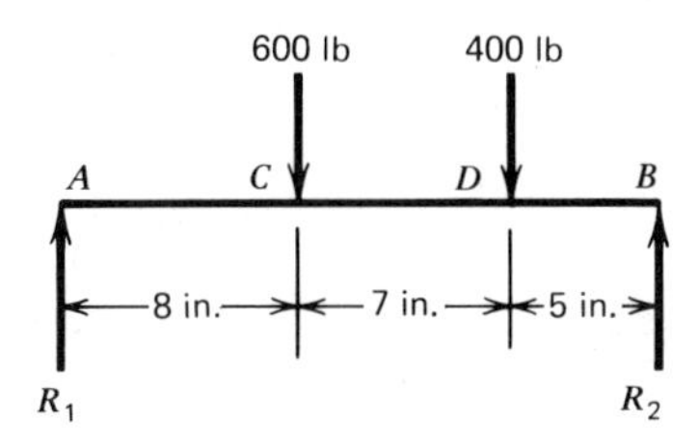

Fig. 11-1. Free-body diagram of beam.

The first step in the beam problem is to find the reactions, and we do this just as we did before. We write a moment equation with center at A ($\Sigma M_a = 0$) to find R_2, and then a moment equation with center at B ($\Sigma M_b = 0$) to find R_1.

The next thing is the bending moment. You remember that the bending moment at any point in the beam is the moment, with respect to that point, of all forces to the left of the point. If we think of point C, for instance, the only force to the left of C is the force R_1, and so the bending moment at C is just the moment of R_1 with respect to C. We say that a counterclockwise moment is *negative* and a clockwise moment is *positive.*

You remember that the moment diagram is a diagram that shows the value of the bending moment at each point along the beam, and you should remember, also, that when there is no distributed load on a beam, the moment diagram consists of straight lines between the points where the concentrated forces are applied. (You are fortunate that in this chapter there are no distributed loads in the type of problem we are going to study.) You have only to calculate the bending moments at points where loads are applied to the bar; then you plot the bending moment at these points, and connect the plotted points with straight lines. Now let's take an example.

Illustrative Example 1. Draw the moment diagram for the beam in Fig. 11-1.

Solution:

A. The reactions.

1. The reaction R_2 is found by writing a moment equation with center at A. (You should remember that the moment of a force with respect to a point is equal to the product of the force and the perpendicular distance from the point to the force. We choose point A because the force R_1 goes through A.) So,

$$\Sigma M_A = 0$$

$$600 \times 8 + 400 \times 15 - 20 R_2 = 0$$

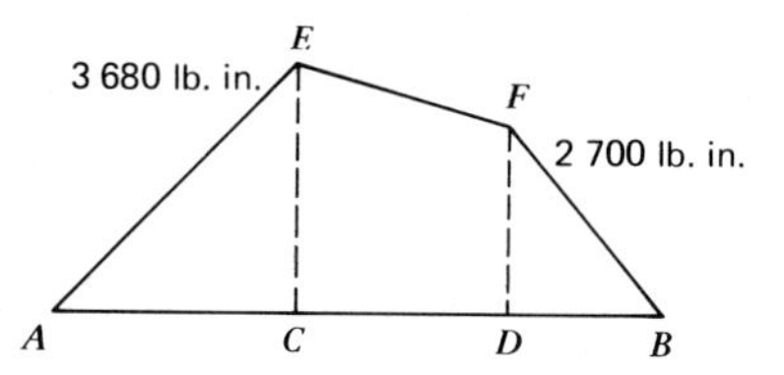

Fig. 11-2. Moment diagram of beam.

$$20R_2 = 600 \times 8 + 400 \times 15 = 4\,800 + 6\,000 = 10\,800$$

$$R_2 = 540 \text{ lb}$$

2. The reaction R_1 is found by writing a moment equation with center at B. Thus

$$\Sigma M_B = 0$$

$$20R_1 - 600 \times 12 - 400 \times 5 = 0$$

We transpose,

$$20R_1 = 600 \times 12 + 400 \times 5 = 7\,200 + 2\,000 = 9\,200$$

$$R_1 = 460 \text{ lb}$$

B. The moment diagram.

1. We start by drawing the base line *ACDB* in Fig. 11-2.
2. The bending moment at A is zero because there are no forces to the left of A.
3. The bending moment at C is the moment, with respect to C, of all forces to the left of C. This is just the moment of R_1 which is 460 lb. Thus

$$M_c = 460 \times 8 = 3\,680 \text{ lb in.}$$

The value of 3 680 lb in. is plotted upward from C to locate point E in the moment diagram, and the straight line AE is drawn.

4. The bending moment at D is the moment with respect to D of all forces to the left of D. So,

$$M_D = 460 \times 15 - 600 \times 7$$

$$= 6\,900 - 4\,200 = 2\,700 \text{ lb in.}$$

The value of 2 700 lb in. is plotted upward from D to locate point F in the moment diagram, and the straight line EF is drawn.

5. The bending moment at B is zero, because the moment of all forces to the left of B, with respect to B, is the moment of

all of the forces on the beam. The moment of all of the forces is zero, because the beam is in equilibrium. The straight line *FB* is drawn to complete the moment diagram.

11-2. REVIEW OF TORQUE

Let's do a thorough job of reviewing while we are at it and look at torque again. Remember that *torque* is twisting moment; that is, a moment that tends to twist the shaft. Let's look at Fig. 11-3, which shows the free-body diagram of a shaft that carries the pulleys *A* and *B* and is supported in bearings at *C* and *D*. The heavy line represents the shaft; a belt over pulley *A* exerts the forces of 240 lb and 480 lb, and a belt over pulley *B* exerts the forces *P* and 200 lb; the bearings exert the reactions R_1 (at *C*) and R_2 (at *D*).

The first thing to do is to find the force *P*, because it is one of the forces that has moment about the center of the shaft. We do this from Fig. 11-3. We can find *P* (as we did in several problems before) by writing a moment equation about the *x* axis.

The *torque,* or twisting moment, at any point in the shaft is the moment with respect to the *x* axis of all forces to the left of the point. Forces such as R_1 and R_2, which go through the center of the shaft, have no moment with respect to the *x* axis and do not contribute to the torque. For any point between the pulleys in Fig. 11-3, the torque is the moment of the forces of 240 lb and 480 lb with respect to the center of the shaft, because these are the only forces to the left of any such point that have moment about the center of the shaft.

Illustrative Example 1. Find the force *P* on the shaft in Fig. 11-3.

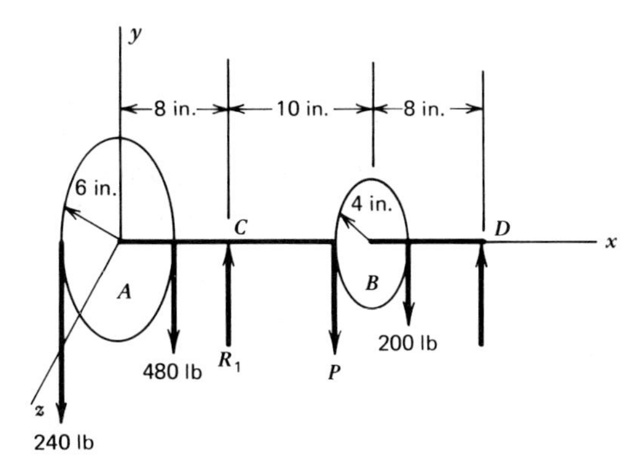

Fig. 11-3. Free-body diagram of shaft.

Solution:
The force P is found by writing a moment equation about the x axis. Thus

$$\Sigma M_x = 0$$

$$-240 \times 6 - 4P + 200 \times 4 + 480 \times 6 = 0$$

Transposing,

$$4P = -240 \times 6 + 200 \times 4 + 480 \times 6$$

$$= -1\,440 + 800 + 2\,880 = 2\,240$$

$$P = 560 \text{ lb}$$

Illustrative Example 2. Calculate the torque at point C in Fig. 11-3.

Solution:
The torque at C is the moment with respect to the x axis of the forces to the left of C. The forces to the left are the forces of 240 lb and 480 lb. From Fig. 11-3 the moment arm of each force is 6 in., so

$$T_c = -240 \times 6 + 480 \times 6 = -1\,440 + 2\,880$$

$$= 1\,440 \text{ lb in.}$$

11-3. TORQUE DIAGRAMS

A *torque diagram* is a diagram that shows the value of the torque at each point along the shaft. We like to have a torque diagram, because we can look at it and see where the torque is a maximum and how much it is at each point along the shaft. Then we can see where the torsional shearing stress is a maximum and calculate the stress at that point.

You already know how to calculate the torque at any point in a shaft. Now, to draw a torque diagram, you start by drawing a base line, just as you did for a moment diagram. Then you calculate the torque at different points along the shaft. You plot these torques and connect the plotted points. It isn't hard, and you can learn it if you follow a couple of examples.

Illustrative Example 1. Figure 11-4*a* shows the free-body diagram of a shaft that carries the pulleys A and B and is supported in bearings at C and D. Draw the torque diagram.

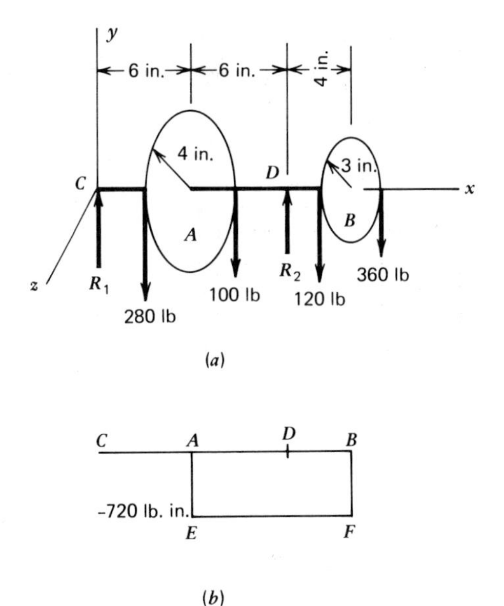

Fig. 11-4. Shaft for Illustrative Example 1. (*a*) Free-body diagram. (*b*) Torque diagram.

Solution:

1. The first step is to draw the base line *CADB* in Fig. 11-4*b*.
2. The torque is zero for any point between C and A because the force R_1, which is the only force to the left of any point in this length, goes through the center of the shaft and does not cause any twisting moment.
3. For any point between A and B the torque is the moment with respect to the x axis of the forces on pulley A.

$$T = -280 \times 4 + 100 \times 4 = -1\,120 + 400$$
$$= -720 \text{ lb in.}$$

This value of T is plotted downward from A to locate point E in the torque diagram. Then the line EF is drawn; there is no change in the torque between the pulleys A and B, because for any point in this length we have the same forces on pulley A to cause the torque. The line FB is drawn to finish the torque diagram.

A SPECIAL POINT

Torque diagrams are easy to draw, and you can see why when you look at Fig. 11-4*b*. The torque doesn't change between a pulley such as A and the next pulley B, and the same thing would be true if the shaft carried gears instead of pulleys. This means that you only have to calculate the torque between each pair of gears or pulleys. Let's try it again.

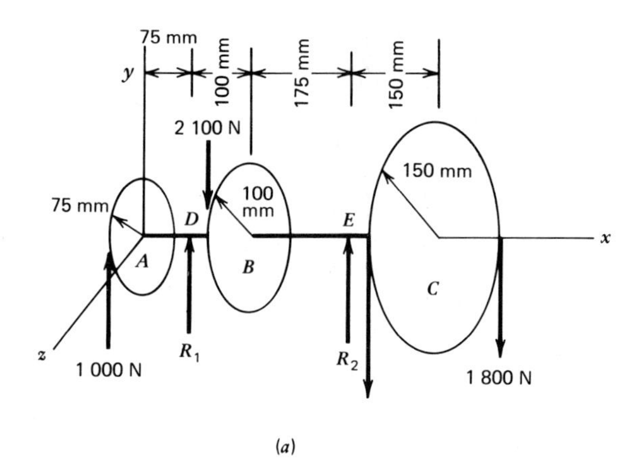

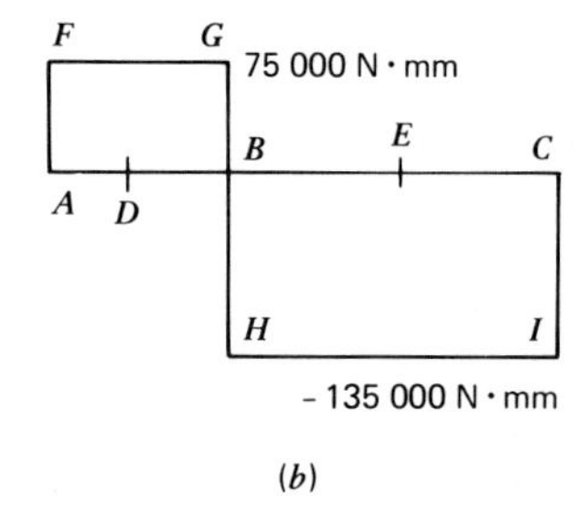

Fig. 11-5. Shaft for Illustrative Example 2. (*a*) Free-body diagram. (*b*) Torque diagram.

Illustrative Example 2. Figure 11-5*a* shows the free-body diagram of a shaft that carries the gears *A* and *B* and the pulley *C*; the shaft is supported in bearings that exert the reactions R_1 and R_2. Draw the torque diagram.

Solution:

1. The first step is to draw the base line *ADBEC* in Fig. 11-5*b*.
2. Now look at Fig. 11-5*a* to see which forces are to the left of a certain point. For any point between the gears *A* and *B*, we only have to consider the 1 000-N force on *A*, because it is the only force to the left of the point that has any torque. So

$$T = 1\ 000 \times 75 = 75\ 000 \text{ N·mm}$$

   This torque is plotted upward from *A* to locate point *F* in the torque diagram. Then the line *FG* is drawn.
3. For any point between the gear *B* and the pulley *C*, we must consider the 1 000-N force on *A* and the 2 100-N force on *B*. Then,

$$T = 1\,000 \times 75 - 2\,100 \times 100 = 75\,000 - 210\,000$$
$$= -135\,000 \text{ N·mm}$$

The value of $-135\,000$ is plotted downward from B to locate point H in the torque diagram. The lines BH, HI, and IC are drawn to complete the diagram.

Practice Problems (Section 11-3). Just draw the torque diagrams for the shafts shown in the figures in this list.

1. Figure 11-6.
2. Figure 11-7.
3. Figure 11-8.
4. Figure 11-9.

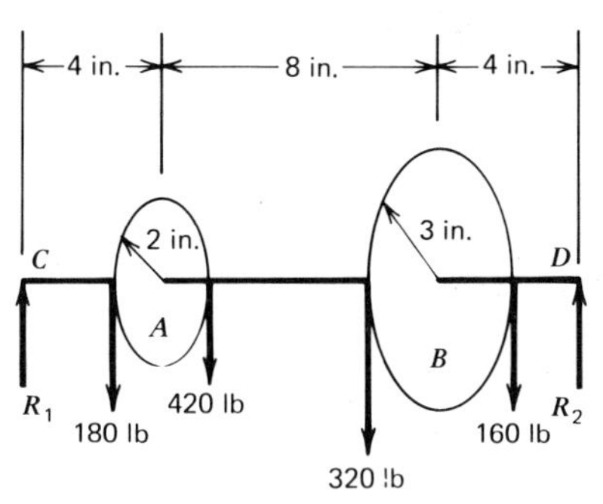

Fig. 11-6. Shaft for Problem 1.

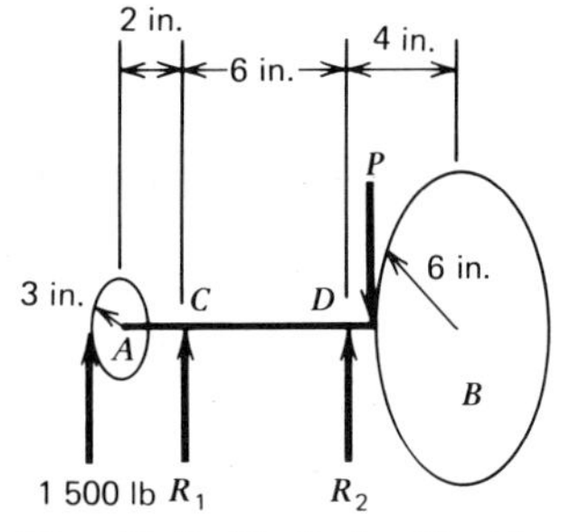

Fig. 11-7. Shaft for Problem 2.

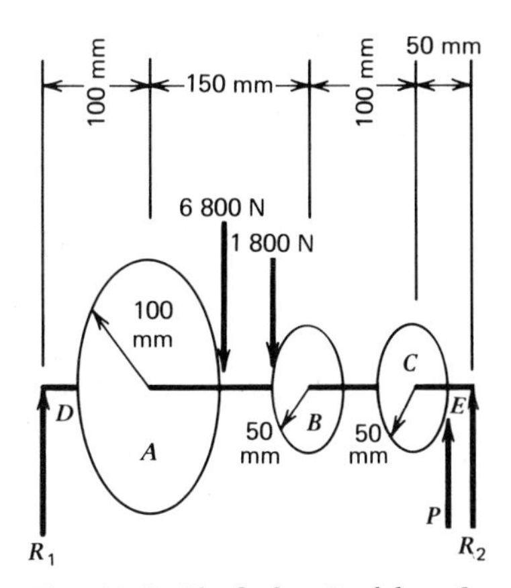

Fig. 11-8. Shaft for Problem 3.

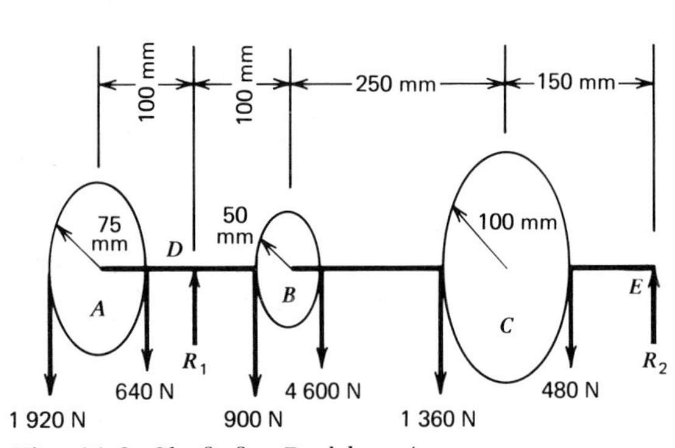

Fig. 11-9. Shaft for Problem 4.

11-4. MOMENT DIAGRAMS FOR SHAFTS

We drew moment diagrams for beams, and we can draw moment diagrams for shafts. We calculate the reactions so that we know all the forces, and then we calculate the bending moments and plot them. For a shaft, we look at the forces in a plane parallel to the shaft and work with what we see there. An example will show you how to do it. The first step is to draw this projection, the second is to calculate the reactions, and the last is to draw the moment diagram.

Illustrative Example 1. Figure 11-10*a* shows a shaft that carries the pulleys *A* and *B* and is supported in bearings at *C* and *D*; the bearings exert the reactions R_1 and R_2. Draw the moment diagram.

Solution:

A. The projection in a plane parallel to the shaft. First we draw the projection of the shaft in a plane parallel to the shaft, as in Fig. 11-10*b*. At *A* we see the sum of the belt pulls on pulley *A*, and this is

$$260 + 140 = 400 \text{ lb}$$

At *B* we see the sum of the belt pulls on pulley *B*, and this is

$$90 + 162 = 252 \text{ lb}$$

We see the reaction R_1 at *C* and the reaction R_2 at *D*. Now we have the picture to work with.

B. The reactions.

1. We find the reaction R_2 by writing a moment equation with center at *C*. Thus

$$\Sigma M_c = 0$$

$$400 \times 4 - 12R_2 + 252 \times 17 = 0$$

Transposing,

$$12R_2 = 400 \times 4 + 252 \times 17 = 1\,600 + 4\,284 = 5\,884$$

$$R_2 = 490 \text{ lb}$$

2. We find the reaction R_1 by writing a moment equation with center at *D*. (Why at *D*?) So,

$$\Sigma M_d = 0$$

$$12R_1 - 400 \times 8 + 252 \times 5 = 0$$

Transposing,

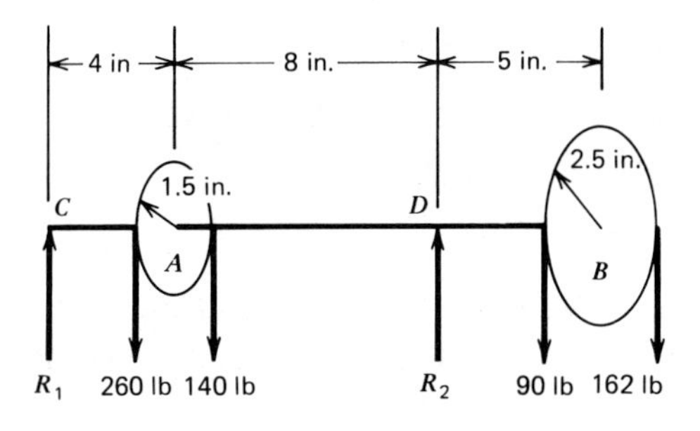

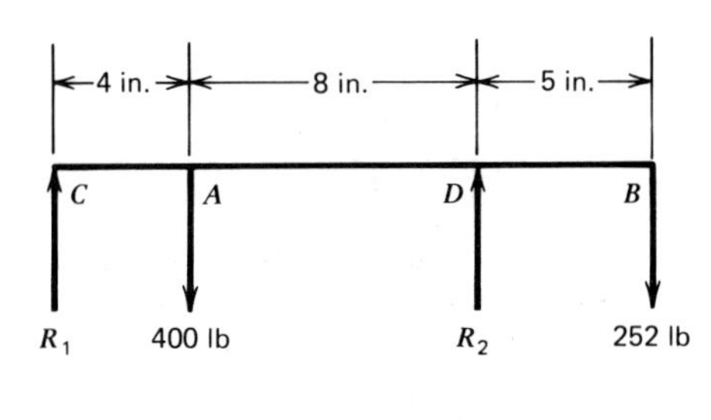

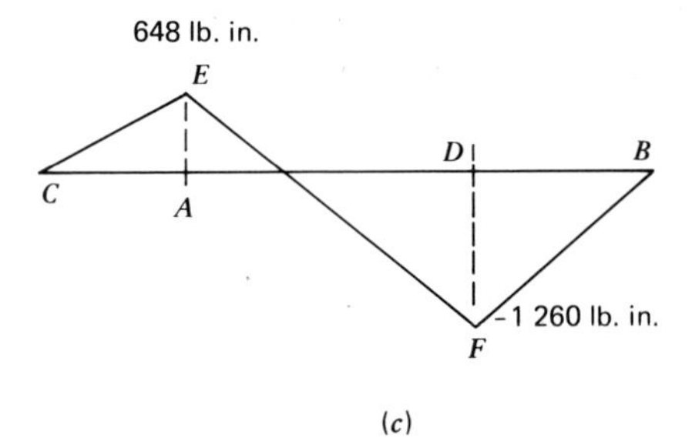

Fig. 11-10. Shaft for Illustrative Example 1. (*a*) Free-body diagram. (*b*) Projection in plane parallel to shaft. (*c*) Moment diagram.

$$12R_1 = 400 \times 8 - 252 \times 5 = 3\,200 - 1\,260 = 1\,940$$

$$R_1 = 162 \text{ lb}$$

C. The moment diagram.

1. We start the moment diagram by drawing the base line *CADB* in Fig. 11-10*c*. Remember now that the *bending moment* at any point is equal to the moment, with respect to that point, of all forces to the left of the point.
2. The bending moment at *C* is zero, because there are no forces to the left of *C*.

3. The bending moment at A is

$$M_a = 4\,R_1$$

and, since $R_1 = 162$ lb

$$M_a = 4 \times 162 = 648 \text{ lb in.}$$

The value of 648 lb in. is plotted upward (upward because it is positive) from A to locate point E in the moment diagram, and the straight line CE is drawn.

4. The bending moment at D is

$$M_d = 12R_1 - 400 \times 8$$

Using $R_1 = 162$ lb

$$M_d = 12 \times 162 - 400 \times 8 = 1\,940 - 3\,200$$

$$= -1\,260 \text{ lb in.}$$

This value of $-1\,260$ lb in. is plotted downward (downward because it is negative) from D to locate point F in the moment diagram. Then the straight lines EF and FB are drawn to complete the moment diagram.

Practice Problems (Section 11-4). Practice by drawing moment diagrams for the shafts in the figures in this list.

1. Figure 11-6.
2. Figure 11-7.
3. Figure 11-8.
4. Figure 11-9.

11-5. MAXIMUM SHEARING STRESS IN A SOLID CIRCULAR SHAFT SUBJECTED TO BOTH BENDING MOMENT AND TORQUE

Now we are ready to calculate the most important stress in the shaft. We are thinking of a circular shaft of diameter d, which is subjected to both bending moment and torque. You see two views of such a shaft in Fig. 11-11, and you already know that there is a tensile stress S_t (due to the bending moment) on the cross section B-B; also, you know that there is a shearing stress S_s (due to the torque) on the cross section B-B. However, neither of these stresses is the most important stress in the shaft, because neither one of them is the stress that will cause failure. Instead, the most important stress is usually considered to be the maximum shear-

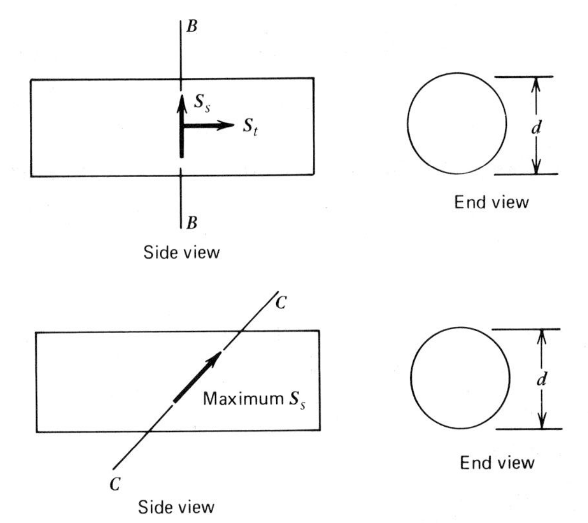

Fig. 11-11. Solid circular shaft.

Fig. 11-12. Maximum shearing stress in shaft.

ing stress, and it occurs on an oblique cross section such as *C-C* in Fig. 11-12. This stress is due to the combination of bending moment and torque. We don't derive the formula for you here because we would have to use advanced mathematics. But we do give you the formula and show you how to use it. Here it is:

$$\max S_s = \frac{16}{\pi d^3}\sqrt{M^2 + T^2}$$

where S_s is the stress in psi or megapascals, d is the diameter of the shaft in inches or millimeters; M is the bending moment in pound inches or newton millimeters, and T is the torque in pound inches or newton millimeters. Let's use it.

Illustrative Example 1. A solid circular shaft 2 in. in diameter is subjected to a bending moment of 7 200 lb in. and a torque of 9 300 lb in. Calculate the maximum shearing stress.

Solution:

1. The bending moment M is 7 200 lb in.
2. The torque T is 9 300 lb in.
3. The diameter d is 2 in.
4. The maximum shearing stress is

$$\max S_s = \frac{16}{\pi d^3}\sqrt{M^2 + T^2}$$

$$= \frac{16}{\pi(2)^3}\sqrt{(7\,200)^2 + (9\,300)^2} = 7\,490 \text{ psi}$$

Illustrative Example 2. What is the maximum shearing stress in a solid shaft 150 mm in diameter that is subjected to a bending moment of 22.4×10^6 N·mm and a torque of 13.8×10^6 N·mm?

Solution:

1. The bending moment M is 22.4×10^6 N·mm.
2. The torque T is 13.8×10^6 N·mm.
3. The diameter d is 150 mm.
4. The maximum shearing stress is

$$\max S_s = \frac{16}{\pi d^3}\sqrt{M^2 + T^2}$$

$$= \frac{16}{\pi(150)^3}\sqrt{(22.4 \times 10^6)^2 + (13.8 \times 10^6)^2} = 39.7 \text{ MPa}$$

THE WHOLE PROBLEM

This is a good start, but it is a long way from the complete problem. Now let's commence at the beginning with the picture of the shaft and go all the way through. We will draw the moment diagram and the torque diagram so that we can pick the point in the shaft where the bending moment and the torque are the greatest, because the larger the bending moment and torque, the larger the maximum shearing stress. You can see this in the formula

$$\max S_s = \frac{16}{\pi d^3}\sqrt{M^2 + T^2}$$

We must use the maximum values of M and T to get the greatest value of the maximum shearing stress.

Illustrative Example 3. Figure 11-13*a* shows a shaft that carries the pulleys A and B and is supported in bearings at C and D. The shaft is 1.25 in. in diameter. Find the maximum shearing stress.

Solution:

A. The free-body diagram.

   1. Figure 11-13*b* shows the free-body diagram. The bearing reactions are R_1 (at C) and R_2 (at D).

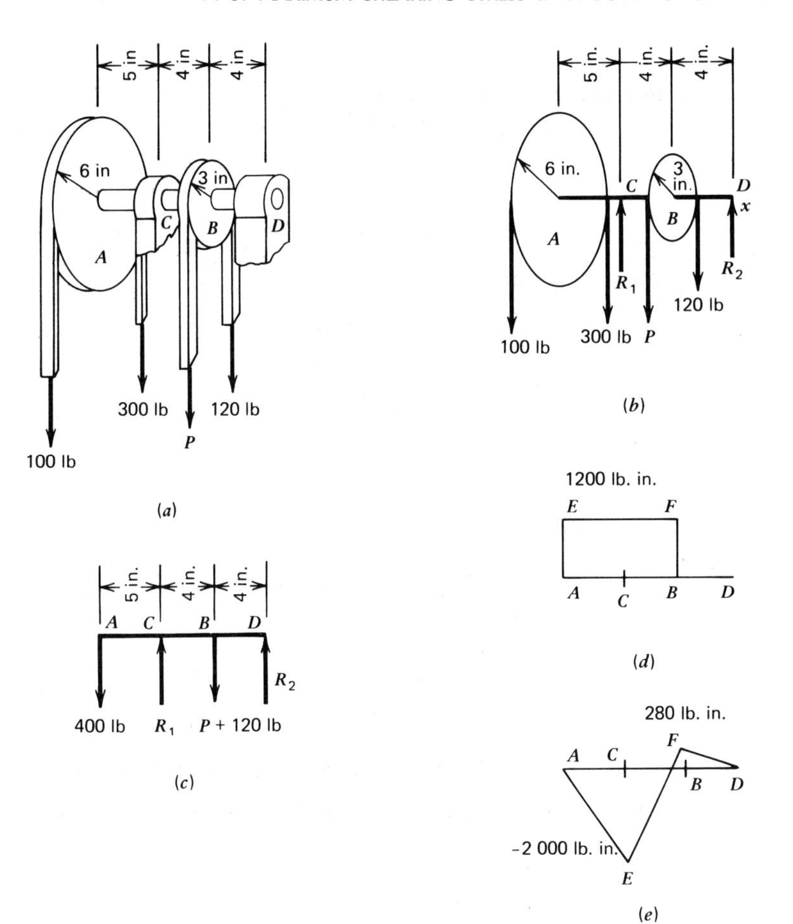

Fig. 11-13. Shaft for Illustrative Example 3. (*a*) Shaft and pulleys. (*b*) Free-body diagram. (*c*) Projection in plane parallel to shaft. (*d*) Torque diagram. (*e*) Moment diagram.

2. Figure 11-13*c* shows the projection of the forces in a plane parallel to the shaft. At A you see the sum of the belt pulls on pulley A, and this is

$$100 + 300 = 400 \text{ lb}$$

At B you see the sum of the forces P and 120 lb.

B. The reactions.
The reactions are the unknown forces P, R_1, and R_2.

1. The force P is found by writing a moment equation about the x axis in Fig. 11-13*b*. Thus

$$\Sigma M_x = 0$$

$$-100 \times 6 - 3P + 120 \times 3 + 300 \times 6 = 0$$

Transposing the term containing P,

$$3P = -600 + 360 + 1\ 800 = 1\ 560$$

$$P = 520 \text{ lb}$$

2. The reaction R_2 is found by writing a moment equation with center at C in Fig. 11-13c. Like this,

$$\Sigma M_c = 0$$

$$-400 \times 5 + (P + 120) \times 4 - 8R_2 = 0$$

Transposing the term containing R_2,

$$8R_2 = -400 \times 5 + (P + 120)4 = -2\ 000 + 4P + 480$$

$$= -1\ 520 + 4P$$

Using $P = 520$ lb

$$8R_2 = -1\ 520 + 4 \times 520 = -1\ 520 + 2\ 080 = 560$$

$$R_2 = 70 \text{ lb}$$

3. The reaction R_1 is found by writing a moment equation with center at D in Fig. 11-13c. So,

$$\Sigma M_d = 0$$

$$-400 \times 13 + 8R_1 - (P + 120)4 = 0$$

Transposing all of the terms except the term containing R_1

$$8R_1 = 400 \times 13 + (P + 120)4 = 5\ 200 + 4P + 480$$

$$= 5\ 680 + 4P$$

Using $P = 520$ lb,

$$8R_1 = 5\ 680 + 4 \times 520 = 5\ 680 + 2\ 080 = 7\ 760$$

$$R_1 = 970 \text{ lb}$$

C. The torque diagram.

1. The torque diagram is started by drawing the base line $ACBD$ in Fig. 11-13d.
2. For any point between A and B the torque is

   $$T = -100 \times 6 + 300 \times 6 = -600 + 1\ 800 = 1\ 200 \text{ lb in.}$$

   This value of 1 200 lb in. is plotted upward from A to locate point E in the torque diagram, and the line EF is drawn.
3. For any point between B and D the torque is zero, because all the forces that have twisting moment are to the left of the point. The moment of all of the forces must be zero, since the shaft is in equilibrium. The line FB is drawn to complete the torque diagram.

D. The moment diagram.

1. The moment diagram is started by drawing the base line $ACBD$ in Fig. 11-13e.
2. The bending moment at A is zero, because there are no forces to the left of point A.
3. The bending moment at C is

$$M_c = -400 \times 5 = -2\,000 \text{ lb in.}$$

This value of $-2\,000$ lb in. is plotted downward from C to locate point E in the moment diagram, and the straight line AE is drawn.

4. The bending moment at B is

$$M_b = -400 \times 9 + 4R_1 = -3\,600 + 4R_1$$

Using $R_1 = 970$ lb,

$$M_b = -3\,600 + 4 \times 970 = -3\,600 + 3\,880 = 280 \text{ lb in.}$$

The value of 280 lb in. is plotted upward from B to locate point F in the moment diagram, and the straight line EF is drawn.

5. The bending moment at D is zero, because all the forces are to the left of D, and the moment of all of the forces is zero. The line FD is drawn to complete the moment diagram.

E. The maximum shearing stress.

1. The maximum bending moment M is 2 000 lb in., as you can see in Fig. 11-13e. Here the numerical value is the most important, and you pay no attention to the negative sign.
2. The maximum torque T is 1 200 lb in., as you can see in Fig. 11-13e. The maximum M and the maximum T both occur at point C.
3. The diameter d is 1.25 in.
4. The maximum shearing stress is

$$\max S_s = \frac{16}{\pi d^3}\sqrt{M^2 + T^2}$$

$$= \frac{16}{\pi(1.25)^3}\sqrt{(2\,000)^2 + (1\,200)^2} = 6\,080 \text{ psi}$$

MAXIMUM SHEARING STRESS IN A HOLLOW CIRCULAR SHAFT

You must know how to find the maximum shearing stress in a hollow circular shaft, too. Figure 11-14 shows the cross section of a hollow

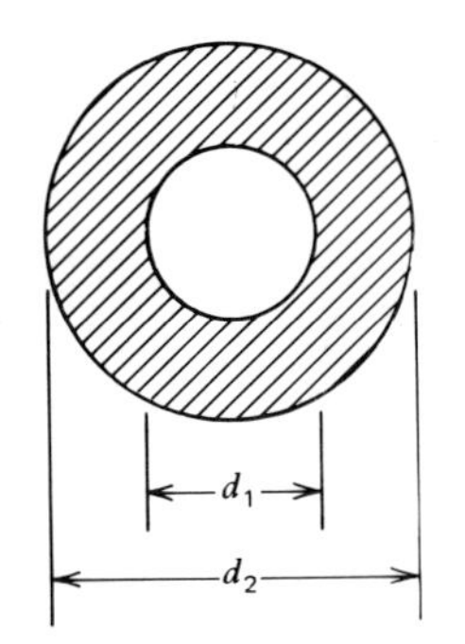

Fig. 11-14. Cross section of hollow shaft.

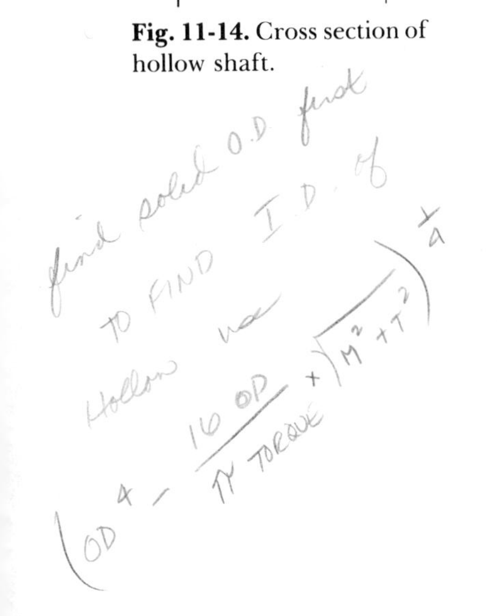

circular shaft; we will call the outer diameter d_2 and the inner diameter d_1. The formula for the maximum shearing stress is

$$\max S_s = \frac{16 d_2}{\pi(d_2^4 - d_1^4)} \sqrt{M^2 + T^2}$$

a little harder than the other formula, but not bad.

Illustrative Example 4. A hollow shaft 72 mm outer diameter and 48 mm inner diameter is subjected to a bending moment of 2.6×10^6 N·mm and a torque of 3.7×10^6 N·mm. Calculate the maximum shearing stress.

Solution:

1. The bending moment M is 2.6×10^6 N·mm.
2. The torque T is 3.7×10^6 N·mm.
3. The outer diameter d_2 is 72 mm.
4. The inner diameter d_1 is 48 mm.
5. The maximum shearing stress is

$$\max S_s = \frac{16 d_2}{\pi(d_2^4 - d_1^4)} \sqrt{M^2 + T^2}$$

$$= \frac{16 \times 72}{\pi[(72)^4 - (48)^4]} \sqrt{(2.6 \times 10^6)^2 + (3.7 \times 10^6)^2}$$

$$= 76.9 \text{ MPa}$$

Illustrative Example 5. What is the maximum shearing stress in a hollow shaft ½ in. outer diameter and ⅜ in. inner diameter that is subjected to a bending moment of 90 lb in. and a torque of 65 lb in.?

Solution:

1. The bending moment M is 90 lb in.
2. The torque T is 65 lb in.
3. The outer diameter d_2 is ½ in.
4. The inner diameter d_1 is ⅜ in.
5. The maximum shearing stress is

$$\max S_s = \frac{16 d_2}{\pi(d_2^4 - d_1^4)} \sqrt{M^2 + T^2}$$

$$= \frac{16 \times \frac{1}{2}}{\pi[(\frac{1}{2})^4 - (\frac{3}{8})^4]} \sqrt{(90)^2 + (65)^2} = 6\,620 \text{ psi}$$

Practice Problems (Section 11-5). Work the following problems:

1. A solid circular shaft 3 in. in diameter is subjected to a bending moment of 40 000 lb in. and a torque of 38 000 lb in. What is the maximum shearing stress?
2. What is the maximum shearing stress in a solid circular shaft 6 mm in diameter, subjected to a bending moment of 180 N·mm and a torque of 370 N·mm?
3. The shaft in Fig. 11-15 carries the gears *A* and *B* and is supported in bearings at *C* and *D*. The shaft is solid and is 1.2 in. in diameter. Calculate the maximum shearing stress.
4. In Fig. 11-16 the shaft is solid and is 20 mm in diameter. Find the maximum shearing stress.
5. A hollow circular shaft, 1.5 in. outer diameter and 1 in. inner diameter, is subjected to a bending moment of 5 300 lb in. and a torque of 3 500 lb in. What is the maximum shearing stress?
6. What is the maximum shearing stress in a hollow circular shaft, 8 in. outer diameter and 6 in. inner diameter, from a bending moment of 580 000 lb in. and a torque of 430 000 lb in.?

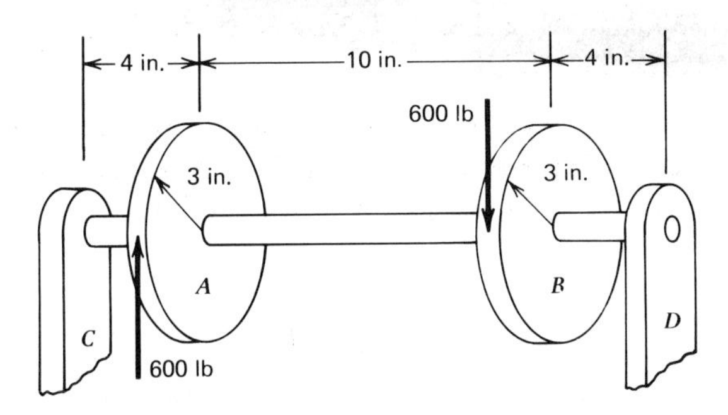

Fig. 11-15. Shaft for Problem 3.

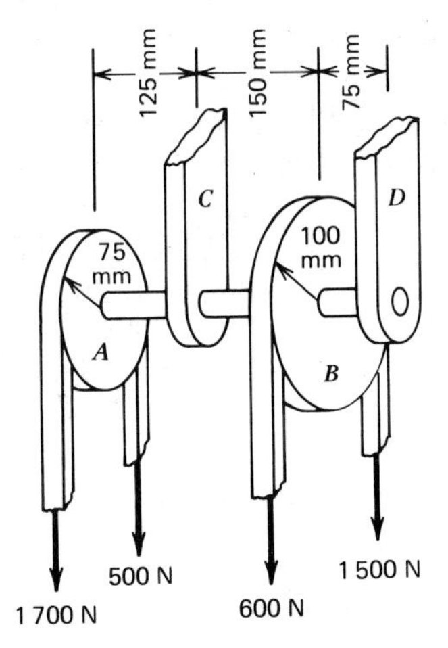

Fig. 11-16. Shaft for Problem 4.

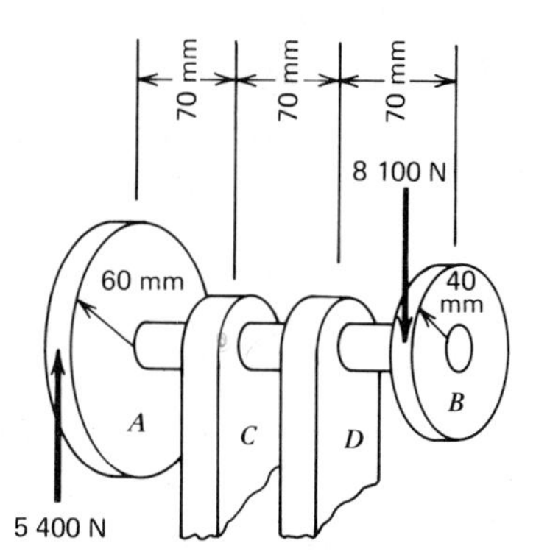

Fig. 11-17. Shaft for Problem 7.

7. The shaft in Fig. 11-17 is hollow, with outer diameter of 52 mm and inner diameter of 28 mm. Calculate the maximum shearing stress.

11-6. HOW TO FIND THE DIAMETER OF A SOLID CIRCULAR SHAFT UNDER BENDING MOMENT AND TORQUE

One of the real problems of machine design is that of finding the diameter of a shaft subjected to both bending moment and torque. We can use the same formula for this, but we will have to turn it around. We wrote it as,

$$\max S_s = \frac{16}{\pi d^3}\sqrt{M^2 + T^2}$$

where S_s is the maximum stress in psi or megapascals, d is the diameter of the solid circular shaft in inches or millimeters, M is the bending moment in pound inches or newton millimeters, and T is the torque in pound inches or newton millimeters. Let's multiply both sides of this equation by d^3 and divide by max S_s. Then we will have

$$\frac{d^3}{\max S_s}\max S_s = \frac{d^3\,16}{\max S_s\,\pi d^3}\sqrt{M^2 + T^2}$$

Next, we will cancel max S_s on the left side of the equation and d^3 on the right side. Then,

$$d^3 = \frac{16}{\pi(\max S_s)}\sqrt{M^2 + T^2}$$

The next step is to use this formula.

Illustrative Example 1. A solid circular shaft is to be subjected to a bending moment of 18 000 lb in. and a torque of 22 000 lb in. The maximum shearing stress is not to exceed 9 000 psi. What diameter of shaft is required?

Solution:

1. The bending moment M is 18 000 lb in.
2. The torque T is 22 000 lb in.
3. The max S_s is 9 000 psi.
4. The cube of the diameter is

$$d^3 = \frac{16}{\pi(\max S_s)}\sqrt{M^2 + T^2}$$

or $d = \left(\frac{16}{\pi\,(\text{SHEAR STRESS})}\sqrt{M^2+T^2}\right)^{\frac{1}{3}}$

$$= \frac{16}{\pi(9\ 000)}\sqrt{(18\ 000)^2 + (22\ 000)^2} = 16.09$$

5. Then, $d = 2.52$ in.

Illustrative Example 2. What diameter of solid circular shaft is needed to withstand a bending moment of 360 000 N·mm and a torque of 180 000 N·mm, with a maximum shearing stress of 82 MPa?

Solution:

1. The bending moment is 360 000 N·mm.
2. The torque is 180 000 N·mm.
3. The maximum S_s is 82 MPa.
4. The cube of the diameter is

$$d^3 = \frac{16}{\pi(82)}\sqrt{(360\ 000)^2 + (180\ 000)^2}$$

$$= 25\ 000$$

5. Then,

$$d = 29.2 \text{ mm}$$

Practice Problems (Section 11-6). Calculate the following shaft diameters:

1. Calculate the diameter of a solid circular shaft to carry a bending moment of 92 000 lb in. and a torque of 53 000 lb in. with a maximum shearing stress of 9 000 psi.
2. A solid circular shaft is to be subjected to a bending moment of

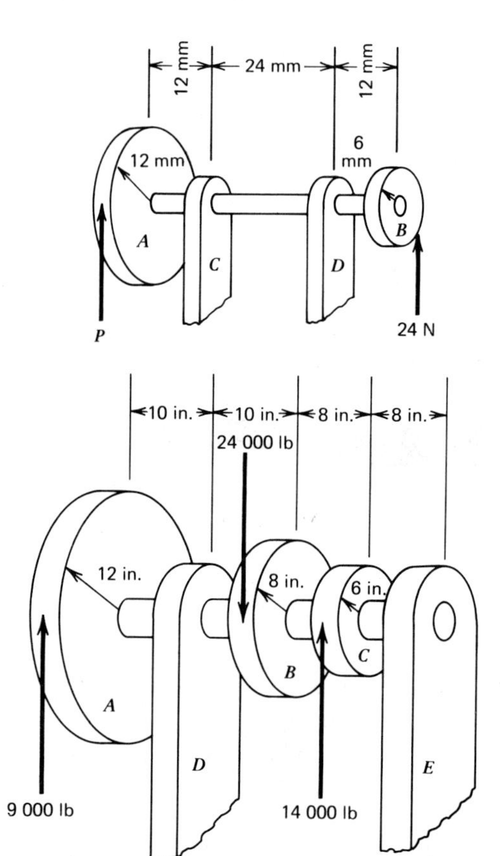

Fig. 11-18. Shaft for Problem 4.

Fig. 11-19. Shaft for Problem 5.

28 000 N·mm and a torque of 41 000 N·mm. The maximum shearing stress is not to exceed 70 MPa. What diameter of shaft is required?

3. What diameter of solid circular shaft is required to carry a bending moment of 62 000 lb in. and a torque of 42 000 lb in. with a maximum shearing stress of 11 600 psi?
4. The maximum shearing stress in the solid circular shaft in Fig. 11-18 is not to exceed 28 MPa. What diameter is required?
5. Find the required diameter for the solid circular shaft in Fig. 11-19 so that the maximum shearing stress will not exceed 8 000 psi.

SUMMARY Here is a summary that covers the main points in this chapter.

1. A bending moment is a moment that tends to bend a bar or shaft.
   (a) The bending moment at any point in a shaft is equal to the

moment with respect to that point of all forces to the left of the point. Counterclockwise moments are negative, and clockwise moments are positive.

(b) The bending moment in a shaft can be calculated most easily from a projection of the forces in a plane parallel to the shaft.

(c) A moment diagram is a diagram that shows the value of the bending moment at each point along the shaft.

2. A torque is a moment that tends to twist a bar or shaft.

(a) The torque at any point in a shaft is equal to the twisting moment with respect to that point of all forces to the left of the point. Counterclockwise torques are negative, and clockwise torques are positive.

(b) A torque diagram is a diagram that shows the value of the torque at each point along the shaft.

3. The maximum shearing stress, which occurs on an oblique plane, is usually considered to be the most important stress in a shaft that is subjected to both bending moment and torque.

(a) The value of the maximum shearing stress in a solid circular shaft is

$$\max S_s = \frac{16}{\pi d^3} \sqrt{M^2 + T^2}$$

Here S_s is the stress in psi or megapascals, d is the diameter in inches or millimeters, M is the bending moment in pound inches or newton millimeters, and T is the torque in pound inches or newton millimeters.

(b) The value of the maximum shearing stress in a hollow circular shaft is

$$\max S_s = \frac{16 d_2}{\pi (d_2^4 - d_1^4)} \sqrt{M^2 + T^2}$$

Here d_2 is the outer diameter of the shaft, and d_1 is the inner diameter.

(c) The diameter required for a solid circular shaft in which a certain value of the maximum shearing stress is not to be exceeded is

$$d = \sqrt[3]{\frac{16}{\pi (\max S_s)} \sqrt{M^2 + T^2}}$$

REVIEW QUESTIONS

Now answer these questions without looking at the preceding pages.

1. What is *torque*?

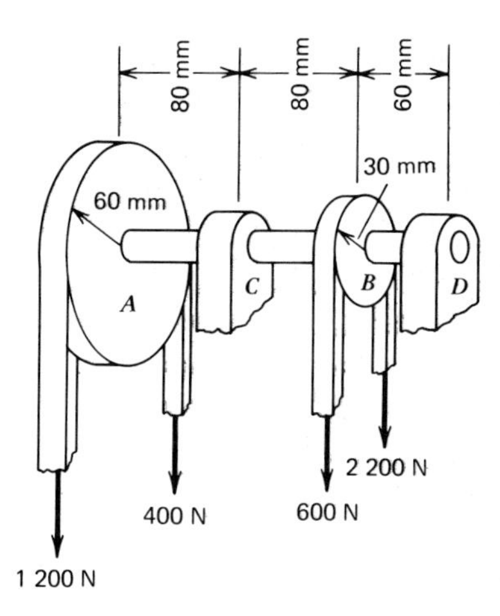

Fig. 11-20. Shaft for Problem 2.

2. What is *bending moment*?
3. What is a *torque diagram*?
4. What is a *moment diagram*?
5. What is the difference between maximum shearing stress and torsional shearing stress?
6. State the formula for the maximum shearing stress in a solid circular shaft subjected to both bending moment and torque.
7. What is the maximum shearing stress in a hollow circular shaft subjected to both bending moment and torque?
8. How would you find the diameter required for a solid circular shaft that is to be subjected to both bending moment and torque?

REVIEW PROBLEMS Try these problems to see whether you are ready to go on to the next chapter.

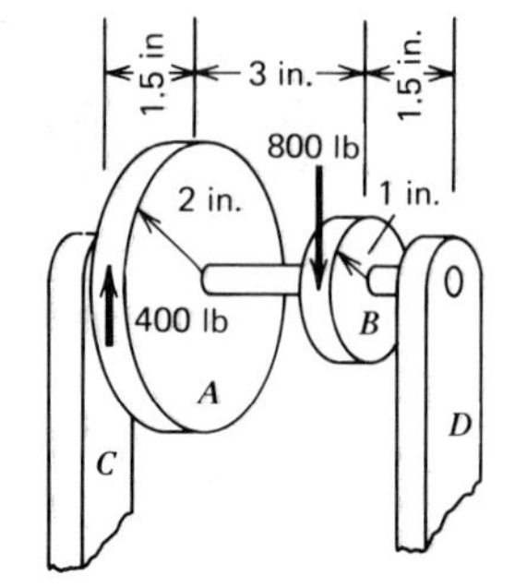

Fig. 11-21. Shaft for Problem 4.

1. A solid circular shaft 12 in. in diameter is subjected to a bending moment of 1 960 000 lb in. and a torque of 2 100 000 lb in. Calculate the maximum shearing stress.
2. The shaft in Fig. 11-20 is 20 mm in diameter. What is the maximum shearing stress?
3. A hollow circular shaft 64 mm outer diameter and 42 mm inner diameter carries a bending moment of 3.2×10^6 N·mm and a torque of 1.4×10^6 N·mm. Find the maximum shearing stress.
4. The shaft in Fig. 11-21 is hollow. The outer diameter is 1.2 in., and the inner diameter is 0.6 in. What is the maximum shearing stress?
5. A solid circular shaft is to be subjected to a bending moment of

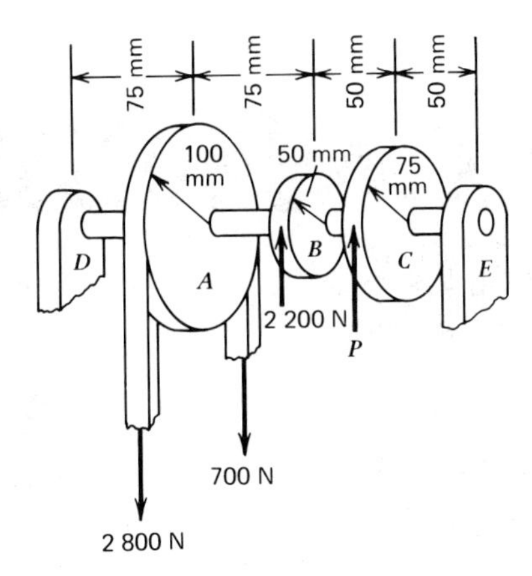

Fig. 11-22. Shaft for Problem 6.

156 000 lb in. and a torque of 124 000 lb in. The maximum shearing stress is not to exceed 8 000 psi. What diameter is required?

6. The maximum shearing stress in the solid circular shaft in Fig. 11-22 is not to exceed 82 MPa. Find the required diameter.

12 COMPRESSION MEMBERS

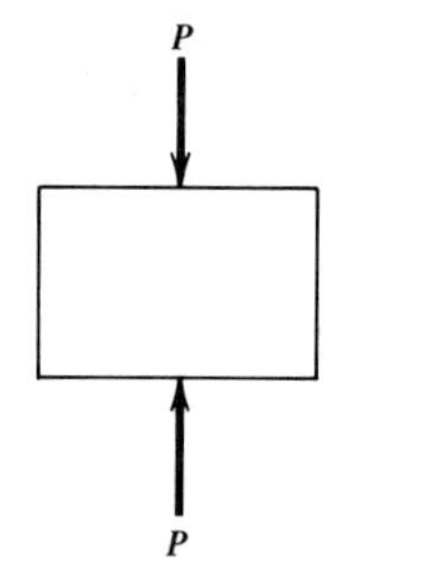

Fig. 12-1. Block in compression.

PURPOSE OF THIS CHAPTER. The purpose of this chapter is to study compression members and to learn how to calculate the strength of a compression member. You remember what compression is: a bar is in *compression* when the forces acting on it tend to squeeze it together or to shorten it. Figure 12-1 shows a block in compression; the forces P act to shorten the block. Now of course you studied compression in earlier parts of this book, but there is a lot more to it than appeared there. This time we will go into the subject more thoroughly, and when we get through, you will be well informed on compression members.

12-1. REVIEW OF CENTROID OF AN AREA

You remember that we studied *centroid of an area* in Chapter 4. We had better review the centroid of a composite area here, because we will need it in this chapter. A composite area, as you recall, is an area that can be divided into simple parts such as rectangles, triangles, and quadrants of circles. You are supposed to know the properties of these simple areas and, just to be sure that you do, we will give them to you again here. Figure 12-2a shows a rectangular area of width b and altitude h. The centroid of the rectangular area is at G; the distance of the centroid from the left side of the rectangle is

$$\bar{x} = \frac{b}{2}$$

and the distance from the base of the rectangle to G is

$$\bar{y} = \frac{h}{2}$$

The area of the rectangle is

$$A = bh$$

Figure 12-2b shows a right triangular area of width b and altitude h. The centroid is at G; the distance from the vertical leg to G is

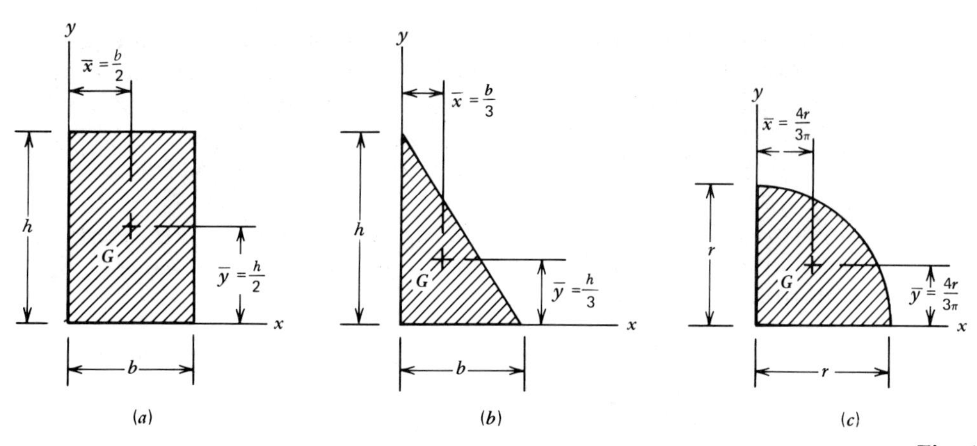

Fig. 12-2. Centroids of areas. (*a*) Rectangle. (*b*) Triangle. (*c*) Quadrant.

$$\bar{x} = \frac{b}{3}$$

and the distance from the horizontal leg to G is

$$\bar{y} = \frac{h}{3}$$

The area of the triangle is

$$A = \frac{bh}{2}$$

Figure 12-2*c* shows a quadrant of radius r. The centroid of the quadrant is at G; the distance from the vertical side to G is

$$\bar{x} = \frac{4r}{3\pi}$$

and the distance from the horizontal side to G is

$$\bar{y} = \frac{4r}{3\pi}$$

The area of the quadrant is

$$A = \tfrac{1}{4}(\pi r^2)$$

You remember that the moment of an area with respect to an axis (say the y axis) is equal to the product of the area (in square inches or square millimeters) and the distance (in inches or millimeters) of the centroid of the area from the axis; the result for the moment of the area is expressed in inches (or millimeters) to the third power. Now, if we know the properties of the simple areas, we can divide a composite area into parts and calculate the moment of each part with respect to the axis. Then we say that the moment of the composite area is equal to the sum of the moments of the parts of the area. (Remember from Chapter 4?) We designate the moment of the composite area with respect to the y axis as M_y, and say that the distance $\bar{x}$ of the centroid from the y axis is

$$\bar{x} = \frac{M_y}{A}$$

where A is the total area. Also, we designate the moment of the composite area with respect to the x axis as M_x and say that the distance $\bar{y}$ of the centroid from the x axis is

$$\bar{y} = \frac{M_x}{A}$$

The centroid of the composite area is the point located by $\bar{x}$ and $\bar{y}$.

12-2. REVIEW OF MOMENT OF INERTIA

Do you remember how to calculate moment of inertia from Chapter 5? If you don't, you had better pay close attention while we review.

Moment of inertia is a property of an area and is used in problems of bending. We need to use moment of inertia in studying compression members because, as we will see, some compression members bend under load. As we said in Chapter 5, don't try to get a picture of moment of inertia, because you can't. The thing to do is just to accept it and use it.

Moment of inertia of area is expressed in length to the fourth power, and we have formulas for the moments of inertia of the simple areas with respect to axes through the centroid. As you should remember, the moment of inertia of a rectangular area with respect to an axis through the centroid and parallel to the base is

$$I = \frac{1}{12}bh^3$$

where b is the base (width) and h is the height (altitude). We take the width as the dimension parallel to the axis and the altitude as the dimension perpendicular to the axis.

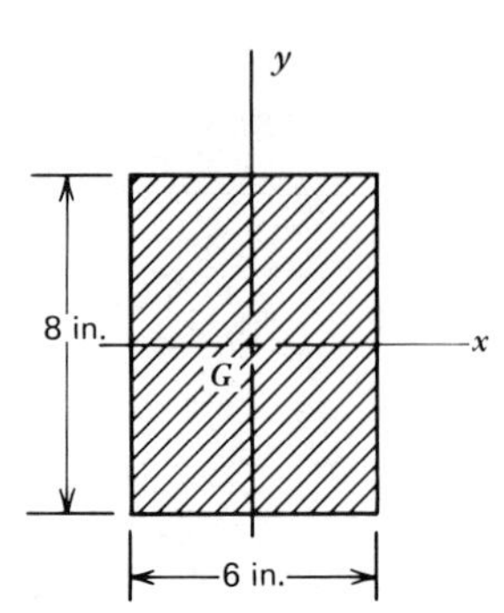

Fig. 12-3. Rectangular area for Illustrative Example 1.

Illustrative Example 1. Calculate the moment of inertia of the rectangular area in Fig. 12-3 with respect to the x axis and with respect to the y axis.

Solution:

1. For the x axis
   (a) The width is the dimension parallel to the x axis, so

$$b = 6 \text{ in.}$$

(b) The altitude is the dimension perpendicular to the x axis. Thus,

$$h = 8 \text{ in.}$$

(c) The moment of inertia of the area with respect to the x axis is

$$I_x = \frac{1}{12}bh^3 = \frac{1}{12} \times 6(8)^3 = 256 \text{ in.}^4$$

2. For the y axis
   (a) The width is the dimension parallel to the y axis

$$b = 8 \text{ in.}$$

   (b) The depth is the dimension perpendicular to the y axis, so

$$h = 6 \text{ in.}$$

   (c) The moment of inertia with respect to the y axis is

$$I_y = \frac{1}{12}bh^3 = \frac{1}{12} \times 8(6)^3 = 144 \text{ in.}^4$$

TRIANGLE

We have to work with the moment of inertia of a right triangle, too, so here's the formula for that. The moment of inertia of the area of a right triangle with respect to an axis through the centroid and parallel to a leg is

$$I = \frac{1}{36}bh^3$$

where b is the width and h is the depth. Here again, we take the width as the dimension parallel to the axis and the depth as the dimension perpendicular to the axis.

Illustrative Example 2. Calculate the moment of inertia of the triangular area in Fig. 12-4 with respect to the x axis and with respect to the y axis.

Solution:

1. For the x axis
   (a) The width b is 12 mm.
   (b) The depth h is 9 mm.
   (c) The moment of inertia is

$$I_x = \frac{1}{36}bh^3 = \frac{1}{36} \times 12(9)^3 = 243 \text{ mm}^4$$

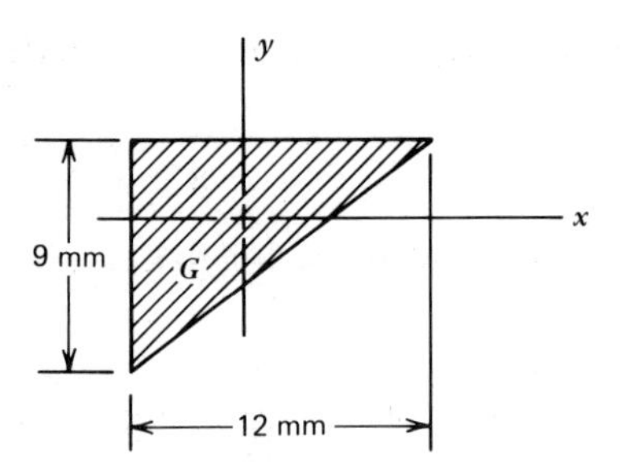

Fig. 12-4. Triangular area for Illustrative Example 2.

2. For the y axis
   (a) The width b is 9 mm.
   (b) The depth h is 12 mm.
   (c) The moment of inertia is

$$I_y = \frac{1}{36}bh^3 = \frac{1}{36} \times 9(12)^3 = 432 \text{ mm}^4$$

CIRCLE

A circle is a simple plane figure. The moment of inertia of a circular area with respect to any axis through the center is

$$I = \frac{\pi}{4}r^4$$

where r is the radius, or

$$I = \frac{\pi}{64}d^4$$

where d is the diameter. Now look at Fig. 12-5. The moment of inertia of each of the four quadrants A, B, C, and D is the same with respect to the x axis as with respect to the y axis, and each is

$$I = \frac{\pi}{16}r^4$$

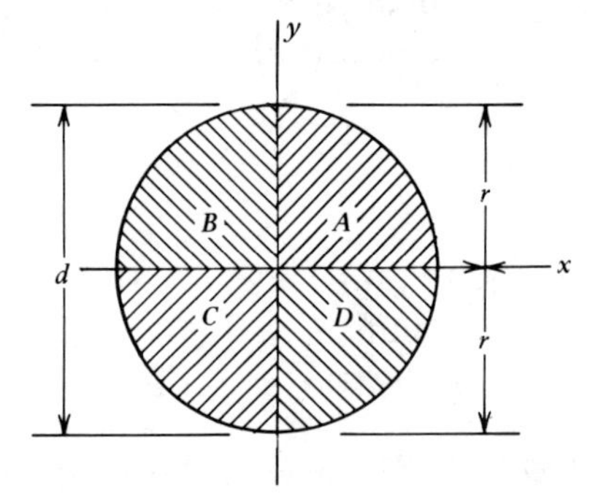

Fig. 12-5. Circular area divided into Quadrants.

which is just one-fourth of the moment of inertia of the whole circle.

We could see four semicircles in Fig. 12-5 if we wanted to see them. They are:

1. The semicircle above the x axis (quadrants A and B).
2. The semicircle below the x axis (quadrants C and D).
3. The semicircle to the right of the y axis (quadrants A and D).
4. The semicircle to the left of the y axis (quadrants B and C).

The moment of inertia of each of the semicircles is the same with respect to the x axis as it is with respect to the y axis. Each moment of inertia is

$$I = \frac{\pi}{8}r^4$$

which is just one-half of the moment of inertia of a whole circular area.

PARALLEL-AXIS THEOREM

You should remember that we use the parallel-axis theorem to calculate the moment of inertia of an area with respect to an axis that does not pass through the centroid of the area. We could apply the theorem to the circular area in Fig. 12-6 if we wanted to calculate the moment of inertia with respect to the x axis. The formula would be

$$I_x = I_{xg} + Ac^2$$

where I_x is the moment of inertia with respect to the x axis, I_{xg} is the moment of inertia with respect to the centroidal axis x_g (the x axis and the x_g axis are parallel), A is the amount of the area, and c is the distance between the two axes.

Illustrative Example 3. Calculate I_x for the area in Fig. 12-6.

Solution:

1. The moment of inertia with respect to the x_g axis is

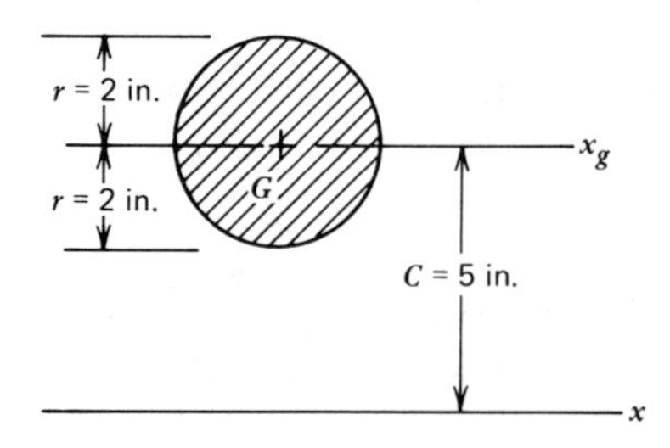

Fig. 12-6. Circular area to illustrate parallel-axis theorem.

$$I_{xg} = \frac{\pi}{4}r^4 = \frac{\pi}{4}(2)^4 = 12.56 \text{ in.}^4$$

2. The area of the circle is

$$A = \pi r^2 = \pi(2)^2 = 12.56 \text{ in.}^2$$

3. The distance c is 5 in.
4. The moment of inertia with respect to the x axis is

$$I_x = I_{xg} + Ac^2 = 12.56 + 12.56(5)^2$$

$$= 12.56 + 314 = 327 \text{ in.}^4$$

12-3. RADIUS OF GYRATION

Are you ready to learn something new? Let's tackle radius of gyration (pronounced *jyration*) because you need it to study compression members. You know how to calculate the moment of inertia of an area, and of course you can calculate the amount of an area. Then the radius of gyration of an area is given by the formula

$$r = \sqrt{\frac{I}{A}}$$

where r is the radius of gyration, I is the moment of inertia of the area, and A is the amount of the area.

The units of the quantities in this equation can be shown in this array

	Gravitational Units	*Metric Units*
r	in.	mm
I	in.^{4}	mm^4
A	in.^{2}	mm^2

We can calculate a moment of inertia with respect to the x axis and a moment of inertia with respect to the y axis. Then we have a radius of gyration with respect to the x axis, and this is

$$r_x = \sqrt{\frac{I_x}{A}}$$

Also, we have a radius of gyration with respect to the y axis. It is

$$r_y = \sqrt{\frac{I_y}{A}}$$

If you want the radius of gyration with respect to any certain axis, all you have to do is to use the amount of inertia with respect to that axis in the formula.

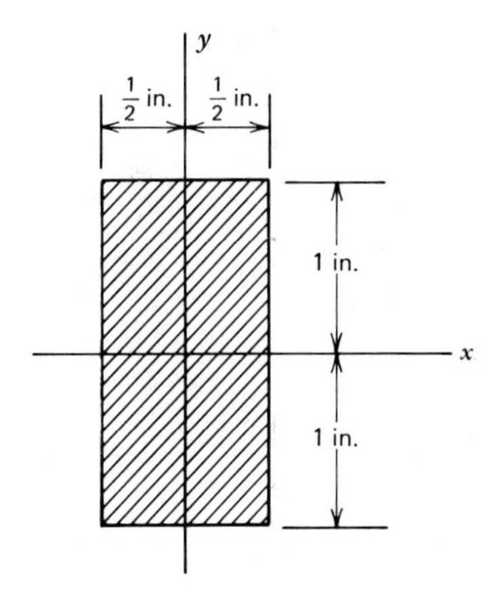

Fig. 12-7. Rectangular area for Illustrative Example 1.

Illustrative Example 1. Calculate the radius of gyration of the rectangular area in Fig. 12-7 with respect to the x axis and the radius of gyration with respect to the y axis.

Solution:

1. The area of the rectangle is

$$A = bh = 1 \times 2 = 2 \text{ in.}^2$$

2. With respect to the x axis
   (a) The moment of inertia is

$$I_x = \frac{1}{12} bh^3 = \frac{1}{12} \times 1(2)^3 = 0.667 \text{ in.}^4$$

   (b) The radius of gyration is

$$r_x = \sqrt{\frac{I_x}{A}} = \sqrt{\frac{0.667}{2}} = 0.577 \text{ in.}$$

3. With respect to the y axis
   (a) The moment of inertia is

$$I_y = \frac{1}{12} bh^3 = \frac{1}{12} \times 2\,(1)^3 = 0.167 \text{ in.}^4$$

   (b) The radius of gyration is

$$r_y = \sqrt{\frac{I_y}{A}} = \sqrt{\frac{0.167}{2}} = 0.289 \text{ in.}$$

NO PICTURE

Don't worry about trying to get a picture of radius of gyration. Instead, just take it as a quantity that is used in studying and designing compression members; it takes some high-powered mathematics to justify its use.

STANDARD SHAPES

Now let's think about standard shapes for awhile. You remember that we studied I beams, channels, and angles in Chapter 6 and that we used tables of the properties of the standard shapes. You will find the radius of gyration of each of the standard shapes in these tables, which means that you don't have to calculate it.

Illustrative Example 2. Find the radius of gyration with respect to each of the x and y axes of a steel beam designated as W 6 × 20.

Solution:

1. The designation W 6 × 20 means that the beam is approximately 6 in. deep, that it is shaped like the letter I, and that it weighs 20 lb/ft of length. Look in Table 6.1 for W-shape beams.
2. Look down column one to the number 6. There is only one 6-in. beam in this table, and it is W 6 × 20.
3. Look across to the ninth column (this gives the radius of gyration with respect to the x axis). Read

$$r = 2.66 \text{ in.}$$

4. Look across to the last column (this gives the radius of gyration with respect to the y axis). Read

$$r = 1.50 \text{ in.}$$

Illustrative Example 3. What is the radius of gyration with respect to each of the x and y axes of an aluminum-alloy channel designated as 4 [2.331?

Solution:

1. The designation 4 [2.331 means that the channel is 4 in. deep and that it weighs 2.331 lb/ft of length. Look in Table 6.2 for aluminum-alloy channels.
2. Look down the column headed d to the numbers 4. Look in the second column for the weight and notice that the first of the 4-in. channels weighs 2.331 lb/ft.
3. Look across to the ninth column for r_x and read

$$r_x = 1.62 \text{ in.}$$

4. Look across to the twelfth column for r_y and read

$$r_y = 0.72 \text{ in.}$$

ANGLES

There is a little more to it for angles. Figure 12-8 shows the cross section of a standard angle, and you remember that we designated such an angle as $L_1 \times L_2 \times t$, where L_1 and L_2 are the lengths of the legs and t is the thickness of each leg. Table 6.3 gives the properties of recommended sizes of angles in the metric system and, when you look at Table 6.3, you see listed not only the x and y axis but also the z axis. The reason for this is that when we study compression members we are usually interested in the least (smallest) radius of gyration for the area, and the least radius of gyration is with respect to the z axis. This figure is given in the last column in Table 6.3.

Illustrative Example 3. What is the least radius of gyration for an angle 120 × 120 × 15?

Solution:

1. The designation 120 × 120 × 15 means that each leg of the angle is 120 mm and the thickness of each leg is 15 mm. Look in Table 6.3 for angles.
2. Look down the column headed Designation until you see 120 × 120 × 15. The second of the 120 × 120 angles is it.
3. Look across to the last column for r_z. It is

$$r_z = 2.33 \text{ cm} \qquad \text{or} \qquad 23.3 \text{ mm}$$

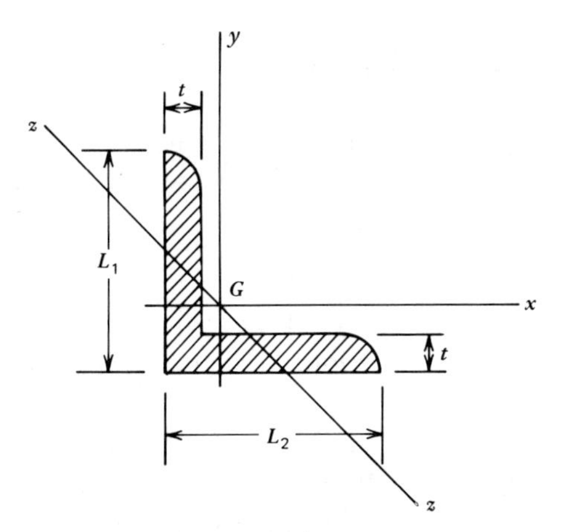

Fig. 12-8. Angle section.

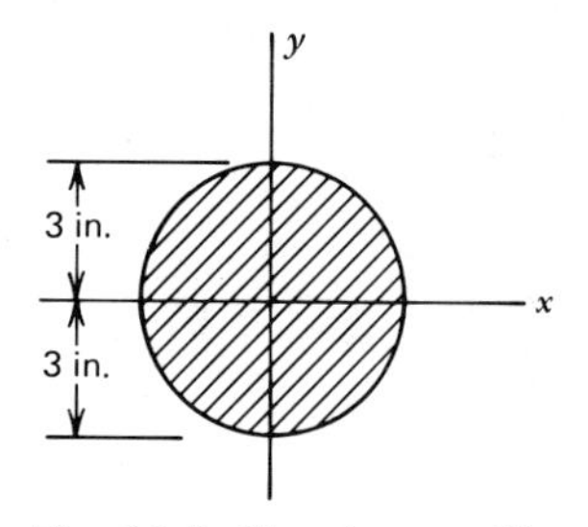

Fig. 12-9. Circular area for Problem 1.

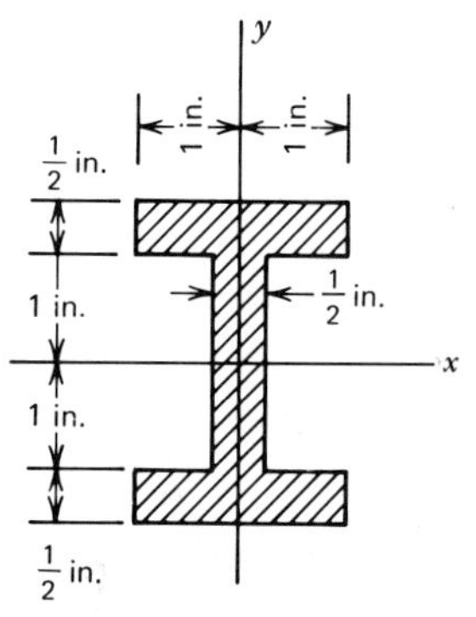

Fig. 12-10. Section for Problem 2.

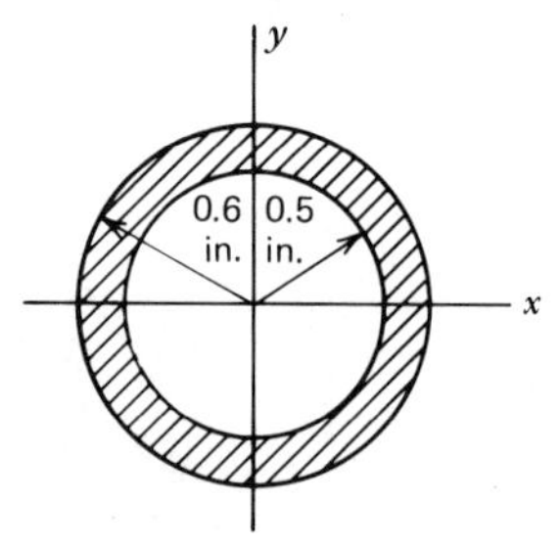

Fig. 12-11. Tube section for Problem 3.

Practice Problems (Section 12-3). You will have to know how to calculate radius of gyration if you are going to study compresssion members.

1. Calculate the radius of gyration of the circular area in Fig. 12-9 with respect to each axis.
2. What is the radius of gyration of the area in Fig. 12-10 with respect to (a) the x axis and (b) the y axis?
3. Find the radius of gyration of the hollow circular area in Fig. 12-11 with respect to the x axis.
4. What is the radius of gyration with respect to the y axis of a steel beam designated as W 10 × 100?
5. Find the radius of gyration with respect to the x axis of an aluminum-alloy channel designated as 3 [1.597.
6. What is the least radius of gyration of an angle designated as 30 × 30 × 3?

12-4. SLENDERNESS RATIO

Now we are ready for what is probably the most important term in the subject of compression members. That term is *slenderness ratio*, and it is equal to the length of the compression member divided by the least radius of gyration of the cross section. We don't use any special symbol for the slenderness ratio but just write it as L/r, where L is the length of the member and r is the least radius of gyration. Here we express L and r in inches, millimeters, or centimeters. Let's learn to calculate the slenderness ratio now and learn how to use it a little later.

We have the radius of gyration as

$$r = \sqrt{\frac{I}{A}}$$

and we know that we can calculate the moment of inertia with respect to either the x or y axis. We are after the least radius of gyration, and so we must use the smallest moment of inertia in the formula. Usually this means that we calculate I_x and I_y both and then use the smaller one to calculate r.

Illustrative Example 1. Figure 12-12 shows the cross section of a compression member 6 ft long. Calculate the slenderness ratio.

Solution:

1. The area of the rectangle is

$$A = bh = 3 \times 2 = 6 \text{ in.}^2$$

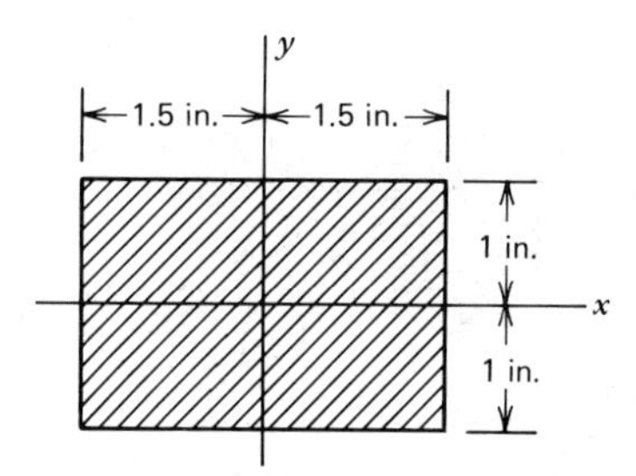

Fig. 12-12. Cross section for Illustrative Example 1.

2. The moment of inertia of the area with respect to the x axis is

$$I_x = \frac{1}{12}bh^3 = \frac{1}{12} \times 3(2)^3 = 2 \text{ in.}^4$$

3. The moment of inertia with respect to the y axis is

$$I_y = \frac{1}{12}bh^3 = \frac{1}{12} \times 2(3)^3 = 4.5 \text{ in.}^4$$

4. We use the smallest moment of inertia to calculate the least radius of gyration, so we use I_x. Then,

$$r = \sqrt{\frac{I_x}{A}} = \sqrt{\frac{2}{6}} = 0.577 \text{ in.}$$

5. The length L is 6 ft, but we must change L to inches by multiplying by 12, so

$$L = 6 \times 12 = 72 \text{ in.}$$

6. The slenderness ratio is

$$\frac{L}{r} = \frac{72}{0.577} = 124.8$$

CENTROIDAL AXIS

When we studied beams, we found we had to calculate the moment of inertia with respect to an axis through the centroid of the area. It's the same way here. The moment of inertia must be taken with respect to a centroidal axis.

Illustrative Example 2. Figure 12-13 shows the cross section of a compression member 3 m long. What is the slenderness ratio?

Solution:

1. The numerical amount of the area is equal to the area of the outer circle minus the area of the inner circle.

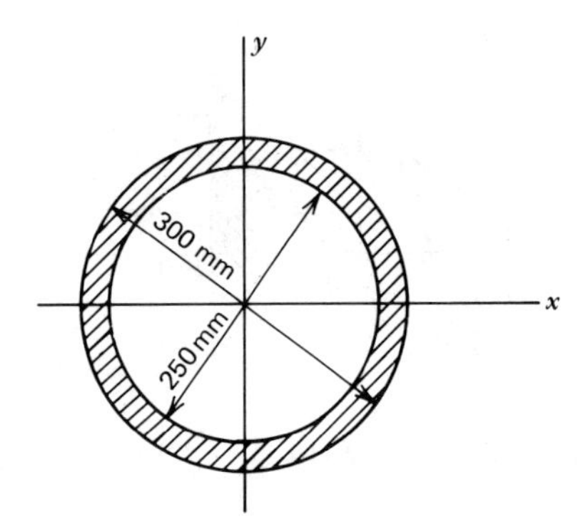

Fig. 12-13. Hollow circular area for Illustrative Example 2.

(a) The area of the outer circle is

$$A_o = \frac{\pi}{4}d^2 = \frac{\pi}{4}(300)^2 = 70\ 700\ \text{mm}^2$$

$$= 7.07 \times 10^4\ \text{mm}^2$$

(b) The area of the inner circle is

$$A_i = \frac{\pi}{4}d^2 = \frac{\pi}{4}(250)^2 = 49\ 100\ \text{mm}^2$$

$$= 4.91 \times 10^4\ \text{mm}^2$$

(c) The hollow circular area is

$$A = 70\ 700 - 49\ 100 = 21\ 600\ \text{mm}^2 = 2.16 \times 10^4\ \text{mm}^2$$

2. For a circular area, the moment of inertia with respect to the x axis is equal to the moment of inertia with respect to the y axis. We can calculate either one. Let's take I_x. We will subtract the moment of inertia of the inner circle from the moment of inertia of the outer circle.
   (a) For the outer circle:

$$I_{xo} = \frac{\pi}{64}d^4 = \frac{\pi}{64}(300)^4 = 3.976 \times 10^8\ \text{mm}^4$$

   (b) For the inner circle:

$$I_{xi} = \frac{\pi}{64}d^4 = \frac{\pi}{64}(250)^4 = 1.918 \times 10^8\ \text{mm}^4$$

   (c) For the hollow circular area:

$$I_x = 3.976 \times 10^8 - 1.918 \times 10^8 = 2.058 \times 10^8\ \text{mm}^4$$

3. The least radius of gyration is

$$r = \sqrt{\frac{I_x}{A}} = \sqrt{\frac{2.058 \times 10^8}{2.16 \times 10^4}} = 97.6 \text{ mm}$$

4. The length of the compression member is 3 m, and we will change it to millimeters. Thus

$$L = 3 \times 10^3 \text{ mm}$$

5. The slenderness ratio is

$$\frac{L}{r} = \frac{3 \times 10^3}{97.6} = 30.73$$

STANDARD SHAPES

The easiest shape to work with is one that is standard, because you just read the radius of gyration from a table. Be sure to use the smallest radius of gyration.

Illustrative Example 3. What is the slenderness ratio of an aluminum-alloy channel designated as 5 [3.089 that is 12 ft long?

Solution:

1. The least radius of gyration is read from Table 6.2 as

$$r = 0.88 \text{ in.}$$

This is the radius of gyration with respect to the y axis.

2. The length L is

$$L = 12 \times 12 = 144 \text{ in.}$$

3. The slenderness ratio is

$$\frac{L}{r} = \frac{144}{0.88} = 163.6$$

Practice Problems (Section 12-4). The rest of the chapter will be easier if you do the following problems.

1. What is the slenderness ratio of a solid circular bar that is 40 in. long and 1 in. in diameter?
2. A rectangular bar 1 in. × ½ in. is 18 in. long. Calculate the slenderness ratio.
3. Figure 12-14 shows the cross section of part of a machine frame. This particular member is 8 ft long. What is the slenderness ratio?

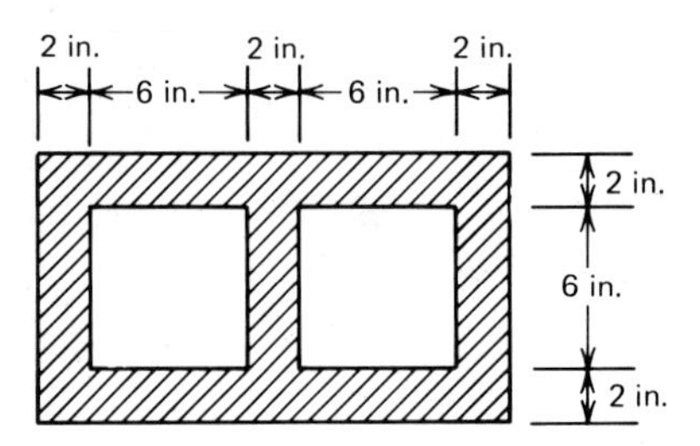

Fig. 12-14. Cross section for Problem 3.

4. What is the slenderness ratio of a steel beam 9 ft long and designated as W 8 × 28?
5. Calculate the slenderness ratio of an aluminum-alloy channel designated as 7 [4.715 and is 90 in. long.
6. What is the slenderness ratio of an angle designated as 50 × 50 × 4 (in mm) and is 800 mm long?

12-5. SHORT-COMPRESSION MEMBERS

We will divide compression members into three classes, and the first of the three classes is the *short-compression member* or block such as you see in Fig. 12-15*a*. A short-compression member fails by crushing of the material; a ductile material (*ductile* means the material can be deformed a lot before breaking) barrels out, as shown in Fig. 12-15*b*, and a brittle material (*brittle* means the material snaps before deforming much) fails by sliding on an oblique plane, as shown in Fig. 12-15*c*. The important property of the material here is its compressive strength, because failure results from straight crushing.

You will be wondering how short a short-compression member is, and that's where the slenderness ratio comes in. If the slenderness ratio of the compression member is below a certain value, we say the member is a short-compression member. For low-carbon steel, a short-compression member is one that has a slenderness ratio below about 40; for

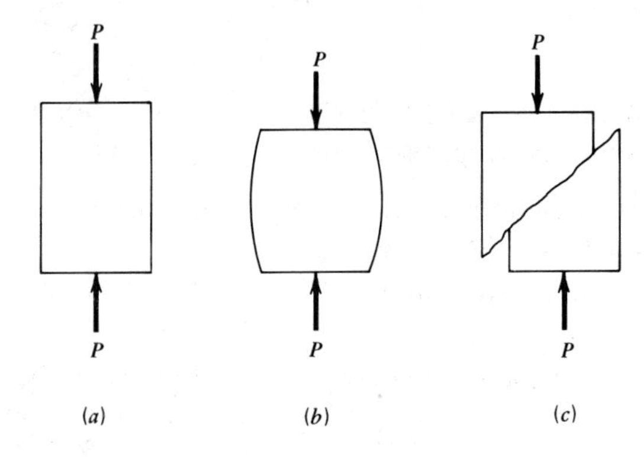

Fig. 12-15. Short compression members. (*a*) Short compression member. (*b*) Failure of ductile material in compression. (*c*) Failure of brittle material in compression.

aluminum and magnesium alloys any compression member that has a slenderness ratio below about 10 is a short-compression member. These values of 40 and 10 have been established from experiments on a great many compression members.

It isn't just the length of the member that determines whether it is a short-compression member. The width is important, too, and the effect of the width is represented in the slenderness ratio. A member may be several feet long and yet be a short-compression member if it is wide enough. On the other hand, a circular bar 6 in. long and ⅛ in. in diameter is not a short-compression member, because it is so narrow. The only safe way to find out for sure whether a compression member is a short-compression member or not is to calculate the slenderness ratio.

It's easy to calculate the stress in a short-compression member. The stress is given by the formula,

$$S = \frac{P}{A}$$

where S is the stress in psi or megapascals, P is the force in pounds or newtons, and A is the cross sectional area in square inches or square millimeters. You probably remember that we have used this formula several times. Before using it for compression members, you should make sure the slenderness ratio is small enough so the member is a short-compression member.

Illustrative Example 1. A solid circular steel bar 2 in. in diameter and 16 in. long is subjected to a compressive force of 34 000 lb. Calculate the compressive stress.

Solution:

Figure 12-16 shows the cross section of the circular bar.

A. The slenderness ratio.

1. The area of the cross section is

$$A = \frac{\pi}{4}d^2 = \frac{\pi}{4}(2)^2 = 3.14 \text{ in.}^2$$

2. The moment of inertia of the circular area is the same with respect to the x axis as with respect to the y axis, so only one moment of inertia has to be calculated.

$$I_x = \frac{\pi}{64}d^4 = \frac{\pi}{64}(2)^4 = 0.785 \text{ in.}^4$$

3. The radius of gyration is

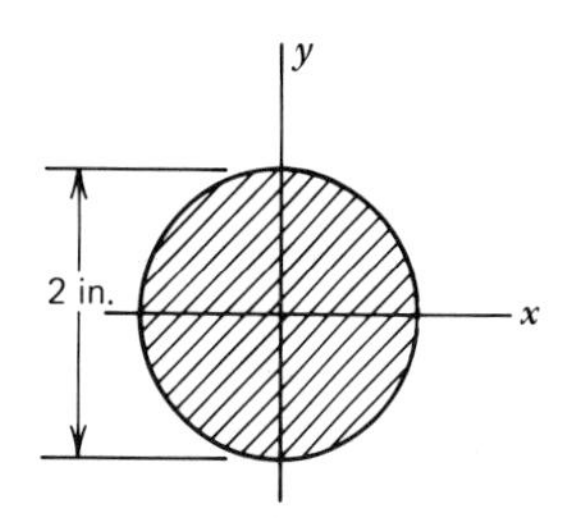

Fig. 12-16. Cross section for Illustrative Example 1.

$$r = \sqrt{\frac{I_x}{A}} = \sqrt{\frac{0.785}{3.14}} = 0.5 \text{ in.}$$

4. The length L is 16 in.
5. The slenderness ratio is

$$\frac{L}{r} = \frac{16}{0.5} = 32$$

The slenderness ratio is below 40, so this steel member is a short-compression member, and it is all right to use the formula $S = P/A$.

B. The compressive stress.
1. The force P is 34 000 lb.
2. The area A is 3.14 in.^{2}
3. The compressive stress is

$$S = \frac{P}{A} = \frac{34\ 000}{3.14} = 10\ 800 \text{ psi}$$

CALCULATING THE FORCE

There will be lots of times when you will want to turn the formula around and write it as

$$P = AS$$

Then you can use it to calculate the force if you know the area and the stress. However, you will have to be sure that the slenderness ratio is small enough so that the member is really a short-compression member.

Illustrative Example 2. An aluminum-alloy angle designated as 80 × 80 × 8 is used as a compression member 14 cm long. The compressive stress is not to exceed 70 MPa. What is a reasonable load?

Solution:

The properties of the cross section are found in Table 6.3.

A. The slenderness ratio.
1. The least radius of gyration is read from the table as

$$r = 1.56 \text{ cm}$$

This is the radius of gyration with respect to the z axis.

2. The length L is 14 cm.
3. The slenderness ratio is

$$\frac{L}{r} = \frac{14}{1.56} = 8.97$$

The slenderness ratio is below 10, so the aluminum-alloy angle is considered a short-compression member.

B. The load.

1. The area of the channel is read from the table as

$$A = 12.3 \text{ cm}^2 = 1\,230 \text{ mm}^2$$

2. The stress S is 70 MPa.
3. The load P is $P = AS = 1\,230 \times 70 = 86\,100$ N.

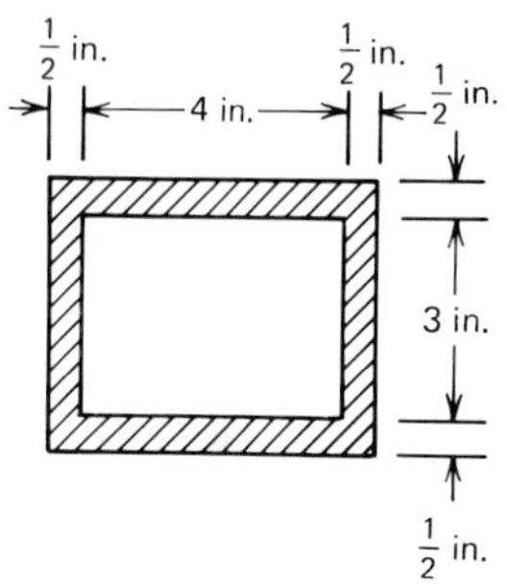

Fig. 12-17. Cross section for Problem 2.

Practice Problems (Section 12-5). Short-compression members are easy to figure, but you had better practice if you want to remember.

1. A hollow circular steel washer that has an outer diameter of 3 in. and an inner diameter of 2 in. is ½ in. long. The washer is subjected to a compressive force of 87 000 lb. Calculate the compressive stress.
2. Figure 12-17 shows the cross section of a magnesium-alloy casting 12 in. long subjected to a compressive load of 23 000 lb. What is the compressive stress?
3. A magnesium-alloy angle, designated as 50 × 50 × 4, is 160 mm long and carries a compressive force of 18 000 N. Find the compressive stress.
4. A steel beam, designated as W 8 × 18, is 36 in. long and subjected to a compressive force; the compressive stress is not to exceed 12 000 psi. Find the load to which the member can be subjected.
5. Figure 12-18 shows the cross section of an aluminum-alloy casting 4 in. long subjected to a compressive force. What load will cause a compressive stress of 10 000 psi?
6. A compression member in the frame of a heavy machine is 10 ft long and has the cross section shown in Fig. 12-19; the member is built up by welding steel plates together. The compressive stress is not to exceed 18 000 psi. Find the load.

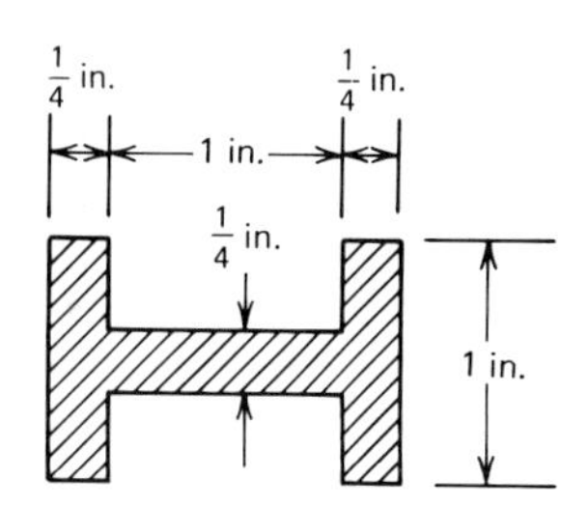

Fig. 12-18. Cross section for Problem 5.

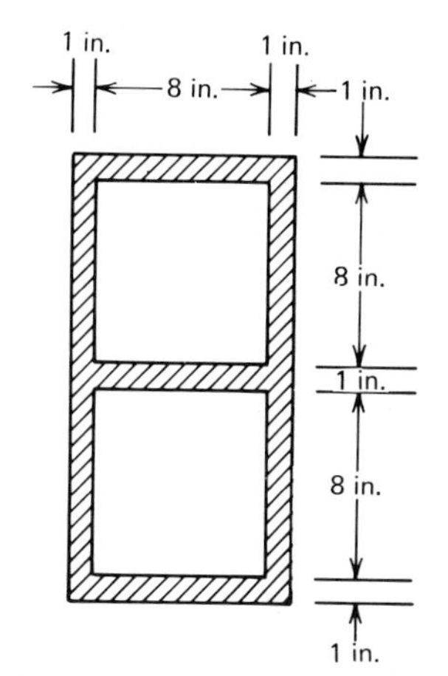

Fig. 12-19. Cross section for Problem 6.

12-6. INTERMEDIATE COLUMNS

The second type of compression member we are going to study is the *intermediate-compression member*; or, as it is often called, the intermediate column. You may wonder how to tell whether a compression member is an intermediate column, and you can tell from the slenderness ratio.

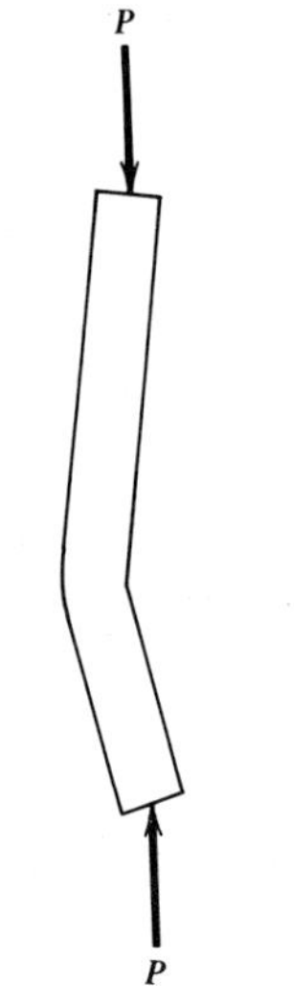

Fig. 12-20. Failure of intermediate column.

A steel compression member is an intermediate column if its slenderness ratio is between about 40 and 150; an aluminum- or magnesium-alloy compression member is an intermediate column if its slenderness ratio is between 10 and 70. You already know how to calculate the slenderness ratio for a member, so you shouldn't have any trouble in finding whether a member is an intermediate column.

An intermediate column usually fails by bending rather sharply at one point, as shown in Fig. 12-20. The bending causes stresses beyond the yield point (Do you remember yield point from Chapter 3?) and the column remains bent after the load is removed. The failure of an intermediate column involves both bending and the yield point of the material; the process is so complicated it is hard to derive an exact formula for the strength of the column. Many kinds of formulas have been proposed, but there is no one type that everyone agrees is the best. This could lead to a confusing picture, and it is profitless for us to enter the controversy. We will just give you several types that are in common use and show you how to use them. Later when you are doing real engineering design, you will probably be using formulas that are specified by design codes.

12-7. STRAIGHT-LINE COLUMN FORMULAS

First, we are going to show you a type of formula that is called the straight-line formula, because if you plotted it on coordinate paper (with P/A vertically and L/r horizontally) it would appear as a straight line. The formula is usually written as

$$\frac{P}{A} = S - k\frac{L}{r}$$

where P is the load on the column, A is the area of the cross section, S is a stress, k is a constant in the formula, and L/r is the slenderness ratio (we will give you values of S and k). You would use the formula to calculate the load P when you already knew everything else in the formula. It may make it a little simpler if we multiply the formula by A and write it as

$$P = A\left(S - k\frac{L}{r}\right)$$

For intermediate columns of low-carbon steel, two examples of straight-line formulas are

$$P = A\left(15\,000 - 50\frac{L}{r}\right)$$

$$P = A\left(16\,000 - 70\frac{L}{r}\right)$$

where P is the working load, that is, the load the column would carry safely in service. The first number in each formula is a stress in psi and the second is in the same units as stress.

In metric units these formulas would be (with some rounding off of numbers)

$$P = A\left(103 - 0.345\frac{L}{r}\right)$$

$$P = A\left(110 - 0.483\frac{L}{r}\right)$$

Here the first number in each formula is stress in megapascals.

Illustrative Example 1. What is a reasonable working load for a steel beam 10 ft long, designated as W 8 × 28, when used as a compression member? Use $P = A\left(15\,000 - 50\frac{L}{r}\right)$.

Solution:

A. The slenderness ratio.

1. The length is 10 ft, but we must convert it to inches by multiplying by 12. So

$$L = 10 \times 12 = 120 \text{ in.}$$

2. The least radius of gyration is obtained from Table 6.1 as

$$r = 1.62 \text{ in.}$$

3. The slenderness ratio is

$$\frac{L}{r} = \frac{120}{1.62} = 74.1$$

The value of 74.1 for the slenderness ratio shows that the compression member is an intermediate column.

B. The working load.

1. The area is read from Table 6.1 as

$$A = 8.25 \text{ in.}^2$$

2. The working load is

$$P = A\left(15\,000 - 50\frac{L}{r}\right) = 8.25(15\,000 - 50 \times 74.1)$$

$$= 93\,200 \text{ lb}$$

ANOTHER FORMULA

The subject of compression members is comprehensive, and we could give you a different formula for each different alloy that might be used in a machine or structure. This would take too long, though, so the best we can do is to give you a few sample formulas and show you how to use them.

For intermediate columns of several aluminum-alloys 6061, the straight-line formula can be written as

$$P = A\left(20.2 - 0.126\frac{L}{r}\right)$$

to give a reasonable working value of the load. Here the load P will be in kips. Remember, though, that the slenderness ratio must be between about 10 and 70 for an aluminum bar to be an intermediate column.

Illustrative Example 2. An aluminum-alloy (6061) channel, designated as 6 [4.030 is 35 in. long and is used as a compression member. What is a reasonable working load?

Solution:

A. The slenderness ratio.

1. The length L is 35 in.
2. The least radius of gyration is read from Table 6.2 as

$$r = 1.05 \text{ in.}$$

3. The slenderness ratio is

$$\frac{L}{r} = \frac{35}{1.05} = 33.3$$

The value of 33.3 is between the limits for an intermediate column of an aluminum alloy.

B. The working load.

1. The area is read from Table 6.2 as

$$A = 3.427 \text{ in.}^2$$

2. The working load is

$$P = A\left(20.2 - 0.126\frac{L}{r}\right) = 3.427\,(20.2 - 0.126 \times 33.3)$$

$$= 54.8 \text{ kips}$$

Practice Problems (Section 12-7). Now you can practice using straight-line formulas on intermediate columns.

1. What is a reasonable working load in compression for a circular steel rod 60 mm in diameter and 1.2 m long? Use $P = A(103 - 0.345\ L/r)$.
2. An aluminum-alloy channel is to be used as a compression member. The channel is designated as 3 [1.135 and is 30 in. long. Calculate the working load.
3. A steel beam section 9 ft long and designated as W 14 × 159 is to be used as a compression member. Find the working load. Use $P = A(16\ 000 - 70L/r)$.
4. A steel angle, 60 × 60 × 5, is used as a compression member 700 mm long. What is a reasonable working load? Use the formula

$$P = A\left(110 - 0.483\frac{L}{r}\right)$$

Be sure to take the least radius of gyration here.

5. A cast-iron column 11 ft long has the cross section shown in Fig. 12-21. Calculate the working load. Use the formula

$$P = A\left(9\ 000 - 40\frac{L}{r}\right)$$

6. A rectangular steel bar 4 in. by 1 in. and 4 ft long is used as a compression member. What is a reasonable working load? Use $P = A(15\ 000 - 50L/r)$

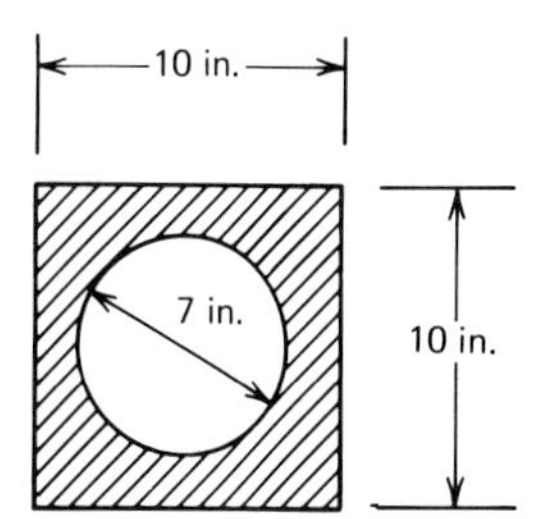

Fig. 12-21. Section of cast-iron column for Problem 5.

12-8. PARABOLIC COLUMN FORMULAS

A second type of formula for intermediate columns is the parabolic column formula. One example is

$$\frac{P}{A} = 21\ 900 - 1.75\left(\frac{L}{r}\right)^2$$

for aluminum-alloy columns. If you plotted P/A against L/r in this formula, the curve would be a parabola.

The AISC (American Institute of Steel Construction, Inc.) recommends an elaboration of the parabolic type of formula, thus

$$\frac{P}{A} = \frac{S_y}{N}\left[1 - \frac{\left(K\frac{L}{r}\right)^2}{2C_c^2}\right]$$

where N is the factor of safety and is

$$N = \frac{5}{3} + \frac{3\left(K\dfrac{L}{r}\right)}{8C_c} - \frac{\left(K\dfrac{L}{r}\right)^3}{8C_c^3}$$

S_y is the yield point of the material.

The constant K depends on the end conditions of the column

$$K = 1 \text{ for pinned or hinged ends}$$

$$K = 0.65 \text{ for fixed ends}$$

As a practical matter, K might as well be taken as 1, because it is very difficult to achieve any degree of fixedness at the ends of a large steel column.

The constant C_c is

$$C_c = \sqrt{\frac{2\pi^2 E}{S_y}}$$

This is a complicated formula, so let's use it in an example and see how it goes.

Illustrative Example 1. Calculate the working load for a steel column, W 16 × 89, 15 ft. long. The yield point of the material is 36 000 psi.

Solution:

1. The length L is 15 × 12 = 180 in.
2. The least radius of gyration (see Table 6.1) is 2.49 in.
3. The slenderness ratio is

$$\frac{L}{r} = \frac{180}{2.49} = 72.3$$

4. The area of the cross section is 26.2 in.^{2}
5. The yield point is 36 000 psi.
6. E = 30 000 000 psi.
7. The constant C_c is

$$C_c = \sqrt{\frac{2\pi^2 E}{S_y}} = \sqrt{\frac{2\pi^2\, 30\,000\,000}{36\,000}} = 128.3$$

8. Take K as 1.
9. The constant N is

$$N = \frac{5}{3} + \frac{3\left(K\frac{L}{r}\right)}{8C_c} - \frac{\left(K\frac{L}{r}\right)^3}{8C_c^3}$$

$$= \frac{5}{3} + \frac{3(1 \times 72.3)}{8 \times 128.3} - \frac{(1 \times 72.3)^3}{8(128.3)^3}$$

$$= 1.667 + 0.211 - 0.022 = 1.856$$

10. Finally,

$$\frac{P}{A} = \frac{S_y}{N}\left[1 - \frac{\left(K\frac{L}{r}\right)^2}{2C_c^2}\right]$$

$$\frac{P}{26.2} = \frac{36\,000}{1{,}856}\left[1 - \frac{(1 \times 72.3)^2}{2(128.3)^2}\right]$$

$$= 19\,400\,[1 - 0.159] = 16\,300$$

$$P = 427\,000 \text{ lb.}$$

and there you are.

Practice Problems (Section 12-8). A few problems will help you to remember how to use parabolic column formulas.

1. Use the formula $P/A = 21\,900 - 1.75\,(L/r)^2$ to calculate the working load for an aluminum-alloy channel, 12 [11.822, that is 7.5 ft long and is used as a column.
2. Calculate the working load for a steel column, W 18 × 106, that is 16 ft long. Use the formula $P/A = 17\,000 - 0.485\,(L/r)^2$.
3. Use the AISC formula to calculate the working load for a steel column, W 14 × 159, that is 21 ft long. The yield point of the material is 42 000 psi.
4. A steel angle is designated as 200 × 200 × 18 where the dimensions are in millimeters. The length is 3 m. The yield point of the material is 300 MPa. Use the AISC formula to calculate the working load.

12-9. GORDON–RANKINE COLUMN FORMULAS

A third type of column formula that has had wide use is the Gordon–Rankine formula. This might be written generally as

$$\frac{P}{A} = \frac{S}{1 + B\left(\frac{L}{r}\right)^2}$$

where S is in units of stress and B is a dimensionless constant. A particular example for steel columns is

$$\frac{P}{A} = \frac{18\,000}{1 + \frac{1}{18\,000}\left(\frac{L}{r}\right)^2}$$

Let's try it.

Illustrative Example 1. What is the working load for a steel tube column 10 in. outer diameter, 8 in. inner diameter, and 17 ft long? Use the formula just above.

Solution:

1. The moment of inertia of the cross section is equal to the moment of inertia of the outer circular area minus the moment of inertia of the inner circular area:

$$I = \frac{\pi}{64}(10)^4 - \frac{\pi}{64}(8)^4 = 490.9 - 201.1 = 289.8 \text{ in.}^4$$

2. The area of the cross section is

$$A = \frac{\pi}{4}(10)^2 - \frac{\pi}{4}(8)^2 = 78.54 - 50.27 = 28.27 \text{ in.}^2$$

3. The radius of gyration is

$$r = \sqrt{\frac{I}{A}} = \sqrt{\frac{289.8}{28.27}} = 3.20 \text{ in.}$$

4. The slenderness ratio is

$$\frac{L}{r} = \frac{17 \times 12}{3.20} = 63.75$$

5. Then

$$\frac{P}{A} = \frac{18\,000}{1 + \frac{1}{18\,000}\left(\frac{L}{r}\right)^2}$$

$$\frac{P}{28.27} = \frac{18\,000}{1 + \dfrac{1}{18\,000}(63.75)^2} = \frac{18\,000}{1 + 0.226} = 14\,700$$

$$P = 415\,600 \text{ lb}$$

Practice Problems. (Section 12-9). Now try a few problems with Gordon–Rankine formulas.

1. Calculate the working load on a steel column, W 10 × 100, that is 11 ft long. Use

$$\frac{P}{A} = \frac{18\,000}{1 + \dfrac{1}{18\,000}\left(\dfrac{L}{r}\right)^2}$$

2. A rectangular steel tube column has outer dimensions of 200 mm × 120 mm and inner dimensions of 170 mm × 90 mm. It is 2.5 m long. Use

$$\frac{P}{A} = \frac{126}{1 + \dfrac{1}{126}\left(\dfrac{L}{r}\right)^2}$$

   to calculate the working load.

3. A circular steel tube has an outer diameter of 4 in. and an inner diameter of 3 in. The tube is 7 ft long and is used as a column. Calculate the working load using the formula.

$$\frac{P}{A} = \frac{16\,250}{1 + \dfrac{1}{11\,000}\left(\dfrac{L}{r}\right)^2}$$

12-10. SLENDER COLUMNS

We told you there were three classes of compression members, and so far we have studied two classes. That leaves one to go, and it is the *slender column*, that is, a compression member that is slender and has a small width compared to the length. Just as before, we use the slenderness ratio to find in which class the compression member belongs. A steel-compression member that has a slenderness ratio above about 150 is classified as a slender column, and an aluminum- or magnesium-alloy compression member that has a slenderness ratio above about 70 is a slender column.

Fig. 12-22. Bending of slender column.

A slender column fails by bending in a smooth curve, as shown in Fig. 12-22. An interesting fact about this bending is that the stresses are below the yield point of the material, and the column will straighten out if the load is removed. (Try this for yourself with a thin bar of steel or wood.) The load that will cause a slender column to bend is called the *critical load* and is given by the formula,

$$P_{cr} = \frac{\pi^2 EI}{L^2}$$

where P_{cr} is the critical load in pounds or newtons, E is the modulus of elasticity of the material in psi or megapascals, I is the least moment of inertia of the cross section in inches or millimeters to the fourth power, and L is the length of the column in inches or millimeters. (This formula is usually called the *Euler formula*, after the man who derived it first.) The column holds the critical load while it is bent, but any increase in the load causes the column to fail. Now we don't want to load columns up to the point of failure, so we must use a factor of safety f and write the formula as

$$P = \frac{\pi^2 EI}{fL^2}$$

We'll give you the factor of safety for each problem.

You can look up the modulus of elasticity E in Table 3.1; just be sure to take the modulus for tension and compression. You are supposed to be able to calculate the moment of inertia I by now. Remember that the first step in finding the working load for a compression member is to calculate the slenderness ratio to see what type of compression member it is.

Illustrative Example 1. An aluminum-alloy tube is to be used as a compression member. The tube is 1.5 in. outer diameter, 1 in. inner diameter, and 6 ft long. What is the working load with a factor of safety of 2.5?

Solution:

Figure 12-23 shows the cross section.

A. The slenderness ratio.

1. The moment of inertia is the same with respect to the x axis as with respect to the y axis, so we only have to calculate one. We will subtract the moment of inertia of the small circle from the moment of inertia of the large circle.

   (a) The moment of inertia of the outer circle is

$$I = \frac{\pi}{64} d^4 = \frac{\pi}{64}(1.5)^4 = 0.2485 \text{ in.}^4$$

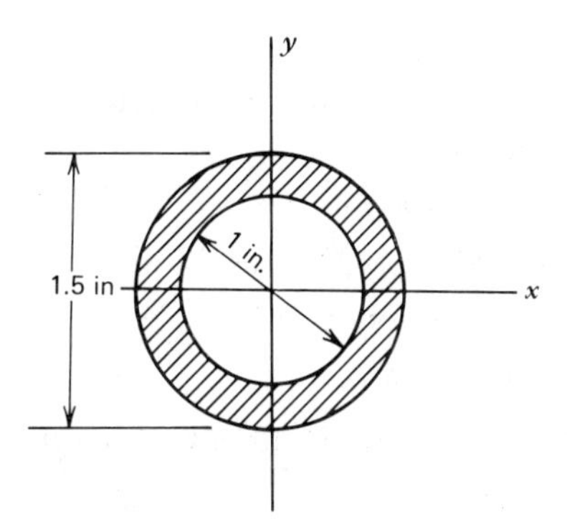

Fig. 12-23. Tube section for Illustrative Example 1.

(b) The moment of inertia of the inner circle is

$$I = \frac{\pi}{64} d^4 = \frac{\pi}{64}(1)^4 = 0.0491 \text{ in.}^4$$

(c) The moment of inertia of the hollow circular area is

$$I = 0.2485 - 0.0491 = 0.1994 \text{ in.}^4$$

2. The area of the cross section is equal to the area of the outer circle minus the area of the inner circle.
   (a) The area of the outer circle is

$$A = \frac{\pi}{4} d^2 = \frac{\pi}{4}(1.5)^2 = 1.767 \text{ in.}^2$$

   (b) The area of the inner circle is

$$A = \frac{\pi}{4} d^2 = \frac{\pi}{4}(1)^2 = 0.785 \text{ in.}^2$$

   (c) The area of the cross section is

$$A = 1.767 - 0.785 = 0.982 \text{ in.}^2$$

3. The radius of gyration is

$$r = \sqrt{\frac{I}{A}} = \sqrt{\frac{0.1994}{0.982}} = 0.451 \text{ in.}$$

4. The length of the tube is 6 ft or

$$L = 6 \times 12 = 72 \text{ in.}$$

5. The slenderness ratio is

$$\frac{L}{r} = \frac{72}{0.451} = 160$$

The value of 160 shows that the aluminum-alloy tube must be considered as a slender column.

B. The working load.

1. The modulus of elasticity is read from Table 3.1. E = 10 000 000 psi.
2. We calculated the moment of inertia I as 0.1994 in.^{4}
3. We figured the length L as 72 in.
4. The factor of safety is 2.5.
5. The working load is

$$P = \frac{\pi^2 EI}{fL^2} = \frac{\pi^2 \times 10\ 000\ 000 \times 0.1994}{2.5\ (72)^2} = 1\ 519 \text{ lb}$$

LEAST I
When the moment of inertia with respect to the x axis is not the same as the moment of inertia with respect to the y axis, you must take the smaller moment of inertia.

Illustrative Example 2. A rectangular steel bar 12 mm by 3 mm and 200 mm long is to be used as a compression member. Find the working load, using a factor of safety of 2.

Solution:

Figure 12-24 shows the cross section of the bar.

A. The slenderness ratio.

1. You should see at a glance that I_x is smaller than I_y

$$I_x = \frac{1}{12} bh^3 = \frac{1}{12} \times 12(3)^3 = 27 \text{ mm}^4$$

2. The area is

$$A = 12 \times 3 = 36 \text{ mm}^2$$

3. The radius of gyration is

$$r = \sqrt{\frac{I_x}{A}} = \sqrt{\frac{27}{36}} = 0.866 \text{ mm}$$

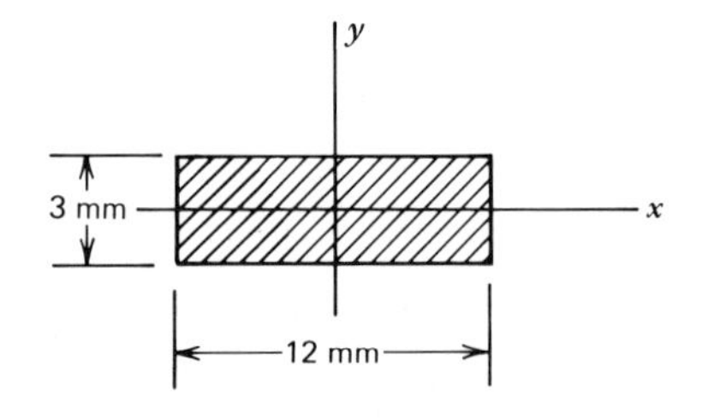

Fig. 12-24. Bar section for Illustrative Example 2.

4. The length L is 200 mm.
5. The slenderness ratio is

$$\frac{L}{r} = \frac{200}{0.866} = 231$$

A steel compression member that has a slenderness ratio of 231 is certainly a slender column.

B. The working load.

1. The modulus of elasticity of steel is 207 GPa or 207×10^3 MPa.
2. The moment of inertia is 27 mm.^{4}
3. The length L is 200 mm.
4. The factor of safety is 2.
5. The working load is

$$P = \frac{\pi^2 EI}{fL^2} = \frac{\pi^2 \times 207 \times 10^3 \times 27}{2(200)^2} = 689.5 \text{ N}$$

Practice Problems (Section 12-10). Here are problems on slender columns but you had better calculate the slenderness ratio for each to be sure that it is a slender column.

1. Find the working load for a circular magnesium-alloy rod 12 mm in diameter and 0.5 m long when used as a compression member. Take the factor of safety as 3.
2. A steel angle designated as $120 \times 120 \times 8$ (mm) is 8 m long and is used as a compression member. Calculate the working load. Let $f = 2$.
3. An aluminum-alloy channel designated as 3 [1.597 is 4 ft long and is used as a compression member. What is a reasonable working load? Let $f = 1.8$.
4. A steel tube 1 in. outer diameter, 0.8 in. inner diameter, and 60 in. long is to be used as a compression member. Find the working load. Let $f = 2.2$.
5. What is a reasonable working load for an aluminum-alloy compression member that has the cross section shown in Fig. 12-25 and is 30 in. long? Let $f = 2.4$.

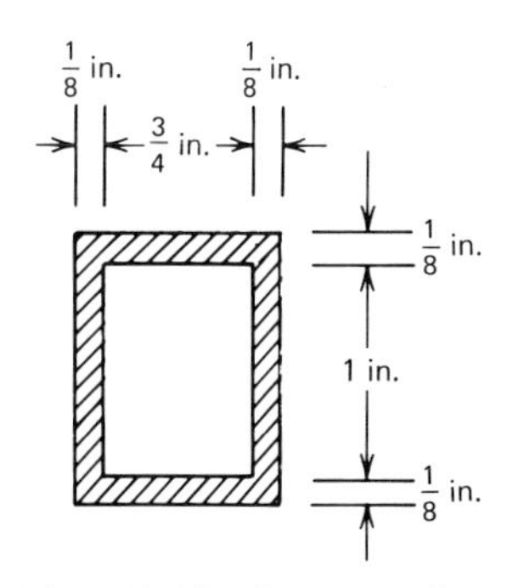

Fig. 12-25. Cross section for Problem 5.

SUMMARY

This summary covers the main new points of the chapter.

1. The least radius of gyration of a compression member is equal to

$$r = \sqrt{\frac{I}{A}}$$

where I is the smallest moment of inertia of the cross section and A is the area of the cross section.

2. The slenderness ratio of a compression member is equal to the length divided by the least radius of gyration.
3. There are three classes of compression members.
   (a) The working load for a short-compression member is

$$P = AS$$

   where A is the area of the cross section and S is the working stress for the material in compression.
   (1) A steel compression member is considered *short* if its slenderness ratio is below about 40.
   (2) An aluminum- or magnesium-alloy compression member is considered to be short if its slenderness ratio is below about 10.
   (b) There are several formulas for calculating the working load for an intermediate column. A designer should be able to use whatever is specified by a design code.
   (1) One type of column formula is the straight-line formula, for example

$$\frac{P}{A} = 16\,000 - 70\frac{L}{r}$$

   in gravitational units, or

$$\frac{P}{A} = 110 - 0.483\frac{L}{r}$$

   in metric units.
   (2) A second type of column formula is the parabolic formula; thus

$$\frac{P}{A} = 21\,9000 - 1.75\left(\frac{L}{r}\right)^2$$

   in gravitational units, or

$$\frac{P}{A} = 151 - 1.21\left(\frac{L}{r}\right)^2$$

   in metric units.
   (3) A third type of column formula is the Gordon–Rankine formula, for example

$$\frac{P}{A} = \frac{18\,000}{1 + \dfrac{1}{18\,000}\left(\dfrac{L}{r}\right)^2}$$

in gravitational units.

4. A steel-compression member is an intermediate column if its slenderness ratio is between 40 and 150.
5. An aluminum- or magnesium-alloy compression member is an intermediate column if its slenderness ratio is between 10 and 70.
6. A working load for a slender column is

$$p = \frac{\pi^2 EI}{fL^2}$$

where E is the modulus of elasticity, I is the moment of inertia, L is the length, and f is the factor of safety.

(a) A steel-compression member is a slender column if its slenderness ratio is above about 150.

(b) An aluminum- or magnesium-alloy compression member is a slender column if its slenderness ratio is above about 70.

REVIEW QUESTIONS Can you answer these questions without looking at the preceding pages?

1. What is *compression*?
2. What is *moment of inertia*?
3. What is *radius of gyration*?
4. What is *slenderness ratio*?
5. You can calculate moment of inertia with respect to the x axis and with respect to the y axis. How do you know which to use to calculate the radius of gyration?
6. How does a short-compression member fail?
7. When is a steel-compression member considered to be a short-compression member?
8. When is an aluminum-alloy compression member considered to be a short-compression member?
9. How does an intermediate column fail?
10. What are the limits of slenderness ratio for an intermediate steel column?
11. What kinds of formulas can you use to calculate a working load for an intermediate column?

12. What is the formula for the working load on a slender column?
13. How does a slender column behave when the critical load is applied?

REVIEW PROBLEMS These problems will test what you have learned in this chapter.

1. Use the formula $P/A = 21\,900 - 1.75(L/r)^2$ to calculate the working load for an aluminum-alloy channel 10 [8.360 6 ft long and is used as a column.
2. Use the formula $P/A = 110 - 0.483\,L/r$ to calculate the working load for a steel angle 150 × 150 × 10 that is 3 m long and is used as a column.
3. A steel beam section, W 10 × 100, is used as a column 14 ft long. The yield point of the material is 40 000 psi. Use the AISC formula to calculate the working load.
4. A circular steel tube 6 in. outer diameter, 5 in. inner diameter, and 32 in. long is loaded in compression. A reasonable working stress in compression is 20 000 psi, and a suitable column formula for this material is

$$\frac{P}{A} = \frac{18\,000}{1 + \dfrac{1}{18\,000}\left(\dfrac{L}{r}\right)^2}$$

   Calculate a reasonable working load.
5. A square magnesium-alloy bar, 60 mm on a side, is 2 m long. Calculate the working load in compression using a factor of safety of 1.6.
6. A steel wire 1 mm in diameter and 50 mm long is used as a compression member. What is a reasonable working load if the factor of safety is 3?

13 RIVETED, BOLTED, AND WELDED JOINTS

PURPOSE OF THIS CHAPTER. In this chapter we will study the common methods of fastening machine and structural parts together. A machine or structure is usually composed of many separate parts, and they must be fastened to one another. We are going to examine the types of connections that are used most commonly and calculate the stresses in the connections. We will find that some parts are stressed in tension (stretching), some in compression (shortening), and some in shear (sliding). You learned something about these stresses in Chapter 3, and you know how to use the formula

$$P = AS$$

where P represents force, A represents area, and S represents stress. We will find more use for the formula in this chapter.

It is easy to see that it is important to design connections properly, because if the connections aren't strong enough, the machine or structure will fall apart when it is placed in operation. The usual procedure in designing a machine or structure is to design each member in it and then to design the connections that hold the members together. The purpose of this chapter is to teach you how to find stresses in the connections and how to design them.

13-1. RIVETS AND BOLTS

Let's look at rivets and bolts now and be sure we know what they are. Figure 13-1 shows two views of a bolted connection (joint) between two plates. The shank (center part) of the bolt is circular. One end has a head which is usually hexagonal (six sided) and the other end is threaded to receive the nut. The function of the bolt in Fig. 13-1 is to transmit the force P from one plate to the other. We will look at this function in detail a little later and see how it is done and what the stresses are. Bolts

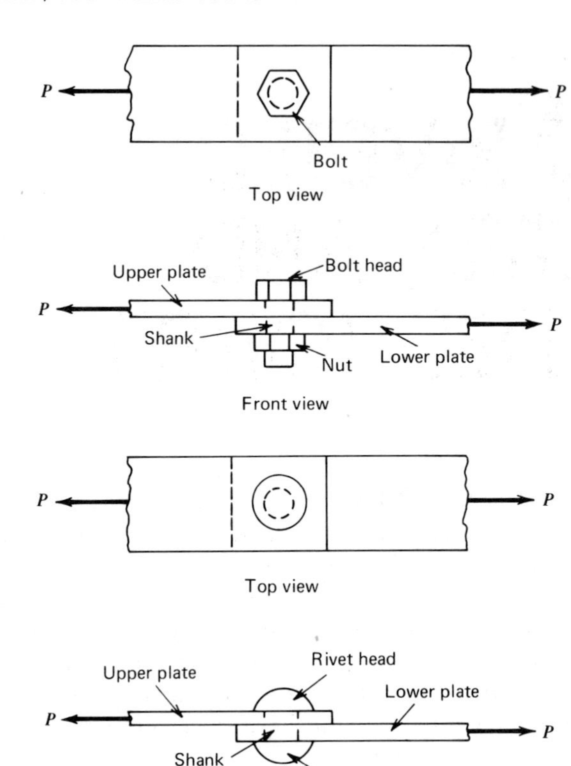

Fig. 13-1. Bolted joint.

Fig. 13-2. Riveted joint.

are usually made of steel but are sometimes made of brass or an aluminum alloy.

A *rivet* is a circular rod with a head on each end. Figure 13-2 shows two views of a riveted connection (joint) between two plates. The shank of the rivet is circular and the heads may be of almost any shape. Figure 13-3 shows a rivet before and after driving. The rivet is made in the shape shown in Fig. 13-3*a* and is inserted in the hole in the plates (the diameter of the rivet must be somewhat less than the diameter of the

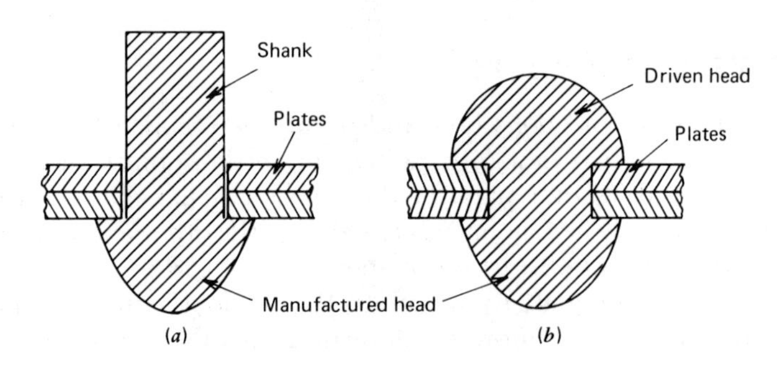

Fig. 13-3. Cross section of rivets and plates. (*a*) Before driving. (*b*) After driving.

hole in the plates to make this possible). Then the manufactured head of the rivet is supported while the driven head is formed by striking the end of the rivet with a hammer or by squeezing with a press. During the process of forming the driven head of the rivet, the metal flows, so that the shank of the rivet expands and makes a tight fit in the hole in the plates. The function of a rivet is the same as that of a bolt, that is, to transmit force from one piece of metal to another; in fact, there are many design problems in which a rivet would do as well as a bolt and vice versa. In general, a rivet makes a tighter connection than a bolt and is more permanent. It is also likely to be cheaper when there is occasion to use a large number of fasteners. One advantage of a bolted connection is that it can be dismantled easily for repair or adjustment. A disadvantage of a bolted connection is that the nuts tend to jiggle loose if there is any shaking or vibration of the machine or structure. (The tendency to loosen can be reduced by the use of special locking devices and by cotter pins.) Rivets are usually made of soft steel or iron, but are sometimes made of soft copper or of an aluminum alloy. Steel and iron rivets of 3/8 in. in diameter or larger are usually heated before driving, whereas steel and iron rivets of smaller diameter and copper or aluminum rivets can be driven cold.

Calculation of stresses is the same for a bolted joint as for a riveted joint. We use the formula,

$$P = AS$$

and the area is the same for a bolted joint as for a riveted joint.

13-2. SHEARING STRESSES IN RIVETS AND BOLTS

Remember, we learned in Chapter 3 that there is shearing stress in a body whenever one part of the body tends to slip over another part. Let's see where the shearing stress occurs in a rivet or bolt and see how to find it. Figure 13-4*a* shows a riveted joint between two steel plates. The force P that acts on the upper plate tends to pull the upper plate to the left, and the force P that acts on the lower plate tends to pull the lower plate to the right. As a result, the upper plate tends to slide to the left with respect to the lower plate, but it is prevented from doing so by

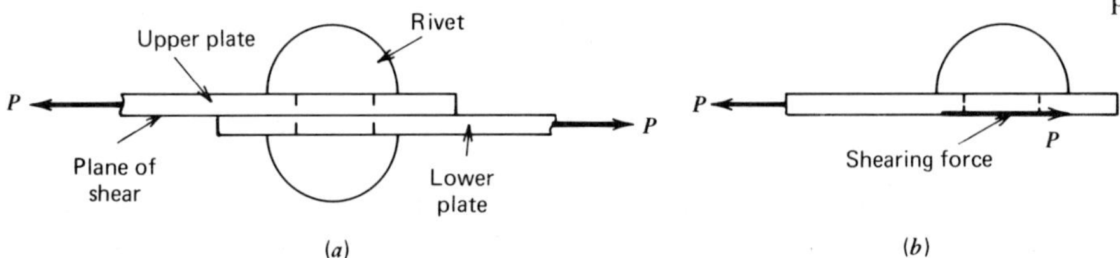

Fig. 13-4. Rivet in single shear. (*a*) Free-body diagram of joint. (*b*) Free-body diagram of upper plate.

the rivet. The upper plate tends to pull the upper half of the rivet to the left and the lower plate tends to pull the lower half to the right. The plates tend to shear the rivet on the plane of contact of the two plates, and this is the plane on which the shearing stress occurs. Figure 13-4*b* shows the free-body diagram of the upper plate and the upper half of the rivet. Notice the shearing force P exerted on the upper half of the rivet by the lower half. This force must be equal to P for the upper half of the plate to be in equilibrium; because, if a body is in equilibrium, the sum of the x components of forces acting on it must be zero, that is, $\Sigma F_x = 0$ (remember how we used this equation in Chapter 2). The only way for ΣF_x to equal zero is for the two forces to be equal, so that

$$\Sigma F_x = 0$$

$$P - P = 0$$

The area in shear is simply the area of the cross section of the rivet; we represent the area by A, and it is just $\pi/4d^2$, where d is the diameter of the rivet. The shearing stress S_s is equal to the force divided by the area.

$$S_s = \frac{P}{A_S}$$

Here the subscript s indicates that the stress and the area are in shear.

Illustrative Example 1. In Fig. 13-4, the force P is 7 300 lb, and the rivet diameter is 7/8 in. Calculate the shearing stress in the rivet.

Solution:

1. The magnitude of the force P is 7 300 lb.
2. The area in shear is the circular area of the cross section of the rivet. So

$$A = \frac{\pi}{4}d^2 = \frac{\pi}{4}\left(\frac{7}{8}\right)^2 = 0.601 \text{ in.}^2$$

3. Then the stress is

$$S_s = \frac{P}{A} = \frac{7\,300}{0.601} = 12\,100 \text{ psi}$$

(Do you remember that we introduced psi as an abbreviation for *pounds per square inch* in Chapter 3?)

DOUBLE SHEAR

Figure 13-5*a* shows a different type of riveted joint. Here a center plate A is fastened to two outer plates B and C by means of a rivet. The rivet

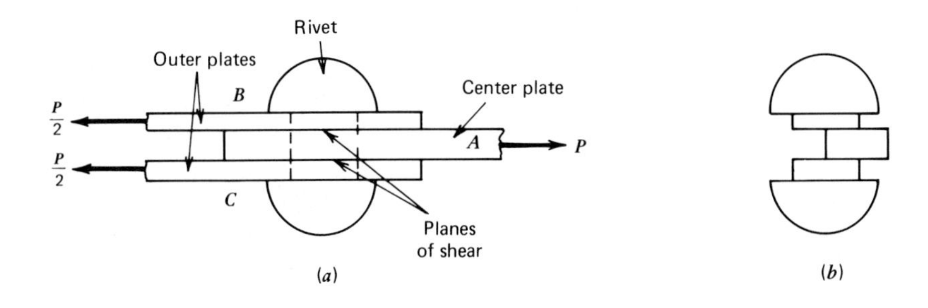

Fig. 13-5. Rivet in double shear. (*a*) Free-body diagram of joint. (*b*) Partly sheared rivet.

transmits the force P from the center plate to the two outer plates and is subjected to shearing stress. The force P tends to slide the center plate out from between the two outer plates and develops shearing stress in the rivet on each plane of contact between the center plate and the outer plates. Figure 13-5*b* shows the rivet as it would appear when partially sheared in two. You can see that there are two planes on which shearing stress occurs. The shearing area on each plane is the area of cross section of the rivet, so the total area in shear is

$$A_s = 2\frac{\pi}{4}d^2$$

We say that a rivet is in *double shear* when shearing stress is developed on two planes, as in Fig. 13-5. When there is shearing stress on only one plane, as in Fig. 13-4, we say that the rivet is in *single shear*. There is twice as much shear area for a rivet in double shear as for a rivet of the same diameter in single shear, because there are two cross sections in shear instead of one.

Illustrative Example 2. The force P in Fig. 13-5*a* is 5 600 N, and the diameter of the rivet is 6 mm. What is the shearing stress in the rivet?

Solution:

1. The force is $P = 5\,600$ N.
2. The rivet is in shear on two planes, so the area in shear is twice the area of the circular cross section. Thus

$$A_s = 2\frac{\pi}{4}d^2 = 2\frac{\pi}{4}(6)^2 = 56.5 \text{ mm}^2$$

3. Then the shearing stress is

$$S_s = \frac{P}{A} = \frac{5\,600}{56.5} = 99.1 \text{ MPa}$$

RIVETS OR BOLTS

We have looked at a rivet in single shear and at a rivet in double shear (the shearing stress in a bolt is calculated in the same manner as the shearing stress in a rivet); practically all rivets and bolts are stressed either in single shear or in double shear, and the only way to make a riveted or bolted connection more complicated is to use more than one rivet or bolt. This doesn't present any difficulty. You can tell whether the rivets or bolts are in single shear or in double shear. If they are in single shear, just multiply the number of rivets or bolts by the area of one cross section to get the area in shear. If they are in double shear, each rivet or bolt has two areas in shear, so multiply two times the number of rivets or bolts by the area of one cross section to get the area in shear. Let's try this in a pair of examples.

Illustrative Example 3. The bar A in Fig. 13-6 is fastened to the machine frame B by three ⅜ in. cap screws. Find the shearing stress in the cap screws. (A cap screw is just like a bolt, but is screwed into a piece of metal instead of going through and having a nut on the other end.)

Solution:

1. The force P is 1 700 lb.
2. Each cap screw is in single shear, so each has one cross section in shear. There are three cap screws, so the total area in shear is three times the area of one cap screw. The area of one is $\pi d^2/4$, so

$$A_s = 3 \times \frac{\pi}{4} d^2 = 3 \times \frac{\pi}{4}\left(\frac{3}{8}\right)^2 = 0.331 \text{ in.}^2$$

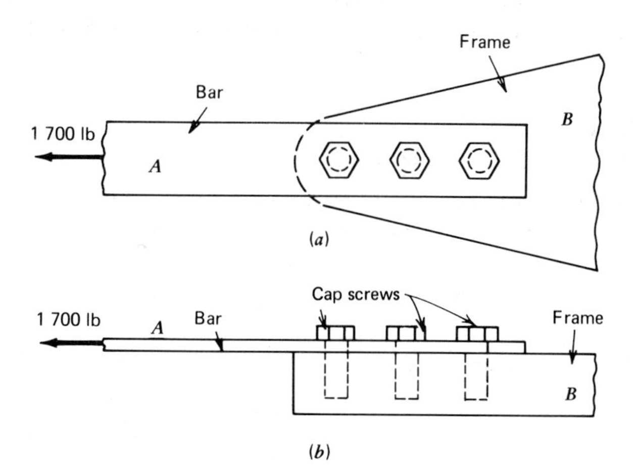

Fig. 13-6. Cap screws in shear. (*a*) Top view. (*b*) Front view.

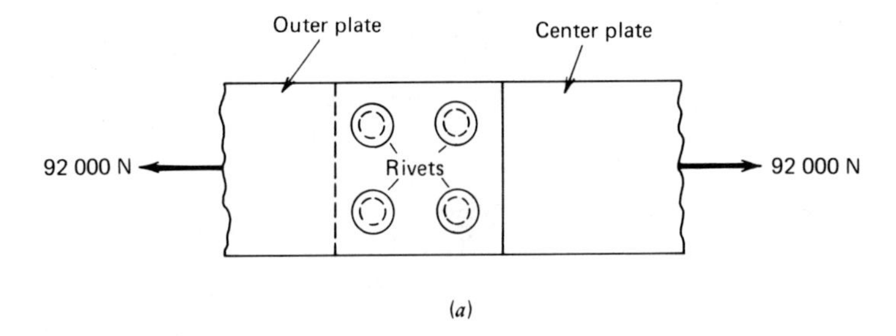

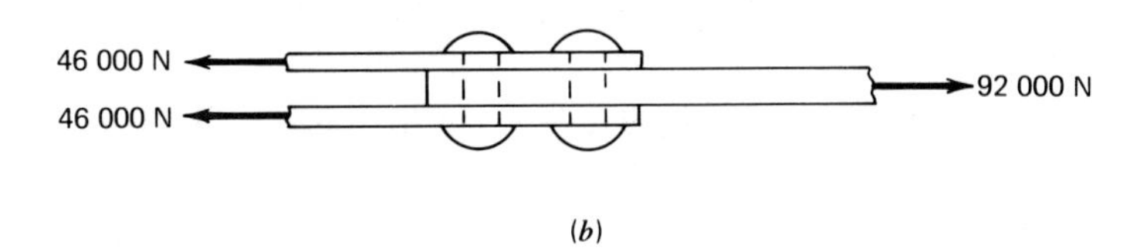

Fig. 13-7. Riveted joint for Illustrative Example 4. (*a*) Top view. (*b*) Front view.

3. The shearing stress is

$$S_s = \frac{P}{A_s} = \frac{1\ 700}{0.331} = 5\ 140 \text{ psi}$$

Illustrative Example 4. The center plate in Fig. 13-7 is fastened to the outer plates by four 20-mm rivets as shown. Find the shearing stress in the rivets.

Solution:

1. The force P is 92 000 N.
2. Each rivet is in double shear, so each has two cross sections in shear. There are four rivets, so the total area in shear is 2 × 4 × the area of one rivet. Thus the area in shear is

$$A_s = 2 \times 4 \times \frac{\pi}{4}d^2 = 8 \times \frac{\pi}{4}(20)^2 = 2\ 510 \text{ mm}^2$$

3. The shearing stress is

$$S_s = \frac{P}{A_s} = \frac{92\ 000}{2\ 510} = 36.7 \text{ MPa}$$

Practice Problems (Section 13-2). It is now time to practice again. Here are six problems on rivets and bolts that you should be able to do quickly.

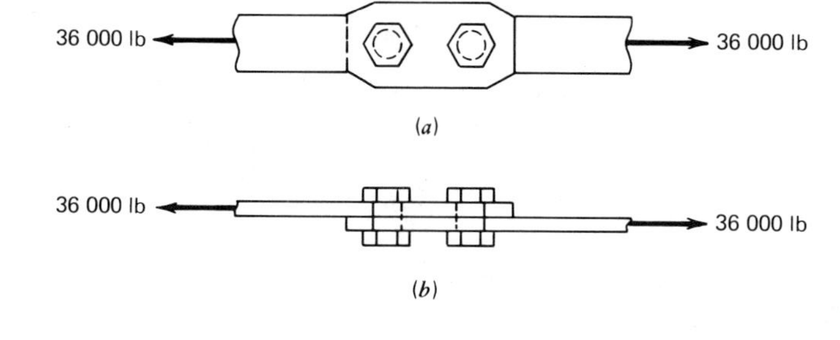

Fig. 13-8. Bolted joint for Problem 1. (*a*) Top view. (*b*) Front view.

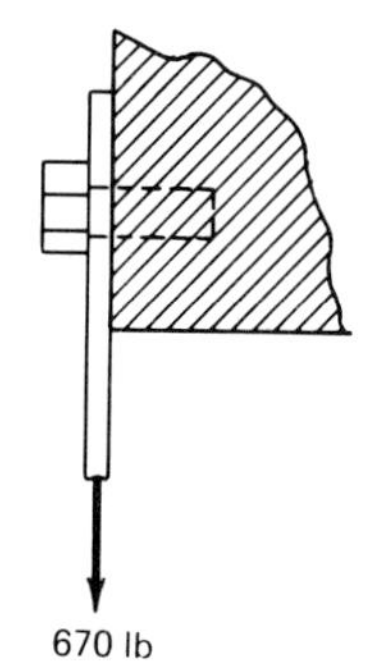

Fig. 13-9. Cap screw in shear for Problem 2.

1. The bolts in Fig. 13-8 are ¾ in. in diameter. Find the shearing stress.
2. The cap screw in Fig. 13-9 is ½ in. in diameter. Find the shearing stress.
3. The rivets in Fig. 13-10 are ¼ in. in diameter. Find the shearing stress.
4. The rivets in Fig. 13-11 are 8 mm in diameter. Find the shearing stress.

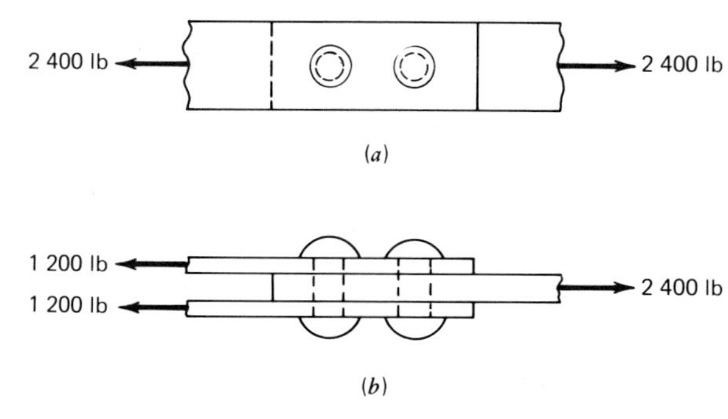

Fig. 13-10. Riveted joint for Problem 3. (*a*) Top view. (*b*) Front view.

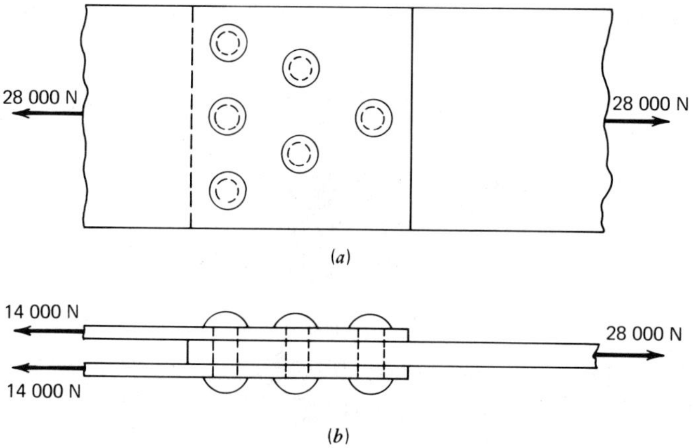

28 000 N

28 000 N

(*a*)

14 000 N

14 000 N

28 000 N

(*b*)

Fig. 13-11. Joint with six rivets for Problem 4. (*a*) Top view. (*b*) Front view.

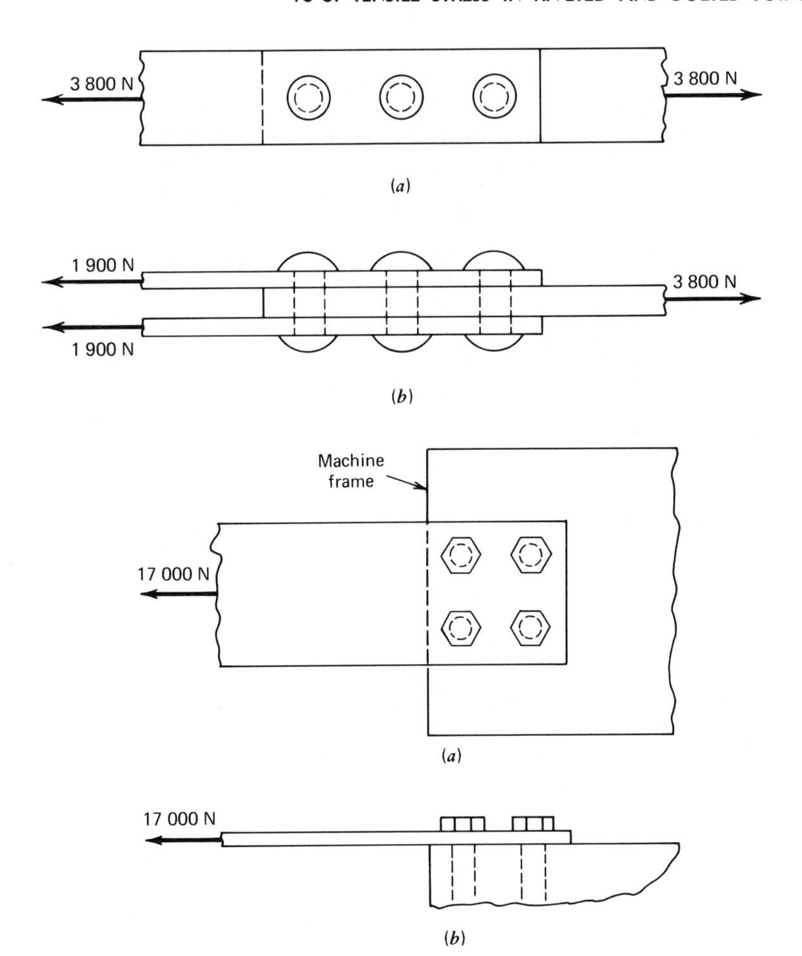

Fig. 13-12. Three rivets in shear for Problem 5. (*a*) Top view. (*b*) Front view.

Fig. 13-13. Cap screws in shear for Problem 6. (*a*) Top view. (*b*) Front view.

5. Calculate the shearing stress in the rivets in Fig. 13-12. The diameter of the rivets is 3 mm.
6. Figure 13-13 shows a plate fastened to a machine frame by four cap screws, each 12 mm in diameter. Find the shearing stress.

13-3. TENSILE STRESS IN RIVETED AND BOLTED JOINTS

We come now to the tensile stress in a riveted or bolted joint. Let's look at Fig. 13-14 while we explain it. Figures 13-14*a* and 13-14*b* show two views of a pair of plates fastened together by a bolt of diameter d and subjected to a load P. Each cross section of each plate carries tensile stress, but we are interested in the maximum tensile stress, because if the plate breaks, it will break where the stress is greatest. The maximum

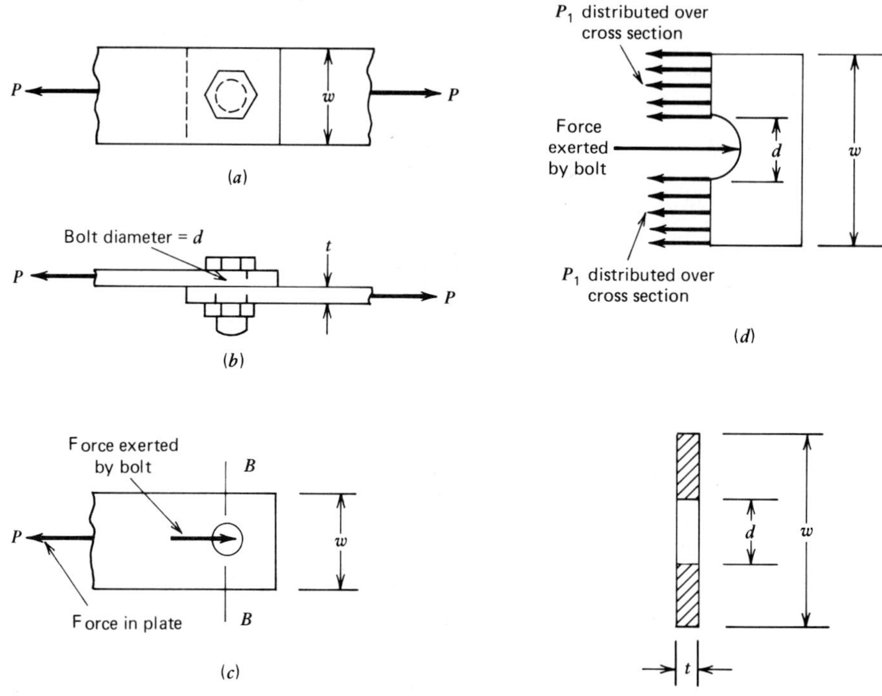

Fig. 13-14. Tensile stress in a joint. (*a*) Top view of joint. (*b*) Front view of joint. (*c*) Free-body diagram of upper plate. (*d*) Free-body diagram of part of upper plate. (*e*) Net section.

stress occurs where the cross section is the smallest, and this is the cross section through the center of the hole. Figure 13-14*c* shows the free-body diagram of part of the upper plate with the force exerted by the bolt. The plate is subjected to the force P (to the left), and the bolt must exert the force P to the right to hold the plate in equilibrium. The cross section B-B through the center of the hole is the smallest cross section, so we will look at it in more detail. Figure 13-14*d* shows the free-body diagram of the part of the upper plate to the right of the cross section B-B. Here we see the force to the right exerted by the bolt, and this is balanced by an equal force to the left which is distributed over the cross section. The cross section through the center of the hole is usually called the *net section* and is shown in Fig. 13-14*e*. The width of the plate is w, the diameter of the rivet is d, and the thickness of the plate is t. We can think of this area as a rectangle of length w with a smaller rectangle of length d cut out of it. The thickness of each rectangle is t, so the area is

$$A_t = wt - dt = (w - d)t$$

The area, then, equals the plate thickness times the difference between the width and the diameter. The subscript t for A indicates that it is an area subjected to tensile stress, and we will write the formula as

$$S_t = \frac{P}{A_t}$$

to indicate that the stress is a tensile stress.

Illustrative Example 1. In Fig. 13-14, let $w = 3$ in., $d = 1$ in., and $t = \frac{1}{2}$ in. $P = 8\ 500$ lb. Find the tensile stress on the net section.

Solution:

1. The force P is 8 500 lb.
2. The area of the net section is

$$A_t = (w - d)t = (3 - 1)\tfrac{1}{2} = 2 \times \tfrac{1}{2} = 1 \text{ in.}^2$$

3. The tensile stress is

$$S_t = \frac{P}{A_t} = \frac{8\ 500}{1} = 8\ 500 \text{ psi}$$

ROWS OF RIVETS

There may be several rivets in a row perpendicular to the load, as in Fig. 13-15, but we can proceed in the same way to find the area of the net section. The smallest cross section of the plate is still the section through the center of the holes, and we see this in Fig. 13-15*c*. Here there are three small rectangles (each of length d and thickness t) cut out of the large rectangle of length w and thickness t, so the area of the net section is equal to the area of the large rectangle minus the area of the three small rectangles. This is

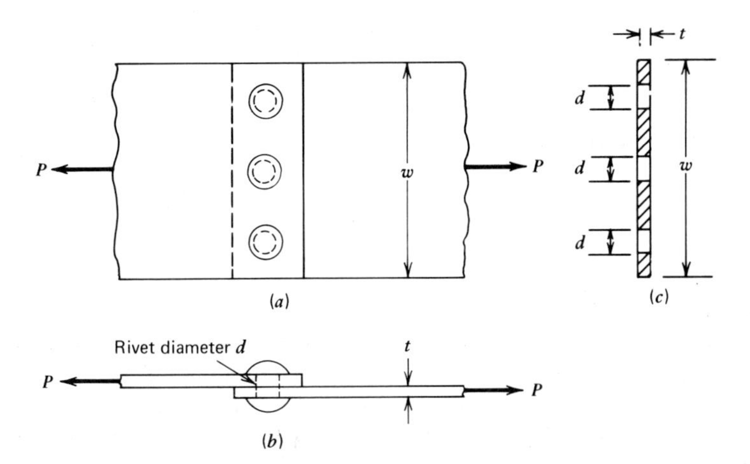

Fig. 13-15. Riveted joint and net section. (*a*) Top view. (*b*) Front view. (*c*) Net section.

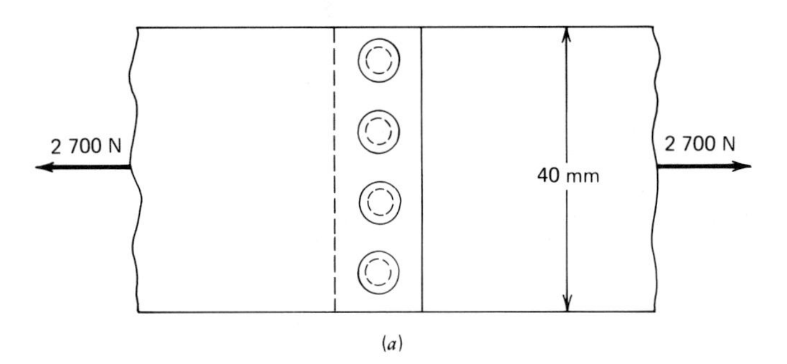

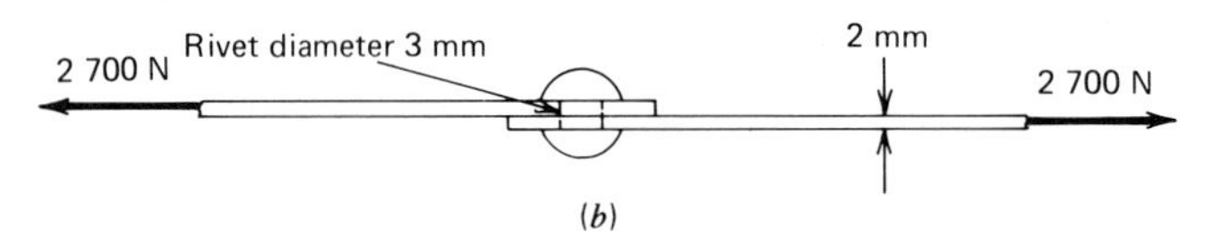

Fig. 13-16. Riveted joint for Illustrative Example 2. (*a*) Top view. (*b*) Front view.

$$A_t = wt - 3dt = (w - 3d)t$$

We have here the basis for a rule.

Rule 1. *Take the width of the plate, minus the number of rivets in the row times the diameter, and multiply by the thickness of the plate to find the area of the net section.*

Let's try it.

Illustrative Example 2. Find the tensile stress on the net section in Fig. 13-16.

Solution:

1. The force P is 2 700 N.
2. There are four rivet holes in the net section; consequently

$$A_t = (w - 4d)t = (40 - 4 \times 3)2 = (40 - 12)2$$

$$= 28 \times 2 = 56 \text{ mm}^2$$

3. The tensile stress is

$$S_t = \frac{P}{A_t} = \frac{2\,700}{56} = 48.2 \text{ MPa}$$

COVER PLATES

We have passed over one question so far, but now we are ready to settle it. Forces on a riveted joint are shown in Fig. 13-17. Look at Fig. 13-17*b*, which shows the front view of a riveted joint. The force on the center plate is P, and you might wonder how much of this is balanced by the

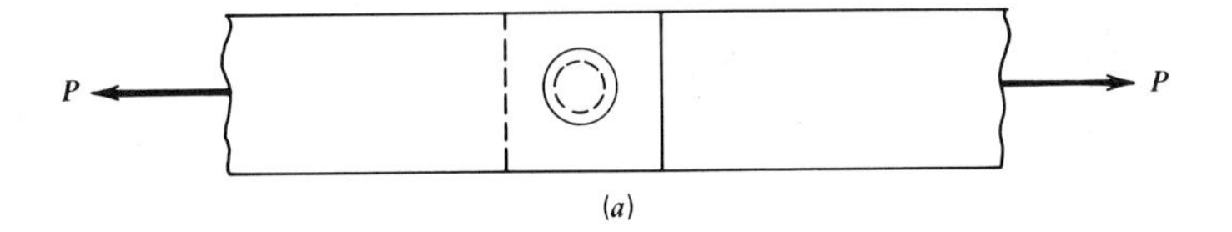

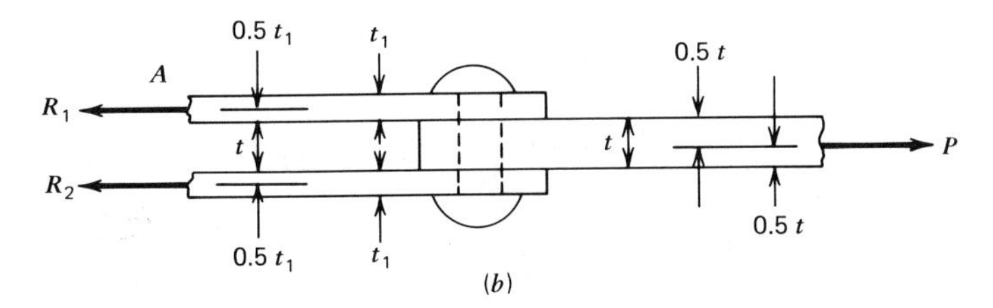

Fig. 13-17. Forces on riveted joint. (*a*) Top view. (*b*) Front view.

upper one of the outer plates and how much by the lower one. Well, let's find out. We will represent the thickness of the center plate by t and the thickness of each outer plate by t_1 (it is common practice to make the two outer plates of the same thickness). Next, let's designate the force in the upper outer plate as R_1 (unknown) and the force in the lower outer plate as R_2 (also unknown). Now we have a set of forces in equilibrium, and we will find R_1 and R_2 by writing moment equations (as we did in Chapter 2). Each force acts at the center of its plate, so we know where the forces are. Let's start by writing a moment equation with the moment center at point A on the line of action of R_1. The moment arm of the force P is $0.5t_1 + 0.5t$ (This is $0.5t_1$ from A to the plane of contact of the two plates plus $0.5t$ from this plane to P), and the moment arm of R_2 is $t_1 + t$ (twice as much as the moment arm of P). Then

$$\Sigma M_A = 0$$

$$-(0.5t_1 + 0.5t)P + (t_1 + t)R_2 = 0$$

$$-0.5(t_1 + t)P + (t_1 + t)R_2 = 0$$

Let's cancel $t_1 + t$, and then we will have

$$-0.5P + R_2 = 0$$

$$R_2 = 0.5P$$

Now that we know the value of R_2, let's write the equilibrium $\Sigma F_x = 0$ to find R_1. Thus

$$\Sigma F_x = 0$$

$$-R_1 - R_2 + P = 0$$

We will substitute $R_2 = 0.5P$, and

$$-R_1 - 0.5P + P = 0$$

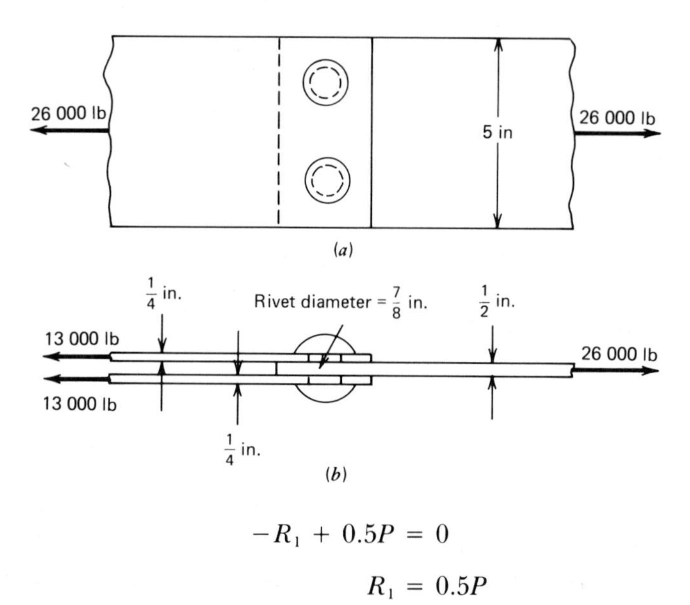

Fig. 13-18. Riveted joint for Illustrative Example 3. (*a*) Top view. (*b*) Front view.

$$-R_1 + 0.5P = 0$$

$$R_1 = 0.5P$$

So we have proved that each outer plate carries half of the total force when the outer plates are of equal thickness. Why not use this knowledge in an example?

Illustrative Example 3. Find the tensile stress on the net section of the outer plate in the riveted joint in Fig. 13-18.

Solution:

1. The force on each outer plate is one-half of the total, so

$$P = 13\,000 \text{ lb}$$

2. The width of the plate is 5 in., and there are two rivet holes to be deducted, so the area of the net section is

$$A_t = (w - 2d)t = \left(5 - 2 \times \frac{7}{8}\right)\frac{1}{4} = (5 - 1.75)\frac{1}{4}$$

$$= 3.25 \times \frac{1}{4} = 0.8125 \text{ in.}^2$$

3. The tensile stress is

$$S_t = \frac{P}{A_t} = \frac{13\,000}{0.8125} = 16\,000 \text{ psi}$$

Practice Problems (Section 13-3). Let's practice now on calculating tensile stress. The important thing is to find the area of the net section. Then just divide the force by the area.

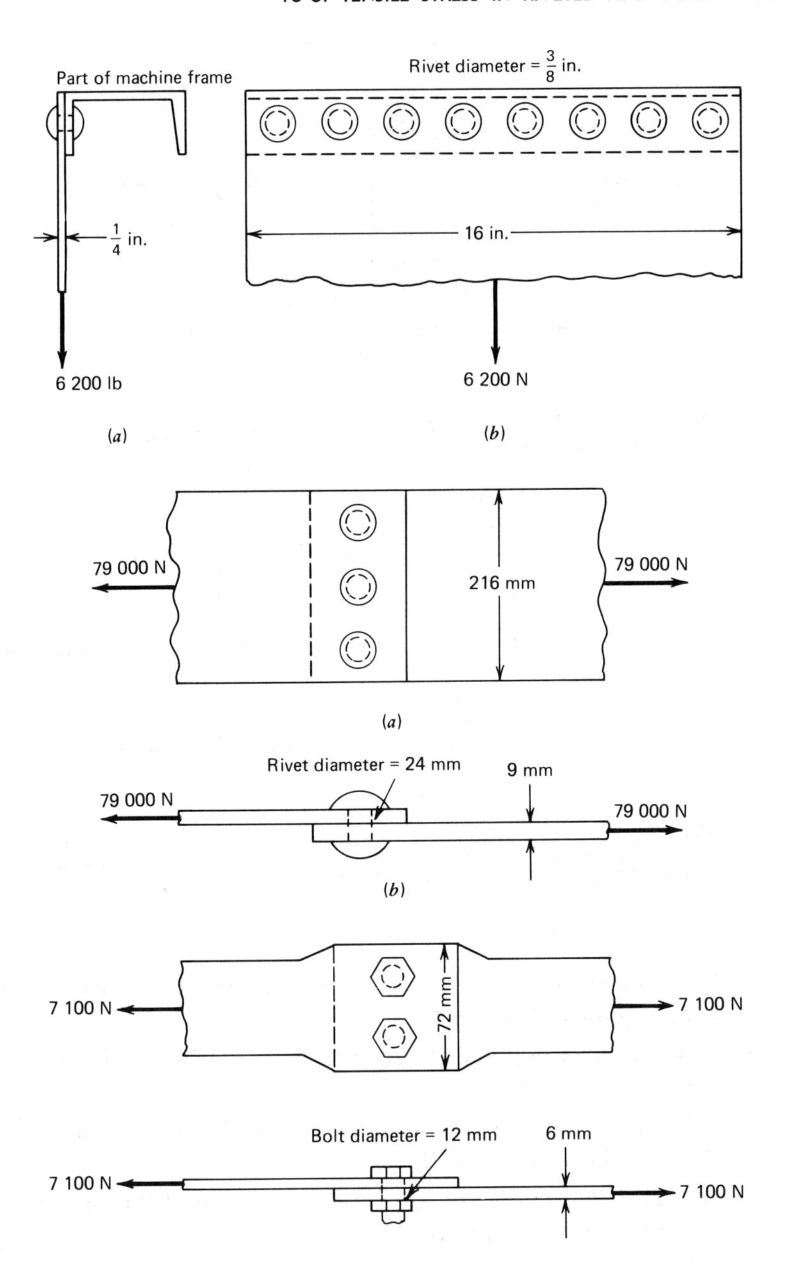

Fig. 13-19. Plate riveted to a machine frame for Problem 1. (*a*) Front view. (*b*) Side view.

Fig. 13-20. Riveted joint for Problem 2. (*a*) Top view. (*b*) Front view.

Fig. 13-21. Bolted joint for Problem 3.

1. Find the maximum tensile stress in the plate in Fig. 13-19.
2. Find the maximum tensile stress in the riveted joint in Fig. 13-20.
3. Calculate the tensile stress on the net section of the bolted joint in Fig. 13-21.

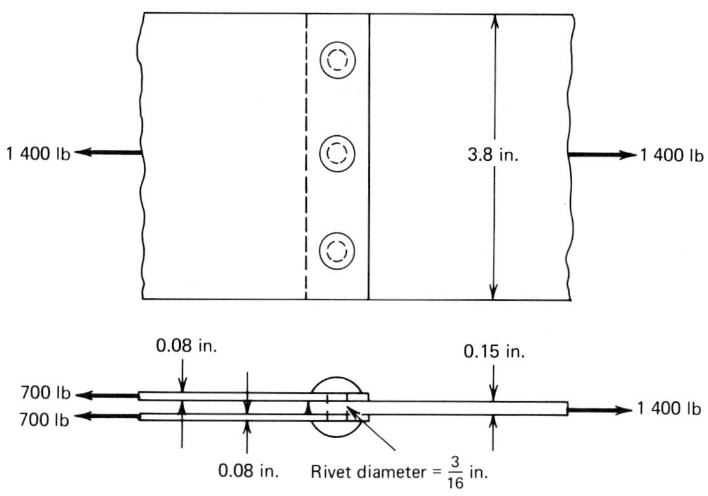

Fig. 13-22. Riveted joint for Problems 4 and 5.

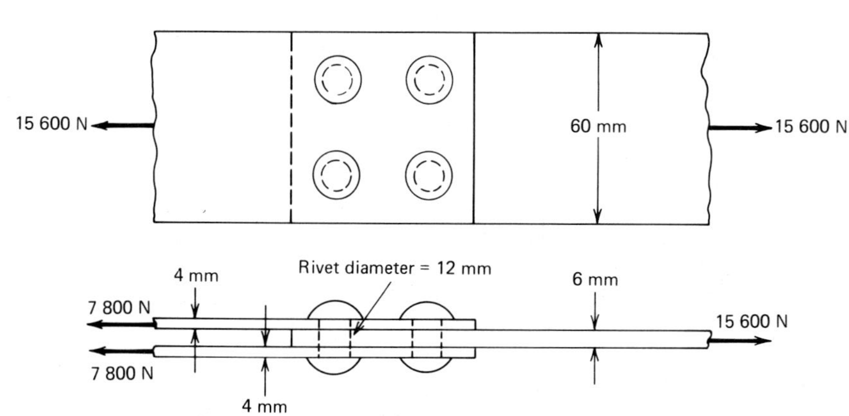

Fig. 13-23. Riveted joint for Problems 6 and 7.

4. Calculate the maximum tensile stress on the outer plates of the riveted joint in Fig. 13-22.
5. Find the maximum tensile stress in the center plate of the riveted joint in Fig. 13-22.
6. Calculate the maximum tensile stress in the outer plates of the riveted joint in Fig. 13-23.
7. Calculate the tensile stress on the net section of the center plate of the riveted joint in Fig. 13-23.

13-4. BEARING STRESS ON RIVETS AND BOLTS

Bearing stress is a special form of compressive stress. (A member shortens when it is subjected to a compressive stress.) When the compressive stress is at the surface of a body and is exerted there by another body,

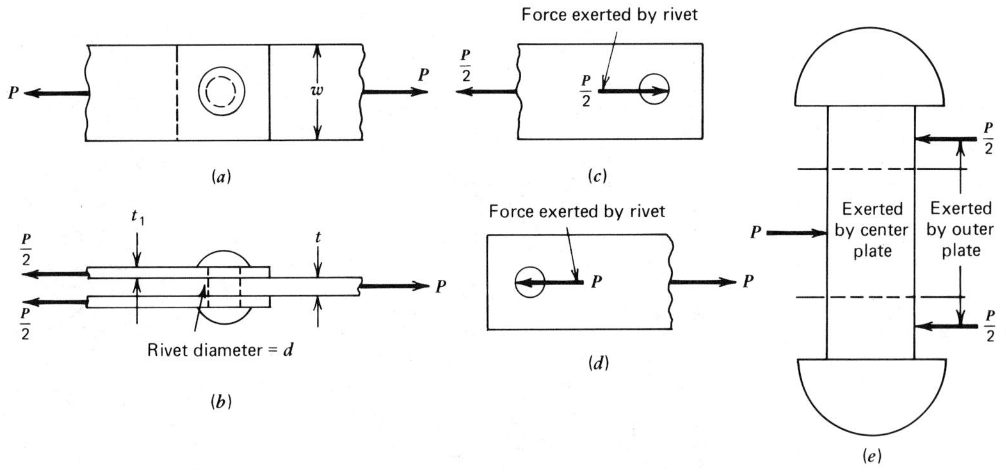

Fig. 13-24. Forces in riveted joint. (*a*) Top view. (*b*) Front view. (*c*) Free-body diagram of outer plate. (*d*) Free-body diagram of center plate. (*e*) Free-body diagram of rivet.

we call it bearing stress. Let's look closely at a simple riveted joint and see how and where the bearing stress occurs. Figures 13-24*a* and 13-24*b* show two views of the joint, and all of the dimensions are shown. Figure 13-24*c* shows the free-body diagram of one of the outer plates; the rivet must exert a force $P/2$ to the right upon each of the outer plates to hold it in equilibrium. Figure 13-24*d* shows the free-body diagram of the center plate; the rivet exerts a force P to the left on the plate. Then Fig. 13-24*e* shows the free-body diagram of the rivet; each outer plate exerts a force of $P/2$ to the left and the center plate exerts the force P to the right. These are compressive forces exerted on the rivet by the plates, and so there is compressive stress between the rivet and the plates, but since the compressive stress in on the surface between the rivet and the plates we call it *bearing stress.*

Figure 13-25*a* shows the shank (cylindrical part) of the rivet with the forces exerted on it by the plates. Here we see the areas over which the forces are distributed, and each area is half of a cylindrical surface. The diameter of each cylinder is the rivet diameter d, and the height of each cylinder is equal to the thickness of the plate that exerts the force; the height is t_1 for each cover plate and is t for the center plate. In Fig. 13-25*a* you see the actual areas over which the compressive stress is distributed, and because it is at the surface between the rivet and plates, we call it *bearing stress*. However, *bearing stress* is calculated in a peculiar way (probably because someone started doing it that way and other people followed the example). A nominal area is used instead of the actual area over which the force is distributed, and this nominal area is the projection of the actual area on a central plane of the rivet (the central plane is the plane *ABCD* in Fig. 13-25*a*). Figure 13-25*b* shows the nominal bearing areas; each is a rectangle with width equal to the rivet diameter d and height equal to the plate thickness, so the amount of the area is equal to the product of the rivet diameter and the plate thickness.

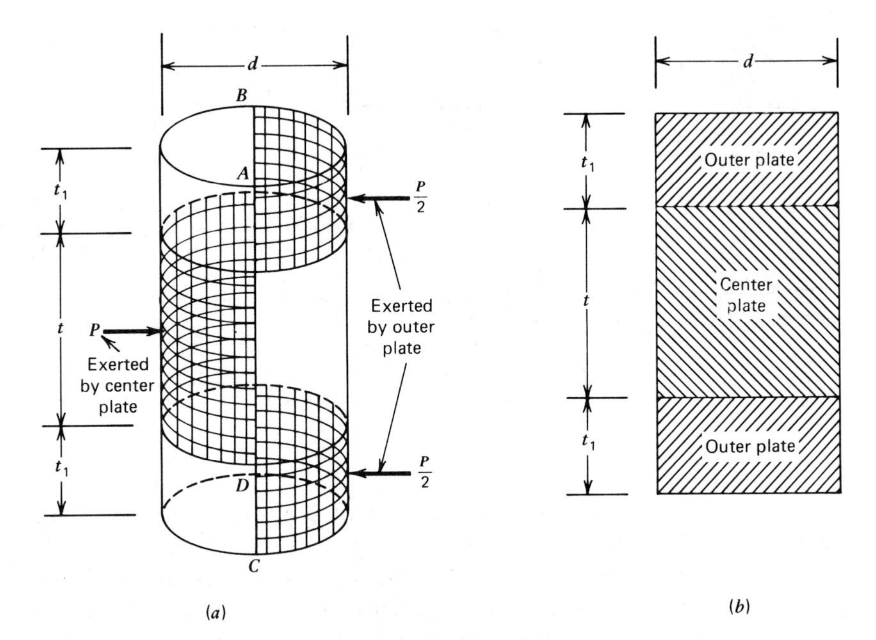

Fig. 13-25. Bearing areas on rivet. (*a*) Actual bearing areas. (*b*) Nominal bearing areas.

Hence the area of bearing for one of the cover plates is dt_1, and the area of bearing for the center plate is dt. We write the formula for stress as

$$S_b = \frac{P}{A_b}$$

where the subscript b stands for bearing.

Illustrative Example 1. Find the bearing stress between the rivet and plates in the riveted joint in Fig. 13-26.

Solution:

1. The force P is 2 930 lb.

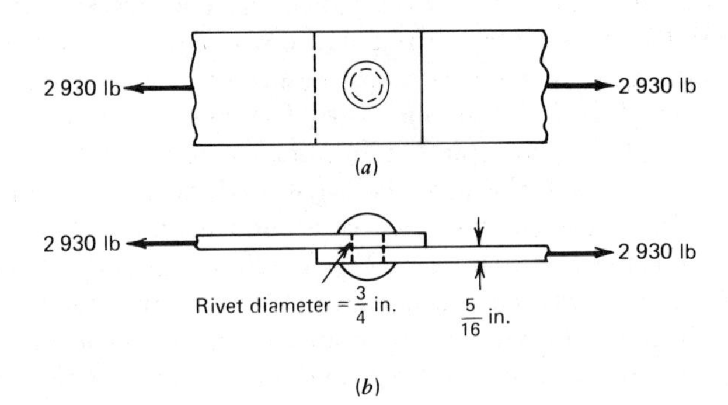

Fig. 13-26. Riveted joint for Illustrative Example 1. (*a*) Top view. (*b*) Front view.

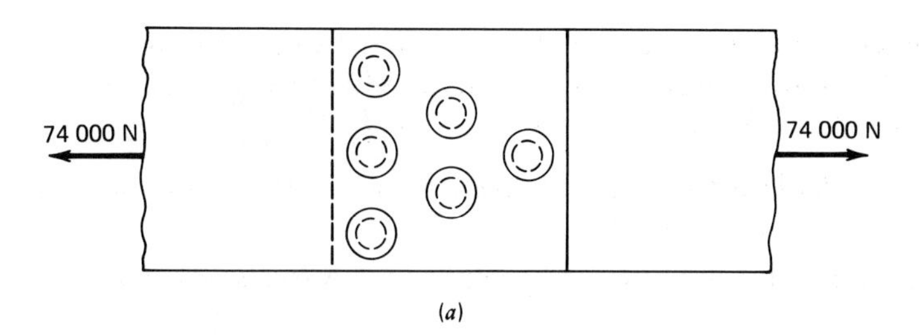

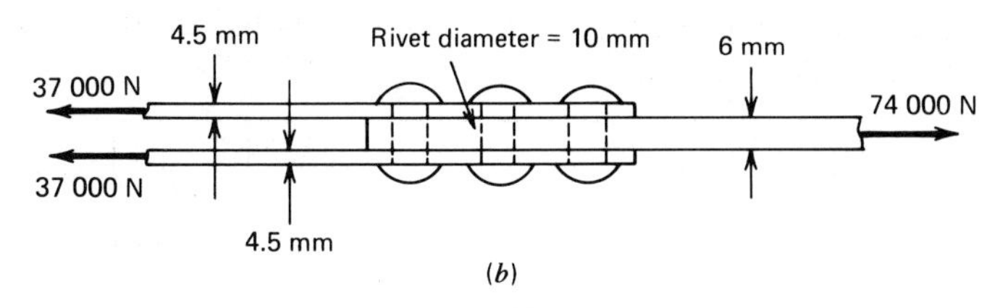

Fig. 13-27. Riveted joint for Illustrative Example 4 and Problem 1. (*a*) Top view. (*b*) Front view.

2. The area in bearing is equal to the product of the rivet diameter and the plate thickness; hence

$$A_b = dt = \frac{3}{4} \times \frac{5}{16} = 0.234 \text{ in.}^2$$

3. The bearing stress is

$$S_b = \frac{P}{A_b} = \frac{2\,930}{0.234} = 12\,500 \text{ psi}$$

SEVERAL RIVETS

When there are several rivets or bolts in the connection, the total bearing area is equal to the number of rivets or bolts times the bearing area of one. Just multiply dt by the number of rivets or bolts.

Illustrative Example 2. Calculate the bearing stress between the rivets and the outer plates in the riveted connection in Fig. 13-27.

Solution:

1. The force on one cover plate is 37 000 N.
2. There are six rivets, so the bearing area between the rivets and each cover plate is six times the rivet diameter times the plate thickness. Thus

$$A_b = 6dt = 6 \times 10 \times 4.5 = 270 \text{ mm}^2$$

3. The bearing stress between the rivets and the cover plates is

$$S_b = \frac{P}{A_b} = \frac{37\,000}{270} = 137 \text{ MPa}$$

Practice Problems (Section 13-4). Now let's practice on bearing stress. This is something that everyone should know and the more problems you do, the better you will know it.

1. Calculate the bearing stress between the rivets and the center plate in Fig. 13-27.
2. Find the bearing stress between the rivets and plate in Fig. 13-28.
3. Calculate the bearing stress between the bolts and the outer plates in the bolted joint in Fig. 13-29.

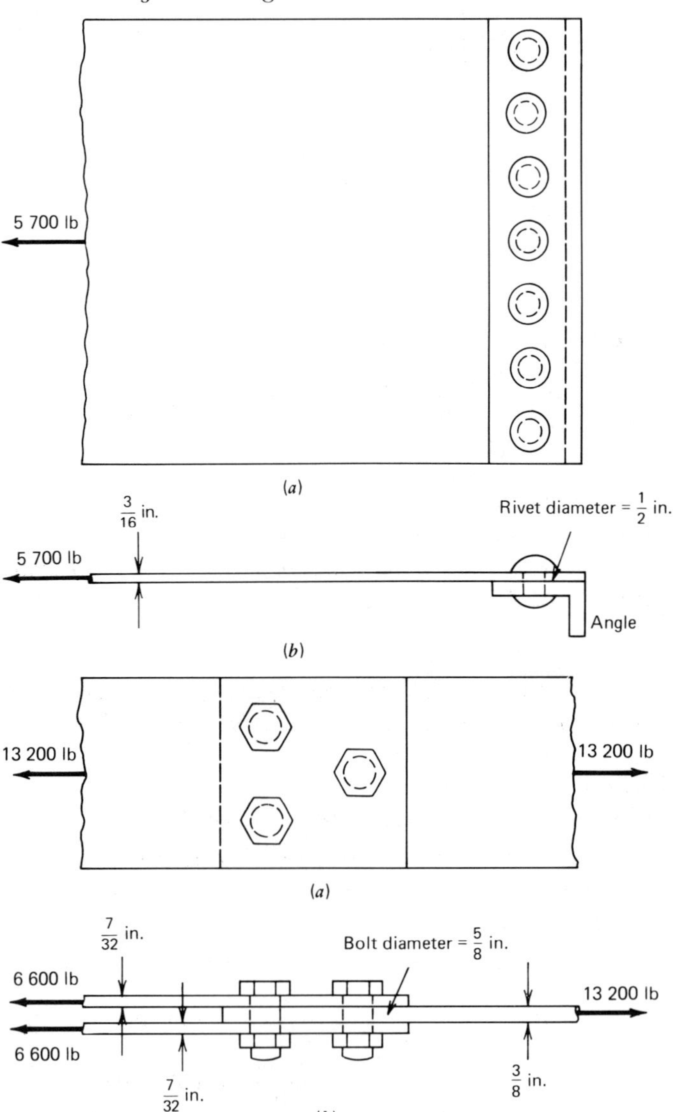

Fig. 13-28. Riveted joint between plate and angle for Problem 2. (*a*) Top view. (*b*) Front view.

Fig. 13-29. Bolted joint for Problems 3 and 4. (*a*) Top view. (*b*) Front view.

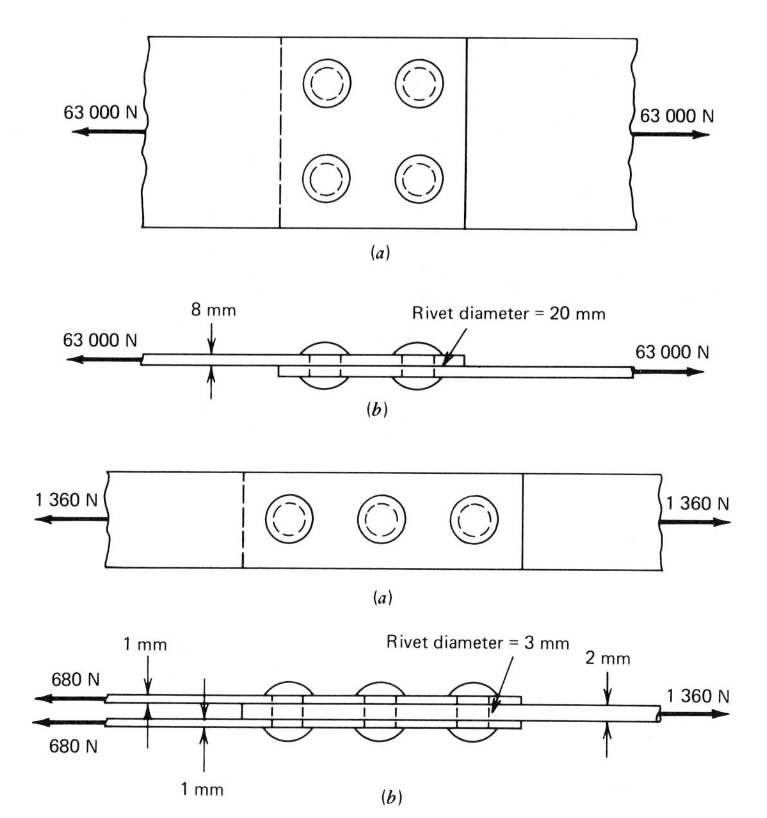

Fig. 13-30. Riveted joint for Problem 5. (*a*) Top view. (*b*) Front view.

Fig. 13-31. Riveted joint for Problem 6. (*a*) Top view. (*b*) Front view.

4. Calculate the bearing stress between the bolts and center plate in Fig. 13-29.
5. Find the bearing stress between the rivets and plates in Fig. 13-30.
6. Calculate the bearing stress between the rivets and center plate in Fig. 13-31.

13-5. STANDARD SIZES OF RIVETS AND BOLTS

The next step in studying connections is to become acquainted with the standard sizes of rivets and bolts. A designer has to know what the standard sizes are, because practically all bolts and rivets in machines are of standard size. You see, the standard sizes are made in great quantity, usually by automatic machines, and are available at low cost. To make any odd size, it would be necessary to make expensive adjustments on an automatic machine, and the cost of the odd size of bolt or rivet would be high. It is cheaper to use a larger size of bolt or rivet than is necessary for strength of the machine part just to get a standard size. Table 13.1

TABLE 13.1. STANDARD SIZES OF RIVETS AND BOLTS

Diameter (in.)[a]		*Area of Cross Section (in.²)*
Fraction	*Decimal*	
1/16	0.0625	0.00307
3/32	0.0938	0.0069
1/8	0.125	0.0123
5/32	0.1562	0.0192
3/16	0.1875	0.0276
1/4	0.250	0.0490
5/16	0.3125	0.0767
3/8	0.375	0.1104
7/16	0.4375	0.150
1/2	0.500	0.196
5/8	0.625	0.307
3/4	0.750	0.442
7/8	0.875	0.601
1	1.000	0.785
1 1/8	1.125	0.994
1 1/4	1.250	1.228

[a]These dimensions are also the sizes of rectangular bars.

shows the standard sizes of rivets and bolts; the table gives the diameter as a common fraction in gravitational units in the first column and as a decimal fraction in the second column. The area of the cross section appears in the third column.

Table 13.2 gives preferred standard diameters in the metric system.

The designer's calculations usually give either the diameter or area of cross section that is needed for the bolt or rivet. This diameter or area won't ordinarily be a standard size exactly, so the designer usually takes the next larger standard size. Shall we try it?

Illustrative Example 1. The diameter found to be necessary for a certain steel bolt is 0.34 in. What diameter should be used?

Solution:
The calculated diameter of 0.34 in. is between 0.3125 in. for a 5/16-in. bolt and 0.375 in. for a 3/8-in. bolt. See Table 13.1. The 3/8-in. bolt is the next larger standard size and is the one to be used. Then

$$d = 3/8 \text{ in.}$$

TABLE 13.2. PREFERRED RIVET AND BOLT DIAMETERS IN MILLIMETERS

Diameter (mm)	*Area of Cross Section* (mm^2)
3	7.069
4	12.57
5	19.63
6	28.27
8	50.27
10	78.54
12	113.10
16	201.06
20	314.16
24	452.39
30	706.86
36	1017.88

Illustrative Example 2. The calculations for the required area of cross section of a certain aluminum rivet give 62 mm^2 as required. What diameter should be used?

Solution:
We can see in Table 13.2 that an 8-mm rivet has an area of 50.27 mm^2 and a 10-mm rivet has an area of 78.54 mm^2. The 10-mm rivet is the next larger standard size from what we need, so we choose it:

$$d = 10 \text{ mm}$$

Practice Problems (Section 13-5). What diameter of rivet or bolt should be used in each of the following cases? The calculations give:

1. A required diameter of 0.618 in.
2. A required diameter of 0.173 in.
3. A required diameter of 0.487 in.
4. A required diameter of 7.45 mm.
5. A required diameter of 22.7 mm.
6. A required area of cross section of 0.337 $in.^2$
7. A required area of cross section of 0.695 $in.^2$
8. A required area of cross section of 0.156 $in.^2$
9. A required area of cross section of 24.8 $mm.^2$
10. A required area of cross section of 287 $mm.^2$

13-6. STANDARD SIZES OF BARS, STRIP, SHEET, AND PLATE

A designer must know something about the standard sizes of bars, sheets, and plates, because many members can be made most cheaply from standard sizes. Pieces of standard size are made in great quantity and, as a result, their cost is low. Usually it is cheaper to use a standard size a little larger than is required for strength, even though this means extra material. The cost of the extra material in the piece of standard size is less than the cost of shaping a piece that is not of standard size.

Strip, sheet, and plate are all pieces of rectangular cross section. *Strip* usually means material that is less than 1/4 in. thick and less than 12 in. wide. *Sheet* usually means material that is less than 1/4 in. thick and more than 12 in. wide. *Plate* usually means material that is more than 1/4 in. thick and more than 6 in. wide. These statements give you an idea of what is meant by strip, sheet, and plate, but you will find some variation from them. So many companies manufacture material, and there are so many materials and uses for materials, that it is impossible to find any simple system followed by all. Any size not covered by strip, sheet, or plate (as well as some that are) can be called a *bar*, although a bar is not necessarily rectangular in cross section. Right now, you probably feel that this is a confusing picture, and you are exactly right. It is! There just isn't any simple system of classifying different sizes and shapes of pieces of material.

The thickness of strip and sheet is usually given by a gage (also spelled *gauge*) number. A *gage* is a system for stating sizes in which each number in gage represents a certain size. Table 13.3 shows the thicknesses included in the Manufacturer's Standard Gage for Sheet Steel.

Metal strip can be obtained in many different standard widths. Usually the smallest is 3/8 in. and it goes from 3/8 in. to 1 1/2 in. by steps of 1/8 in., from 1 1/2 in. to 3 1/2 in. by steps of 1/4 in., from 3 1/2 in. to 6 in. by steps of 1/2 in., from 6 in. to 10 in. by steps of 1 in., and finally there is the 12-in. width. There are additional widths in some alloys and by some manufacturers, but those given here are the widths you can rely on. Plates can be obtained in standard widths from 24 in. up to as much as 12 ft for some thicknesses of steel plates.

Rectangular metal bars can usually be obtained in sizes ranging from 1/16 in. to 1/2 in. by steps of 1/16 in., from 1/2 in. to 1 in. by steps of 1/8 in., and from 1 in. to 2 in. by steps of 1/4 in. Above 2 in., the standard dimensions are about the same as the standard widths of strip. The dimensions in Table 13.1 can be used for small bars. You will find the decimal fractions of value.

Figure 13-32 shows an ISO (International Organization for Standardization) recommendation for standard sizes of hot-rolled flat steel bars. Here each horizontal line represents a thickness in millimeters, and each vertical line represents a width in millimeters. Each x mark indicates

TABLE 13.3. MANUFACTURER'S STANDARD GAGE FOR SHEET STEEL

Standard Gage	*Thickness (in.)*	*Standard Gage*	*Thickness (in.)*
38	0.0060	20	0.0359
37	0.0064	19	0.0418
36	0.0067	18	0.0478
35	0.0075	17	0.0538
34	0.0082	16	0.0598
33	0.0090	15	0.0673
32	0.0097	14	0.0747
31	0.0105	13	0.0897
30	0.0120	12	0.1046
29	0.0135	11	0.1196
28	0.0149	10	0.1345
27	0.0164	9	0.1495
26	0.0179	8	0.1644
25	0.0209	7	0.1793
24	0.0239	6	0.1943
23	0.0269	5	0.2092
22	0.0299	4	0.2242
21	0.0329	3	0.2391

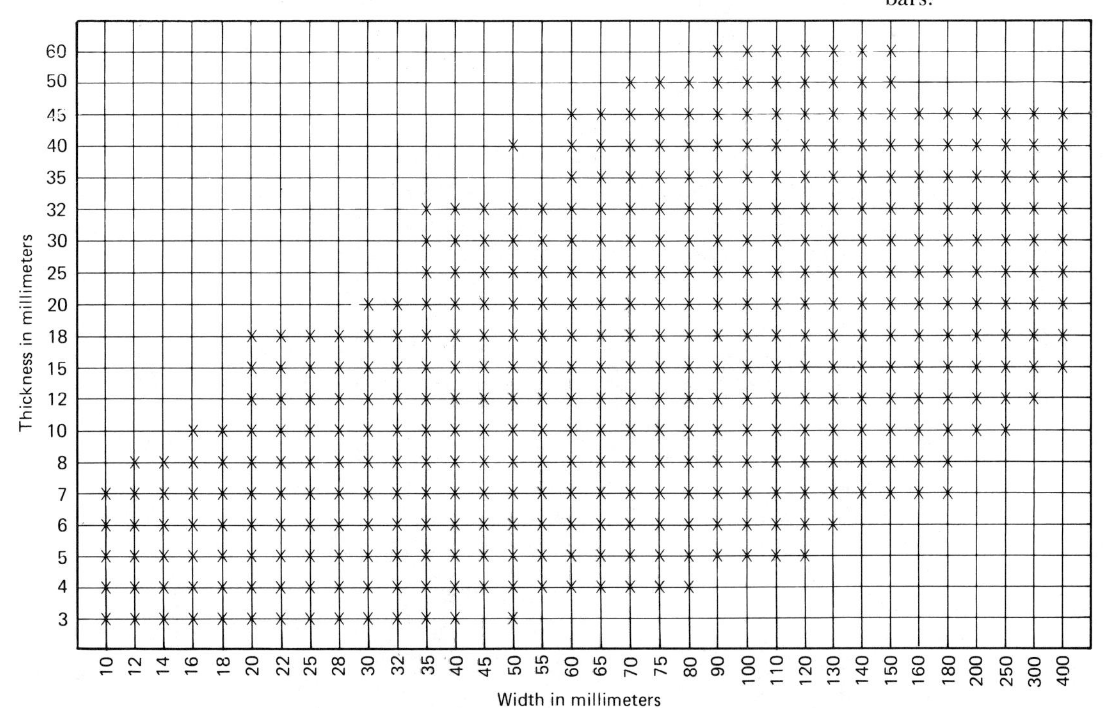

Fig. 13-32. Chart showing standard metric sizes of bars.

the availability of a certain size bar. If you choose the vertical line numbered 12 (indicating a width of 12 mm), you can see *x* marks on this line to show that bars of 12-mm width are available in thicknesses ranging from 3 mm to 8 mm.

The foregoing information is not the whole story on standard sizes but only a part of the story. However, enough information is given for the purpose of this book, which is to cover the subject of strength of materials from the point of view of the designer. Each manufacturer issues a handbook or reference book which gives complete information on his product and a designer should be well supplied with such reference material.

13-7. DESIGN OF SIMPLE CONNECTIONS

Now let's see how we go about designing a simple connection, for instance the one in Fig. 13-33. We want to fasten the bar to the machine frame by means of rivets. We will know the force P, the kind of material we are going to use, and the working stresses for the material in tension, shear, and bearing. We will need to find the diameter of the rivets, the number of the rivets required, and the thickness and width of the bar. We can set up a definite procedure to follow. (There are other possible procedures, and some people prefer them, but the one we are going to use is reliable.) The first step is to decide what rivet diameter is to be used. This seems arbitrary, and it is. As a matter of fact, you could make a good design with any one of several different sizes of rivets. Experience is a great help in choosing the rivet diameter, both your own experience and the experience of the company you work for; frequently, the chief engineer makes the decision. After choosing the rivet diameter, you calculate the force that one rivet can carry when it is stressed to working stress in shear, using the formula,

$$P_1 = A_s S_s$$

where P_1 is the force for one rivet, A_s is the area of one rivet in shear, and S_s is the working stress in shear. Then you divide the total force P

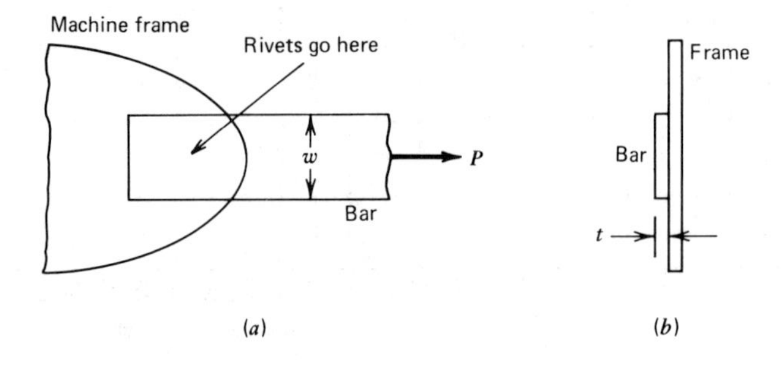

Fig. 13-33. Bar connected to a machine frame with rivets. (*a*) Front view. (*b*) End view.

by the force P_1 that one rivet can carry to get the number of rivets n. Thus

$$n = \frac{P}{P_1}$$

Ordinarily, you won't obtain a whole number for n, so you will take the next larger whole number as the number of rivets to be used. For instance, if you get 2.7 for n, you will use three rivets. If n comes out as 1.63, you will use two.

The next step is to find the thickness of the plate, and this depends on bearing stress between the rivets and the bar. You must choose a bar thick enough so that the bearing stress won't be more than the working stress in bearing. Here is the way to do it. The bearing area for one rivet is dt, where d is the rivet diameter and t is the thickness of the bar. The bearing area for n rivets is n times as much, or $n\,dt$. Then you can use the formula,

$$A_b = \frac{P}{S_b}$$

where A_b is the area in bearing, P is the total force, and s_b is the working stress in bearing. We can substitute $n\,dt$ for A_b and we have

$$n\,dt = \frac{P}{S_b}$$

or

$$t = \frac{P}{ndS_b}$$

Use this formula to find the thickness of the bar. You probably won't come out with a dimension that represents any standard size, so take the next larger standard size from the thickness calculated. For instance, if you get 0.603 in. as the required thickness, when the next smaller standard thickness is ½ in., which is 0.500 in., and the next larger standard thickness is ⅝ in., which is 0.625 in., you will use a bar ⅝ in. thick. (You can use Table 13.1 for the decimal equivalents of the common fractions, even though this table shows rivet and bolt diameters. The fraction ⅝ in. is always equivalent to the decimal 0.625 in. whether it represents the diameter of a rivet or the thickness of a bar.)

The only thing left to do is to find the width of the bar, and here we have to make the net section (through a rivet hole) strong enough in tension. If we place the rivets in a row parallel to the force, there is only one rivet hole in any cross section, and the net section is as represented in Fig. 13-34. Let's look at the net area as a large rectangle (of height w and width t) with a small rectangle (of height d and width t) taken away. Then the area of the net section is

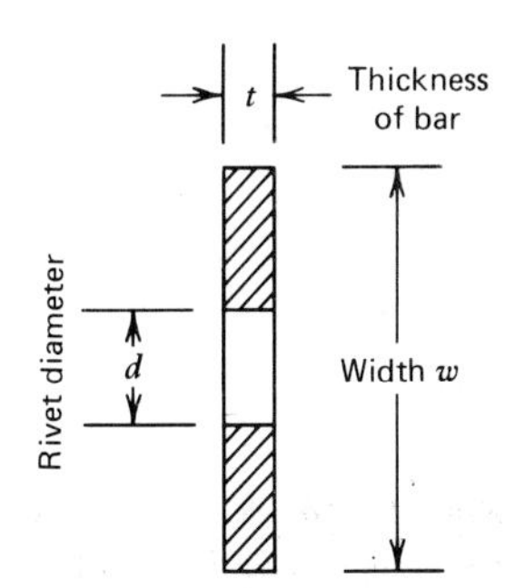

Fig. 13-34. Net section of bar.

$$A_t = wt - dt = (w - d)t$$

We have the formula,

$$A_t = \frac{P}{S_t}$$

where A_t is the area in tension, P is the force, and S_t is the working stress in tension. If we substitute $(w - d)t$ for A_t, we have

$$(w - d)t = \frac{P}{S_t}$$

$$w - d = \frac{P}{tS_t}$$

$$w = d + \frac{P}{tS_t}$$

This formula gives the required width of bar, and since it probably won't be a standard width, we choose the next larger standard width. Now let's try an example.

Illustrative Example 1. In Fig. 13-35, let the load P be 2 100 lb. The bar is to be made of a magnesium alloy, and the rivet of a magnesium alloy. Working stresses are 8 000 psi in tension, 11 000 psi in shear, and 14 000 psi in bearing. Choose a rivet diameter of 5/16 in. and find the number of rivets, the thickness of the bar, and the width of the bar.

Solution:

1. The force that one rivet can carry in shear is found first.

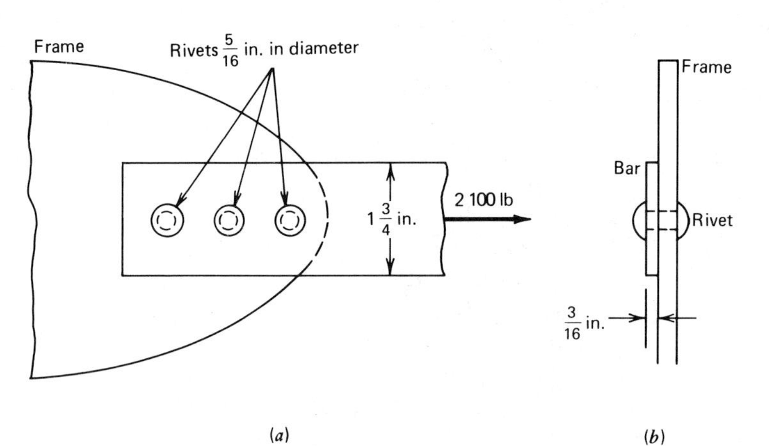

Fig. 13-35. Riveted joint for Illustrative Example 1. (*a*) Front view. (*b*) End view.

(a) The area of cross section of a 5⁄16-in. rivet is found from Table 13.1 as 0.0767 in.[2]
(b) The working stress in shear is 11 000 psi.
(c) The strength of one rivet in shear is

$$P_1 = A_s S_s = 0.0767 \times 11\,000 = 844 \text{ lb}$$

2. Next we find the number of rivets.
(a) The total force P is 2 100 lb.
(b) The strength P_1 of one rivet is 844 lb.
(c) The number of rivets needed is

$$n = \frac{P}{P_1} = \frac{2\,100}{844} = 2.49$$

(d) We take the next larger whole number from 2.49, and this is 3. So we use $n = 3$.

3. Then we calculate the bar thickness.
(a) The total force P is 2 100 lb.
(b) $n = 3$. NUMBER OF RIVETS
(c) The rivet diameter is 5⁄16 in.
(d) The working stress in bearing is 14 000 psi.
(e) The required thickness of bar is

$$t = \frac{P}{ndS_b} = \frac{2\,100}{3 \times 5/16 \times 14\,000} = 0.16 \text{ in.}$$

(f) Using the decimal equivalents in Table 13.1, the next larger standard thickness is 3/16 in., which is 0.1875 in., so we take

$$t = \frac{3}{16} \text{ in.}$$

4. Finally, let's calculate the width of the bar.
(a) The load P is 2 100 lb.
(b) d = 5⁄16 in. = 0.3125 in. (See Table 13.1).
(c) t = 3⁄16 in. = 0.1875 in. (See Table 13.1).
(d) The working stress in tension is 8 000 psi
(e) The required width of bar is

$$w = d + \frac{P}{tS_t} = 0.3125 + \frac{2\,100}{3/16 \times 8\,000}$$

$$= 0.3125 + 1.40 = 1.7125 \text{ in.}$$

(f) From 1 in. to 2 in. the standard widths go in steps of ¼ in.. The next larger standard size from 1.7125 in. is 1¾ in., which is 1.75 in.. So we choose

$$w = 1\frac{3}{4} \text{ in.}$$

Now the problem is done, and you see the dimensions in Fig. 13-35.

MANY RIVETS

When a large number of rivets or bolts is needed in a joint, it is better to arrange them in more than one row. Figure 13-36*a* shows a joint with eight rivets in one row parallel to the force. This is not a good arrangement, because of the extra length of plate needed to make one plate overlap the other. Also, the joint is so long that it may be hard to make room for it. As a practical rule, it is better not to have more than four rivets or bolts in a row parallel to the force. Figure 13-36*b* shows a joint with eight rivets arranged in two rows, and this is a better arrangement. The joint occupies less space, and less extra material is needed for the overlap. However, you must be careful in figuring the net section. On a section *B-B* (Fig. 13-36*b*) there are two rivet holes, and the two holes must be deducted when you calculate the net section. We can change a previous formula so that we can use it in such a case. You remember that we used

$$w = d + \frac{P}{tS_t}$$

to find the width of plate (d is the rivet diameter, P is the total force, t is the plate thickness, and S_t is the working stress in tension). All we have to do is write the formula as

$$w = qd + \frac{P}{tS_t}$$

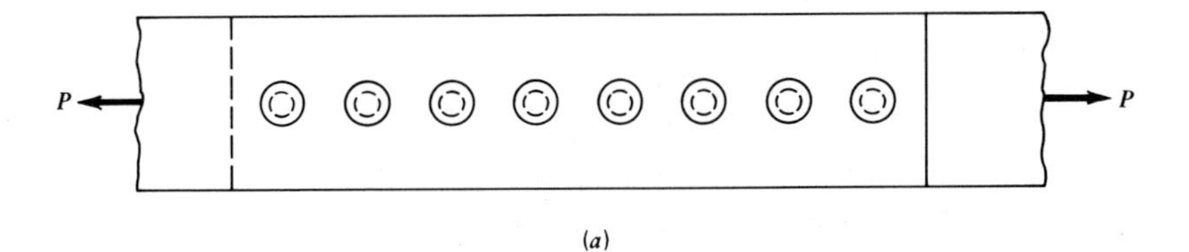

B
P
P
B
(b)

Fig. 13-36. Arrangements of rivets in joints. (*a*) Poor arrangement of rivets. (*b*) Good arrangement of rivets.

where q is the number of rivet or bolt holes in the net section. (When there is only one hole in the net section, we have the simple formula that we used first.) Let's try an example.

Illustrative Example 2. A steel lap joint (like the ones in Fig. 13-36) is to carry a load of 45 000 N. The working stresses are 70 MPa in shear, 105 MPa in tension, and 140 MPa in bearing. The plates are to be held together by 12-mm bolts. Design the joint.

Solution:

1. Find the force that one bolt can carry in shear.
   (a) The area of cross section for a 12-mm bolt is

$$A_s = \frac{\pi}{4}(12)^2 = 113.1 \text{ mm}^2$$

   (b) The working stress in shear is 70 MPa.
   (c) The strength of one bolt in shear is

$$P_1 = A_s S_s = 113.1 \times 70 = 7\,920 \text{ N}$$

2. Now find the number of bolts.
   (a) The total force P is 45 000 N.
   (b) The force P_1 that one bolt can carry is 7 920 N.
   (c) The number of bolts required is

$$n = \frac{P}{P_1} = \frac{45\,000}{7\,920} = 5.68$$

   (d) The next larger whole number from 5.68 is 6, then

$$n = 6 \qquad \text{(6 bolts)}$$

3. Calculate the plate thickness.
   (a) The total force P is 45 000 N.
   (b) $n = 6$.
   (c) The bolt diameter is 12 mm.
   (d) The working stress in bearing is 140 MPa.
   (e) The required plate thickness is

$$t = \frac{P}{ndS_b} = \frac{45\,000}{6 \times 12 \times 140} = 4.46 \text{ mm}$$

   (f) The next larger standard thickness (Fig. 13-32) from 4.46 mm is 5 mm. So

$$t = 5 \text{ mm}$$

4. Now arrange the bolts and find the width of the plate.
   (a) Six bolts are too many for one row parallel to the force, so

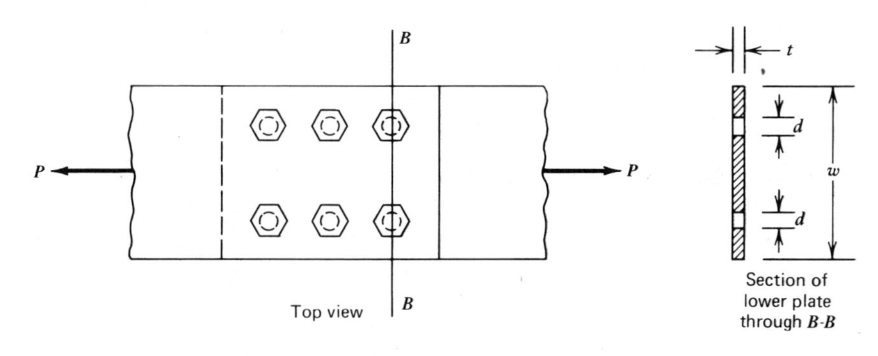

Fig. 13-37. Good arrangement for bolts.

let's make it two rows. Figure 13-37 shows a good arrangement. Section B-B is the net section.

(b) The total force P is 45 000 N.

(c) The number of bolt holes in the net section is 2 ($q = 2$).

(d) $d = 12$ mm.

(c) $t = 5$ mm.

(f) The working stress in tension is 105 MPa.

(g) The required width of plate is

$$w = qd + \frac{P}{tS_t} = 2 \times 12 + \frac{45\,000}{5 \times 105}$$ TENSION

$$= 24 + 85.7 = 109.7 \text{ mm}$$

(h) The next larger standard width from 109.7 mm is 110 mm. See Figure 13-32. Then

$$w = 110 \text{ mm}$$

Practice Problems (Section 13-7). Now you know how to do it, so practice on a few problems. If there are more than four rivets or bolts, arrange them in two rows parallel to the force. If you need an odd number of rivets or bolts, you can put one in a row by itself, as in Fig. 13-38. You are to find the number of rivets or bolts, the thickness of the plate, and width of the plate.

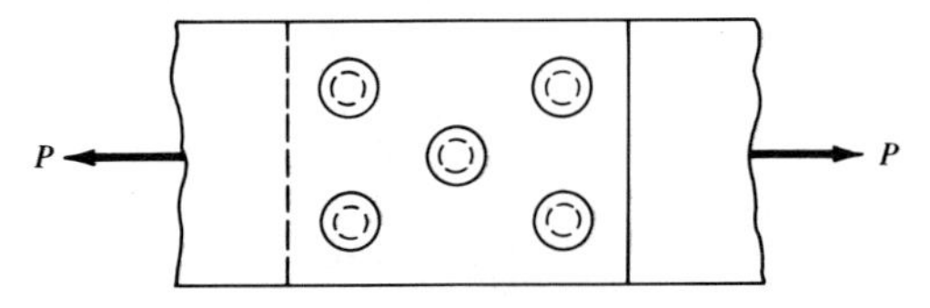

Fig. 13-38. Joint with five rivets.

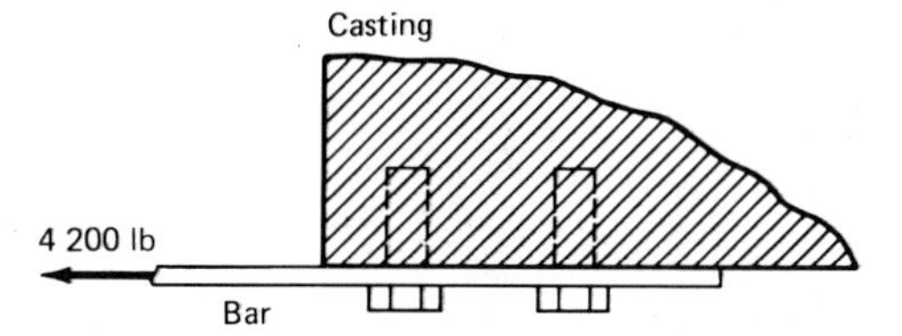

Fig. 13-39. Joint with cap screws for Problem 1.

1. A steel bar is fastened to a heavy casting by steel cap screws, as in Fig. 13-39, and carries a load of 4 200 lb. The cap screws are to be 3/8 in. in diameter. Design the connection. For working stresses use S_t = 12 000 psi, S_s = 8 000 psi, S_b = 16 000 psi.

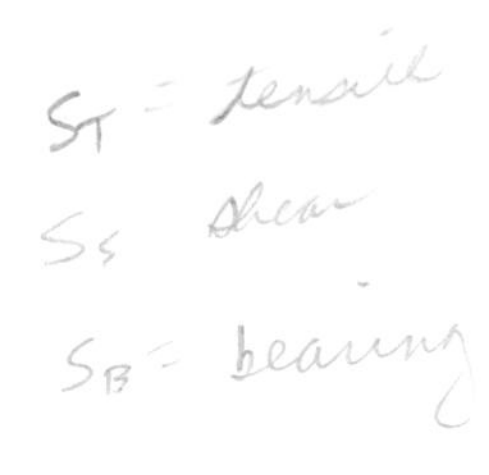

2. Design a lap joint between two aluminum-alloy bars, using aluminum rivets 15 mm in diameter. The total force is 36 000 N. For working stresses use S_t = 100 MPa, S_s = 60 MPa, S_b = 160 MPa.
3. Design a lap joint between two steel plates to carry a total force of 32 000 lb. Use 3/4 in. rivets. For working stresses use S_t = 20 000 psi, S_s = 13 000 psi, S_b = 24 000 psi.
4. Design a lap joint between two magnesium alloy plates, using 6-mm aluminum-alloy rivets. The total force is 14 000 N. For working stresses use S_t = 80 MPa, S_s = 60 MPa, S_b = 120 MPa.

13-8. HOW TO SPACE RIVETS AND BOLTS

After you have found how many rivets or bolts you need and have found the dimensions of the bar or plate, there is another question. It is, *how are you going to space the rivets?* This isn't a hard question. There are a couple of simple rules to follow, and beyond that you can space the rivets as you wish. These rules have to do with minimum pitch and minimum edge distance. Figure 13-40 illustrates pitch and edge distance, and you can see that the *pitch* is the distance from the center of one rivet to the center of the next rivet; the *edge distance* is the distance from the center of a rivet to the edge of the plate. As general rules we have

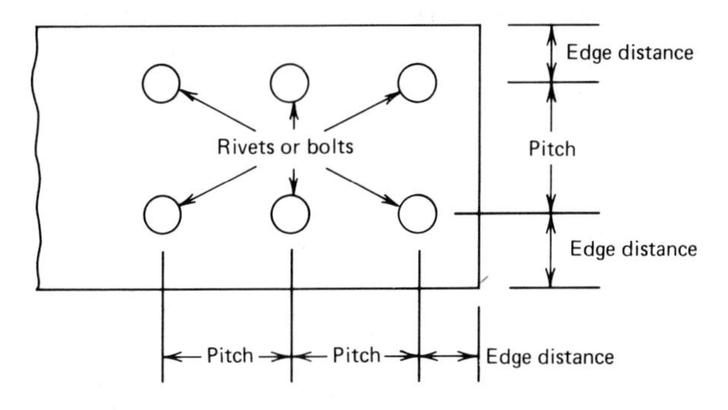

Fig. 13-40. Pitch and edge distance of rivets or bolts.

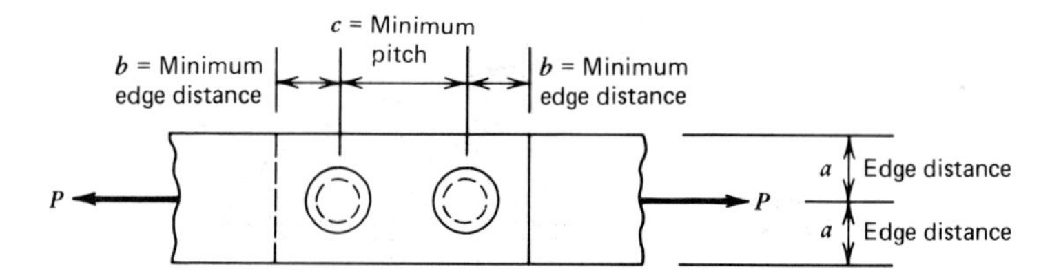

Fig. 13-41. Spacing of rivets or bolts.

Rule 1. *The minimum pitch is three times the diameter of rivet or bolt and the minimum edge distance is one and one-half times the diameter for a steel or aluminum-alloy plate.*

Rule 2. *The minimum ptich is four times the diameter of rivet or bolt and the minimum edge distance is two and one-half times the diameter for a magnesium-alloy plate.*

The foregoing are general rules, and you can deviate from them at times. However, they are good rules to follow, and you will be safe in doing so. It is all right, though, to make the pitch or edge distance greater than the rules specify.

Now let's suppose that you have calculated the rivet diameter, the plate thickness, and the width of the plate, and you are ready to space the rivets. If the rivets are in one row parallel to the pull, as in Fig. 13-41, the edge distance a is half of the width. If the width is more than the minimum spacing, you are not in error. If this is not the case, then increase the width to twice the minimum edge distance. The distance b should be made equal to the minimum edge distance, and the distance c should be made equal to the minimum pitch.

Illustrative Example 1. Determine the rivet spacing for the riveted connection of Section 13-7, Illustrative Example 1.

Solution:

Figure 13-42 shows the joint. There are three rivets.

1. The distance a is ⅞ in. (one-half of 1¾ in.). The minimum edge distance is two and one-half times the rivet diameter (for a magnesium-alloy plate), and this is

$$2.5\text{d} = 2.5 \times \frac{5}{16} = 0.781 \text{ in.}$$

   ⅞ in. is 0.875 in. (Table 13.1), and this is greater than 0.781, so the edge distance a is all right.

2. The distance b is made equal to the minimum edge distance of two and one-half times the rivet diameter. This was calculated as

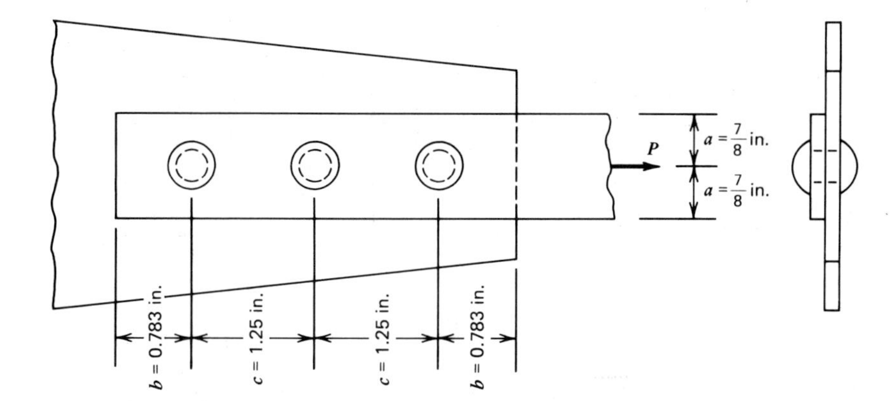

Fig. 13-42. Riveted joint for Illustrative Example 1.

0.781 in. in Solution 1, so

$$b = 0.781 \text{ in.}$$

3. The distances c are made equal to the minimum pitch, which is four times the rivet diameter (for a magnesium-alloy plate). This is

$$c = 4d = 4 \times \frac{5}{16} = 1.25 \text{ in.}$$

SEVERAL ROWS OF RIVETS

When there are two rows of rivets or bolts, as in Fig. 13-43, the edge distance a is made equal to the minimum edge distance, and the distance e is whatever is leftover from the width of the plate. If this gives a value of e as large as the minimum pitch, everything is fine; if not, the width of the plate must be increased. This doesn't happen often. The distances b and c, parallel to the force, are determined as before.

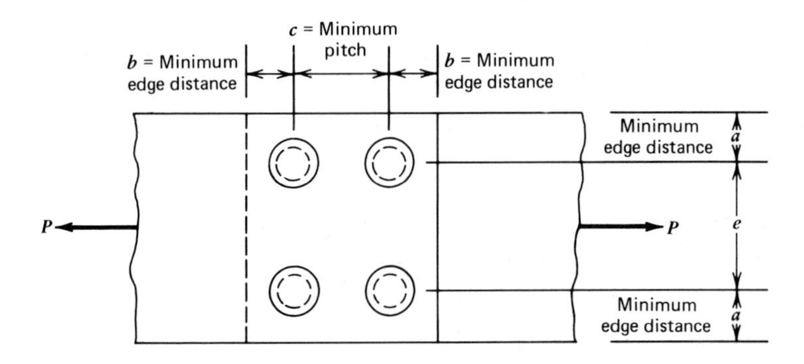

Fig. 13-43. Joint with two rows of rivets.

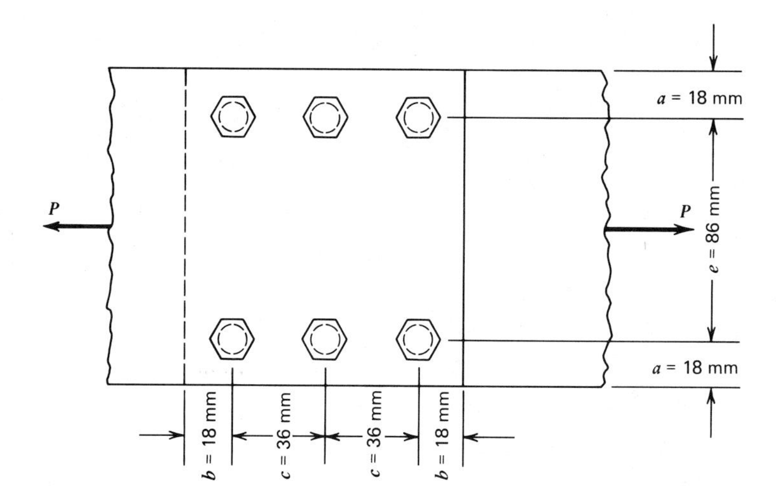

Fig. 13-44. Bolted joint for Illustrative Example 2.

Illustrative Example 2. Determine the spacing of the bolts of Section 13-7, Illustrative Example 2.

Solution:

Figure 13-44 shows the top view of the joint. There are six bolts (two rows of three bolts each).

1. The edge distance a is one and one-half times the bolt diameter (for a steel plate); hence

$$a = 1.5d = 1.5 \times 12 = 18 \text{ mm}$$

2. The distance e is what is left after subtracting two distances a from the width of 110 mm.

$$e = 110 - 2a = 110 - 2 \times 12 = 110 - 24 = 86 \text{ mm}$$

3. The distances b are each the minimum edge distance of 1.5 diameters for a steel plate. This minimum edge distance was calculated to be 18 mm in Solution 1, so

$$b = 18 \text{ mm}$$

4. The distances c are each equal to the minimum pitch of three times the diameter for a steel plate. Thus

$$c = 3 \times 12 = 36 \text{ mm}$$

Practice Problems (Section 13-8). Why not practice a bit on spacing rivets and bolts? Look up the answers to the last set of practice problems and space the rivets or bolts.

1. Problem 1 from Section 13-7.

2. Problem 2 from Section 13-7.
3. Problem 3 from Section 13-7.
4. Problem 4 from Section 13-7.

13-9 WELDING

Welding is the process of joining two or more pieces of metal by the application of heat and sometimes of pressure also. The process is called *fusion welding* when the pieces are heated to the melting point, because then the separate parts fuse together and solidify as one member. *Forge welding* is welding in which the pieces of metal are heated to a temperature below the melting point, but in forge welding the pieces of metal must be squeezed or hammered together by pressure. Both fusion and forge welding are used extensively in joining metal parts and even in building up a metal part from several pieces, so we are going to look at several of the common methods of welding and see how to go about designing welded connections.

Most of the materials used for machine parts can be welded satisfactorily, although it is much easier to weld some metals than others. Low-carbon steel is one of the easiest metals to weld, and for this reason it is used more for welded parts than all other metals together. However, the alloy steels, copper alloys, nickel alloys, aluminum and its alloys, and the magnesium alloys can be welded by using suitable techniques. The designer needs to know something of how welds are made and how to design welds for strength. He or she does not have to know all of the techniques for making welds in various alloys but can leave that up to the shop that is to make the machine or structure after is it designed.

13-10. ARC WELDING

Arc welding is one of the common methods of welding pieces of metal together. Two pieces are shown ready for welding in Fig. 13-45*a*, and Fig. 13-45*b* shows a partly completed arc weld. Important elements here are the base metal (the pieces being welded together) and the weld rod. The weld rod supplies the material to fill the groove between the two pieces to be welded. In arc welding the *base metal* and the *weld rod* form part of an electric circuit. An *electric arc* between the weld rod and the base metal reaches a high temperature (6 000 to 7 000°F), and melts the weld rod and the base metal. The material from the weld rod and the base metal mix while they are in the molten state and solidify as one piece. Enough metal is melted from the weld rod to fill the space between the pieces of base metal, and the finished weld has the appearance shown

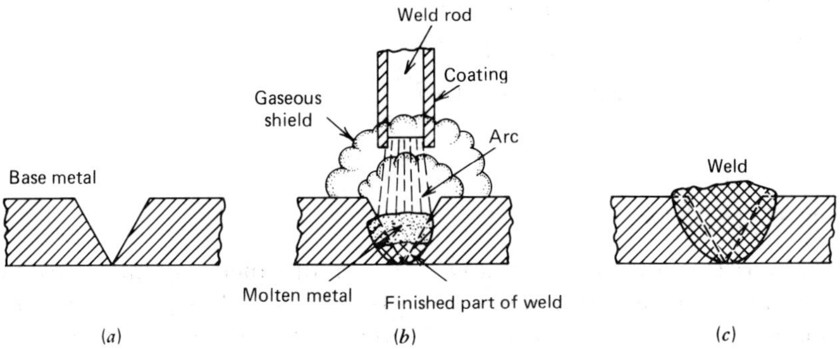

Fig. 13-45. Illustration of arc welding. (*a*) Pieces ready for welding. (*b*) Weld in process. (*c*) Finished weld.

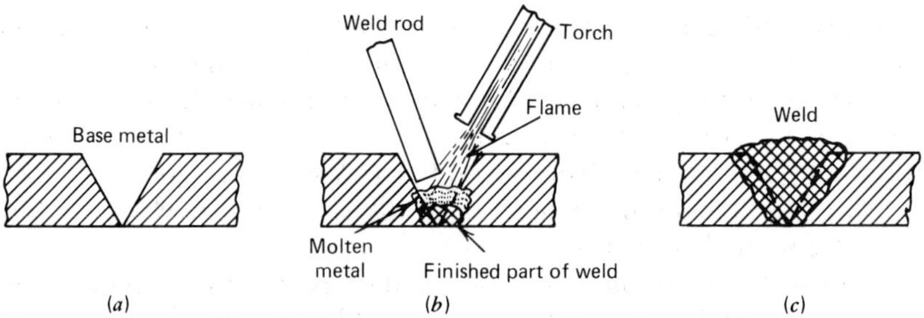

Fig. 13-46. Illustration of gas welding. (*a*) Pieces ready for welding. (*b*) Weld in process. (*c*) Finished weld.

in Fig. 13-45*c*. The finished weld extends into the base metal and is a strong joint if properly made.

13-11. GAS WELDING

Gas welding is welding in which the heat is supplied by a gas torch. Ordinarily, acetylene gas and oxygen are mixed in the torch and the acetylene is burned at the tip of the torch. The torch is called an oxy-acetylene torch and produces a hot flame (about 6 300°F) which melts the base metal and the weld rod. Figure 13-46*a* shows two pieces ready to be welded, and Fig. 13-46*b* shows the partially completed weld. The flame is played on the base metal and the weld rod to melt them; they mix in the molten state and solidify as one piece of metal. Figure 13-46*c* shows the finished weld.

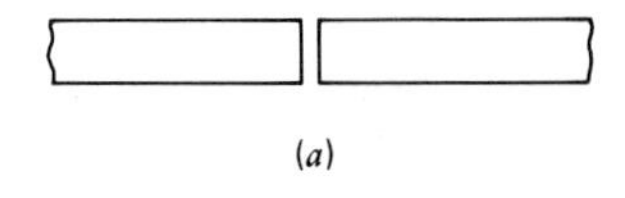

(*a*)

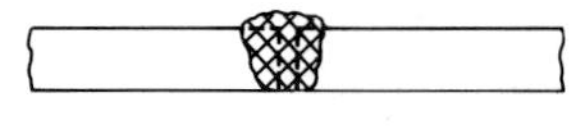

(*b*)

Fig. 13-47. Plain butt weld in thin plate. (*a*) Pieces ready for welding. (*b*) Finished weld.

13-12. BUTT WELDS

A *butt weld* is a weld in which the plates are lined up, butt to butt, and welded. Butt welds can be made with an arc or with a torch.

Plain butt welds are made in thin plates without any special shaping of the ends. Figure 13-47*a* shows the pieces before welding, and Fig. 13-47*b* shows the finished weld. *Thin* is a word of broad interpretation in

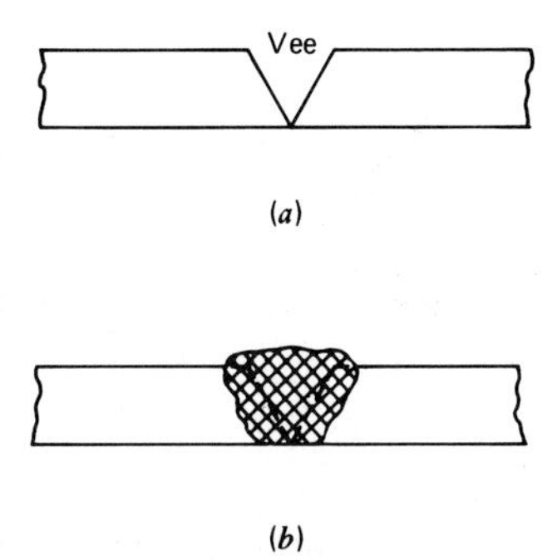

Fig. 13-48. Single vee butt weld in medium plate. (*a*) Pieces ready for welding. (*b*) Finished weld.

this connection, since it may mean up to $\frac{1}{16}$ in. in an aluminum alloy, up to $\frac{1}{8}$ in. in a magnesium alloy, and up to $\frac{3}{8}$ in. in low-carbon steel. The weld is usually made from only one side of the plate.

Single *vee-butt welds* are made in plates of medium thickness. The plates are planed on the ends so that the pieces form a *vee* before welding (see Fig. 13-48*a*). The finished weld is shown in Fig. 13-48*b*. It is made from one side of the plate only. Whether a plate is to be considered as *thin, medium,* or *heavy* depends on the material and special conditions of the job. It is a good idea to get recommendations from the manufacturer of the material unless you have had a lot of experience with this particular type of work.

Double vee-butt welds are made in thick plates. The plates are planed on the ends so they form a double vee, Fig. 13-49*a*, when ready for welding. The weld is made from both sides and the finished weld is shown in Fig. 13-49*b*.

It is easy to calculate stresses in a butt weld. Any extra thickness of the weld is neglected, and the thickness is taken to be the same as the thickness of the plate (see Fig. 13-50). Then the area in tension is just the product of the width of the plate and the thickness of the plate. The techniques of welding are so far advanced that a butt weld can be made as strong as the material to be welded, so all that is necessary for design is to calculate the size of the pieces to be welded. You already know how to do this.

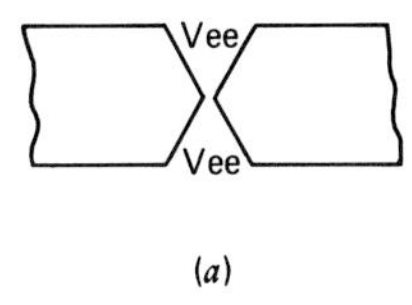

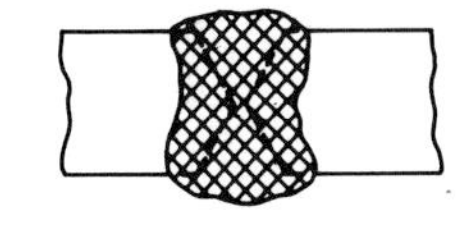

Fig. 13-49. Double vee butt weld in thick plate. (*a*) Pieces ready for welding. (*b*) Finished weld.

13-13. FILLET WELDS

Material used to fill in a corner is called a *fillet.* For example, Fig. 13-51*a* shows the cross section of a *circular fillet* and Fig. 13-51*b* the cross section of an *angle fillet.* When weld material is deposited in such a corner, it is

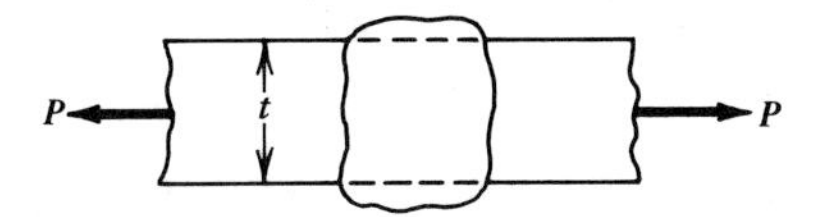

Fig. 13-50. Thickness of weld.

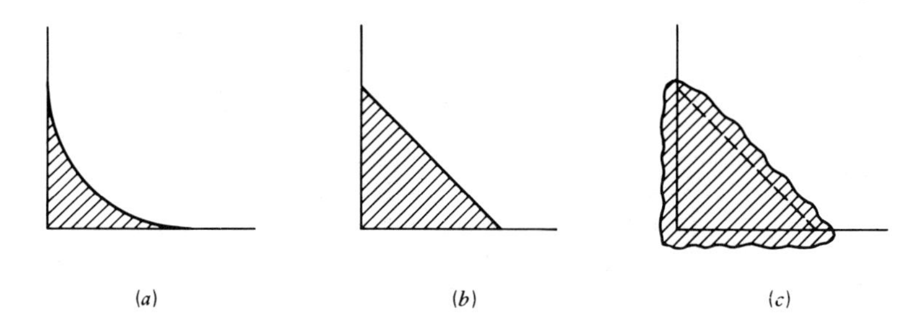

Fig. 13-51. Types of fillets. (*a*) Circular fillet. (*b*) Angle fillet. (*c*) Fillet weld.

called a *fillet weld.* Figure 13-51*c* shows the cross section of a fillet weld. Fillet welds are used a great deal in making machines and structures and can be made either by arc welding or gas welding.

Figure 13-52 shows fillet welds as they are commonly used to fasten one member to another. A fillet weld at the side of a bar is called a *side-fillet weld,* and a fillet weld at the end of a bar is called an *end-fillet weld.* The cross section of a bar with two side-fillet welds is shown in Fig. 13-53; the force acting on the bar is perpendicular to the plane of the cross section. The force on the bar tends to make the bar slip out from between the welds and causes shearing stress on different planes in the welds. Part of the weld might cling to the bar and part to the plate if the weld broke, so that there is a possibility of the weld failing on any one of such planes, as *A, B, C,* or *D* in Fig. 13-53. The stress is highest on the plane where the thickness of the weld is smallest. The thickness of the weld is smallest on a 45° plane, and this smallest thickness is called the *throat.*

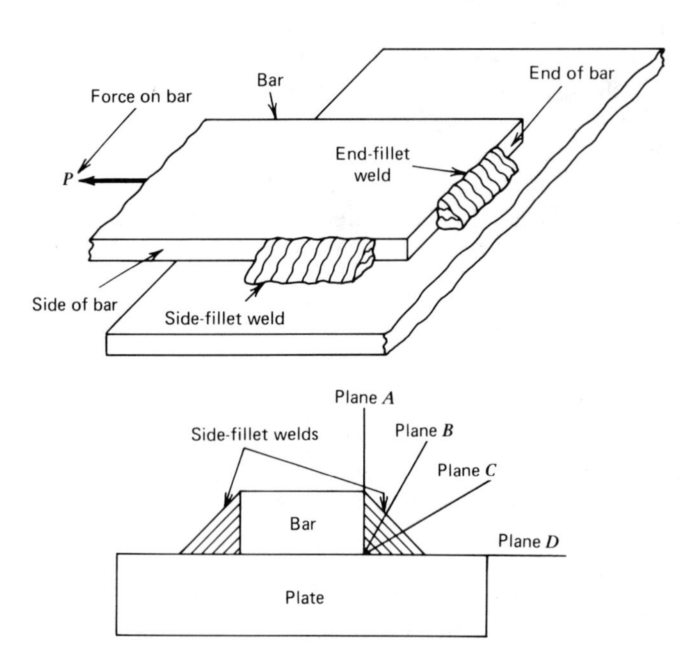

Fig. 13-52. Fillet welds on sides and ends of bar.

Fig. 13-53. Cross section of welds and members.

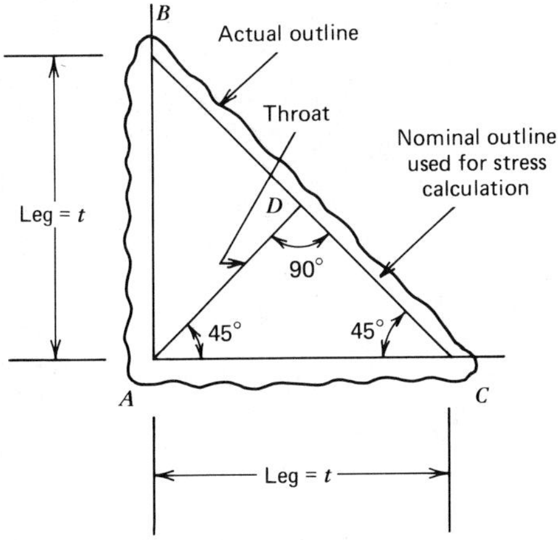

Fig. 13-54. Cross section of a fillet weld.

Figure 13-54 shows the cross section of a fillet weld in detail; the actual outline is shown and the nominal outline as it is assumed to be in calculating stresses. Fillet welds are nearly always in the shape of a 45° right triangle, so the throat is the length AD perpendicular to the hypotenuse. We can look at the right triangle ADC, where AC is the hypotenuse and see that

$$AD = AC \cos 45° = 0.707\ AC$$

The length AC is called the *leg* of the weld, and we designate it by t, because it is the same as the thickness of the bar which is welded to the plate. We designate the thickness of the throat as t_t, so

$$t_t = 0.707t$$

Figure 13-55 shows a bar and half of a side-fillet weld. The other half of the weld is cut away so that the stress exerted by it can be seen. The area in shear is a rectangle with length equal to the length L of the weld

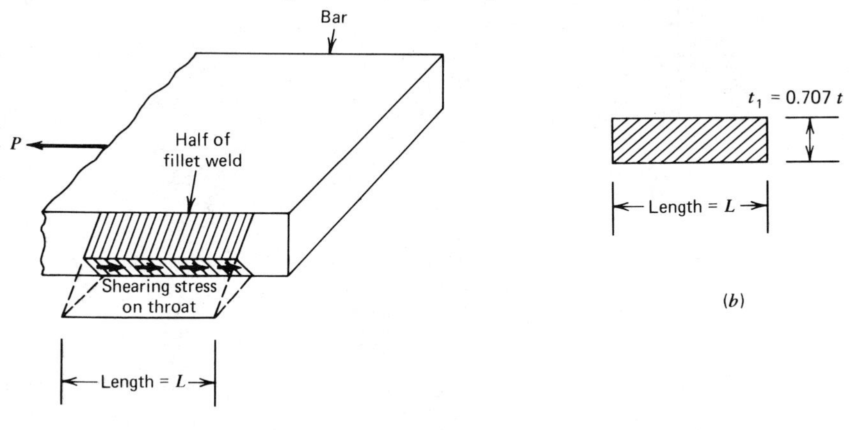

Fig. 13-55. Stresses on throat of weld. (*a*) Forces on bar. (*b*) Area of threat.

and the width equal to the thickness t_t of the throat. The amount of the area is

$$A = Lt_t = 0.707Lt$$

where t is the leg of the weld. The force a fillet weld can withstand is equal to the product of the area and the stress. Thus

$$P = AS$$

The area A is $0.707Lt$, and when this is substituted for A

$$P = 0.707LtS_s$$

This formula can be solved for the length L and gives

$$L = \frac{P}{0.707tS_s}$$

This is the lenth of side-fillet weld needed to fasten one steel bar to another. For a rectangular bar, half of the length is placed on each side.

Illustrative Example 1. The steel bar in Fig. 13-56 is to be welded to the steel plate with two side-fillet welds as shown. Find the length L_1 for each weld. Use a working stress of 16 000 psi.

Solution:

1. The formula

$$L = \frac{P}{0.707tS_s}$$

is reasonable for fillet welds between pieces of steel.

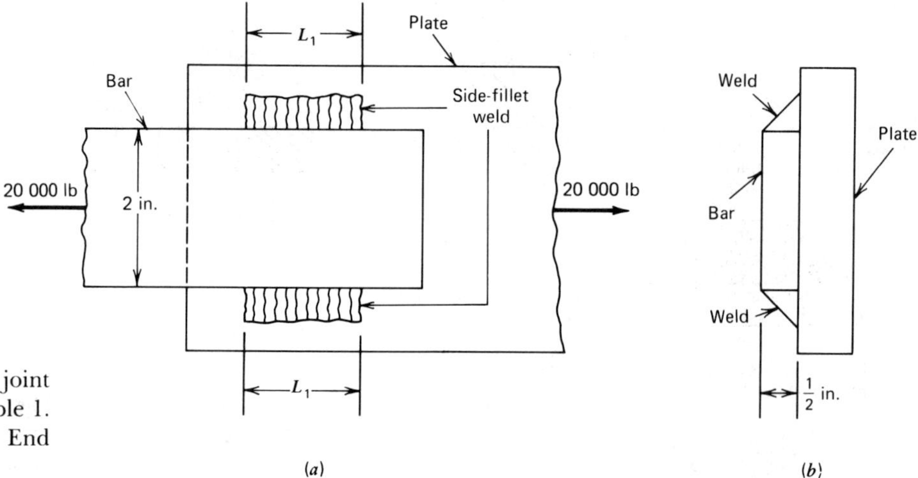

Fig. 13-56. Welded joint for Illustrative Example 1. (*a*) Front view. (*b*) End view.

2. The force P is 20 000 lb.
3. The leg t of the fillet weld is always made the same as the thickness of the bar, so

$$t = \tfrac{1}{2} \text{ in.}$$

4. The total length of the weld is

$$L = \frac{P}{0.707tS_s} = \frac{20\ 000}{0.707 \times \frac{1}{2} \times 16\ 000} = 3.536 \text{ in.}$$

5. Half of the total length is placed on each side, so

$$L_1 = \frac{L}{2} = \frac{3.536}{2} = 1.768 \text{ in.}$$

END-FILLET WELDS

End-fillet welds have been shown by test to be somewhat stronger than side-fillet welds, but they are usually figured on the same basis, without any allowance for the increased strength. The total length of the fillet weld is calculated and then distributed between the sides and the ends. It is up to the individual designer to decide how much of the total length is to be placed at the sides and how much at the ends. A fair rule would be to weld all the way across the end and place what is left on the sides.

Illustrative Example 2.
A steel bar 72 mm by 8 mm is to be welded to a steel plate, as shown in Fig. 13-57. There is to be a fillet weld across the end and a fillet

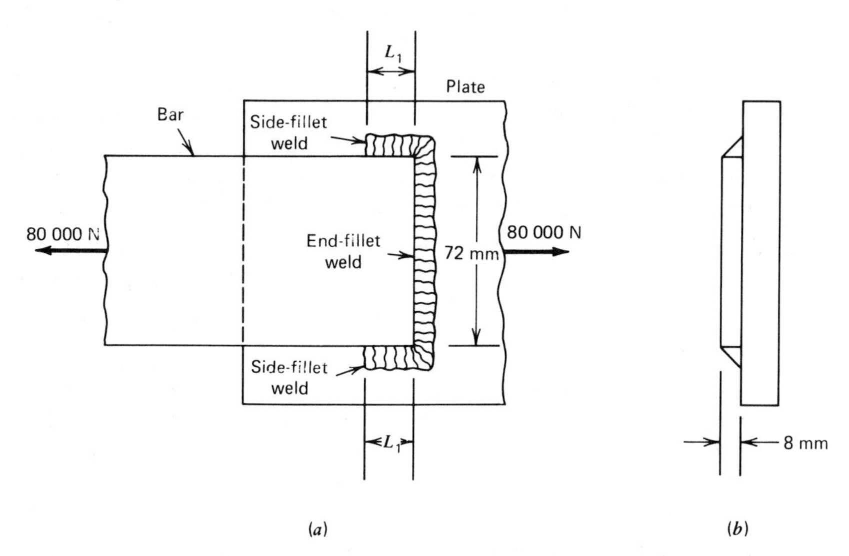

Fig. 13-57. Welded joint for Illustrative Example 2. (*a*) Front view. (*b*) End view.

weld on each side. Find the length of each side-fillet weld. The working stress is 140 MPa.

Solution:

1. The force P is 80 000 N.
2. The leg t of the weld is the same as the thickness of the bar, so

$$t = 8 \text{ mm}$$

3. The total length of weld is

$$L = \frac{P}{0.707tS_s} = \frac{80\,000}{0.707 \times 8 \times 140} = 101 \text{ mm}$$

4. The length of the end-fillet weld is 72 mm. The remainder is left for the sides, and this is

$$101 - 72 = 29 \text{ mm}$$

The length L_1 of each side-fillet weld is half of the remainder, so

$$L_1 = \frac{29}{2} = 14.5 \text{ mm}$$

Practice Problems (Section 13-13). You will now have an opportunity to practice by working the following problems.

1. A rectangular steel bar 1 in. by ⅛ in. is to be welded to a steel machine frame by side-fillet welds. The tensile force in the bar is 2 000 lb. Find the length of fillet weld for each side. The working stress is 18 000 psi.
2. A rectangular steel bar 6 in. by ¾ in. is to carry a tensile force of 90 000 lb. The bar is to be welded to a steel plate by means of an end-fillet weld, which goes clear across the end, and by side-fillet welds. Find the length of each side-fillet weld. The working stress is 18 000 psi.
3. A rectangular steel bar of 12 mm thickness is to be welded to a steel plate by side-fillet welds. The tensile force in the bar is 70 000 N. Find the length of each side-fillet weld. The working stress is 130 MPa.
4. A rectangular aluminum-alloy bar of ¼ in. thickness is to be welded to an aluminum plate by means of side-fillet welds. The bar is to be subjected to a tensile force of 4 600 lb, and the working stress in the fillet welds is 4 000 psi. Find the length of each side-fillet weld.
5. A rectangular magnesium-alloy bar 50 mm × 10 mm is to be

welded to a plate of the same alloy by an end-fillet weld that goes clear across the end and side-fillet welds. The tensile force in the plate is 12 000 N, and the working stress in the weld is 22 MPa. Find the length of each side-fillet weld.

13-14. SPOT WELDS

Spot welding is an electrical method of welding thin sheets of metal together. Figure 13-58 shows the general set-up. The sheets to be welded are clamped together by two copper or copper-alloy electrodes, and an electric current is passed between the electrodes. The current passes through the sheets to be welded and, because of the electrical resistance of the sheets, develops a high temperature in the sheets. The combination of high temperature and pressure produces a weld. Figure 13-59 shows a cross section of the finished spot weld.

Spot welding is a ticklish process and, if good spot welds are to be produced, it is best to use machines with automatic controls for the pressure between the plates and for the electric current. The amount of the current and the length of time that it is applied must be carefully controlled. (These matters are really the responsibility of the shop, but the designer should have some idea of what is involved.) Another important item is that the contact surfaces should be clean of paint, oxide, and grease.

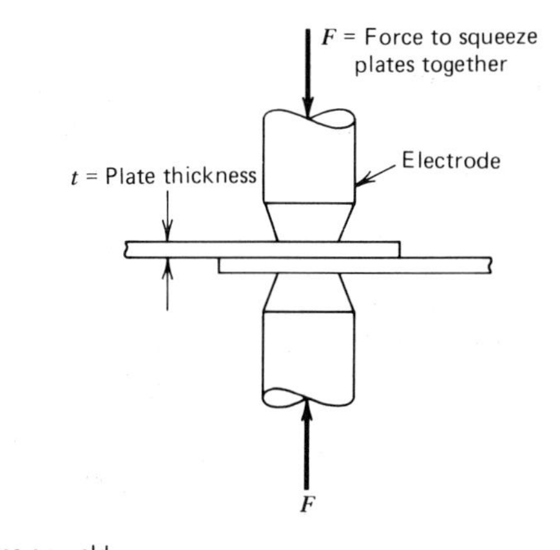

Fig. 13-58. Electrode and plates for spot welding.

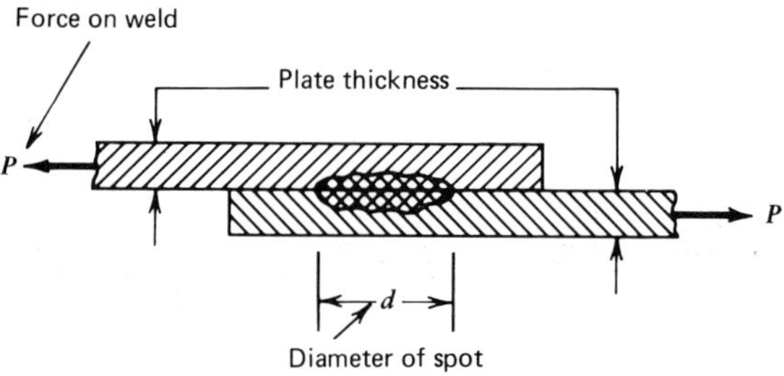

Fig. 13-59. Cross section of spot weld.

How thick is a thin sheet? Well, *thin*, as we use it in speaking of spot welding, means up to ⅛ in. for aluminum and magnesium alloys, and up to ⅜ in. for low-carbon steel.

Now that we know something about the process, let's see how to design spot-welded connections. Spot welds are usually loaded in shear, as shown in Fig. 13-59. You would know the force P and the thickness t of the sheets to be welded. You would want to find out how many spots are necessary to carry the force P. Here's how it's done. The strength of a single spot weld depends on the thickness of the sheet to be welded (the thicker the plate, the bigger the spot) and on the material of the sheets. Let's call the working strength of a single spot R. The working strength is the load we would be willing to apply in service. Then if the number of spots in the connection is n, the working strength of the connection is nR. The working strength is is equal to the load P; consequently

$$nR = P$$

We divide this equation by R to get

$$n = \frac{P}{R}$$

as a formula for the number of spots.

The value of n probably won't come out as a whole number, and you just take the next larger whole number as the number of spots.

Illustrative Example 1. The two sheets in Fig. 13-60 are to be spot welded. They are made of aluminum alloy 52S-O. How many spots should be used? The working strength for one spot is 85 lb.

Solution:

1. The total force P is 500 lb.
2. The working strength R of a single spot is 85 lb.
3. The number of spot welds required is

$$n = \frac{P}{R} = \frac{500}{85} = 5.88$$

4. The next larger whole number from 5.88 is 6, so we will use six spot welds.

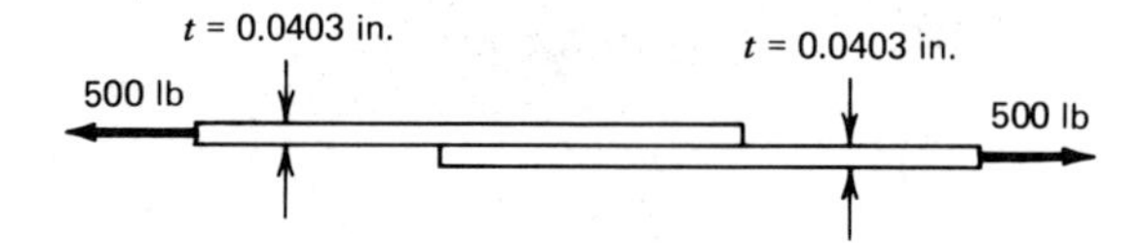

Fig. 13-60. Sheets to be spot welded.

Illustrative Example 2. Two steel sheets, each of 3 mm thickness, are to be spot welded together. The total force is 16 000 N. Find the number of spot welds. The working strength of one spot is 6 200 N.

Solution:

1. The total force P is 16 000 N.
2. The working strength R of a single spot weld is 6 200 N.
3. The number of spots required is

$$n = \frac{P}{R} = \frac{16\,000}{6\,200} = 2.58$$

4. The next larger whole number from 2.58 is 3, so we use three spot welds.

Practice Problems (Section 13-14). Just a few practice problems on spot welds. Be sure to raise the answer to the next larger whole number (you can't make half a spot weld).

1. Two steel sheets, each of ⅛ in. thickness, are to carry a force of 2 400 lb and are to be spot welded together. How many spot welds would you use? Let R = 1 400 lb.
2. Two steel sheets, each of ¼ in. thickness, are to be spot welded together. The total force on the joint is 10 600 lb. Find the number of spot welds. R = 3 000 lb.
3. Two aluminum-alloy sheets, each of 0.201 in. thickness, carry a total force of 240 lb and are to be spot welded together. How many spot welds should be used? Use R = 18 lb.
4. Two steel sheets, each of 5 mm thickness, are to be spot welded together. The total force is 12 000 N. Find the number of spot welds. R = 9 000 N.
5. Two magnesium-alloy sheets, each of 3 mm thickness, are to carry a total force of 2 300 N and are to be spot welded together. Find the number of spot welds. R = 800 N.

13-15. THIN-WALLED PRESSURE VESSELS

A pressure vessel is just a container that holds a liquid or gas under pressure. Most pressure vessels are cylindrical or spherical in shape. Many are made of metal and some are made of plastic. The stresses in a *thin-walled* pressure vessel can be calculated from single formulas. These formulas are easy to derive, so we will do that for you. Besides, it is excellent practice in using free-body diagrams and equations of equilibrium.

CYLINDERS

Figure 13-61*a* shows a cylindrical pressure vessel. The diameter of the vessel is d and the wall thickness is t. The cylinder contains a liquid or gas under pressure p (pressure is compressive stress).

There are two stresses of interest in a cylindrical pressure vessel. First, we will derive a formula for the longitudinal stress. This is the stress in the direction of the length. We will cut the cylinder and work with the free-body diagram of the left part of it, as shown in Fig. 13-61*b*, including the pressurized gas or liquid in that part. The contents of the other part of the vessel exert the pressure p on the circular area of diameter d, so there is a total force directed to the left of $\frac{\pi}{4}d^2p$. This is balanced by a force directed to the right, which is made up of the longitudinal tensile stress S_L acting on a circular band of diameter d and thickness t. The area is the product of the circumference πd and the thickness t. So the total force to the right is πdtS_L.

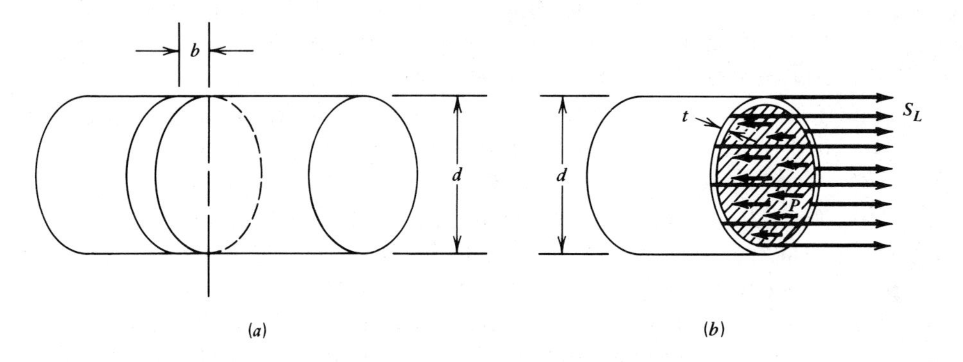

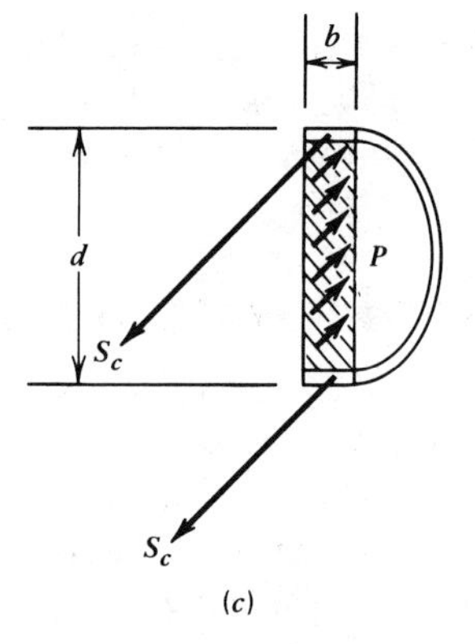

Fig. 13-61. Cylindrical pressure vessel. (*a*) Cylindrical pressure vessel. (*b*) Free-body diagram of part of cylinder. (*c*) Free-body diagram of part of cylinder.

The force directed to the right must be equal to the force directed to the left, thus,

$$\pi dtS_L = \frac{\pi}{4}d^2p$$

from which we calculate the longitudinal stress as

$$S_L = \frac{pd}{4t}$$

Let's use this to calculate a stress.

Illustrative Example 1. A polyethelene sprayer tank is 8 in. in diameter and has a wall thickness of ⅛ in. Calculate the longitudinal stress due to an internal pressure of 20 psi.

Solution:

1. The pressure p is 20 psi.
2. The diameter d is 8 in.
3. The wall thickness t is ⅛ in.
4. Then the longitudinal stress is

$$S_L = \frac{pd}{4t} = \frac{20 \times 8}{4 \times 1/8} = 320 \text{ psi}$$

CIRCUMFERENTIAL STRESS

There is also a circumferential stress in a cylindrical pressure vessel. We will derive a formula for it by using a free-body diagram of half of the length b of the cylinder shown in Fig. 13-61*a*. Figure 13-61*c* shows this free-body diagram, including the pressurized contents of this part of the cylinder. The pressure p is applied to a rectangular area of width b and height d, so there is a total force perpendicular to the plane of the paper of pbd, and this is directed inward. There is an outward force due to the circumferential stress S_c acting on the shell. The area to which this stress is applied consists of two rectangles, each of width b and height t. So, the total outward force is

$$S_c\,(2bt) = 2btS_c$$

The two forces must be equal for this part of the cylinder to be in equilibrium. Then

$$2btS_c = pbd$$

and

$$S_c = \frac{pd}{2t}$$

This is the circumferential stress. It is twice as large as the longitudinal stress so it would be considered to be more important. Let's calculate it in an example.

Illustrative Example 2. A cylindrical steel tank is 3 m in diameter and has a wall thickness of 6 mm. The tank is full of water at a pressure of 500 Pa (pascals). What is the circumferential stress in the wall of the tank?

Solution:

1. The pressure is 500 Pa, and this should be converted to megapascals. Thus,

$$p = 500 \times 10^{-3} = 0.500 \text{ MPa}$$

2. The diameter d is 3 m and should bc converted to millimeters. So,

$$d = 3 \times 10^3 = 3\,000 \text{ mm}$$

3. The wall thickness t is 6 mm.
4. Then,

$$S_c = \frac{pd}{2t} = \frac{0.500 \times 3\,000}{2 \times 6} = 125 \text{ MPa}$$

THIN-WALLED SPHERES

Figure 13-62*a* shows a sphere which is a thin-walled pressure vessel. The diameter is d, the wall thickness is t, and the internal pressure is p. Let's see how we can derive a formula for the stress.

Figure 13-62*b* shows the free-body diagram of half of the sphere and half of its contents. The contents of the other half exert the pressure p against this half. Also, the other half of the sphere exerts the stress S on this half. We can get a formula for S by applying the condition that the sum of the horizontal forces is zero.

d

(a)

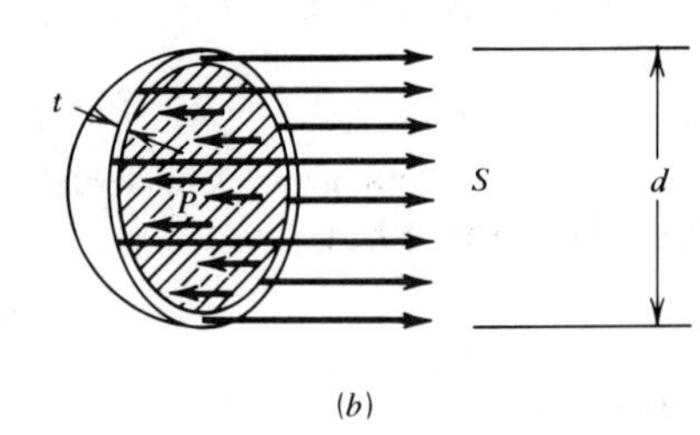

(b)

Fig. 13-62. Spherical pressure vessel. (*a*) Sphere. (*b*) Free-body diagram of half of sphere.

The total horizontal force to the left is equal to the pressure p times the circular area of diameter d. So it is

$$p\frac{\pi d^2}{4}$$

The total horizontal force to the right is equal to the stress S times the area of a circular ring of diameter d and thickness t. It is

$$S\pi dt$$

These two forces must be equal for the half sphere to be in equilibrium. Then

$$S\pi dt = \pi p\frac{d^2}{4}$$

which comes down to

$$S = \frac{pd}{4t}$$

Let's use this formula.

Illustrative Example 3. A spherical steel tank is 22 ft in diameter. The wall thickness is ⅜ in. The internal pressure is 60 psi. Calculate the stress in the wall of the tank.

Solution:

1. The pressure p is 60 psi.
2. The diameter is 22 ft, but this must be converted to inches.

$$d = 22 \times 12 = 264 \text{ in.}$$

3. The wall thickness t is ⅜ in.
4. The stress is

$$S = \frac{pd}{4t} = \frac{60 \times 264}{4 \times 3/8} = 10\,560 \text{ psi}$$

WORDS OF CAUTION

Be sure to have the diameter d and the thickness t in the same units whenever you use these formulas. You would probably choose inches if you are working with gravitational units and millimeters if you are working with the metric system. If you follow this advice, the stress will come out in the same units as the pressure, that is, psi or megapascals.

You might wonder how you are to know when a pressure vessel is a thin-walled pressure vessel. The answer is that it is thin-walled if the ratio of diameter to thickness is 20 or more.

CALCULATING THICKNESS

Such formulas as

$$S = \frac{pd}{2t} \text{ and } S = \frac{pd}{4t}$$

can be rewritten as

$$t = \frac{pd}{2S}$$

to calculate the wall thickness of a cylindrical tank and

$$t = \frac{pd}{4S}$$

to calculate the wall thickness of a spherical tank, if the diameter, pressure, and stress are known.

Illustrative Example 4. What wall thickness is required for a cylindrical steel tank that is 40 ft in diameter and is filled with oil at a pressure of 30 psi. The working stress is 18 000 psi.

Solution:

1. The pressure p is 30 psi.
2. The diameter must be converted to inches.

$$d = 40 \times 12 = 480 \text{ in.}$$

3. The working stress is 18 000 psi.
4. Then

$$t = \frac{pd}{2S} = \frac{30 \times 480}{2 \times 18\,000} = 0.4 \text{ in.}$$

5. As a practical matter, the next larger standard thickness of steel plate is 7⁄16 in., which is 0.4375 in., and this would probably be chosen.

Practice Problems. (Section 13-15). Your next task is to work these problems of thin-walled pressure vessels.

1. A cylindrical tank is 24 in. in diameter and has a wall thickness

of ¼ in. The internal pressure is 100 psi. Calculate the longitudinal and circumferential stresses.

2. A cylindrical tank is 5.2 m in diameter and has a wall thickness of 12 mm. The internal pressure is 350 Pa. Calculate the longitudinal and circumferential stresses.
3. A cylindrical tank is 6 ft in diameter and is subjected to an internal pressure of 80 psi. The working stress is 20 000 psi. What wall thickness is needed?
4. A cylindrical tank is 0.6 m in diameter and is subjected to an internal pressure of 520 Pa. The working stress is 120 MPa. What wall thickness is needed?
5. A spherical pressure vessel is 24 in. in diameter and has a wall thickness of 0.1 in.. The internal pressure is 90 psi. Calculate the stress in the vessel.
6. A spherical pressure vessel is 4 m in diameter and has a wall thickness of 10 mm. The internal pressure is 280 Pa. What is the stress in the vessel?
7. A spherical pressure vessel is 12 ft in diameter and is subjected to an internal pressure of 42 psi. The working stress is 20 000 psi. What wall thickness is needed?
8. A spherical pressure vessel is 2.6 m in diameter and is subjected to an internal pressure of 600 Pa. The working stress is 90 MPa. What wall thickness is needed?

13-16. BEAM BEARING PLATES

There are many cases in which one end of a steel beam is supported by a masonry wall. Then it is usually necessary to place a steel bearing plate between the beam and the wall; Fig. 13-63 shows two views of a bearing plate between a steel beam and a wall.

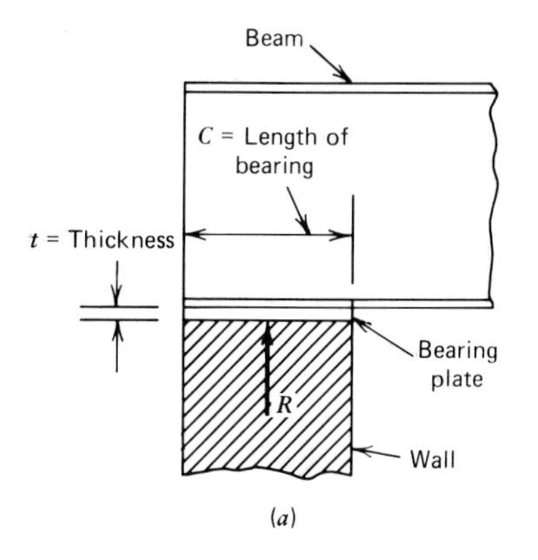

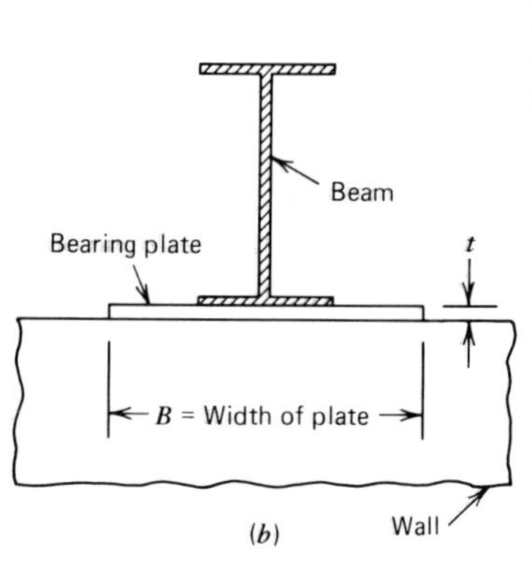

Fig. 13-63. Steel bearing plate for beam. (*a*) Front view. (*b*) End view.

You can see the reaction, R, of the wall on the beam in Fig. 13-63*a*. (You learned to calculate beam reactions in Chapter 2 and worked beam problems later.) The force R causes bearing stress between the wall and the beam. If no bearing plate were used, the bearing stress on the wall would be too high for the masonry. The bearing plate provides more bearing area against the wall and reduces the bearing stress on the masonry to a reasonable value.

Notice the dimensions of the bearing plate in Fig. 13-63. The dimension C is the length of the bearing, and this is often the same as the thickness of the wall. In any event, C is known when you start to design the bearing plate.

The dimension B is the width of the bearing plate. We can derive a formula for B. The area of bearing between the plate and the wall is the product of the dimensions C and B. Thus

$$A_b = CB$$

Then, the working stress in bearing against the wall is S_b, and the load is R. The bearing stress is equal to the load divided by the area; that is

$$S_b = \frac{R}{A_b} = \frac{R}{CB}$$

and we can use a little algebra to change this to

$$B = \frac{R}{CS_b}$$

We can solve this formula for the width B when we know the reaction R, the length of bearing C, and the bearing stress S_b.

With B known, the only unknown left to find is the thickness t. We could go through the derivation of a formula for t, but we will give it to you to save time. It is

$$t = \sqrt{\frac{3}{4}\frac{RB}{CS}}$$

where R is the reaction, B and C are dimensions of the steel plate, and S is the working stress for the steel in bending. (The steel plate does bend.)

You need to use two working stresses to design a steel bearing plate: the working stress S_b for the masonry in bearing and the working stress S for the steel plate in bending. We will give you the working stresses in our problems, and, in practical work, you will find them in the building codes.

The dimension B will usually come out as a number such as 9.73 in. or 7.57 in., which we will round off to the next larger inch. Thus if we calculate B as 9.73 in., we will increase it to 10 in. If we calculate B as 7.57 in., we will increase it to 8 in. Then we will use the whole number for B when we calculate t. The dimension t will probably come out as something like 0.692 in., but we will increase it to the next larger eighth of an inch. Thus 0.692 in. would be increased to 0.75 in., which is ¾ in.

Illustrative Example 1. A steel beam is to be supported by a concrete wall 8 in thick; the reaction of the wall on the beam is 29 700 lb. The working stresses are 500 psi for bearing against the concrete and 20 000 psi for bending in the steel. Find the width and thickness of the bearing plate.

Solution:

1. The length of bearing is $C = 8$ in.
2. The reaction R is 29 700 lb.
3. The working stress in bearing is 500 psi.
4. The required width of the plate is

$$B = \frac{R}{CS_b} = \frac{29\,700}{8 \times 500} = 7.425 \text{ in.}$$

and we increase this to the next larger inch, which is 8 in.

5. The working stress in bending is 20 000 psi.
6. The thickness needed is

$$t = \sqrt{\frac{3}{4}\frac{RB}{CS}} = \sqrt{\frac{3}{4}\frac{29\,700 \times 8}{8 \times 20\,000}} = 1.055 \text{ in.}$$

and we increase this to the next larger eighth of an inch, which is 1.125 in. or 1⅛ in.

7. Then $B = 8$ in.; and $t = 1⅛$ in.

Practice Problems (Section 13-16). You can learn how to design bearing plates by working problems. In each of the following problems find the width and thickness of the bearing plate.

1. A concrete wall 6 in. thick supports a steel beam and exerts a reaction of 18 000 lb on the beam. The working stress in bearing is 700 psi, and the working stress in bending is 20 000 psi.
2. Figure 13-64 shows an arrangement in which the length of bearing is only 4 in. The reaction is 16 200 lb. Working stresses are 600 psi in bearing and 18 000 psi in bending.
3. A concrete wall 9 in. thick exerts a reaction of 46 600 lb on a steel beam. Working stresses are 600 psi in bearing and 18 000 psi in bending.

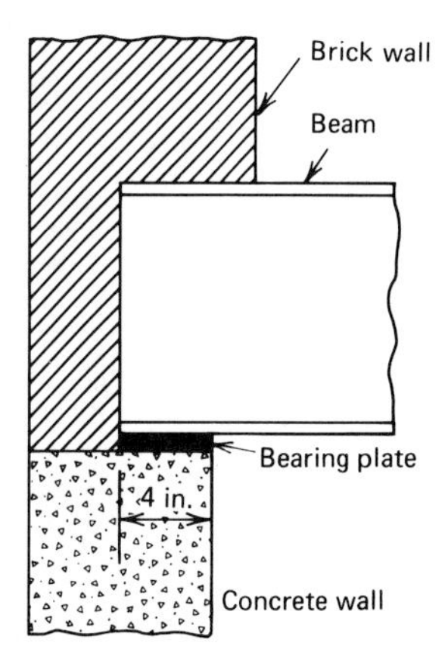

Fig. 13-64. Short bearing plate for Problem 2.

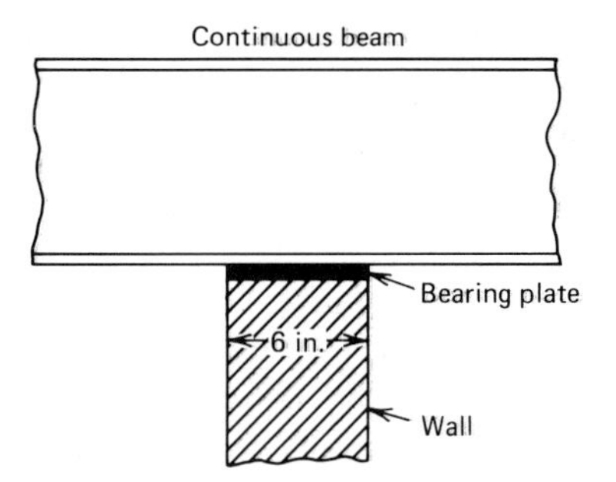

Fig. 13-65. Bearing plate under continuous beam for Problem 4.

4. Figure 13-65 shows a continuous beam (remember *continuous beams* from Chapter 9) supported by a wall and bearing plate. The reaction is 24 800 lb. Working stresses are 750 psi in bearing and 20 000 psi in bending.

SUMMARY We have covered a lot of ground in this chapter, and now it's time to summarize.

1. Rivets and bolts are essentially circular rods used to transmit force from one member to another.
   (a) A rivet has a formed head on each end.
   (b) A bolt has a formed head on one end and threads on the other end for a nut.
   (c) A cap screw has a formed head on one end, and the other end is threaded so that the screw can be screwed into a threaded hole.
2. Rivets and bolts are stressed in shear. The stress in shear is equal to the force divided by the area.

$$S_s = \frac{P}{A_s}$$

   (a) The rivet or bolt is in *single shear* when there is shear on only one cross section.
   (b) The rivet or bolt is in *double shear* when there is shear on two cross section.
   (c) The area of the cross section is

$$\frac{\pi d^2}{4} \ (d = \text{rivet diameter})$$

3. The maximum tensile stress in a riveted or bolted joint occurs on the net section, which is a section through the holes.

$$S_t = \frac{P}{A_t}$$

(a) The area of the net section is

A_t = (width of plate − number of rivets in net section × rivet diameter) × thickness of plate

$$A_t = (w - nd)t$$

4. Bearing stress occurs between the rivet and plate

$$S_b = \frac{P}{A_b}$$

(a) The bearing area for one rivet is equal to the product of the rivet diameter and the plate thickness.

5. Most designs are made with standard sizes of material.
   (a) Standard diameters of rivets and bolts are given in Table 13.1.
   (b) Standard gages of strip and sheet are given in Table 13.2.
6. The procedure in designing a simple riveted or bolted connection is
   (a) Decide on the diameter d.
   (b) Calculate the shearing strength P_1 of one rivet or bolt.

$$P_1 = A_s S_s$$

   (c) Find the number of rivets n.

$$n = \frac{P}{P_1}$$

   (d) Calculate the thickness t of the plate.

$$t = \frac{P}{ndS_b}$$

   (e) Find the width of the plate.

$$w = qd + \frac{P}{tS_t}$$

(q is the number of rivets in a row perpendicular to the force)

7. A *weld* is a joint made by applying heat, or heat and pressure, to pieces of metal.
   (a) An arc weld is made by the heat of an electric arc.
   (b) A gas weld is made by the heat from a torch.
   (c) A spot weld is made by the heat from an electric current while the sheets of metal are squeezed between electrodes.
8. The plates in a butt weld are placed butt to butt.
   (a) The cross section of a butt weld is rectangular.
   (b) If the width is known, the thickness is found from the formula,

$$t = \frac{P}{wS}$$

(c) If the thickness is known, the width is found from the formula

$$w = \frac{P}{tS}$$

9. A *fillet weld* is a weld that fills a corner.
   (a) A *side-fillet weld* is a fillet weld along the side of a bar.
   (b) An *end fillet* is a fillet weld at the end of a bar.
   (c) The total length of a fillet weld is found from the formula,

$$L = \frac{P}{0.707St}$$

10. The number of spot welds in a joint between two sheets of metal is found from the formula,

$$n = \frac{P}{R}$$

where P is the total force and R is the working strength of a single spot.

11. When a thin-walled cylinder is used as a pressure vessel.
    (a) The circumferential stress is

$$S_c = \frac{pd}{2t}$$

    (b) The longitudinal stress is

$$S_L = \frac{pd}{4t}$$

12. The stress in a thin-walled spherical pressure vessel is

$$S = \frac{pd}{4t}$$

13. Beam bearing plates.
    (a) A steel bearing plate is used to distribute the reaction over a large area when a steel beam is supported by a masonry wall.
    (b) The length of bearing C is determined by the space available.
    (c) The width B of the plate is given by the formula

$$B = \frac{R}{CS_b}$$

    where R is the reaction, C is the length of bearing, and S_b is the working stress in bearing.
    (d) The thickness of the bearing plate is determined from the formula

$$t = \sqrt{\frac{3}{4}\frac{RB}{CS}}$$

Here S is working stress in bending.

REVIEW QUESTIONS

Test your knowledge of this chapter with the following questions. Answer without looking back at the preceding pages.

1. What is a *rivet*?
2. What is a *bolt*?
3. Where does the shearing stress occur on a rivet or bolt?
4. What is the *area of cross section* of a rivet or bolt?
5. Where does the maximum tensile stress occur in a riveted joint?
6. What is *bearing stress*?
7. What is the *area of bearing* for a rivet?
8. How do you calculate the area of the net section?
9. Why are standard sizes of material used so extensively?
10. What is the difference between *single shear* and *double shear*?
11. What is a *weld*?
12. What is a *butt weld*?
13. What is a *fillet weld*?
14. What is a *spot weld*?
15. Where does the *maximum stress* occur in a fillet weld?
16. What is the kind of stress in a spot weld?
17. What are the stresses in a thin-walled cylindrical pressure vessel?
18. What are the stresses in a thin-walled spherical pressure vessel?
19. What is a beam bearing plate?
20. Why is a bearing plate used?
21. What is *length of bearing* for a beam bearing plate?
22. What is the *width of a bearing plate*?
23. How do you find the thickness required for a beam bearing plate?

REVIEW PROBLEMS

1. Find the shearing stress in the cap screws in Fig. 13-66.
2. Find the maximum tensile stress in the plate in Fig. 13-66.

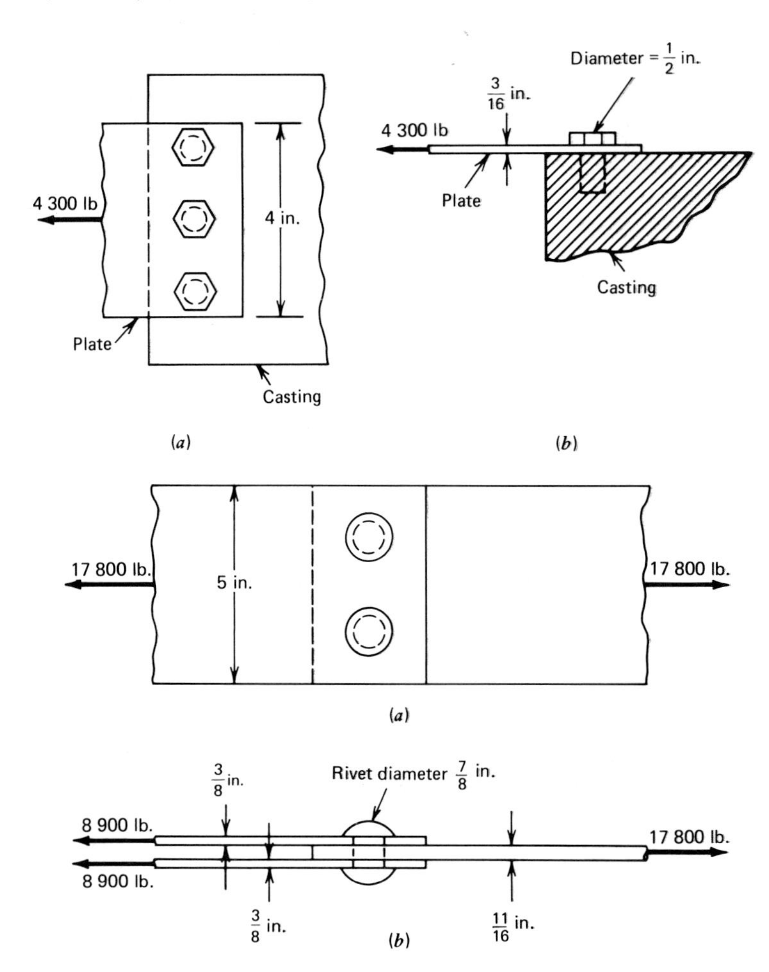

Fig. 13-66. Joint with cap screws for Problems 1, 2, and 3. (*a*) Top view. (*b*) Side view.

Fig. 13-67. Riveted joint for Problems 4 through 8. (*a*) Top view. (*b*) Front view.

3. Find the bearing stress between the cap screws and plate in Fig. 13-66.
4. Find the shearing stress in the rivets in Fig. 13-67.
5. Find the maximum tensile stress in the main plate in Fig. 13-67.
6. Find the maximum tensile stress in the cover plates in Fig. 13-67.
7. Find the bearing stress between the rivets and the main plate in Fig. 13-67.
8. Find the bearing stress between the rivets and cover plates in Fig. 13-67.
9. Figure 13-68 shows two plates that are to be riveted together; the rivets are to be placed in one row *A-A* and are to be ¾ in. in diameter. The working stresses are 10 000 psi in shear, 15 000 psi in tension and 20 000 psi in bearing. Find the number of rivets, the thickness of the plate, and the width of the plate.

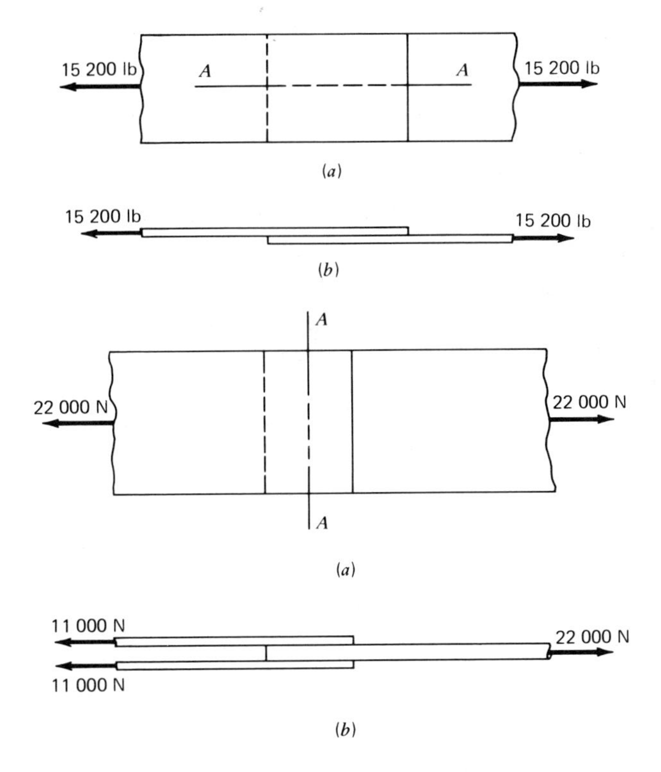

Fig. 13-68. Joint to be riveted for Problem 9. (*a*) Top view. (*b*) Front view.

Fig. 13-69. Joint to be bolted for Problem 10. (*a*) Top view. (*b*) Front view.

10. Figure 13-69 shows a joint that is to be bolted; the bolts are to be 12 mm in diameter and placed in one row *A-A*. The working stresses are 60 MPa in shear, 90 MPa in tension, and 120 MPa in bearing. Find the number of bolts, the thickness of the main plate, and the width of the main plate.
11. A welded-butt joint between two steel plates is to carry a force of 18 000 lb. with a working stress of 13 000 psi. The plates are $3/16$ in. thick. How wide should they be?
12. A welded-butt joint between two magnesium-alloy plates is to be subjected to a tensile force of 38 000 N. The plates are 75 mm wide, and the working stress is 40 MPa. How thick should the plates be?
13. A rectangular steel bar 2 in. × $1/2$ in. is to be welded to a steel plate with side-fillet welds. The total force is 14 000 lb. Find the length of the fillet weld on each side. The working stress is 16 000 psi.
14. A rectangular steel bar of aluminum-alloy 1.5 in. × $1/8$ in. is to be welded to an aluminum-alloy plate with a fillet weld across the end and a fillet weld on each side. The working stress for the welds is 6 000 psi and the total force is 1 940 lb. Find the length of each side-fillet weld.
15. Two steel sheets, each 0.109 in. thick, are to be spot welded together.

The total force is 4 200 lb. How many spot welds are to be used? Let $R = 800$ lb.

16. Two magnesium-alloy sheets, each 0.0201 in. thick, are to be spot welded together to carry a force of 165 lb. Find the number of spot welds. Let $R = 34$ lb.
17. Calculate the longitudinal and circumferential stresses in a cylindrical pressure vessel that is 42 in. in diameter and has a wall thickness of ¼ in. The internal pressure is 34 psi.
18. Calculate the tensile stress in a spherical pressure vessel that is 600 mm in diameter and has a wall thickness of 8 mm. The internal pressure is 2 MPa.
19. Design a beam bearing plate for a steel beam that has a reaction of 22 800 lb. The length of bearing is 6 in., the working stress in bearing against the support is 600 psi, and the working stress in bending for the plate is 18 000 psi.

14 REPEATED STRESS AND STRESS CONCENTRATION

PURPOSE OF THIS CHAPTER. This chapter has two purposes. First, we are going to see what happens to a machine part when a load is applied a great many times instead of just once. We will find that a member will fail under a smaller force if the load is repeated than if the force is applied only once. Now we know that force causes stress, and if the force is repeated, the member is subjected to repeated stress. We want to see how much the strength of a material is reduced by the repetition of stress, and we want to know how to take the reduced strength into account in designing members. The topic of repeated stress is important, because so many machine parts are subjected to repeated stress.

The second purpose of this chapter is to learn about a phenomenon called *stress concentration*. You have already learned how to calculate stresses due to axial forces, bending moments, and torques. The formulas for these stresses apply only to bars of constant cross section. At any point at which there is a change in the cross section of a bar, the stress is higher than that given by the formula. We say that there is a stress concentration at such a point. We will show you some examples of stress concentration and show you how to take account of it in design. And you will see that there is a good reason for studying repeated stress and stress concentration at the same time.

14-1. EXAMPLES OF REPEATED STRESS

Let's look at a few examples of repeated stress to get an idea of how this can happen and how many times the stress might be repeated. Figure 14-1 shows part of a machine for breaking concrete. Part A is the housing, and part B is the hammer. A rotating hook (not shown in the figure)

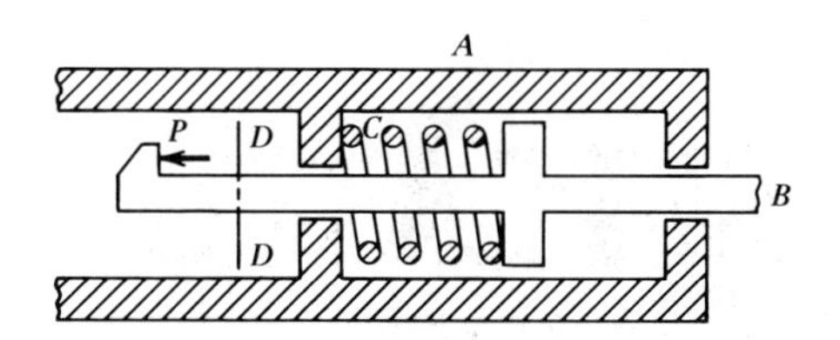

Fig. 14-1. Part of machine for breaking concrete.

catches the projection at the left end of B; the hook exerts the force P and pulls the hammer back against the spring C and compresses the spring. Then the hook turns on around and releases the hammer B; the compressed spring C drives the hammer to the right, and the right end of the hammer delivers the blow on the concrete. This operation is repeated over and over, and each time the hook exerts the force P on the hammer, it causes stress on such a section as D-D. Usually, in such cases the machine operates at a definite speed, and it is possible to make a fair prediction of the amount of time the machine will be used, so that a person can make a reasonable estimate of the number of times that the machine part will be subjected to stress.

Illustrative Example 1. The machine in Fig. 14-1 is expected to operate at 900 strokes/min for an average of 6 hr/day for 100 days/yr for 5 yr. On this basis, how many times is the stress repeated on such a section as D-D?

Solution:

A. There are 60 min in 1 hr, so the number of strokes per hour is

$$900 \times 60 = 54\ 000$$

B. With 6 hr of operation per day

$$54\ 000 \times 6 = 324\ 000 \text{ strokes/day}$$

C. At 100 days use per year

$$324\ 000 \times 100 = 32\ 400\ 000 \text{ strokes/yr}$$

D. During 5 yr of operation there will be

$$32\ 400\ 000 \times 5 = 162\ 000\ 000 \text{ strokes}$$

E. There is one application of stress for each stroke, so the number of times that the stress is repeated is 162 000 000.

REPEATED BENDING STRESS

A bar may be subjected to repetitions of bending stress, even though the load doesn't change, because the bar may rotate under the load. For

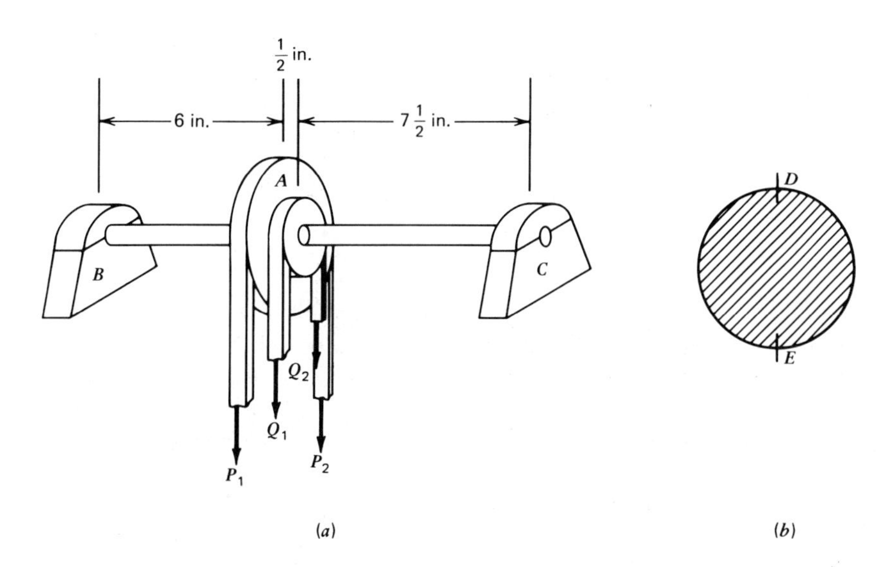

Fig. 14-2. Circular shaft in bearings. (*a*) Compound pulley on shaft. (*b*) Cross section.

example, Fig. 14-2*a* shows a circular shaft that carries a compound pulley *A* and is supported in bearings at *B* and *C*. A belt that passes over the larger part of the compound pulley exerts the forces P_1 and P_2, and a belt that passes over the smaller part of the compound pulley exerts the forces Q_1 and Q_2. You are supposed to know by now that these forces cause a positive bending moment at the center of the shaft. Now let's look at Fig. 14-2*b*, which shows the cross section of the shaft. The positive bending moment causes compressive stress at point *D* at the top of the section and tensile stress at point *E* at the bottom of the section. However, this shaft is rotating, or turning; as it turns through one half of a revolution, a point starting at *D* moves to *E*, and the stress changes from compression to tension; then as the shaft turns through the next half revolution, the point moves from *E* to *D*, and the stress changes from tension to compression. Thus for each revolution of the shaft there is a complete reversal of stress, from compression to tension and back to compression. This occurrence is called *completely reversed bending stress*, and it happens often in machine parts, because so many machine parts are rotating parts. Be sure to realize here that completely reversed bending stress can happen even when the load remains steady.

In manufacturing operations there are lots of machines that are started and stopped frequently. The load is applied each time the machine is started, so each starting operation causes one application of stress in parts of the machine. This can lead to a great many repetitions of stress during the life of the machine.

Metal-cutting machines often contain shafts that reverse their direction of rotation frequently. These shafts are likely to be driven by an electric motor that exerts first a counterclockwise torque on the shaft to turn the shaft counterclockwise, and then a clockwise torque to turn the

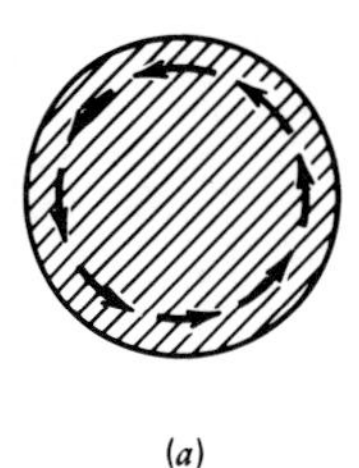
(a)

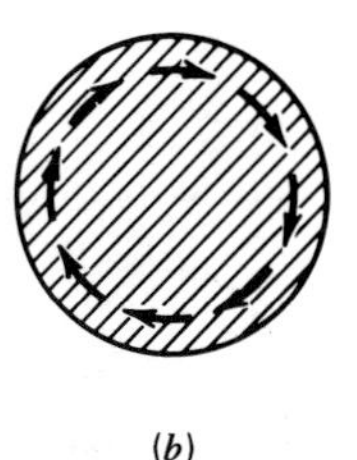
(b)

Fig. 14-3. Reversed torsional shearing stress. (*a*) Direction of shearing stresses due to counterclockwise torque. (*b*) Direction of shearing stresses due to clockwise torque.

shaft clockwise. Figure 14-3*a* shows the direction of the torsional shearing stresses on a section of the shaft when the torque is counterclockwise, and Fig. 14-3*b* shows the direction of the torsional shearing stresses when the torque is clockwise. You can see that the shearing stress is reversed in direction, and the name of the stress reversal is reversed torsional shearing stress.

In general, each application of stress, during which the stress changes from a certain value to another value and then changes back to the first value, is called a *cycle of stress*. For example, a stress might change from zero to 10 000 psi and then change back to zero. We would call the process a cycle of stress, and it might be repeated many times. For another example, the stress might change from a torsional shearing stress of 8 000 psi in one direction to a torsional shearing stress of 8 000 psi in the opposite direction, and then change back to the torsional shearing stress of 8 000 psi in the first direction. This whole process is also called a cycle of stress.

Practice Problems (Section 14-1). Here is your opportunity to calculate the number of cycles of stress for several machine parts.

1. A crankshaft in an engine is subjected to a cycle of stress with each revolution of the engine. The engine is expected to operate at 3 000 r.p.m. for an average of 3 hr/day for 300 days/yr for 12 yr. How many cycles of stress will the crankshaft be subjected to?
2. The shaft in Fig. 14-2 is expected to rotate at 900 rpm (revolutions per minute) for an average of 8 hr/day for 260 days/yr for 20 yr. How many cycles of completely reversed bending stress should the shaft be designed for?
3. It is expected that a certain open-end wrench will be used to tighten 30 nuts per day and that the mechanic will need to apply a torque six times to each nut. On the basis that the wrench is to be used for an average of 200 days/yr for 5 yr, how many cycles of stress will the wrench be subjected to?
4. It is planned that the main shaft on a certain metal-cutting machine will be reversed in direction five times a minute. It is ex-

pected that the machine will be used 16 hr/day for an average of 250 days/yr for 8 yr. For how many cycles of completely reversed torsional shearing stress would you design the shaft?

14-2. EFFECT OF REPETITION OF STRESS ON THE STRENGTH OF A MATERIAL

We are ready now to see how the strength of a material is affected by repeating the stress, and we are going to find that a material may fail under a rather small stress if the stress is applied enough times. You have probably had some experience of this sort already. For instance, you have probably broken a wire or paperclip by bending it back and forth several times. What you really did here was to apply the stress a number of times; the stress was too small to cause failure the first time it was applied, but the stress was large enough to cause failure when repeated several times.

The usual way of showing the effect of the number of cycles of stress on the strength of a material is to plot what is called an *S-N diagram*, such as you see in Fig. 14-4. In this diagram, the stress that will cause failure is plotted vertically, and the number of cycles of this stress is plotted horizontally. (You notice that the horizontal plotting is unusual; this is a logarithmic plotting which makes it possible to show a large range of values in small space.) You can take the *S-N* diagram in Fig. 14-4 as representing the behavior of iron and steel alloys, but the behavior of nonferrous (nonferrous means that the metal contains no iron) metals is not so simple.

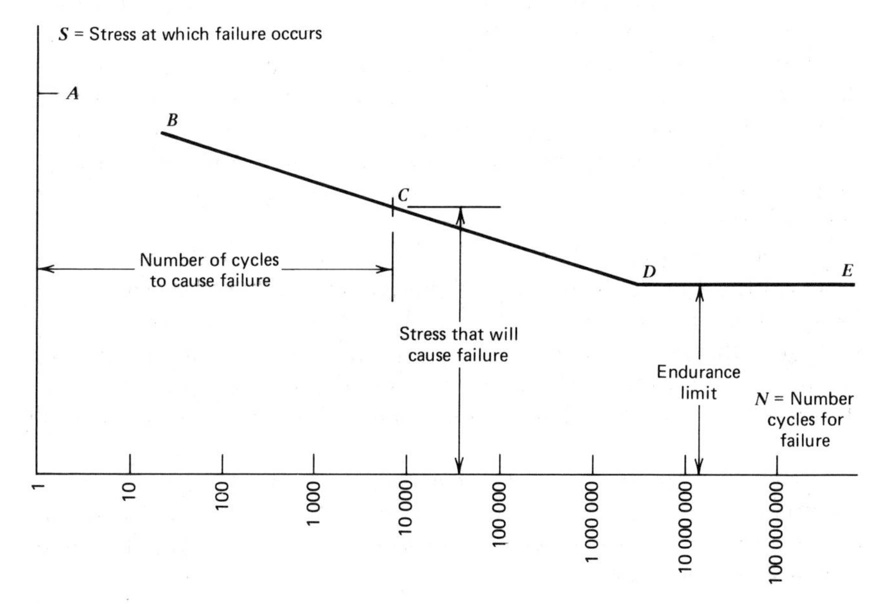

Fig. 14-4. S–N diagram.

The point A in Fig. 14-4 represents the ultimate strength of the material; that is, the stress that will cause failure if applied only once. Then for any point C on the curve the vertical distance or coordinate of C represents the stress that will cause failure, and the horizontal distance to C represents the number of times that the stress must be applied. You can see that the stress goes down as the number of cycles goes up until point D is reached (point D usually represents between 1 000 000 and 10 000 000 cycles of stress, for steel). Beyond point D the stress that will cause failure seems to be the same, no matter how many times it is applied. The value of the stress, for any point on the line DE is called the *endurance limit*, and the importance of the endurance limit is that any stress smaller than the endurance limit will not cause failure, no matter how many times it is applied.

It used to be thought that steel grew tired under repeated stress, and the whole business was called *fatigue of metals*; the endurance limit was called the *fatigue strength*; also, it was said that the steel crystallized. We know now, though, that it isn't a matter of steel getting tired and that steel is always composed of crystals. Today, we just call it the *effect* of repeated stress.

14-3. VALUES OF THE ENDURANCE LIMIT FOR COMPLETELY REVERSED BENDING STRESS

It is time now for you to see some numerical values for the endurance limits of various metals. Most of the experimental work on repeated stress has been done with machines that subject a specimen to cycles of completely reversed bending stress, so we will have to give you values of endurance limits for that condition. As we said before, there is a definite endurance limit for iron and steel alloys, but nonferrous metals, such as aluminum and magnesium alloys, are not so regular in their behavior. The value of the endurance limits for aluminum and magnesium alloys is usually taken as the stress that will cause failure after 500 000 000 cycles.

Table 14.1 gives values of the endurance limit under completely reversed bending stress for several metals. The table is not intended to be so complete that you would never have to look elsewhere for information; rather, it is intended to give sample information so that you can get an idea of how these things go. You should not attempt to memorize the table, but you should remember where it is and what information it contains.

There is a great deal of information available on the behavior of materials under repeated stress (more than we have room for in this book), and you can usually get the information you need from the manufacturer of the material.

TABLE 14.1. SAMPLE VALVES OF ENDURANCE LIMIT FOR COMPLETELY REVERSED BENDING STRESS

Material	*Treatment*	*Endurance Limit* *MPa*	*psi*
Steel 0.2% Carbon 0.7% Manganese	Normalized	193	28 000
Steel 0.58% Carbon 0.65% Manganese	Hardened and tempered	293	42 600
Steel 0.40% Carbon 0.80% Manganese 3.5% Nickel	Hardened and tempered	525	76 000
Aluminum alloy 4% Copper 0.5% Manganese 0.5% Magnesium	Extruded	103	15 000
Aluminum alloy 4% Copper 1.5% Magnesium 2% Nickel	Forged	117	17 000
Magnesium alloy 1.5% Manganese	Extruded	83	12 000
Magnesium alloy 8% Aluminum 0.4% Zinc	Sand cast	86	12 500
Titanium alloy 11% Tin 4% Molybdenum 2.25% Aluminum 0.2% Silicon	Quenched and aged	710	103 000

14-4. FACTOR OF SAFETY-WORKING STRESS

The endurance limit of a material is the greatest stress that can be applied an indefinite number of times without causing failure. Now you don't want failure to occur in a machine part and so the working stress (the working stress is the stress that is expected to be developed in service) should be well below the endurance limit. You should use a factor of safety based on the endurance limit, when the stress is to be repeated, and find the working stress this way:

$$\text{working stress} = \frac{\text{endurance limit}}{\text{factor of safety}}$$

A reasonable factor of safety is 3 and, if you use 3, you get

$$\text{working stress} = \frac{\text{endurance limit}}{3}$$

Whether the factor of safety should be 3 is a matter of judgment; it isn't anything that can be proved exactly. If you think that 2.5 or 3.5 would be better for a particular case, go ahead and use it. You don't want to use too low a factor of safety, because then the machine part might fail. If you use too high a factor of safety, the design is not economical. The factor of safety is supposed to take care of all of the uncertainties in the problem, such as the possible error in estimating the load, the possibility that the particular piece of material is weaker than the average, the question of just how well the formulas used apply. The factor of safety should be high enough so that there is no reasonable possibility that the machine part will fail.

The values of endurance limit in Table 14.1 are for completely reversed bending stress, and you remember that bending stress is either tension or compression. There is not a great deal of information available on endurance limits for tension or compression due to an axial force or for tension or compression due to an axial force combined with bending. However, you would probably be safe in using the values for completely reversed bending stress for any case of repeated tension or compression, and you could rely on the factor of safety to take care of any error involved. New let's practice with factors of safety.

Illustrative Example 1. What is a reasonable working stress for a steel bar that is to be subjected to completely reversed bending stress if the endurance limit is 48 000 psi?

Solution:

A. The endurance limit is 48 000 psi.

B. A reasonable factor of safety is 3.

C. The working stress is

$$\text{working stress} = \frac{\text{endurance limit}}{\text{factor of safety}} = \frac{48\ 000}{3}$$

$$= 16\ 000 \text{ psi}$$

TORSION

If you do much machine designing, you will encounter problems of repeated torsional shearing stress, and you will need some information on endurance limits for this type of stress. It has been fairly well established for steel that the endurance limit for completely reversed torsional shearing stress is about one-half of the endurance limit for completely reversed bending stress. There isn't much known about the values of endurance limit for torsional shearing stress that is not completely reversed, even for steel, and there isn't much information at all for repeated torsional shearing stress for nonferrous metals. However, it would seem to be reasonable to just take the endurance limit for any case of torsional shearing stress as one-half of the endurance limit for completely reversed bending stress.

Illustrative Example 2. A bar of aluminum alloy is to be subjected to repeated torsional shearing stress. Determine a reasonable working stress. The endurance limit for completely reversed bending stress is 5 500 psi.

Solution:

A. The endurance limit for completely reversed bending stress is 5 500 psi.

B. We will multiply by one-half to get a value of endurance limit for torsional shearing stress. So

$$\tfrac{1}{2} \times 5\,500 = 2\,750 \text{ psi}$$

C. A reasonable factor of safety is 3.

D. The working stress is

$$\text{working stress} = \frac{\text{endurance limit}}{\text{factor of safety}} = \frac{2\,750}{3} = 917 \text{ psi}$$

Practice Problems (Section 14-4). Figure out a few working stresses for yourself.

1. What is a reasonable working stress for a bar of aluminum alloy which is to be subjected to repeated tensile stress? The endurance limit for completely reversed bending stress is 8 000 psi.
2. A bar is to be subjected to completely reversed torsional shearing stress. What is a reasonable working stress if the endurance limit for completely reversed bending stress is 60 000 psi?
3. A bar of aluminum alloy is to be subjected to completely reversed bending stress. Calculate a reasonable working stress if the en-

durance limit for completely reversed bending stress is 108 MPa.

4. What is a reasonable working stress for a bar of magnesium alloy that is to be subjected to repeated torsional shearing stress? The endurance limit for completely reversed bending stress is 92 MPa.

14-5. DESIGN OF MEMBERS SUBJECTED TO REPEATED STRESS

Let's see now how to use what we know about repeated stress in designing members, and let's review part of the subject of strength of materials at the same time. We will do it this way. You will have to determine the size of a beam or shaft; to do this you will have to determine a suitable working stress, and that's where you need to know something about repeated stress; also, you will have to use the proper formula to get the size of member, and that's where you will get the review.

You aren't supposed to remember every formula in this book, but you are supposed to know what formulas are here and how to use them. You can always look up the formula if you have forgotten just how it goes, but it won't do you any good unless you know how to use it. We aren't going to tell you just how to do all of these problems nor what formulas to use. You can be sure that you can do the problems if you really know strength of materials, and you can be sure that the formulas you need are in this book. Now let's do some examples.

Illustrative Example 1. A solid circular bar is subjected to a repeated torque of 23 000 lb in. What is a suitable diameter for the bar? The endurance limit for completely reversed bending stress is 56 000 psi.

Solution:

A. The torque T is 23 000 lb in.

B. The endurance limit for completely reversed bending stress is 56 000 psi.

C. The endurance limit for repeated torsional shearing stress can be taken as one-half of the endurance limit for completely reversed bending stress. This is

$$\tfrac{1}{2} \times 56\,000 = 28\,000 \text{ psi}$$

D. The working stress is equal to the endurance limit divided by the factor of safety of 3. So,

$$\text{working stress} = \frac{28\,000}{3} = 9\,330 \text{ psi}$$

E. A suitable diameter is (using the formula from Section 10-6)

$$d = \sqrt[3]{\frac{16}{\pi}\frac{T}{S}} = \sqrt[3]{\frac{16}{\pi}\frac{23\ 000}{9\ 330}} = 2.32 \text{ in.}$$

Illustrative Example 2. A square aluminum-alloy bar is to be subjected to a repeated tensile force of 34 000 N. What size should the bar be? The endurance limit for completely reversed bending stress is 78.5 MPa.

Solution:

A. The force P is 34 000 N.

B. The endurance limit for completely reversed bending stress is 78.5 MPa. We will use this value for repeated tensile stress.

C. The working stress is equal to the endurance limit divided by the factor of safety. So,

$$\text{working stress} = \frac{78.5}{3} = 26.2 \text{ MPa}$$

D. The area required (formula from Section 3-10) is

$$A = \frac{P}{S} = \frac{34\ 000}{26.2} = 1\ 298 \text{ mm}^2$$

E. Figure 14-5 shows the section of the square bar. We will let the length of the side of the square be d. Then the area is

$$A = d^2 = 1\ 298$$

Taking the square root of each side of the equation

$$d = \sqrt{1\ 298} = 36 \text{ mm}$$

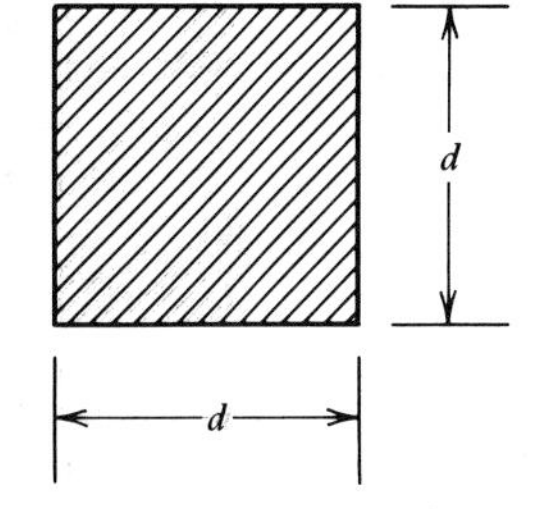

Fig. 14-5. Cross section of square bar.

Illustrative Example 3. Figure 14-6*a* shows a beam which is supported at A and B. The beam carries a bearing at C which supports a shaft and cam, the shaft being driven by a motor. As the shaft turns, the cam is subjected to a load as shown, which varies from zero to a maximum of 1 800 lb. This force is transmitted through the bearing to the beam at C. Thus the beam is subjected to a repeated load of 1 800 lb. The beam is a rectangular steel bar and is ¾ in. deep. What is a suitable width for the beam? The endurance limit for completely reversed bending stress is 34 000 psi.

Solution:

A. The free-body diagram of the beam is shown in Fig. 14-6*b*.

B. The reactions.

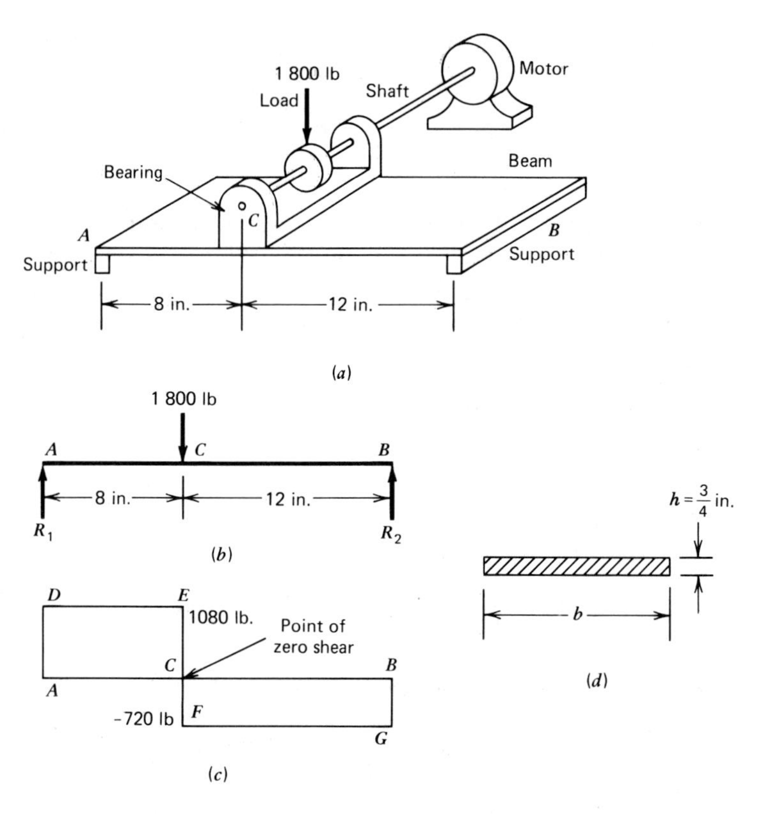

Fig. 14-6. Beam supporting a shaft and cam. (*a*) Beam in use. (*b*) Free-body diagram. (*c*) Shear diagram. (*d*) Cross section.

1. The reaction R_2 is found by writing a moment equation with center at A

$$\Sigma M_a = 0$$

$$1\,800 \times 8 - 20R_2 = 0$$

$$20R_2 = 1\,800 \times 8 = 14\,400 \qquad R_2 = 720 \text{ lb}$$

2. The reaction R_1 is found by writing a moment equation with center at B

$$\Sigma M_b = 0$$

$$20R_1 - 1\,800 \times 12 = 0$$

$$20R_1 = 1\,800 \times 12 = 21\,600; \qquad R_1 = 1\,080 \text{ lb}$$

C. The shear diagram is shown in Fig. 14-6*c*.

1. The first step in drawing the shear diagram is to draw the base line *ACB*.
2. The reaction R_1 (1 080 lb) is plotted upward from *A* to locate

point D in the shear diagram and the line DE is drawn.

3. The load of 1 800 lb is plotted downward from E to locate point F in the shear diagram. The lines FG and GB are drawn to complete the shear diagram.

D. The maximum bending moment.

1. The maximum bending moment occurs at C, where the shear is zero.
2. The bending moment at C is

$$M = 1080 \times 8 = 8\,640 \text{ lb in.}$$

E. The working stress.

1. The endurance limit is 34 000 psi.
2. The working stress is

$$\text{working stress} = \frac{\text{endurance limit}}{\text{factor of safety}} = \frac{34\,000}{3}$$

$$= 11\,300 \text{ psi}$$

F. The width of the beam. Figure 14-6*d* shows the section view.

1. The bending moment M is 8 640 lb in.
2. The working stress S is 11 300 psi.
3. The height h is ¾ in.
4. The section modulus required is

$$\frac{I}{c} = \frac{M}{S} = \frac{8\,640}{11\,300} = 0.765$$

5. The width is found from the formula

$$bh^2 = 6Z$$

$$b\left(\frac{3}{4}\right)^2 = 6 \times 0.765$$

$$b \times \frac{9}{16} = 4.59 \qquad b = 8.16 \text{ in.}$$

Practice Problems (Section 14-5). Now you can practice and review at the same time.

1. A rectangular bar is to be subjected to a repeated tensile force of 6 600 N. The bar is to be 10 mm thick. How wide should it be? TELFCRBS is 84 MPa. TELFCRBS = the endurance limit for completely reversed bending stress.
2. A solid circular aluminum-alloy bar is to be subjected to a com-

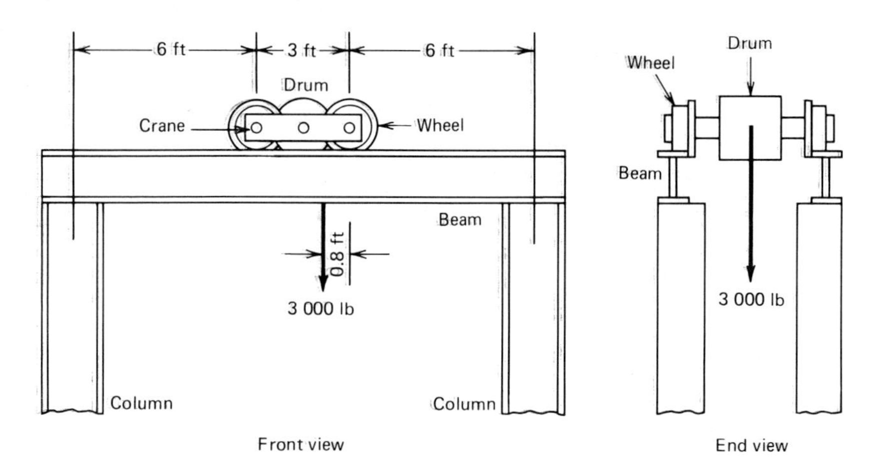

Fig. 14-7. Small crane supported by steel beams.

pletely reversed torque of 2 100 lb in. What is a suitable diameter for the bar? TELFCRBS is 15 000 psi.

3. Figure 14-7 shows two views of a small crane that moves on and is supported by a pair of steel W-shape beams. Assuming that the crane is to be used 100 times/day for an average of 250 days/yr for 40 yr, what size beam should be used? TECFCRBS is 35 000 psi.
4. Figure 14-8 shows the free-body diagram of a solid circular beam that rotates under the loads as the loads remain steady. Find the diameter required for the beam. TELFCRBS is 336 MPa.
5. A solid circular shaft is to be subjected to a repeated torque of 6 000 000 N·mm. Calculate a reasonable diameter for the shaft. TELFCRBS is 280 MPa.
6. A solid circular bar is to be subjected to a completely reversed tensile force of 29 000 lb. What diameter of bar is required? TELFCRBS is 39 000 psi.

14-6. MORE ABOUT STRESS CONCENTRATION

A stress concentration is a larger stress than the nominal stress given by a simple formula such as $S = P/A$ for an axial stress, or $S = Mc/I$ for a bending stress. Stress concentrations occur where there are changes

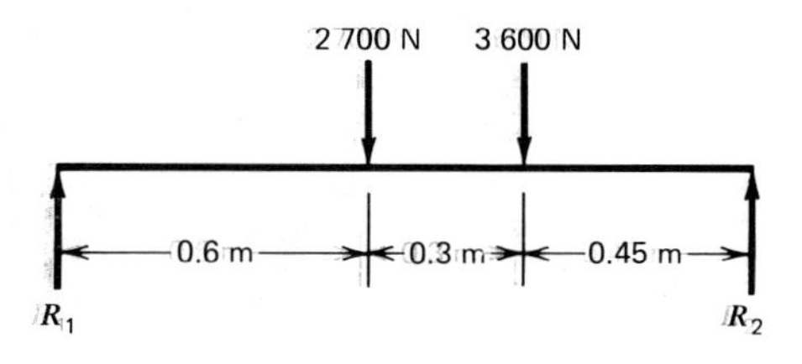

Fig. 14-8. Beam for Problem 4.

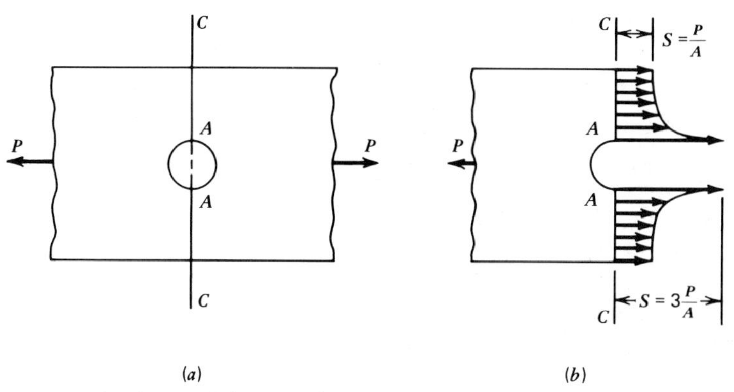

Fig. 14-9. Stress distribution in bar with hole. (*a*) Bar with hole. (*b*) Stress distribution.

in the cross sections of bars. These changes of cross section are often called *stress raisers.*

One well-known example of a stress raiser is a small circular hole in a plate, such as you see in Fig. 14-9*a*. There is a stress concentration at each edge of the hole on the cross section *C-C*, that is, at points *A*. It is possible to show mathematically that the stress at point *A* is *three* times the average stress on the cross section. Figure 14-9*b* shows how the stress varies over the cross section. Here it is evident that the stress diminishes rapidly as the distance from the hole increases. This rapid diminishing is typical of stress concentrations. They are localized phenomena.

The ratio of the maximum stress to the nominal stress is called the stress concentration factor *K*. Thus there is a stress concentration factor of $K = 3$ for the hole in the plate in Fig. 14-9.

There is another important fact about stress concentration. The ratios that we give you for stress concentration factors are only valid as long as the maximum stress does not exceed the proportional limit of the material, that is, as long as stress is proportional to strain. The stress concentration tends to disappear as the material is stressed beyond the yield point, and the stress tends to become distributed in accordance with the simple formulas that you already know. Thus, in Fig. 14-9*b*, the stress would approach a uniform value of P/A if the plate were to be stressed beyond the yield point.

Stress concentration is most important when the stress is repeated, because the endurance limit of a material is often below the proportional limit. Thus the piece may break from repeated stress without ever reaching the yield point and having a chance for the stress to be redistributed.

14-7. TWO EXAMPLES OF STRESS CONCENTRATION

Now we're ready for some examples of stress concentration and some problems. These examples are only samples of a great deal of information about stress concentration, but they are enough to show you how

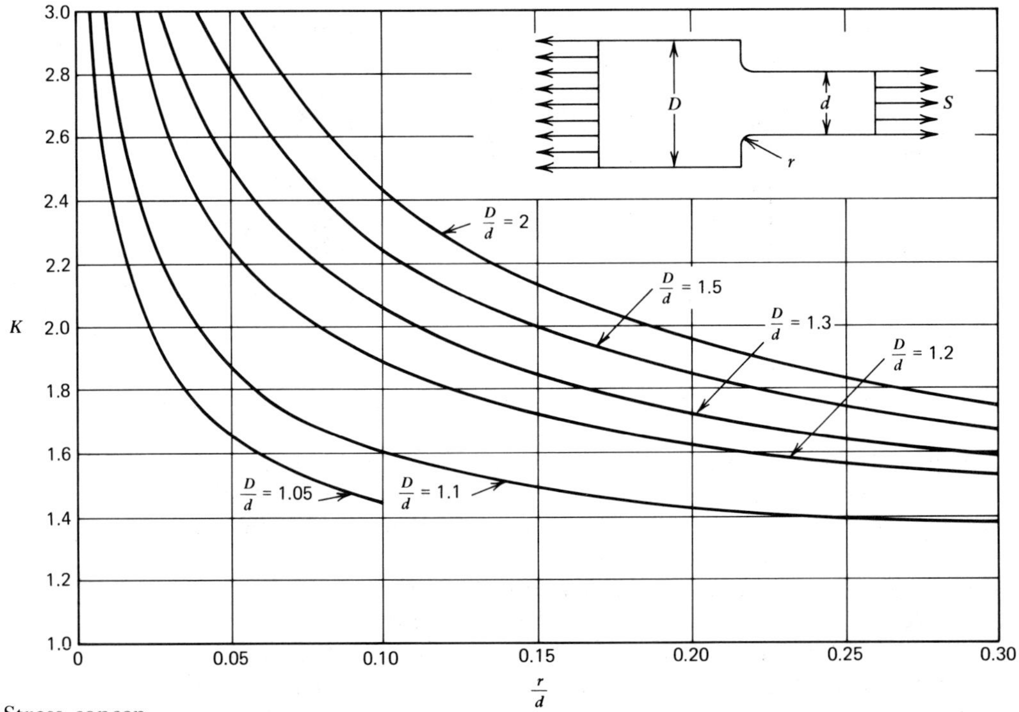

Fig. 14-10. Stress concentration factors for bar with fillets.

to work with stress concentration. Figure 14-10 shows a flat bar with a shoulder fillet. The bar is loaded in tension. Figure 14-10 also shows values of the stress concentration factor K. Here, D is the width of the larger end of the bar, d is the width of the smaller end of the bar, and r is the radius of the fillet. You can see that the stress concentration factor K is dependent on the ratio r/d and the ratio D/d. The maximum stress in the fillet is obtained by multiplying the nominal stress S by the stress concentration factor K.

There is an important point to be deduced from Fig. 14-10. A stress concentration factor is independent of the system of units; because it depends on ratios of dimensional quantities rather than on separate dimensional quantities. So, it is the same whether you are working in gravitational units or metric units.

The way to use Fig. 14-10 is to calculate the stress S, then calculate the ratio D/d to find which curve to use, next calculate the ratio r/d and read the stress concentration factor K, and finally multiply S by K to get the maximum stress. Now let's try some examples.

Illustrative Example 1. What is the maximum stress in the fillet of the flat bar in Fig. 14-11? The bar is ¼ in. thick.

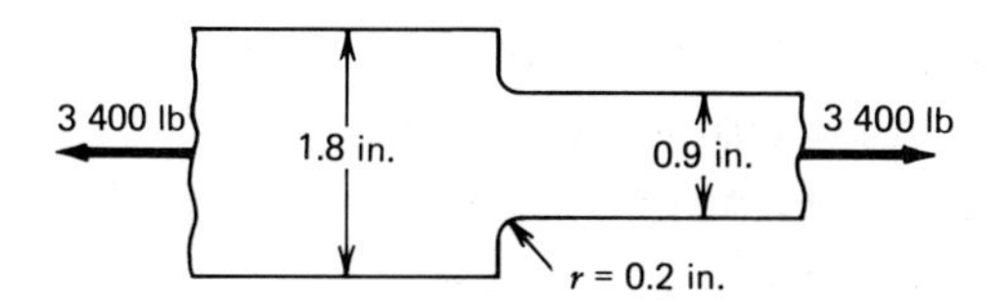

Fig. 14-11. Bar for Illustrative Example 1.

Solution:

1. The load P is 3 400 lb.
2. The smallest area of the cross section is

$$A = 0.9 \times 1/4 = 0.225 \text{ in.}^2$$

3. The nominal stress is

$$S = \frac{P}{A} = \frac{3\,400}{0.225} = 15\,100 \text{ psi}$$

4. The larger width is $D = 1.8$ in.
5. The smaller width is $d = 0.9$ in.
6. The ratio D/d is $1.8/0.9 = 2$.
7. The fillet radius is $r = 0.2$.
8. The ratio r/d is $0.2/0.9 = 0.222$.
9. Now look at the curve labeled $D/d = 2$ in Fig. 14-10. You must estimate where $r/d = 0.222$ is between $r/d = 0.2$ and $r/d = 0.25$. You might take K to be 1.89.
10. Thus maximum stress is $KS = 1.89 \times 15\,100 = 28\,500$ psi.

Illustrative Example 2. Figure 14-12 shows a flat bar 10 mm thick. Calculate the maximum stress in the fillet.

Solution:

1. The load P is 9 800 N.
2. The smallest area of the cross section is

$$A = 20 \times 10 = 200 \text{ mm}^2$$

3. The nominal stress is

$$S = \frac{P}{A} = \frac{9\,800}{200} = 49 \text{ MPa}$$

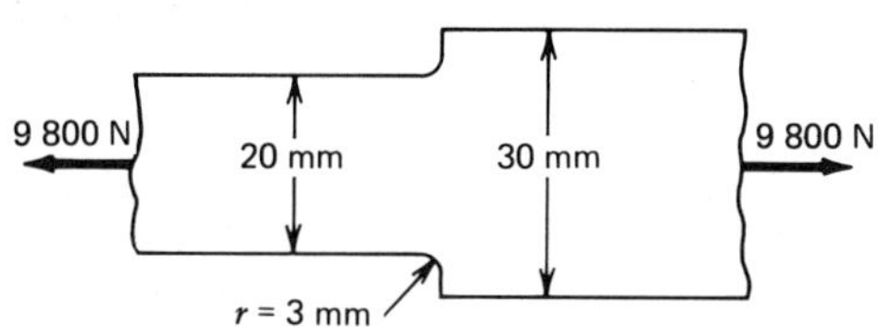

Fig. 14-12. Bar for Illustrative Example 2.

4. The larger width is $D = 30$ mm.
5. The smaller width is $d = 20$ mm.
6. The ratio D/d is $30/20 = 1.5$.
7. The fillet radius is $r = 3$ mm.
8. The ratio r/d is $3/20 = 0.15$.
9. Look at the curve labeled $D/d = 1.5$ in Fig. 14-10. The stress concentration factor K is about 1.98.
10. The maximum stress is $KS = 1.98 \times 49 = 97$ MPa.

A STRESS CONCENTRATION FACTOR IN TORSION

There is a stress concentration at any place where the cross section of a bar changes, whether the bar is subjected to an axial force, or bending, or torsion. Figure 14-13 shows a circular bar loaded with torque T. The diameter of the bar changes from D (larger) to d (smaller), and there is a fillet of radius r at the transition. The nominal stress in the smaller end of the bar is

$$S_s = \frac{16T}{\pi d^3}$$

Fig. 14-13. Stress concentration factors for circular bar with fillet.

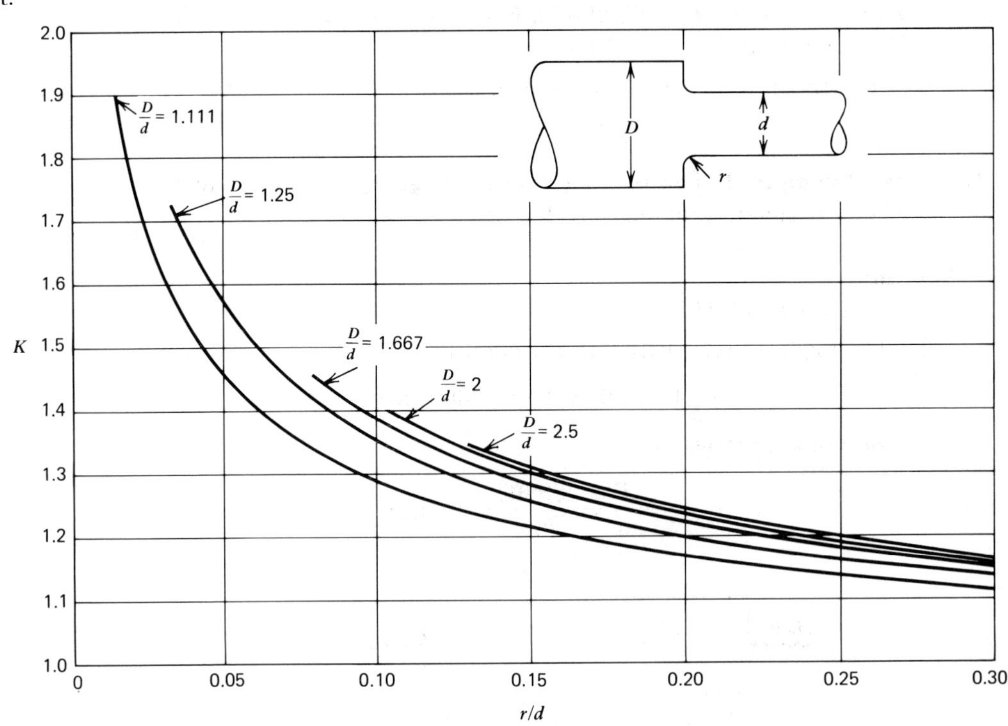

as you should remember from Chapter 10. The maximum stress is KS, and you can read K from the chart if you know D, d, and r. Let's try using this chart in a pair of examples.

Illustrative Example 3. Figure 14-14 shows a shaft which is subjected to a torque of 14 800 lb in. Calculate the maximum stress.

Solution:

1. The torque T is 14 800 lb in.
2. The smaller diameter d is 2.2 in.
3. The nominal stress is

$$S_s = \frac{16T}{\pi d^3} = \frac{16 \times 14\,800}{\pi \times (2.2)^3} = 7\,080 \text{ psi}$$

4. The larger diameter D is 3 in.
5. The smaller diameter d is 2.2 in.
6. The ratio D/d is $3/2.2 = 1.36$.
7. The fillet radius is $r = 0.4$ in.
8. The ratio r/d is $0.4/2.2 = 0.182$.
9. Now it is necessary to interpolate between the curve labeled $D/d = 1.25$ and the curve labeled 1.67. The value of K can be read as 1.22.
10. The maximum stress is $KS = 1.22 \times 7\,080 = 8\,640$ psi.

Illustrative Example 4. The circular bar in Fig. 14-15 is subjected to a torque of 380 N·mm. What is the maximum shearing stress?

Solution:

1. The torque T is 380 N·mm.
2. The smaller diameter d is 4 mm.
3. The nominal shearing stress is

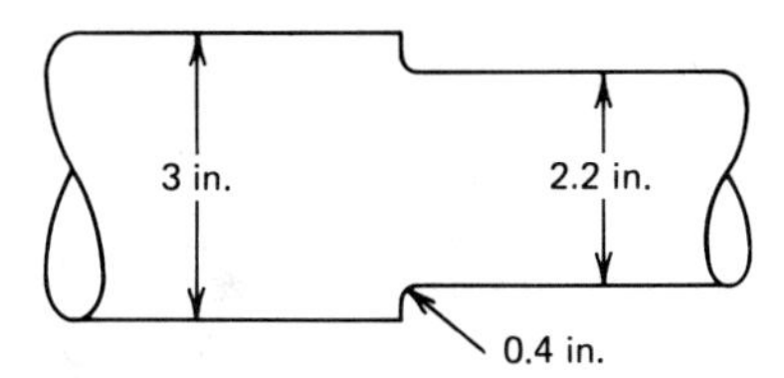

Fig. 14-14. Shaft for Illustrative Example 3.

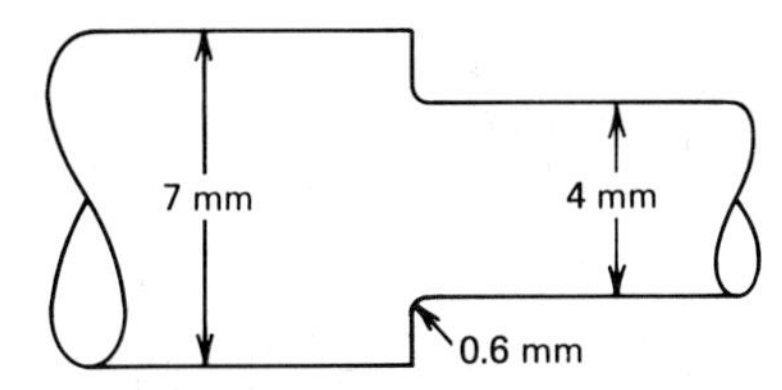

Fig. 14-15. Shaft for Illustrative Example 4.

$$S_s = \frac{16T}{\pi d^3} = \frac{16 \times 380}{\pi(4)^3} = 30.2 \text{ MPa}$$

4. The larger diameter is $D = 7$ mm.
5. The smaller diameter is $d = 4$ mm.
6. The ratio D/d is $7/4 = 1.75$.
7. The fillet radius is 0.6 mm.
8. The ratio r/d is $0.6/4 = 0.15$.
9. It is necessary to interpolate between the curve labeled $D/d = 1.67$ and the curve labeled $D/d = 2.5$. Then the stress concentration factor K can be read as 1.31.
10. The maximum shearing stress is $KS = 1.31 \times 30.2 = 39.6$ MPa

Practice Problems (Section 14-7). It is time now to see how well you can do in reading stress concentration factors and calculating concentrated stresses.

1. The flat bar in Fig. 14-16 is ½ in. thick. What is the maximum stress?
2. The flat bar in Fig. 14-17 is 6 mm thick. Calculate the maximum stress.
3. The circular bar in Fig. 14-18 is subjected to a torque of 2 600 lb in. What is the maximum shearing stress?
4. The circular bar in Fig. 14-19 is subjected to a torque of 4 200 000 N·mm. What is the maximum shearing stress?

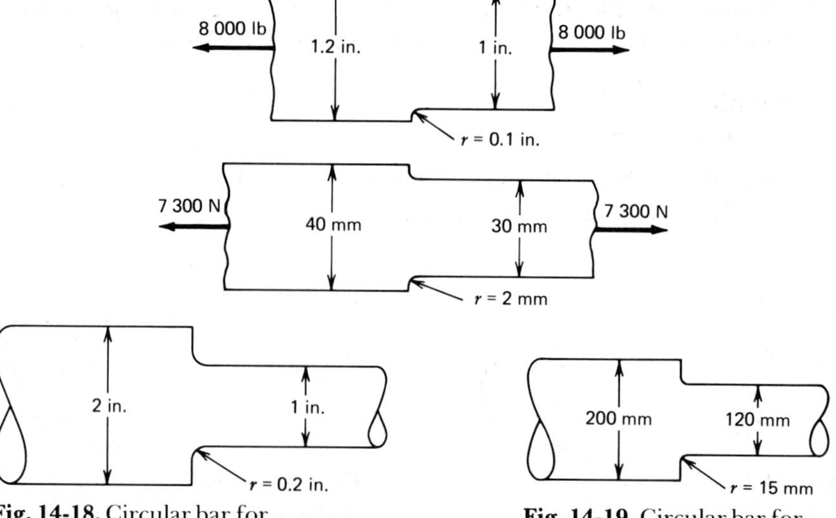

Fig. 14-16. Bar for Problem 1.

Fig. 14-17. Bar for Problem 2.

Fig. 14-18. Circular bar for Problem 3.

Fig. 14-19. Circular bar for Problem 4.

SUMMARY Here are the main new points of this chapter.

1. Material breaks at a lower stress if the stress is repeated than if the stress is applied only once.
   (a) The behavior of a material under repeated stress can be shown by means of an *S-N* diagram.
   (b) The endurance limit is the greatest stress that can be applied an indefinite number of times without causing failure.
   (c) Most of the available information on endurance limits is for completely reversed bending stress. It is considered safe to use these values for any repeated bending stress or for repeated axial stress.
   (d) It is considered safe to take the endurance limit for repeated torsional shearing stress as one-half the endurance limit for completely reversed bending stress.
2. The working stress, for machine parts that are to be subjected to repeated loads, should be based on the endurance limit of the material.
   (a) The working stress is

$$\text{working stress} = \frac{\text{endurance limit}}{\text{factor of safety}}$$

   (b) A reasonable factor of safety for repeated loads is 3.
3. At any point where there is a change in cross section of a bar, the stress is higher than the nominal stress, as given by one of the simple formulas.
   (a) The ratio of this higher stress to the nominal stress is the stress concentration factor K.
   (b) The maximum stress is equal to K times the nominal stress.
   (c) Stress concentration factors usually depend on the ratio of one dimension to another.
   (d) Stress concentrations in a ductile material tend to disappear as the material is stressed beyond the proportional limit.
   (e) A change in cross section is called a stress raiser.

REVIEW QUESTIONS Here is the last set of review questions. See if you can answer them.

1. Is material stronger under repeated stress than under a stress which is applied only once?
2. What is an *S-N diagram*?
3. What is an endurance limit?
4. What is a cycle of stress?
5. What is meant by completely reversed stress?

6. How does the endurance limit for completely reversed torsional shearing stress compare with the endurance limit for completely reversed bending stress?
7. How do you determine a working stress for a machine part that is to be subjected to repeated loads?
8. What is a reasonable factor of safety for a machine part under repeated loads?
9. Why use a factor of safety?
10. What is a stress raiser?
11. What is a stress concentration factor?
12. How is a stress concentration factor used?
13. How can you recognize a point where a stress concentration occurs?
14. Why are stress concentrations especially important when stresses are repeated?

REVIEW PROBLEMS

1. A shaft in an automatic machine is to be reversed in direction during each 10 sec of operation. It is expected that the machine will operate for an average of 12 hr/day for 250 days/yr for 5 yr. The maximum torque is 630 000 N·mm. What is a reasonable diameter for the shaft? TELFCRBS is 340 MPa.
2. Figure 14-20 shows the free-body diagram of a rectangular steel beam, 80 mm wide. The loads shown are to be repeated at least 2 000 000 times. Find the depth required for the beam. TELFCRBS is 220 MPa.
3. An American W-shape steel beam is subjected to a great many complete reversals of a bending moment of 42 000 lb ft. What size beam would you use? TELFCRBS is 34 000 psi.
4. An extruded aluminum-alloy channel is to be subjected to a tensile force of 22 000 lb, which is to be repeated about 1 800 times/min for 6 hr/day for 300 days/yr for an indefinite number of years. What size channel do you think is necessary?

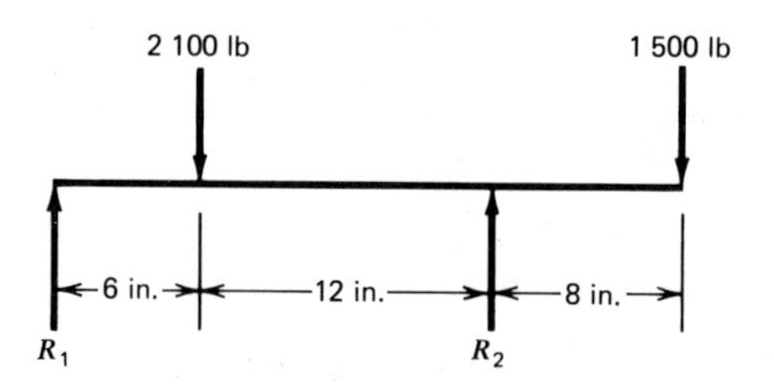

Fig. 14-20. Beam for Problem 2.

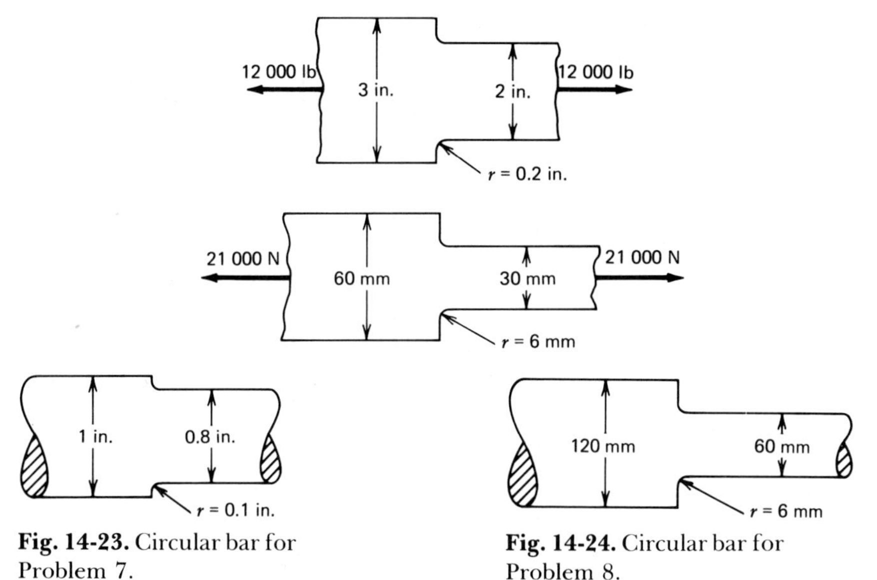

Fig. 14-21. Bar for Problem 5.

Fig. 14-22. Bar for Problem 6.

Fig. 14-23. Circular bar for Problem 7.

Fig. 14-24. Circular bar for Problem 8.

5. The flat bar in Fig. 14-21 is ¾ in. thick. What is the maximum stress?
6. The flat bar in Fig. 14-22 is 10 mm thick. Calculate the maximum stress.
7. The circular bar in Fig. 14-23 is subjected to a torque of 1 960 lb in. What is the maximum shearing stress?
8. The circular bar in Fig. 14-24 is subjected to a torque of 386 000 N·mm. What is the maximum shearing stress?

ANSWERS TO PROBLEMS

Answers to odd-numbered Practice Problems and all Review Problems are included here.

CHAPTER 1

Practice Problems. *Section 1-6.* **1.** 55.3 lb; 65.9 lb **3.** −6.19 lb; −3.87 lb **5.** −530 N; 407 N **7.** −3.87 lb; −8.46 lb **9.** −328 lb; 317 lb

Practice Problems. *Section 1-6 Continued.* **1.** 47.7 lb; 33° **3.** 810 lb; 69.8° **5.** 18 400 N; 29.4° **7.** 12.04 N; 24°

Practice Problems. *Section 1-7.* **1.** 930 lb in. **3.** 540 lb in. **5.** −2 170 lb in. **7.** 168 kip ft **9.** 1 920 N·m **11.** −3 080 N·m

Practice Problems. *Section 1-8.* **1.** 481 lb ft **3.** − 3 620 lb ft **5.** −763 N·m **7.** 307 000 N·m **9.** 107 kip ft

Review Problems. **1.** 11.6 lb; 13.8 lb **2.** −52.8 N; −39.6 N **3.** −5.12 kip; 11.5 kip **4.** 229 lb; −84.7 lb **5.** −1244 N; − 1 244 N **6.** 9 000 lb; 24.3° **7.** 814 N; 25.3° **8.** 4.33 kip; 28.1° **9.** 1 390 lb; 39.8° **10.** 6 330 N; 68.7° **11.** −2 700 lb ft **12.** 1 750 lb ft **13.** −136 000 N·m **14.** −66 600 N·m **15.** 126 kip ft **16.** 1 840 lb in. **17.** −2 000 lb in. **18.** −350 lb in. **19.** 360 N·m **20.** 941 N·m

CHAPTER 2

Practice Problems. *Section 2-3.* **1.** 880 lb; 1 320 lb **3.** 6 000 lb; 4 800 lb **5.** 129 N; 202 N; 68.3° **7.** 14 000 lb; 14 000 lb **9.** 1 250 lb; 2 150 lb

Practice Problems. *Section 2-4.* **1.** 1 273 N; 45° **3.** 1 000 lb; 433 lb; 233 lb **5.** 1 240 lb; 950 lb **7.** 197 lb; 30.4 lb; 9.46°

Practice Problems. *Section 2-5.* **1.** 258 N; 75° **3.** 198 lb; 199 lb; 5.19° **5.** 308 N; 3 100 N

Practice Problems. *Section 2-6.* **1.** 267 N; 1 890 N; 1 280 N **3.** 1 080 lb; 1 080 lb; 1 080 lb **5.** 420 lb; 930 lb; 70.2°; 825 lb; 27°

Review Problems. *Part A.* **1.** 7 710 lb; 4 290 lb **2.** 1 600 lb; 2 400 lb **3.** 31 000 N; 20 000 N **4.** 4 800 lb; 6 790 lb; 45° **5.** 2 670 N; 4 270 N **6.** 3 000 N; 3 120 N; 2 080 N **7.** 1 640 lb; 2 290 lb; 75.4° **8.** 384 N; 1 200 N; 73.9° **9.** 540 lb; 412 lb; 58.4°; 324 lb; 14° **10.** 750 N; 1 575 N; 825 N

Practice Problems. *Section 2-9.* **1.** $F_{ab} = F_{fh}$ = 36 k C; $F_{ac} = F_{gh}$ = 45 k T; $F_{bc} = F_{fg}$ = 27 k C; $F_{bd} = F_{df}$ = 48 k C; $F_{be} = F_{ef}$ = 15 K T; $F_{ce} = F_{eg}$ = 36 k T **3.** F_{ab} = 16 000 N T; F_{ac} = 13 800 N C; F_{ce} = 13 800 N C; F_{bc} = 8 000 N T; F_{be} = 8 000 N C; F_{bd} = 24 000 N T **5.** F_{ab} = 4 350 lb C; F_{ac} = 2 360 lb T; F_{bd} = 2 360 lb C; F_{bc} = 3 670 lb T; F_{cd} = 4 350 lb C; F_{ce} = 4 720 lb T; F_{de} = 11 000 lb T; F_{ef} = 4 720 lb T; F_{df} = 8 700 lb C

Practice Problems. *Section 2-11.* **1.** 28.8 lb; 2.39° **3.** 18 500 lb; 5.45° **5.** 771 N; 10°

Practice Problems. *Section 2-14.* **1.** 14 lb **3.** 25 lb **5.** 52.5 lb **7.** 2.25 ft **9.** 0.289

Practice Problems. *Section 2-15.* **1.** 1 950 N **3.** 0.323

Review Problems. *Part B.* **1.** F_{ab} = 8.41 k T; F_{ac} = 4.67 k C; F_{bd} = 4.67 k T; F_{bc} = 7 k C; F_{cd} = 8.41 k T; F_{ce} = 9.34 k C. **2.** F_{ab} = 9 190 N C; F_{ac} = 7 650 N T; F_{bd} = 10 800 N C; F_{bc} = 12 700 N T; F_{ce} = 12 700 N T; F_{cd} = 7 650 N T; F_{de} = 0; F_{df} = 7 650 N C; F_{ef} = 0 **3.** 1 760 lb **4.** 9 000 N **5.** 99 lb **6.** 43 N **7.** 0.340 **8.** 4.16 m **9.** 231 lb **10.** 96 N

CHAPTER 3

Practice Problems. *Section 3-2.* **1.** 11 300 psi **3.** 4 800 psi **5.** 32 MPa **7.** 11 500 lb

Practice Problems. *Section 3-3.* **1.** 8 770 psi. **3.** 66.7 psi **5.** 25 MPa **7.** 412 000 N

Practice Problems. *Section 3-4.* **1.** 9 500 psi **3.** 53.1 MPa **5.** 2 290 psi **7.** 156 000 N

Practice Problems. *Section 3-5.* **1.** 0.000375 **3.** 4 **5.** 0.00082

Practice Problems. *Section 3-6.* **1.** 0.000789 **3.** 0.072 **5.** 0.5

Practice Problems. *Section 3-7.* **1.** 0.00241 **3.** 208 lb **5.** 571 mm^2 **7.** 1.06×10^6 lb

Practice Problems. *Section 3-10.* **1.** 1.08 in. **3.** 2.5 in. **5.** 6.65 in. **7.** 2.57 in.

Practice Problems. *Section 3-11.* **1.** − 0.0342 in. **3.** − 0.226 mm **5.** 11 700 psi **7.** 122°C

Review Problems. *Chapter 3.* **1.** 14 600; 0.000487; 0.00877 in. **2.** 52.5 mm; 22.5 mm **3.** 1.40 in. **4.** 0.261 mm **5.** 10.8 in. **6.** 30.1 MPa **7.** 0.024 in. **8.** 16.7 mm **9.** 4.11 in. **10.** 10.7 mm **11.** 265°F **12.** 174°C

CHAPTER 4

Practice Problems. *Section 4-1.* **1.** 4 in.; 2 in. **3.** − 5.5 in.; − 4.5 in. **5.** −125 mm; 125 mm

Practice Problems. *Section 4-2.* **1.** 4.33 in.; 5.67 in. **3.** −3.73 in.; −6.73 in. **5.** 80 mm; 30 mm **7.** −85.3 mm; 23.3 mm **9.** − 12.4 mm; 3.6 mm

Practice Problems. *Section 4-3.* **1.** 96 in.^{3}; 204 in.^{3} **3.** −59.2 in.^{3}; −46.6 in.^{3} **5.** 352 in.^{3}; 110 in.^{3} **7.** 100 000 mm^3; 20 000 mm^3 **9.** 5 625 mm^3; 6 750 mm^3

Practice Problems. *Section 4-4.* **1.** 176 in.^{3}; 0 **3.** 4.69 in.^{3}; 9.44 in.^{3} **5.** 1 465 000 mm^3; 2 411 000 mm^3

Review Problems. *Chapter 4.* **1.** 0.904 in.; 0.904 in. **2.** 1.90 in.; 0.485 in. **3.** 0; 6.08 in. **4.** 0; 1.90 in. **5.** 11.4 mm; 6.54 mm **6.** 14.4 mm; 0 **7.** 16 mm; 12.6 mm **8.** 0; 1.91 mm

CHAPTER 5

Practice Problems. *Section 5-2.* **1.** 128 in.^{4}; 18 in.^{4} **3.** 201 in.^{4} **5.** 0.0425 in.^{4}; 0.232 in.^{4} **7.** 31 400 mm^4 **9.** 54 mm^4; 121.5 mm^4

Practice Problems. *Section 5-3.* **1.** 512 in.^{4}; 72 in.^{4} **3.** 0.0319 in.^{4} **5.** 1 184 in.^{4} **7.** 1.75×10^6 mm^4; 15.75×10^6 mm^4 **9.** 296 000 mm^4; 129 000 mm^4

Practice Problems. *Section 5-4.* **1.** 17.35 in.^{4} **3.** 82.4 in.^{4} **5.** 245.9 in.^{4} **7.** 890 000 mm^4; 19 900 mm^4

Review Problems. *Chapter 5.* **1.** 7.29 in.^{4}; 1.541 in.^{4} **2.** 1.3 × 10^6 mm^4 **3.** 172 in.^{4}; 436 in.^{4} **4.** 214 000 mm^4; 67 600 mm^4 **5.** 111 in.^{4}; 125 in.^{4} **6.** 39 800 mm^4; 88 000 mm^4 **7.** 27.2 in.^{4}; 68.8 in.^{4} **8.** 566 000 mm^4; 180 000 mm^4

CHAPTER 6

Practice Problems. *Section 6-3.* **1.** 533 lb **3.** −667 lb **5.** 2 050 N **7.** −1 950 N

Practice Problems. *Section 6-4.*

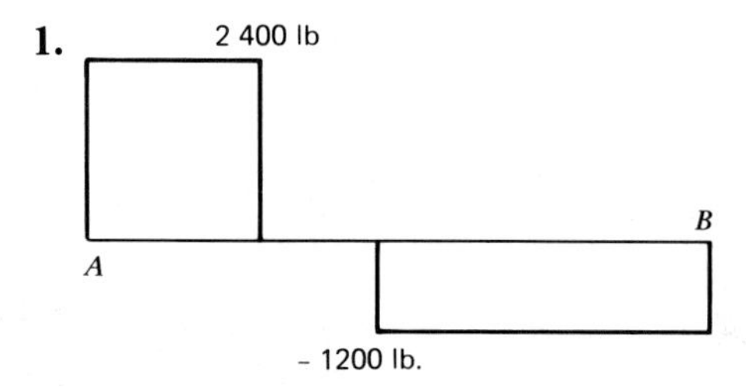

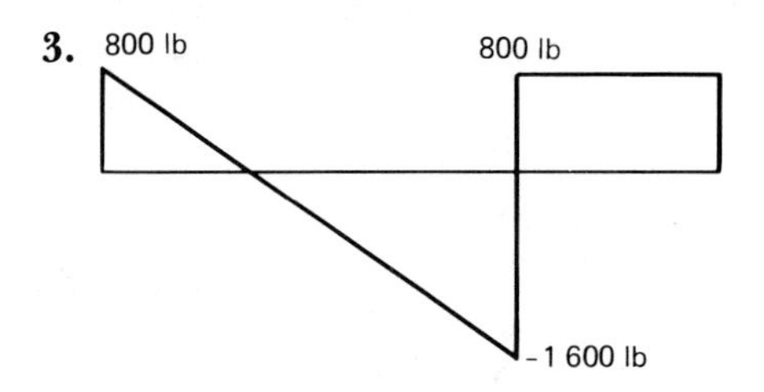

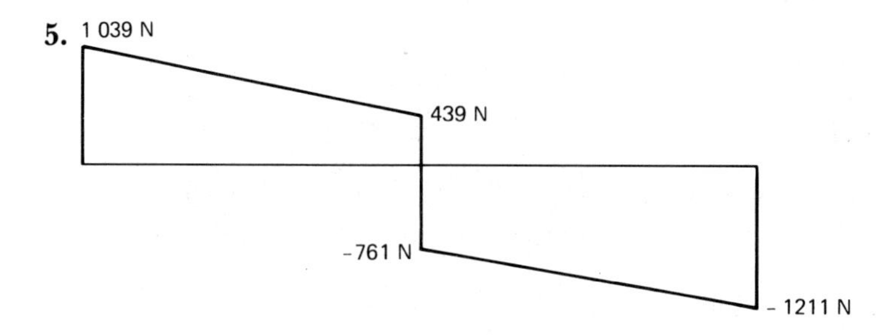

Practice Problems. *Section 6-5.* **1.** −2 820 lb in. **3.** 10 100 N·m **5.** 8 800 lb ft **7.** 5 120 lb ft **9.** 315 N·m

Practice Problems. *Section 6-6.*

1.

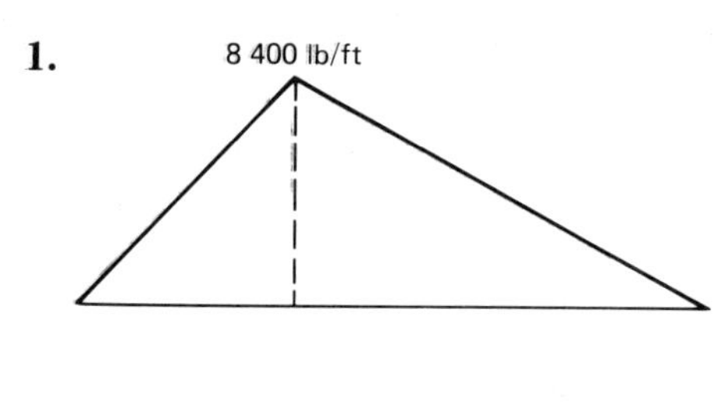

3.

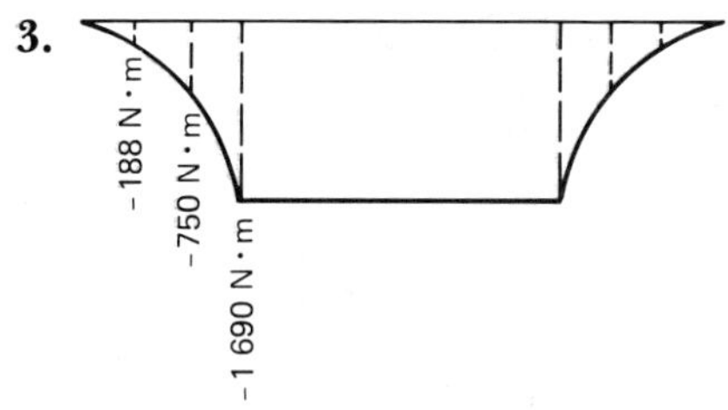

5.

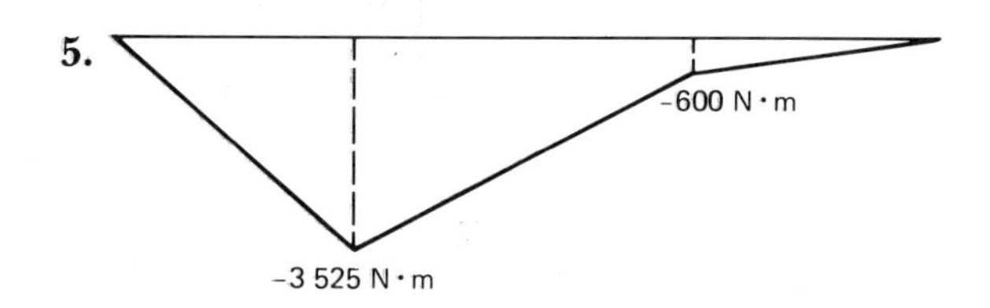

Practice Problems. *Section 6-7.* **1.** 332 lb in. **3.** −6 400 lb ft **5.** −4 500 N·cm

Practice Problems. *Section 6-8.* **1.** 206 MPa **3.** 11 000 psi **5.** 7 460 psi **7.** 18 400 psi **9.** 23 800 psi

Practice Problems. *Section 6-10.* **1.** 0.793 in. **3.** 12.5 mm **5.** 0.588 in.

Practice Problems. *Section 6-11.* **1.** 2.45 in. **3.** 1 160 mm **5.** 0.955 in.

Practice Problems. *Section 6-12.* **1.** 5.87 in.^{2} **3.** 103 in.^{3} **5.** 1.88 in.^{2} **7.** 3.08 cm^2 **9.** 52.4 cm^3

Practice Problems. *Section 6-13.* **1.** W 12 × 30 **3.** 40 × 40 × 6 or 50 × 50 × 4 **5.** 4 [2.331 or 5 [2.212 **7.** 4 [2.331

Practice Problems. *Section 6-17.* **1.** 1 470 psi **3.** 2.88 MPa **5.** 22.5 MPa **7.** 9 670 psi **9.** 2.1 MPa

Review Problems. *Chapter 6.*

1.

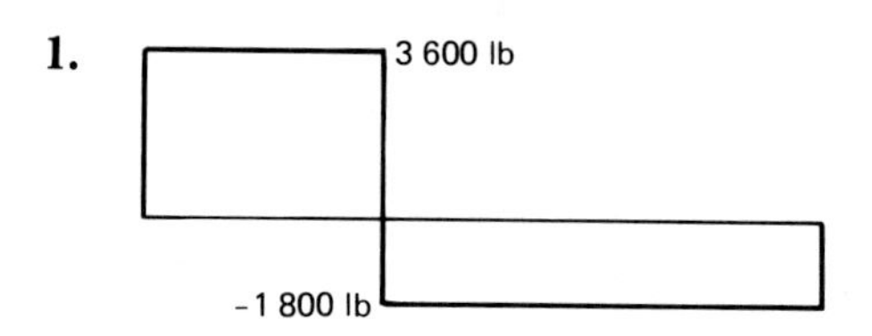

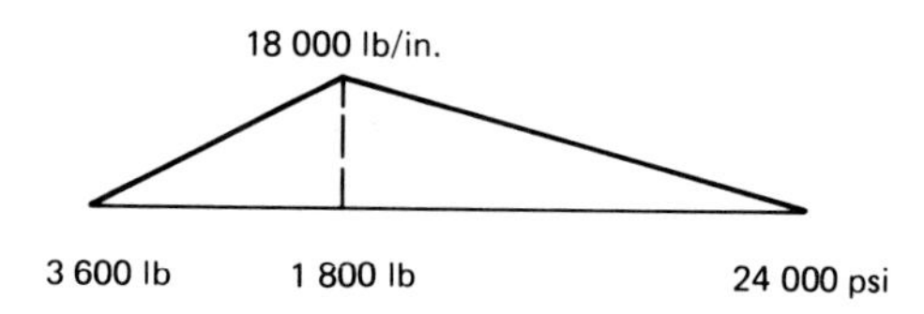

3 600 lb 1 800 lb 24 000 psi

2.

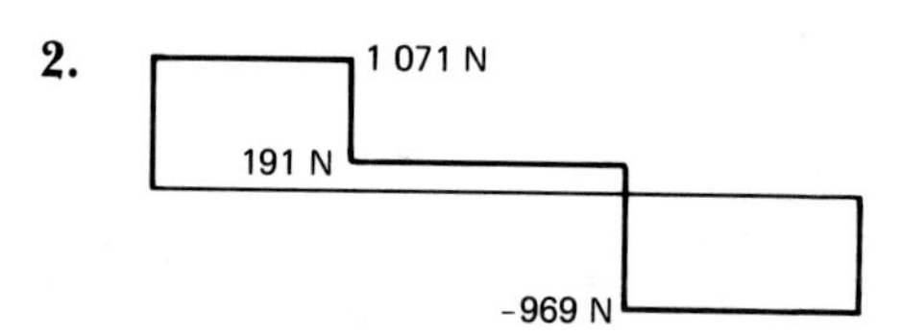

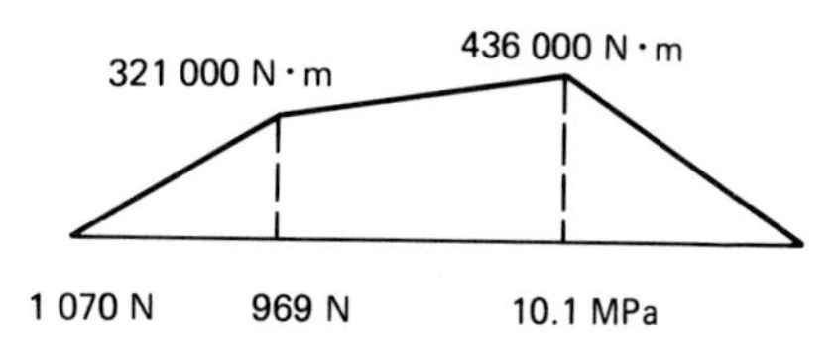

1 070 N 969 N 10.1 MPa

3.

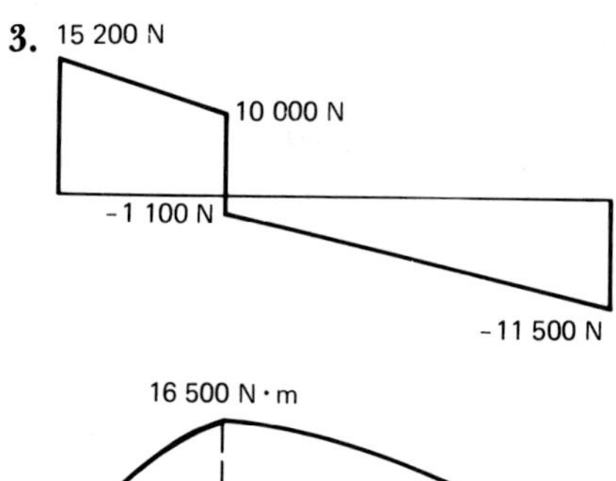

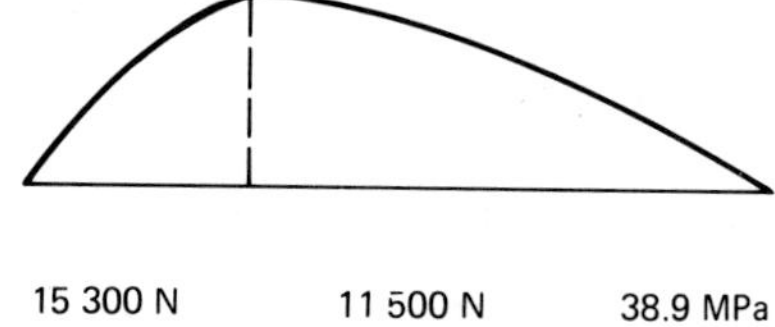

15 300 N 11 500 N 38.9 MPa

4.

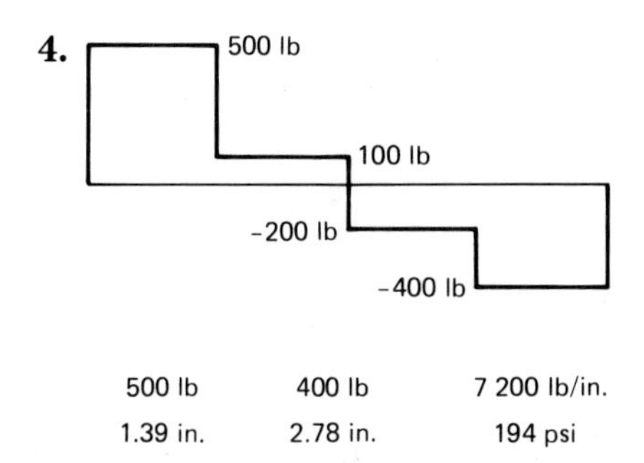

5. **6.**

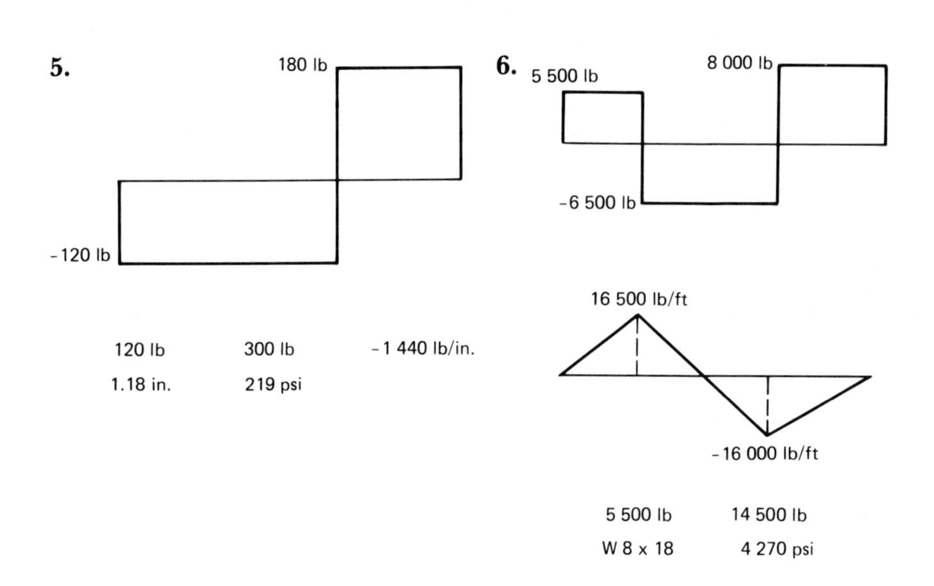

CHAPTER 7

Practice Problems. *Section 7-2.* **1.** 1 253 psi C **3.** 933 psi T **5.** 6.86 MPa T

Practice Problems. *Section 7-3.* **1.** 5 990 psi **3.** 1 910 psi **5.** 60.5 MPa top; 114 MPa bottom

Practice Problems. *Section 7-4.* **1.** 9 340 psi ten. top; 14 000 psi comp. bottom **3.** 13 900 psi ten. left; 12 500 psi comp. right **5.** 131.6 MPa ten. left; 112.2 MPa comp. right

Review Problems. *Chapter 7.* **1.** 12 800 psi ten. left; 11 500 psi comp. right **2.** 9 560 psi comp. left; 4 220 psi ten. right **3.** 5 200 psi comp. top; 4 340 psi ten. bottom **4.** 13 400 psi ten. top; 17 600 psi comp. bottom

5. 7.9 MPa ten. left; 64.2 MPa comp. right **6.** 105 MPa comp. top; 99.4 MPa ten. bottom **7.** 38.4 MPa **8.** 9.9 MPa ten. left; 29.3 MPa comp. right

CHAPTER 8

Practice Problems. *Section 8-2 and 8-3.* **1.** 0.600 in. **3.** 0.976 in. **5.** 3.11 mm

Practice Problems. *Section 8-4.* **1.** 0.337 in. **3.** 10.2 mm

Practice Problems. *Section 8-5.* **1.** 0.283 in. **3.** 0.091 in. upward **5.** 30.4 mm

Review Problems. *Chapter 8.* **1.** 12.6 mm **2.** 0.279 in. **3.** 1.71 in. **4.** 179 mm

CHAPTER 9

Practice Problems. *Section 9-2.* **1.** 7 050 psi **3.** W 10 × 22 **5.** 10 [6.136 **7.** 13 300 psi

Practice Problems. *Section 9-3.* **1.** 26 600 psi **3.** W 14 × 30 **5.** 7.97 MPa

Review Problems. *Chapter 9.* **1.** W 5 × 16 **2.** 5.2 MPa **3.** 6 330 psi **4.** W 6 × 20 or W 8 × 18 **5.** 91 mm; 182 mm **6.** 16 800 psi

CHAPTER 10

Practice Problems. *Section 10-1.* **1.** −1 260 lb in. **3.** −1 350 lb in. **5.** −87 500 N·mm

Practice Problems. *Section 10-2.* **1.** 4 720 psi **3.** 14.9 MPa **5.** 12 600 lb in. **7.** 77 000 N·mm

Practice Problems. *Section 10-3.* **1.** 6 420 psi. **3.** 90.9 MPa **5.** 8.95×10^8 N·mm

Practice Problems. *Section 10-4.* **1.** 0.713 in. **3.** 34.4 mm

Practice Problems. *Section 10-5.* **1.** 0.819° **3.** 3.42° **5.** 0.851°

Practice Problems. *Section 10-6.* **1.** 15 100 psi **3.** 32 000 psi **5.** 56.25 MPa

Practice Problems. *Section 10-7.* **1.** 12 700 psi **3.** 58.5 MPa

Practice Problems. *Section 10-8.* **1.** 12 100 psi **3.** 22 MPa

Review Problems. *Chapter 10.* **1.** 1 770 psi **2.** 66.3 MPa **3.** 1.63 in. **4.** 9.90 mm **5.** 18 800 lb in. **6.** 46.4 × 10^6 N·mm **7.** 2 760 psi **8.** 69 MPa **9.** 0.273° **10.** 78.5° **11.** 4.06° **12.** 6.36° **13.** 5 200 psi **14.** 220 MPa **15.** 11 800 psi **16.** 32.2 MPa **17.** 6 560 psi **18.** 128 MPa

CHAPTER 11

Practice Problems. *Section 11-3.*

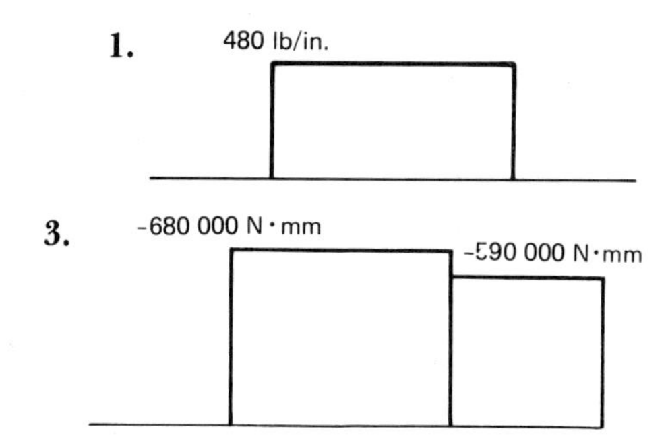

Practice Problems. *Section 11-4.*

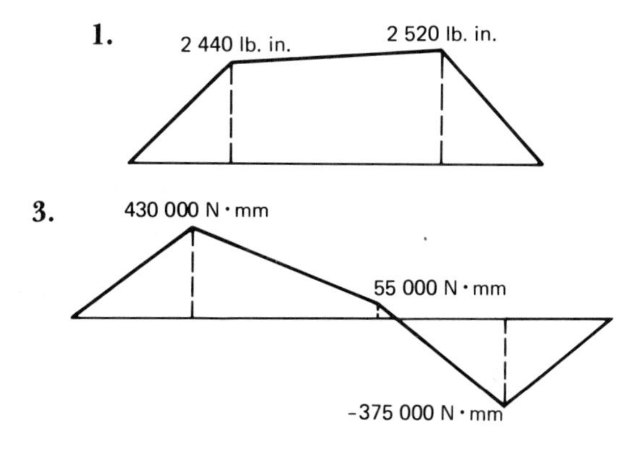

Practice Problems. *Section 11-5.* **1.** 10 400 psi **3.** 6 600 psi **5.** 11 900 psi **7.** 25.8 MPa

Practice Problems. *Section 11-6.* **1.** 3.91 in. **3.** 3.20 in. **5.** 4.97 in.

Review Problems. *Chapter 11.* **1.** 8 470 psi **2.** 87 MPa **3.** 83.2 MPa **4.** 3 450 psi **5.** 5.02 in. **6.** 24.3 mm

CHAPTER 12

Practice Problems. *Section 12-3.* **1.** 1.50 in. **3.** 0.390 in. **5.** 1.20 in.

Practice Problems. *Section 12-4.* **1.** 80 **3.** 27.8 **5.** 79.6

Practice Problems. *Section 12-5.* **1.** 22 200 psi **3.** 46.3 MPa **5.** 7 500 lb

Practice Problems. *Section 12-7.* **1.** 213 000 N **3.** 659 000 lb **5.** 458 000 lb

Practice Problems. *Section 12-8.* **1.** 165 000 lb **3.** 910 000 lb

Practice Problems. *Section 12-9.* **1.** 465 000 lb **3.** 63 300 lb

Practice Problems. *Section 12-10.* **1.** 600 N **3.** 10 000 lb **5.** 3 150 lb.

Review Problems. *Chapter 12.* **1.** 120 000 lb **2.** 179 000 N **3.** 551 000 lb. **4.** 173 000 lb. **5.** 74 600 N **6.** 13.3 N

CHAPTER 13

Practice Problems. *Section 13-2.* **1.** 4 070 psi **3.** 12 200 psi **5.** 89.6 MPa

Practice Problems. *Section 13-3.* **1.** 1 910 psi **3.** 24.7 MPa **5.** 2 880 psi **7.** 72.2 MPa

Practice Problems. *Section 13-4.* **1.** 206 MPa **3.** 16 100 psi **5.** 98.4 MPa

Practice Problems. *Section 13-5.* **1.** $\frac{5}{8}$ in. **3.** $\frac{1}{2}$ in. **5.** 24 mm **7.** 1 in. **9.** 6 mm

Practice Problems. *Section 13-7.* **1.** 5 screws; $\frac{3}{16}$ in.; 2.75 in. **3.** 6 rivets in 2 rows; $\frac{5}{16}$ in.; 7 in.

Practice Problems. *Section 13-8.* **1.** 1 $\frac{1}{8}$ in.; $\frac{13}{16}$ in. **3.** 2.25 in.; 1.75 in.

Practice Problems. *Section 13-13.* **1.** 0.63 in. **3.** 31.8 mm **5.** 13.6 mm

Practice Problems. *Section 13-14.* **1.** 2 **3.** 14 **5.** 3

Practice Problems. *Section 13-15.* **1.** 2 400 psi; 4 800 psi **3.** 0.144 in. **5.** 5 400 psi **7.** 0.0755 in.

Practice Problems. *Section 13-16.* **1.** 5 in.; $\frac{3}{4}$ in. **3.** 9 in.; 1.5 in.

Review Problems. *Chapter 13.* **1.** 7 300 psi **2.** 9 170 psi **3.** 15 300 psi **4.** 7 400 psi **5.** 7 980 psi **6.** 7 300 psi **7.** 14 800 psi **8.** 13 600 psi **9.** 4; 5/16 in.; 4 in. **10.** 2; 8 mm; 72 mm **11.** 7.38 in. **12.** 15 mm **13.** 1.24 in. **14.** 1.08 in. **15.** 6 **16.** 5 **17.** 1 430 psi; 2 860 psi **18.** 75 MPa **19.** 7 in.; 1 1/8 in.

CHAPTER 14

Practice Problems. *Section 14-1.* **1.** $1\,944 \times 10^6$ **3.** 18×10^5

Practice Problems. *Section 14-4.* **1.** 2 670 psi **3.** 36 MPa

Practice Problems. *Section 14-5.* **1.** 23.6 mm **3.** W 5 × 16 **5.** 86.8 mm

Practice Problems. *Section 14-7.* **1.** 30 200 psi **3.** 16 400 psi

Review Problems. *Chapter 14.* **1.** 38.4 mm **2.** 35 mm **3.** W 12 × 45 **4.** 8 [5.789 **5.** 17 900 psi **6.** 137 MPa **7.** 25 200 psi **8.** 12.8 MPa

INDEX